Logic Design and
Switching Theory

Logic Design and Switching Theory

SABURO MUROGA

Professor of Computer Science
and Electrical Engineering
University of Illinois

A WILEY-INTERSCIENCE
PUBLICATION

JOHN WILEY & SONS

New York / Chichester
Brisbane / Toronto

Copyright © 1979 by John Wiley & Sons, Inc.

All rights reserved. Published simultaneously in Canada.

Reproduction or translation of any part of this work
beyond that permitted by Sections 107 or 108 of the
1976 United States Copyright Act without the permission
of the copyright owner is unlawful. Requests for
permission or further information should be addressed to
the Permissions Department, John Wiley & Sons, Inc.

Library of Congress Cataloging in Publication Data

Muroga, Saburo.
   Logic design and switching theory.

   "A Wiley-Interscience publication."
   Includes bibliographical references and index.
     1. Logic circuits. 2. Logic design. 3. Switching
theory. 4. Integrated circuits—Large scale integration.
5. Electronic circuit design—Data processing.
I. Title.
TK7868.L6M87   621.3815'37   78-12407
ISBN 0-471-04418-0

Printed in the United States of America

10 9 8 7 6 5 4 3 2 1

Preface

The enormous progress in integrated circuit technology in the last decade has brought about dramatic changes in logic design, and the pace of change continues to accelerate. Progress in large-scale integration (LSI) has forced many manufacturers having no previous connection with electronics—such as watch, toy, or sewing machine manufacturers—into LSI design and even in-house LSI production. Consequently, logic design is now important for a much wider circle of people, and this circle will continue to grow. Although logic design has been facilitated by off-the-shelf packages, the opportunities for custom LSI design are increasing.

Logic design has also evolved into a more complex discipline, and the logic designer must be familiar with its many aspects. As the number of different integrated circuit (IC) logic families—e.g., ECL, TTL, NMOS, CMOS, VMOS, and IIL—continues to grow, the designer must choose the appropriate family to attain different design objectives, and use different logic design procedures for different families. Compact and inexpensive semiconductor memories, particularly ROMs and PLAs, are now often used as important parts of logic networks. Hence the designer must carefully consider which parts of a network should be realized with gates or memories. Also, as IC costs decline, many complex job functions—such as the computation of trigonometric functions—are being designed in logic networks instead of in software, and the designer must find appropriate algorithms to realize such job functions. As a result of these and many other changes, logic designers must now be familiar with not only switching theory but also architecture, different IC logic families, memories, algorithms for hardware implementation of complex job functions, programming for firmware, diagnosis methods, electronics, and possibly many other related concerns.

The coverage of all these topics in this book is obviously impossible, and instead emphasis is on minimization procedures, a classic topic that retains its importance even in logic design with integrated circuits. Of course, the minimization of an entire system depends on other factors, including the choice of appropriate architectures, algorithms, and electronic circuit designs, but the minimization of logic networks (in particular, key networks such as adders) is still crucial. Unlike conventional minimization with discrete gates, minimization in LSI design does not result directly in cost reduction. Instead, it reduces the size and power consumption of a chip, which in turn reduces the cost of the chip and increases its speed. As designers attempt to pack more and more elements into a single chip, even partial minimization will often dramatically reduce the chip size.

This book emphasizes algebraic minimization procedures for three reasons: (1) Algebraic procedures sometimes can reduce design time drastically. For example, the algebraic procedure for deriving a minimal sum in Section 4.6 is nearly 100 times faster than the conventional tabular method treated in textbooks.

(2) Development of programs for computer-aided design (CAD) has been extensively carried on in industry in regard to every aspect of design and manufacture (including logic design), in order to reduce time, cost, and error. This trend will be even stronger in the future. Algebraic concepts are useful in implementing many CAD programs (also data base management systems, as mentioned in Section 3.7). Like many computer programs in other areas (such as business applications), many CAD programs need continuous maintenance or changes, as technology or usage changes. The computational efficiences of CAD programs are highly dependent on the details of the procedures. Also, many variations of the procedures may be needed for different CAD programs. Thus the computational efficiencies and some variations are discussed in this book so that students can handle different problems efficiently.

(3) When a large number of transistors (soon a quarter million) are to be packed into a single LSI chip, we need to rely more on CAD programs implementing algebraic minimization procedures in order to reduce design time and error. In this case, appropriately specifying or modifying the details of the procedures will result in great differences in the size or speed of the designed networks. This is particularly true when heuristic procedures need to be applied in designing very large networks, giving up on absolute minimization.

In conclusion, it is believed that slightly advanced knowledge of algebraic procedures will greatly enhance students' skill in logic design and also in CAD programming.

This book is intended for a course that follows an introductory course on digital computers (discussion of number representations, elementary switching theory, standard networks, and control logic, typically based on such textbooks as *Computer Logic Design* by M. M. Mano, and *Design of Digital Computers* by H. Gschwind and E. J. McCluskey. To make this book self-contained, some elementary concepts (such as Karnaugh maps) are

Preface vii

included which may overlap with topics in the introductory course on digital computers and hence require only a review.

After the introduction in Chapter 1, this book discusses hardware implementation of logic operations. Analysis of MOS networks is discussed in detail, considering the increasing importance of the MOS in LSI.

Chapters 3 through 5 discuss switching algebra and algebraic procedures. To strengthen students' intuitive grasp, mathematical concepts are pictorially illustrated with Karnaugh maps. Excercises are indispensable for understanding. Students who may be overwhelmed initially by algebraic terminology will understand algebraic concepts as they advance through succeeding chapters. It is believed that Tison's methods in Procedures 3.4.2, 5.1.1, and 5.1.2 are presented here for the first time in a textbook. The author's Procedures 4.6.1, 5.2.1, and 5.2.2 are not available elsewhere in book or paper form. (Procedure 4.6.1, in particular, is highly efficient for hand computation, and useful when a minimal sum cannot be conveniently handled by the Karnaugh map method. Also, Procedure 5.1.1 or Procedure 5.1.2 can yield a minimal sum for a function of many variables very quickly if the given function is not too complex. Where an engineer in industry, as it has come to light, tried to find a minimal sum by working on 16 maps simultaneously, these procedures would have provided a powerful improvement.) By studying these chapters, students can learn techniques to speed up calculation in switching algebra.

Logic design of NAND (or NOR) gate networks is usually not discussed in detail in textbooks despite its vital importance in current IC technology. A major part of Chapter 6 is thus devoted to this subject. To treat this problem, the map-factoring method of G. A. Maley and J. Earle is discussed, with minor modification. Also, the author's extensions (not available elsewhere) are presented as Procedures 6.4.1 and 6.5.1. The rest of Chapter 6 is devoted to problems that are important in design practice; in particular, Section 6.7 discusses the response time of combinational networks.

Chapters 7 and 8 are devoted to sequential networks, the design of which is inherently complex and consequently difficult to teach. If we emphasize easy understanding, as has been done in the past, the complex situation is oversimplified, giving students an unrealistic picture. On the other hand, if we emphasize the realistic picture of sequential network design, this may be harder for students to understand. Facing this dilemma, the author concluded that it may be time to give the realistic picture, since logic design is becoming a more important profession and this book is for an advanced course. On the basis of this decision (this stance is maintained throughout the book—procedures would not be useful if calculation techniques were not detailed), asynchronous sequential networks, considerably more complex than synchronous sequential networks, are first discussed in Chapter 7—because sequential networks in fundamental mode are basic in both asynchronous and synchronous sequential networks (see the next paragraph); because unless the malfunctioning of asynchronous networks due to hazards or races is understood, the advantages of synchronous networks with raceless flip-flops may not be appreciated; and because asynchronous sequential networks are useful for high-speed computers.

Another problem in the discussion of sequential networks is pulse mode, which is widely presented in literature. Since the concept of pulse mode is only vaguely defined or inappropriate in electronic implementation (see the footnote in Section 8.3), skew mode is introduced in conjunction with raceless flips-flops, with the view that the concept is simply a convenient means to analyze or synthesize sequential networks with raceless flip-flops, which can also be analyzed or synthesized by fundamental mode. Also, there are sequential networks that are not in skew mode but in fundamental mode.

No attempt is made to present elaborate mathematical results on sequential networks, which are abundant in literature, because space is limited and such results, though conceptually important and exceedingly interesting academically, are time-consuming in nature and hence usually difficult to apply in design practice.

Chapter 9 tries to bridge the developments in the preceding chapters with logic design practice, discussing computer design practice, various design motivations, design time, important standard networks, logic networks with a mixture of gates and memories, PLAs (programmable logic arrays), flow charts for logic design, and diagnosis methods. Some of these, such as PLAs, are important topics for CAD programs. Since logic design practice is becoming enormously complex, as mentioned in the beginning, this book does not attempt to discuss the entire picture of design practice or all aspects of computer system design. The discussion in Chapter 9 is limited to the aspects related to the preceding chapters. The design procedures in the earlier chapters are useful in LSI design—for example, PLAs and gate arrays, widely used as inexpensive LSI implementations. The author's procedure in Section 9.3(b) is not available elsewhere.

The remarks scattered throughout this book provide pertinent information for advanced readers, with the intention of making this book serve as a reference book. In college courses, it is appropriate that students skip these remarks.

This book is the outcome of lecture notes used for the last several years at the University of Illinois at Urbana. Many people have helped on numerous occasions, for which the author is grateful. In particular, Professors L. L. Dornhoff, the late F. E. Hohn, G. A. Metze, and D. E. Muller, and Dr. J. E. Forbes, used the classnotes in their teaching; and Messrs. R. B. Cutler,

W. T. Dumas, R. Fujimoto, S. Isoda, S. Murai, and T. Ogino, A. Sakurai, Professor D. S. Watanabe, Mrs. M. H. Young, and Dr. R. O. Winder, made valuable discussions and suggestions. Also, especially, Professor L. L. Dornhoff read through the entire manuscript repeatedly and made many corrections and improvements, and Mr. R. B. Cutler's discussions on the procedures in Sections 5.1 and 5.2 were invaluable. The Department of Computer Science at the University of Illinois provided generous support for the preparation of the classnotes, and grants by the National Science Foundation (GJ-503-1, GJ-40221, DCR73-03421 A01, and MSC77-09744) produced research results used throughout this book. Finally, thanks to Mrs. Z. Arbatsky and Mrs. R. Taylor for their excellent and patient typing of many revisions of the classnotes.

Urbana, Illinois SABURO MUROGA
January 1979

How To Use This Book

On the Exercises

A, B, C, ... denote exercises of the same nature, so only one of them should be assigned.

- (R) means that an exercise serves to review what is discussed in text.
- (E) means that an exercise is easy.
- (M) means that an exercise is of medium difficulty. Among those marked with M, exercises easier than the average and those more difficult are designated as $M-$ and $M+$, respectively, the difficulty increasing in the order of $M-$, M, and $M+$.
- (D) means that an exercise is difficult.
- (T) means that an exercise is time-consuming.

On the Sections Labeled with ▲

Sections or topics labeled with ▲ are optional, and skipping them will not impede the understanding of later material. Exercises corresponding to these sections or topics are labeled with ▲.

On the Remarks

Remarks present advanced information, and skipping them will not impede the understanding of later material. It is advised that in college teaching, the remarks be completely skipped.

Contents

CHAPTER 1
INTRODUCTION 1

    1.1 Motivations for Study of Logic Design and Switching Theory 1
    1.2 History 5
    1.3 Introduction to Basic Logic Operations 6
    1.4 Truth Tables 11

CHAPTER 2
IMPLEMENTATION OF LOGIC OPERATIONS AND ANALYSIS OF COMBINATIONAL NETWORKS 15

    2.1 Basic Properties of Relays and Switches 15
    2.2 Analysis of Relay Contact Networks and Electrical Switch Networks 22
    Exercises 29
    2.3 Analysis of Networks of Electronic Gates 36
           Analysis of MOS Networks 36
           Analysis of Bipolar Transistor Networks 46
           Features of Electronic Gate Networks 50
    2.4 Discrete Components and Integrated Circuits 53
    Exercises 61

CHAPTER 3
FUNDAMENTALS OF SWITCHING ALGEBRA 68

    3.1 Theorems of a Few Variables 68
    3.2 Theorems of n Variables 71
    Exercises 82
    3.3 Karnaugh Maps 87

3.4   Implication Relations and Prime Implicants   92
3.5   Algebraic Design Methods and Computer-Aided Design   106
      Exercises   108
▲ 3.6   Boolean Algebra   111
▲ *Exercises*   113
▲ 3.7   Propositional Logic   114
▲ *Exercises*   118
      Exercises   120

CHAPTER 4

SIMPLIFICATION OF SWITCHING EXPRESSIONS   124

4.1   Design Objectives and Minimal Sums   124
4.2   Derivation of Minimal Sums by Karnaugh Map   139
4.3   Prime Implicates, Irredundant Conjunctive Forms, and
      Minimal Products   147
4.4   Design of Two-level Minimal Networks with AND and
      OR Gates   154
      Exercises   156
4.5   Tabular Method to Derive a Minimal Sum   163
4.6   Algebraic Method to Derive Minimal Sums   180
      Exercises   188

CHAPTER 5

ADVANCED SIMPLIFICATION TECHNIQUES AND
BASIC PROPERTIES OF GATES   195

5.1   Derivation of Irredundant Disjunctive Forms without
      Use of Minterms   195
      Exercises   212
5.2   Design of a Two-Level Multiple-Output Network with
      AND and OR gates   214
      Exercises   239
5.3   Comparison of the Different Methods for Derivation of
      a Minimal Sum   245
▲ 5.4   Design of Networks with AND and OR Gates under
      Arbitrary Restrictions   247
5.5   Gate Types   252
      Exercises   263

CHAPTER 6

DESIGN OF NAND (NOR) NETWORKS AND PROPERTIES OF COMBINATIONAL NETWORKS   272

6.1	Introduction to Design of NAND (or NOR) Networks	273
6.2	Switching expressions for NAND (or NOR) Networks	277
6.3	Design of NAND (or NOR) networks in Single-Rail Input Logic by the Map-Factoring Method	281
▲ 6.4	Design of Three-Level NAND (or NOR) Networks in Single-Rail Input Logic	301
▲ 6.5	Design of NAND (or NOR) Networks in Double-Rail Input Logic by the Map-Factoring Method	308
▲ 6.6	Other Design Methods for NAND (or NOR) Networks	310
	Exercises	312
6.7	Transient Response of Combinational Networks	320
▲ 6.8	Classification of Switching Functions and Networks	327
6.9	General Comments on Combinational Networks	336
	Exercises	337

CHAPTER 7

ASYNCHRONOUS SEQUENTIAL NETWORKS   347

7.1	Latches	347
7.2	Sequential Networks in Fundamental Mode	352
7.3	Malfunctions and Performance Descriptions of Sequential Networks	359
	Exercises	367
7.4	Introduction to the Synthesis of Sequential Networks	373
7.5	Translation of Word Statements into Flow-Output Tables	385
	Exercises	391
7.6	Minimization of the Number of States	393
7.7	State Assignment	413
7.8	Design of Sequential Networks in Fundamental Mode	421
7.9	General Comments on Asynchronous Sequential Networks in Fundamental Mode	427
	Exercises	429

CHAPTER 8
SYNTHESIS OF SYNCHRONOUS SEQUENTIAL NETWORKS 440

    8.1 Clocked Networks   440
    8.2 Raceless Flip-Flops   451
    Exercises   460
    8.3 Sequential Networks with Master-Slave Flip-Flops in Skew Mode   463
    8.4 Translation of Word Statements into Flow-Output Tables   472
    8.5 Design of Sequential Networks in Skew Mode   483
    8.6 General Comments on Sequential Networks   490
    Exercises   492

CHAPTER 9
PRACTICAL CONSIDERATIONS IN LOGIC DESIGN 502

    9.1 Diversified Design Motivations and Design Approaches   502
    9.2 Design of Standard and Large Networks   510
▲ 9.3 Design of Sequential Networks with Given Building Blocks   521
    Exercises   534
    9.4 Design of Networks with a Mixture of Memories and Gates   538
    Exercises   559
    9.5 Logic Design with Flow Charts   564
    Exercises   573
    9.6 Simulation, Test, and Diagnosis   574
    Exercises   585

REFERENCES 593
INDEX 613

Logic Design and
Switching Theory

CHAPTER 1

Introduction

Digital systems such as electronic digital computers or electronic telephone exchanges contain digital networks that perform logic operations on digital input signals. Each digital network consists of gates that perform basic logic operations. This book discusses the design of the logic operation aspect of digital systems (in contrast to electronic design or others) as well as a related theory—in other words, logic design of networks with gates and switching theory.

Chapter 1 discusses motivations, history, and some basic concepts of logic design and switching theory.

1.1 Motivations for Study of Logic Design and Switching Theory

Logic design is the design of digital networks with gates so that the networks perform specified logic operations on input signals, in contrast to electronic design of digital networks, which derives electronic circuits with the desired electronic performance, such as the desired speed, power consumption, or waveforms. In a broader sense, "logic design" includes the design of structural aspects of the entire computer systems, that is, the architectural design of computer systems. The distinction between these meanings of "logic design" is often not very clear. This book deals mainly with logic design in the first sense. The design theory of digital networks is called **switching theory**, since it is a theory not only for networks of gates but also for those of switching devices such as relays or electrical switches that are older digital networks. A special case of Boolean algebra that serves as a basic mathematical tool in switching theory is called **switching algebra**.*

The integrated circuit (IC) is a recently developed and still developing approach to implementing gates, connections, or entire networks on a tiny semiconductor chip in an integrated way. The development of IC technology

---

* Very often, "Boolean algebra" is used by authors as a synonym of "switching algebra" because the difference is subtle. Strictly speaking, however, switching algebra is a special case of Boolean algebra, as will be explained in Section 3.6.

2　Introduction

has brought about many changes. The enormous reduction in the cost and size of digital networks due to IC technology is expanding the applications of digital systems from the traditional ones, such as digital computers and electronic telephone exchanges, into many new areas. Examples of such new applications are sewing machines (e.g., Singer's Athena 2000), cameras (e.g., Polaroid SX-70), digital watches, video games, nonvideo games (e.g., Parker Brother's Sector), automobiles, and home computers [Weisbecker 1974]; many others are yet to come. Thus logic design not only is indispensable for computers but also is becoming very important for all products that incorporate digital systems (no matter how small or large these systems may be). As IC technology is still making great strides, it is very difficult at the current stage to foresee what great technological or social impacts it will have. In the following paragraphs we will review the changes in logic design that IC technology is making.

If a designer is not attempting to attain in his or her networks the best electronic performance (e.g., speed) or economy possible with current IC technology, logic design is now much simpler than before, since a large number of ready-made IC packages (i.e., **off-the-shelf packages**), which contain **standard networks**, that is, frequently used networks such as adders or multipliers, are commercially available. What the designer must do is simply to choose appropriate off-the-shelf packages and then interconnect them. In particular, fairly complex digital systems can be quickly and easily implemented by using off-the-shelf microcomputers. Consequently a designer need not design individual networks from scratch, though it may occasionally be necessary to so design some small interface networks which interconnect the off-the-shelf packages. Architectural considerations or clever assembling of IC packages is important.

Custom design of large scale integrated circuit packages, that is, **LSI packages**, including custom design of microcomputer packages such as those for automobiles [Puckett, Marley, and Gragg 1977], is becoming popular, as will be explained in Section 2.4. If a designer does not attempt to exert time-consuming efforts in laying out networks on a chip inside an LSI package, packing as many gates as possible, he or she can take an approach similar to the assembling of off-the-shelf packages explained above. In other words, for each custom LSI design, the designer simply assembles standard layouts, which are previously prepared for frequently used networks, on an LSI chip. With this approach an LSI chip can be easily and quickly custom-designed, though best performance is seldom attained. The short design time reduces design cost. On the other hand, the manufacturing cost of the designed LSI is usually high. This can be justified, however, when the LSI chip is part of much more expensive products such as measurement instruments or intelligent terminals.

If a designer wants to attain the best economy or electronic performance achievable with current IC technology, however, logic design is far more difficult and challenging for the reasons presented below. Excellent LSI designers will become the "superstars" of the profession [Petritz 1977].

First, unlike the days when only two logic families, TTL (transistor–transistor logic) for low cost and ECL (emitter coupled logic) for high speed, were available, there now exist a large number of IC **logic families**, that is, different types of electronic circuits to implement gates or logic operations. As these logic families have different features in regard to economy, performance, and size, the designer must single out the most appropriate family. Design procedures for each logic family are different.

Second, networks can be implemented with discrete components or with IC packages of different integration sizes, that is, SSI (small scale integration), MSI (medium scale integration), LSI, or VLSI (very large scale integration). Also, LSI or VLSI packages can be either custom-designed or off-the-shelf packages. The designer must choose the most appropriate ones (possibly a mixture) by considering production volume, completion time, design change possibilities, cost, size, and performance. If frequent design changes are expected, discrete components are still the common choice. When production volume is high and no design change is expected, custom LSI or VLSI design is becoming popular. In this case, if the designer is trying to attain the best electronic performance or economy, custom design is a highly challenging problem, since it is necessary to pack gates as tightly as possible and to find the most appropriate layout under complex layout rules. Whenever new IC processing methods or layout rules are developed, designers usually must redesign even standard networks from scratch.

Third, IC technology has made possible the replacement of gates by memories (mainly, read-only memories), since memories are becoming compact, inexpensive, and fast, and consequently logic networks are often designed with a mixture of gates and memories, in order to reduce cost. (In this case, memories are not used to store computer programs, data, and intermediate computational results, as is the conventional practice in computers.) Finding the best approaches to using memories is not an easy task for designers.

Fourth, enormously large networks are being designed compactly and economically because of the ever-decreasing size and cost of gates, but it is becoming vitally important to design such very large networks without errors, as well as to determine whether or not designed or manufactured IC packages or digital systems are faulty. Logic design must take care of this problem.

In addition, there are many other problems such as reliability of systems or power supplies.

4   Introduction

In summary, **logic design is now greatly diversified**, especially since it is becoming important in many areas such as digital systems in cameras and sewing machines. Thus **the term "logic design" has very different meanings for different design situations**. Also, depending on the situation, logic design can be an enormously complex and time-consuming task, as explained above, requiring a wide knowledge of IC technology, electronic circuits, architecture, arithmetic methods, microprogramming, microcomputers, software, and computational algorithms. All these subjects* are closely related, though the extent of knowledge required varies depending on where one works in this broad spectrum of subjects in logic design.

A noteworthy important trend is the implementation of software not only by firmware but also by hardware. Implementation by firmware (i.e., implementation of software by read-only memory) is now extensively used; implementation by hardware is done when further speed-up of the operation of computers is desired. (Some notion of this will be given in Chapter 9.) Hardware implementation of software will yield much faster computation than will software in conventional form or firmware [Falk 1974, 1975; Goetz 1975]. Also, this will probably tend to reduce the ever-increasing complexity of software. Therefore even software people need to know something about logic design and switching theory. As was said as early as 1971,† "In the future programmers and hardware engineers will not be [members of] different professions. They will be merged into computer engineers, requiring both software and hardware knowledge." They might be called logico-software engineers.

As mentioned already, logic design, in the narrow sense to which the scope of this book is limited because of space, means the design of networks with logic gates, without going deeply into other aspects, such as electronic circuits or the architecture of digital systems. The major objective of logic design is to design, in as short a time as possible, logic networks with optimum cost or performance which realize given functions under various constraints.

Logic design is still an art rather than a science. Computer-aided design (**CAD**) is extensively used, but tends to be limited to simple aspects of logic

---

* Typical references for each subject are as follows. For electronic circuits: Barna and Porat [1973]; Millman and Halkias [1972]; Luecke, Mize, and Carr [1973]; Carr and Mize [1972]; Taub and Schilling [1977]. For integrated circuit technology: Streetman [1972]; Hamilton and Howard [1975]; Warner [1965]; Hnatek [1973, 1977b]; Engineering Staff of AMI [1972]. For arithmetic methods: Richards [1955, 1971]; Flores [1973]; Schmid [1974]. For computer architecture: Mano [1976]; Abd-Alla and Meltzer [1976]; Hill and Peterson [1973]; Bell and Newell [1971]; Blaauw [1976]. For microcomputers: Peatman [1977]; Hilburn and Julich [1976]; McGlynn [1976]; Ogdin [1976, 1978a, 1978b]; Lin [1977]. For microprogramming: Chu [1972]; Davies [1972]; Husson [1970].

† "Computer in the 70's," *Electronics*, Sept. 13, 1971, p. 62.

design (e.g., checking design errors by simulation) and manufacturing processes (e.g., assemblying IC packages on printed-circuit boards). The key aspects of logic design are so complex that a CAD approach remains only an aid; the key aspects must be done by human designers. Progress in IC technology, however, is making enormously large scale integration feasible. It is extremely important to design such an enormously large network on a single chip without errors in a short time, so efforts to further utilize CAD are being enhanced. In view of this trend, algebraic methods of logic design are emphasized throughout this book, as well as the tabular or map methods conventionally discussed in textbooks. Another important trend is hardware implementation of complex job functions. When hardware was expensive, only simple job functions such as addition were implemented with logic networks. Such simple job functions can be easily and directly expressed in switching expressions and easily implemented with logic networks. Now hardware is inexpensive, and much more complex job functions can be implemented with logic networks. It is usually difficult, however, to express such complex job functions directly in switching expressions. These functions must be treated with appropriate algorithms (which may contain nonswitching algebraic algorithms) before being implemented with logic networks. This trend is another reason why algebraic methods are emphasized in this book.

1.2 History

Boolean algebra was introduced by George Boole more than 100 years ago. It was not used for practical purposes, however, until the late 1930s, although today it is common knowledge, at least for people associated with computers. It may be interesting for us to learn how Boolean algebra was introduced [Bell 1937, 1945]. Boole took a strong interest in an absurd controversy over logic between the Scottish philosopher William Hamilton (not to be confused with the famous mathematician William Rowan Hamilton) and De Morgan. Hamilton was known for his attack on mathematics; he wrote, "Mathematics freeze and parch the mind; an excessive study of mathematics absolutely incapacitates the mind for those intellectual energies which philosophy and life require." He publicly accused De Morgan of plagiarism on the "quantification" of the predicate. In 1848 Boole wrote a pamphlet entitled *The Mathematical Analysis of Logic*, to defend De Morgan, his friend. Six years later he published a book on Boolean algebra, entitled *An Investigation of the Laws of Thought* (reprinted in 1954 by Dover Publications). Then in 1859 he discussed the general symbolic method in the book *Treatise on Differential Equations*. Boolean algebra was thus started under a strange circumstance of hostility.

6 Introduction

The first application of Boolean algebra to the analysis of relay contact networks was attempted by A. Nakashima in 1937. Unaware of the theory already outlined, he tried to develop Boolean algebra by himself. In 1938 C. E. Shannon in the Electrical Engineering Department of MIT wrote a brilliant master's thesis on the application of Boolean algebra to the analysis of relay contact networks. His thesis, which elegantly utilizes the mathematical theory of Boolean algebra, has been the basis for the development of switching theory since then.

After World War II many people extended switching theory to cover not only simple networks called "combinational networks," which respond to only the current values of the inputs, but also "sequential networks," which respond to both the past and current values of the inputs. Synthesis of a sequential network was started independently by D. A. Huffman (Ph.D. thesis, Electrical Engineering Department of MIT) and E. F. Moore, though their papers were published in [1954] and [1956], respectively. Further mathematical sophistication of the theory brought in the finite-state machine theory (i.e., a theory for a machine which has a finite-size memory). The theory was combined with the theory of the Turing machine, which requires an infinite-size memory, and developed into "automata theory." Also, automata theory was extended to "formal language theory," a theory of languages such as natural languages and computer programming languages.

Since World War II many gates based upon new physical phenomena have been devised. A theory that deals with a certain broad class of gates, including conventional ones such as NOR and AND gates, was developed. This theory is called "threshold logic."

Although mathematical theory about abstract aspects of computers has been highly developed (as outlined above), theory for practical logic design does not appear to have progressed much in spite of the fact that the advent of integrated circuitry and an increasing demand for economical manufacture of digital systems make logic design much more complicated than before. However, switching theory is still an important tool for logic design.

1.3 Introduction to Basic Logic Operations

Usually information inside digital systems such as digital computers and electronic telephone exchanges is represented by a binary signal, because of ease of implementation and reliable operation of digital systems. Thus switching algebra, which handles the binary values 0 and 1, is extensively used for designing digital systems.

A digital computer, as well as many other digital systems, consists in general of four basic units: arithmetic unit, control unit, memory unit, and

Introduction to Basic Logic Operations 7

Fig. 1.3.1 Addition of binary numbers.

input–output (or I/O) unit. An adder is the most important part of the arithmetic unit. Let us try to implement an adder (i.e., a network to add two binary numbers) in order to give some idea of how switching algebra is used in computers.

Addition of two binary numbers can be done in a similar way to the decimal case, as shown in Fig. 1.3.1. A carry and a sum digit at the ith digit position are formed from two digits a_i and b_i and carry c_{i-1} of the $(i-1)$st digit position, according to Table 1.3.1. Therefore a network for the addition (i.e., an adder) may be schematically expressed with blocks, as shown in Fig. 1.3.2, if we want to add two numbers, (a_3, a_2, a_1) and (b_3, b_2, b_1). The sum is a binary number expressed by (d_4, d_3, d_2, d_1).

Table 1.3.1 Addition

a_i	b_i	c_{i-1}	SUM DIGIT	CARRY c_i
0	0	0	0	0
0	0	1	1	0
0	1	0	1	0
0	1	1	0	1
1	0	0	1	0
1	0	1	0	1
1	1	0	0	1
1	1	1	1	1

Remark 1.3.1: The rightmost pair of blocks in Fig. 1.3.2 does not have carry inputs. This pair is called a **half adder**, while the pairs in other positions which have carry inputs are called **full adders**. An adder actually used in a computer has carry inputs at the rightmost pair of blocks so that this adder can perform subtraction as well.* The ripple adder in Fig. 1.3.2, which is a cascade of full adders and/or a half adder, is simple, but the simplest adders will be shown in Section 9.2. ∎

---

* For details, see Mano [1976], or Section 8.1 of Gschwind and McClusky [1975], for example.

8 Introduction

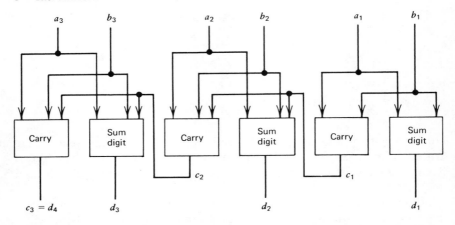

*Fig. 1.3.2   Adder [more specifically, this is called a **ripple adder** (or a carry-ripple adder) in order to differentiate from it adders with more complicated structures].*

Each of the carry and sum bits can be expressed with the three logic operations OR, AND, and NOT. It can also be expressed with other logic operations, as explained in a later chapter.

The **OR operation** of n variables $x_1, x_2, \ldots, x_n$ yields the value 1 whenever at least one of the variables is 1, and 0 otherwise, where each of $x_1, x_2, \ldots, x_n$ assumes the value 0 or 1. This is denoted by $x_1 \vee x_2 \vee \cdots \vee x_n$. In other words, the OR operation defines the function whose value is specified as shown in Table 1.3.2, for example, for $n = 3$. The OR operation defined above is sometimes called "disjunction," "logical sum," "union," "join," or "alternation." (For historical reasons, operations like this have many

Table 1.3.2   OR Operation

x_1	x_2	x_3	$x_1 \vee x_2 \vee x_3$
0	0	0	0
0	0	1	1
0	1	0	1
0	1	1	1
1	0	0	1
1	0	1	1
1	1	0	1
1	1	1	1

different names. Even now, different people use different names.) Throughout this book, **disjunction** will be used very often. Also, some authors use " + " instead of " ∨," but since " + " means the ordinary arithmetic sum in this book, " ∨ " is used to avoid confusion.*

The **AND operation** of n variables yields the value 1 if and only if all variables $x_1, x_2, \ldots, x_n$ are simultaneously 1. This is denoted by $x_1 \cdot x_2 \cdot x_3 \cdots x_n$. An example for $n = 3$ is shown in Table 1.3.3. Since

Table 1.3.3 AND Operation

x_1	x_2	x_3	$x_1 \cdot x_2 \cdot x_3$
0	0	0	0
0	0	1	0
0	1	0	0
0	1	1	0
1	0	0	0
1	0	1	0
1	1	0	0
1	1	1	1

this AND operation is no different from the ordinary multiplication of arithmetic, often these dots are omitted: $x_1 x_2 \ldots x_n$. The AND operation is sometimes called "conjunction," "logical product," or "meet." Throughout this book, **conjunction** will be used often. Some authors use " ∧ " instead of " ⋅ ."

The **NOT operation** of a variable x yields the value 1 if $x = 0$, and 0 if $x = 1$, as shown in Table 1.3.4. This is denoted by $\bar{x}$. Sometimes x' or $\sim x$ is used instead of $\bar{x}$. The NOT operation is sometimes called "complement," "inversion," "negation," or "denial."

Table 1.3.4
NOT Operation

x	$\bar{x}$
0	1
1	0

* In practice, " + " may be preferable, since " ∨ " may be confused with the letter "v," unless, as in this book, " + " is used to mean something else.

10 *Introduction*

Using these operations, AND, OR, and NOT, we can write the sum digit and carry in Table 1.3.1 for the ith digit position of the adder as follows:

$$\text{Sum digit for the } i\text{th bit position} = a_i \bar{b}_i \bar{c}_{i-1} \vee \bar{a}_i b_i \bar{c}_{i-1} \vee \bar{a}_i \bar{b}_i c_{i-1} \vee a_i b_i c_{i-1} \quad (1.3.1)$$

$$\text{Carry for the } i\text{th bit position} = a_i b_i \vee a_i c_{i-1} \vee b_i c_{i-1} \quad (1.3.2)$$

Expressions with logic operations AND, OR, and NOT, such as expressions 1.3.1 and 1.3.2, are called **switching expressions** or **Boolean expressions**. Variables a_i, b_i, and c_{i-1} (also $x_1, x_2, \ldots, x_n$ in the above definitions of logic operations) are sometimes called **switching variables**, and they assume only binary values 0 and 1. When variables are inputs to a network or outputs of a network, they are called **input variables** or **output variables**, respectively.

Readers may check that expressions 1.3.1 and 1.3.2 yield the values shown in Table 1.3.1 for each combination of variable values, by proceeding with the calculation according to Tables 1.3.2 through 1.3.4. When the value of a switching expression for a specific combination of values of variables is to be calculated, **the calculation of AND operations precedes that of OR operations**, conventionally. If there are pairs of parentheses, AND precedes OR in each pair of parentheses. Then we go to the next outer pair of parentheses. For example, when $xy \vee (uz \vee w(x \vee yz))$ is given, the calculation goes in the order indicated by the numbers:

$$xy \vee (uz \vee w(x \vee yz)) \quad (1.3.3)$$

Here the reader may wonder why the term $a_i b_i c_{i-1}$ corresponding to the bottom row in Table 1.3.1 is not contained in the expression for carry, but it can be seen, by comparing the value of expression 1.3.2 for each combination of values of a_i, b_i, c_{i-1} with the value of carry given in Table 1.3.1, that expression 1.3.2 without this term still expresses the carry. Of course the expression with the term included also expresses the carry. (In a later chapter we discuss how to eliminate such redundant terms and how to derive the simplest expression.)

Thus, using AND, OR, and NOT, we can express the carry and sum bits for the ith bit position in the adder as shown in Fig. 1.3.3. This is called a **full adder**.

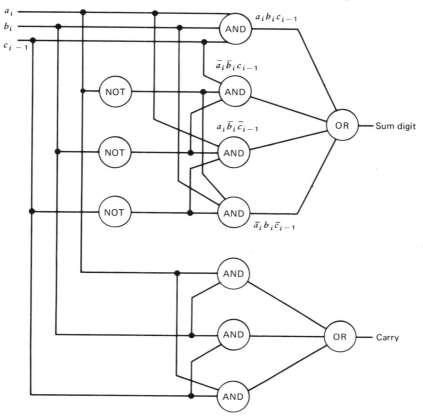

*Fig. 1.3.3 Adder (more specifically, this is called a **full adder**).*

In Chapter 2 we will discuss how the operations OR, AND, and NOT are implemented with physical means, such as relay contacts and electronic gates. A major topic in the succeeding chapters is the design of a network of relays or electronic gates that realizes given functions (such as the sum and carry bits in the above example of the adder). This is called **logic design** (in a narrow sense, in contrast to "logic design" in the broad sense explained in Section 1.1).

1.4 Truth Tables

When we analyze a network, there are three ways to describe its performance:

1. The output of a network is represented in a switching expression in terms of input variables, such as (1.3.1) and (1.3.2). The expression that

12 *Introduction*

results may be complicated. Simplification of such an expression will be discussed in later chapters.

2. The performance of a network may be described in terms of a table such as Tables 1.3.1 through 1.3.4 and also Table 1.4.1, where the output value of the network is shown for each combination of values of input variables. This table is called a **truth table** (or a "combination table"). Since a network of n input variables has 2^n rows in the table, the table size increases rapidly as n increases. A simpler representation of a truth table is discussed later in this section.

3. The performance of a network may be expressed in a word statement. As an example, consider the truth table of Table 1.4.1. The value of f is

Table 1.4.1 A Truth Table (for parity function $x \oplus y \oplus z$)

x	y	z	f
0	0	0	0
0	0	1	1
0	1	0	1
0	1	1	0
1	0	0	1
1	0	1	0
1	1	0	0
1	1	1	1

1 only when the number of variables that assume the value 1 is odd. Thus f may be said to be "the modulo 2 sum of the variables." This f is also called the **parity function** and denoted by $x \oplus y \oplus z$. This statement is much simpler than the truth table. Another example is an adder of two n-bit numbers. If it is represented by a truth table, the table has 2^{2n} rows. For $n = 16$, it means $2^{32} \approx 4 \cdot 10^9$ rows. "Adder" is a much simpler word statement, but some functions may be too complicated for word statements.

Word statements are convenient and logic design is often based on them; however, when we want to design the corresponding networks, word statements must usually be converted to switching expressions. This is sometimes very complicated, particularly in the case of sequential networks, and consequently the desired network performance may be misstated or misunderstood, leading to design errors. Nevertheless word statements are often used. The other two types of descriptions are also used for sequential networks, depending on convenience in each case.

Introduction to Basic Logic Operations

Under certain circumstances, some of the combinations of input variable values never occur. These combinations are called **don't-care conditions** and are denoted by d, as shown in Table 1.4.2. If confusion does not occur, the rows that contain d's may be deleted from the table, thereby reducing its size.

Table 1.4.2 A Truth Table with Don't-Care Conditions

	x	y	z	f
0	0	0	0	0
1	0	0	1	1
2	0	1	0	d
3	0	1	1	0
4	1	0	0	d
5	1	0	1	1
6	1	1	0	1
7	1	1	1	0

Decimal Specifications

As the number of variables increases, the number of rows in a truth table dramatically increases. Writing down all these rows is cumbersome. A concise means of expressing the truth table is to list only rows with $f = 1$, in the following manner. Let us consider the truth table of Table 1.4.3. The set of values of the variables in the second row is the binary number 0 0 1, which represents the decimal number 1. The set in the fourth row is the binary

Table 1.4.3 A Truth Table

	x	y	z	f
0	0	0	0	0
1	0	0	1	1
2	0	1	0	0
3	0	1	1	1
4	1	0	0	0
5	1	0	1	0
6	1	1	0	1
7	1	1	1	0

number 0 1 1, which represents 3, and so on. In other words, we express the binary numbers with $f = 1$ as decimal numbers. Then we list them as

$$f(x, y, z) = \Sigma(1, 3, 6).$$

Notice that the 0's in column f in the table are not considered in this representation. If we change the order of variables (i.e., the order of the columns for variables in the truth table), the rows generally have different corresponding decimal numbers. Hence we need to show the order of variables, x, y, z, as shown above. Only when writing simply f causes no ambiguity, can we write

$$f = \Sigma(1, 3, 6),$$

instead.

Considering only rows with $f = 0$, we can also express the truth table in Table 1.4.3 as

$$f(x, y, z) = \Pi(0, 2, 4, 5, 7),$$

using the different notation Π.

When we have a truth table containing don't-care conditions, such as Table 1.4.2, the table is expressed as

$$f(x, y, z) = \Sigma(1, 5, 6) + d(2, 4)$$
$$= \Pi(0, 3, 7) + d(2, 4).$$

The above **decimal specifications** are often used because of conciseness.

Another concise expression of a truth table is a Karnaugh map, to be discussed in later chapters.

CHAPTER **2**

Implementation of Logic Operations and Analysis of Combinational Networks

In this chapter we discuss first how logic operations such as AND, OR, and NOT, introduced in Section 1.3, are implemented by physical means, such as relays and electronic gates. Then we discuss the analysis of a network, that is, the derivation of a switching expression for the output of a given network.

2.1 Basic Properties of Relays and Switches

Relays are probably the oldest means to realize logic operations which have been widely used in industry. They are still extensively used in telephone exchanges, though telephone exchanges based on relays are being replaced by electronic telephone exchanges. [In electronic telephone exchanges all logic operations are done with electronic gates instead of relays, and relays (reed relays,* which have a somewhat different structure from conventional ones [Torrero 1971]) are used for speech paths but not for logic operations.] Although the use of relays for logic operations appears to be declining† in many areas in industry, it will continue for years to come.

In this book we want to discuss relays for the following reasons:

1. Relays are conceptually simple and appropriate for the introductory discussion of logic operations.
2. The logic performance of a relay contact network is essentially identical to that of the interconnected electrical switches used in our homes. If we know logic design with relays, it is easy to design a network of electrical switches.

---
* Reed relays for speech paths also may be replaced by solid state devices in the future.
† Solid state relays appear to be gaining in popularity over electromechanical relays. Theoretical discussions in this chapter based on electromechanical relays are applicable also to the case of solid state relays.

16 *Implementation of Logic Operations*

3. More importantly, the connection configuration of a relay contact network is very similar to that of a metal oxide semiconductor (MOS) gate, which is an important and widely used integrated circuit logic family of electronic gates. (MOS gates and integrated circuits will be explained later.) Some electronic desk calculators, microcomputers, and minicomputers employ MOS gates for their logic operations.
4. Logic design techniques with relay contacts are useful for designing special types of networks* with ECL gates (more specifically, transistor-level logic design of ECL networks in large-scale integration), where ECL gates are another important integrated circuit logic family of gates for high-speed networks.

In conclusion, logic design with relay contacts is worth learning because it has many important implications, as explained above.

A **relay** (its size is usually a few cubic inches) consists of an armature, a magnet, and a metal contact. An armature is a metal spring made of magnetic material with a metal contact on it. (There are many variations in the structures of armatures.) There are many different structures for relays, which can be classified into a few different types. Most important are a make-contact relay and a break-contact relay.

A **make-contact relay** is a relay such that, when there is no current through the magnet winding, the contact is open. When a direct current is supplied through the magnet winding, the armature is attracted to the magnet and, after a short time delay, the contact is closed. This type of relay contact is called a "make-contact" and is usually denoted by a lower-case x. The current through the magnet is denoted by a capital X, as shown in Fig. 2.1.1.

A **break-contact relay** is a relay such that, when there is no current through the magnet winding, the contact closes. When a direct current is supplied,

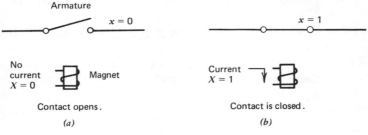

Fig. 2.1.1 A "*make-contact*" relay.

---

* For example, "series gating" (see Motorola [1972]).

Basic Properties of Relays and Switches 17

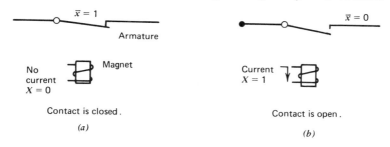

Fig. 2.1.2 A "break-contact" relay.

the armature is attracted to the magnet and, after a short time delay, the contact opens. This type of relay contact is called a "break-contact" and is usually denoted by $\bar{x}$, a lower-case x with a bar. The current through the magnet is again denoted by a capital X, as shown in Fig. 2.1.2.

In the case of a make-contact relay, a magnet having no current is represented by $X = 0$, and then $x = 0$ implies that the contact is open, as shown in Fig. 2.1.1a. A magnet having current is represented by $X = 1$, and then $x = 1$ implies that the contact is closed, as shown in Fig. 2.1.1b. In a break-contact relay, when a magnet has no current, it is represented by $X = 0$ and also $x = 0$. Thus we have $\bar{x} = 1$, which implies that the contact is closed, as shown in Fig. 2.1.2a. When a magnet has a current, we have $X = 1$ and $x = 1$. Thus we have $\bar{x} = 0$ and the contact is open, as shown in Fig. 2.1.2b. Here notice that **the value of X is identical to that of x** in both make-contact and break-contact relays. (Some authors assign $X = 1$ and $x = 1$, respectively, to express no current in the magnet and open contact, contrary to the above. Exercise 2.2.5.)

Let us connect two make-contacts x and y in series, as shown in Fig. 2.1.3. Then we have the combinations of states shown in Table 2.1.1a, which includes only two states, "open" and "close." Let f denote the state of the entire path between terminals a and b, where f is called the **transmission** of the network. Since "open" and "close" of a make-contact are represented by $x = 0$ and $x = 1$, respectively, Table 2.1.1a may be rewritten as shown

Fig. 2.1.3 Series connection of relay contacts.

18 Implementation of Logic Operations

Table 2.1.1 Combinations of States for the Series Connection in Fig. 2.1.3

a.	x	y	ENTIRE PATH BETWEEN a AND b	b.	x	y	f
	Open	Open	Open		0	0	0
	Open	Close	Open		0	1	0
	Close	Open	Open		1	0	0
	Close	Close	Close		1	1	1

Table 2.1.2 Combinations of States for the Parallel Connection in Fig. 2.1.4

a.	x	y	ENTIRE PATH BETWEEN a AND b	b.	x	y	f
	Open	Open	Open		0	0	0
	Open	Close	Close		0	1	1
	Close	Open	Close		1	0	1
	Close	Close	Close		1	1	1

in Table 2.1.1b. This table shows the AND of x and y, defined in Section 1.3 and denoted by $f = xy$. Thus the network of a series connection of make-contacts implements AND operation of x and y.

Let us connect two make-contacts x and y in parallel as shown in Fig. 2.1.4. Then we have the combinations of states shown in Table 2.1.2a. Replacing "open" and "close" by 0 and 1, respectively, we may rewrite Table 2.1.2a as shown in Table 2.1.2b. This table shows the OR of x and y, defined in Section 1.3 and denoted by $f = x \vee y$.

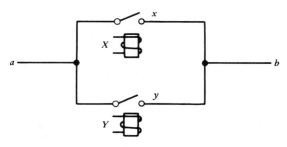

Fig. 2.1.4 Parallel connection of relay contacts.

Basic Properties of Relays and Switches   19

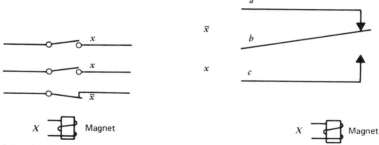

Fig. 2.1.5   *A relay with many contacts.*   Fig. 2.1.6   *Transfer relay.*

Often more than one contact (with armatures) is installed on a single magnet; an example is shown in Fig. 2.1.5. This is called a **multiple-contact relay**. When current is supplied through the magnet winding, all these contacts respond; in other words, make-contacts (x) close and break-contacts ($\bar{x}$) open after a short time delay.

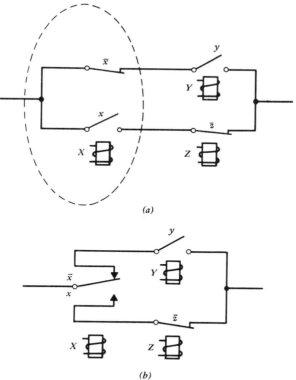

Fig. 2.1.7   *Use of a transfer-contact relay.*

20 Implementation of Logic Operations

A **transfer-contact relay** (or a transfer relay) is a multiple-contact relay which is somewhat different from the ones described above. As shown in Fig. 2.1.6, a make-contact and a break-contact are installed on a single magnet, sharing the central armature. When there is no current through the magnet, the path between terminals a and b is closed and the path between terminals b and c is open, that is, $\bar{x} = 1$ and $x = 0$. When current is supplied, the path between terminals a and b is open and the path between terminals b and c is closed. The transfer-contact relay is often used to simplify a relay contact network when the network contains a pair of contacts, x and $\bar{x}$. For example, the contacts encircled by a dotted line in Fig. 2.1.7a (two contacts, $\bar{x}$ and x, are mounted on magnet X) may be replaced by a transfer-contact relay, yielding the simplified network of Fig. 2.1.7b, where one armature is eliminated. Sometimes more than one transfer-contact is installed on a single magnet.

Since X and x assume identical values at any time, the magnet, along with its symbol X, will henceforth be omitted in drawings unless it is needed for some reason. In the case of sequential networks, magnets must usually be shown, as seen in Fig. 2.2.4, for example.

Electrical Switches

A few different types of electrical switches are available, as shown in Figs. 2.1.8 through 2.1.11. (Notice that none of these switches has magnet windings, as relays do.) Fig. 2.1.8 shows a switch that has only one contact.

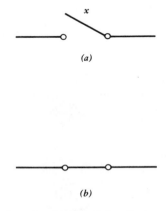

Fig. 2.1.8 *Switch (single-pole).* **(a)** *Switch handle in upper position.* **(b)** *Switch handle in lower position.*

(The handle of this switch is not shown in the figure.) A switch handle has only two positions, "upper" and "lower." If the switch handle is in the upper position, the switch contact is open, as shown in Fig. 2.1.8a, that is, $x = 0$. If the switch handle is in the lower position, the contact is closed, as shown in Fig. 2.1.8b, that is, $x = 1$. (If the upper and lower positions are exchanged, the switch contact may be denoted by $\bar{x}$.) The switch in Fig. 2.1.9 has two contacts, x and $\bar{x}$, corresponding to upper and lower positions of the switch handle (shown in Figs. 2.1.9a and 2.1.9b, respectively). Fig. 2.1.10 shows a set of two switches which are firmly fastened to a single handle. It connects a pair of lines to another pair of lines, x and y, or a third pair of lines, $\bar{x}$ and $\bar{y}$, respectively, corresponding to the upper or lower positions of the switch handle. The switch in Fig. 2.1.11 has an insulator bar (hatched) to which two conductor bars are attached. If the insulator bar is pushed up by its switch handle, as shown in Fig. 2.1.11a, two conductor paths, x and y are formed. If the insulator bar is pushed down, as shown in Fig. 2.1.11b, two other conductor paths, u and v, are formed. Some switches have even more conductor paths.

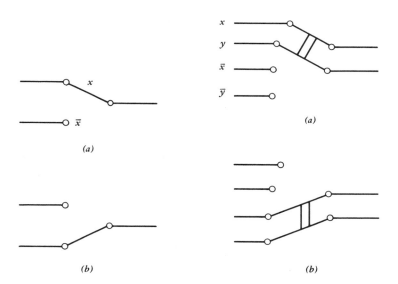

Fig. 2.1.9 Switch (three-way). (a) Switch handle in upper position. (b) Switch handle in lower position.

Fig. 2.1.10 Switch. (a) Switch handle in upper position. (b) Switch handle in lower position.

22 Implementation of Logic Operations

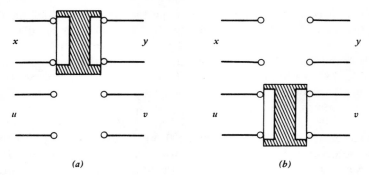

Fig. 2.1.11 (a) *Switch handle in upper position.* (b) *Switch handle in lower position.*

2.2 Analysis of Relay Contact Networks and Electrical Switch Networks

Let us analyze a relay contact network. "Analysis of a network" means the description of the logic performance of the network, usually in terms of a truth table or a switching expression.

To facilitate our discussion, let us define the following terminology. For each variable x, we have to deal with x and $\bar{x}$, as we saw in Section 2.1; x and $\bar{x}$ are called the **literals** of a variable x. Thus each variable has two literals.

Transmission of Series-Parallel Networks

In Section 2.1 we found that the transmission of two contacts, x_1 and x_2, connected in series, is $x_1 x_2$, and the transmission of these contacts in parallel is $x_1 \vee x_2$. Generalization of this is easy to see: when m networks with transmissions $T_1, T_2, \ldots, T_m$ are connected in series, the transmission of the entire new network is $T_1 T_2 \ldots T_m$; and when they are connected in parallel, it is $T_1 \vee T_2 \vee \ldots \vee T_m$. On the basis of this property, we can derive the transmission of any **series-parallel network** of contacts, that is, a network formed by repeatedly interconnecting contacts in series or in parallel. Let us work on an example from which generalization is easily seen.

Assume that the series-parallel network of Fig. 2.2.1 is given. This network can be split into subnetworks in series which have transmissions T_1 and T_2. Thus the transmission f of the entire network is $T_1 T_2$. Each subnetwork is split again into subnetworks in series or in parallel. For example, T_2 is split into subnetworks in parallel which have transmissions T_3 and T_4.

Analysis of Relay Contact Networks and Electrical Switch Networks

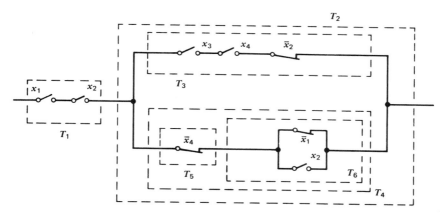

Fig. 2.2.1   Analysis of a series-parallel network.

Thus the transmission of the subnetwork T_2 is $T_2 = T_3 \vee T_4$. Accordingly, the entire network has the transmission

$$f = T_1 T_2 = T_1 (T_3 \vee T_4).$$

We continue to split each subnetwork into smaller subnetworks connected in series or in parallel, until each subnetwork consists of contacts either in series or in parallel only. Thus we can express the transmission of the entire network as

$$f = T_1 T_2 = T_1 (T_3 \vee T_4) = T_1 (T_3 \vee T_5 T_6). \tag{2.2.1}$$

Then, by expressing each transmission with literals that represent contacts, we can express f with literals:

$$f = (x_1 x_2)(\bar{x}_2 x_3 x_4 \vee \bar{x}_4 (\bar{x}_1 \vee x_2)). \tag{2.2.2}$$

Thus we have obtained a switching expression for the given network.

Transmission of Non-Series-Parallel Networks

The above procedure for calculating a transmission is simple, but cannot be applied to a non-series-parallel network such as the network in Fig. 2.2.2. This network can be split into x_5 and the other subnetwork. The latter cannot be split into subnetworks in parallel or in series.

We now discuss general procedures that can be applied to any network configuration. The first general procedure is based on the concept of tie sets, defined as follows.

24   Implementation of Logic Operations

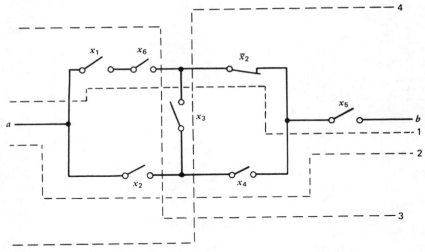

Fig. 2.2.2   *Tie sets of a non-series-parallel network.*

Definition 2.2.1: Consider a path that connects two external terminals, a and b, and no part of which forms a loop. Then the literals that represent the contacts on this path are called a **tie set** of this network.

An example of a tie set is the contacts x_1, x_6, $\bar{x}_2$, and x_5 on the path numbered 1 in Fig. 2.2.2.

Procedure 2.2.1: *Derivation of the Transmission of a Relay Contact Network by Tie Sets*

Find all the tie sets of the network. Form the product of all literals in each tie set. Then the disjunction of all these products yields the transmission of the given network.   □

These tie sets represent all the shortest paths that connect terminals a and b.

As an example, the network of Fig. 2.2.2 has the following tie sets:

For path 1: $x_1, x_6, \bar{x}_2, x_5$.   For path 3: x_1, x_6, x_3, x_4, x_5.
For path 2: x_2, x_4, x_5.   For path 4: $x_2, x_3, \bar{x}_2, x_5$.

Then we get the transmission of the network,

$$f = x_1 x_6 \bar{x}_2 x_5 \vee x_2 x_4 x_5 \vee x_1 x_6 x_3 x_4 x_5 \vee x_2 \bar{x}_2 x_3 x_5. \qquad (2.2.3)$$

Analysis of Relay Contact Networks and Electrical Switch Networks

(The last term, $x_2 \bar{x}_2 x_3 x_5$, may be eliminated since it is identically equal to 0 for any value of x_2. Such simplification of a switching expression will be discussed later.)

Procedure 2.2.1 yields the transmission of the given network because all the tie sets correspond to all the possibilities for making f equal 1. (For example, the first term, $x_1 x_6 \bar{x}_2 x_5$, in expression 2.2.3 becomes 1 for the combination of variables $x_1 = x_6 = x_5 = 1$ and $x_2 = 0$. Correspondingly, the two terminals a and b of the network in Fig. 2.2.2 are connected for this combination.)

The second general procedure is given after the following definition.

Definition 2.2.2: Consider a set of contacts that satisfy the following conditions:

1. If all of the contacts in this set are opened simultaneously (ignore functional relationship among contacts; in other words, even if two contacts, x and $\bar{x}$, are included in this set, it is assumed that contacts x and $\bar{x}$ can be opened simultaneously), the entire network is split into exactly two isolated subnetworks, one containing terminal a and the other containing b.
2. If any of the contacts are closed, the two subnetworks are connected.

Then the literals that represent these contacts are called a **cut set** of this network.

As an example, let us find all the cut sets of the network in Fig. 2.2.2. The network is reproduced in Fig. 2.2.3a. First, let us open contacts x_1 and x_2 simultaneously, as shown in Fig. 2.2.3b. Then terminals a and b are completely disconnected (thus condition 1 of Definition 2.2.2 is satisfied). If either of contacts x_1 and x_2 is closed, the two terminals a and b can be connected by closing the remaining contacts (thus condition 2 of Definition 2.2.2 is satisfied). The literals of these contacts are x_1 and x_2. Therefore x_1 and x_2 are a cut set (numbered cut set 1 in Fig. 2.2.3b). Next, let us consider the two contacts x_6 and x_2, as shown in Fig. 2.2.3c. Since they satisfy the conditions of Definition 2.2.2, x_6 and x_2 are cut set 2. Next, x_1, x_3, and x_4 are similarly cut set 3, as shown in Fig. 2.2.3d. Continuing this process, we will have found all the cut sets in the following list, corresponding to Fig. 2.2.3e:

For cut 1: x_1, x_2.   For cut 5: $\bar{x}_2$, x_3, x_2.
For cut 2: x_6, x_2.   For cut 6: $\bar{x}_2$, x_4.
For cut 3: x_1, x_3, x_4.   For cut 7: x_5.
For cut 4: x_6, x_3, x_4.

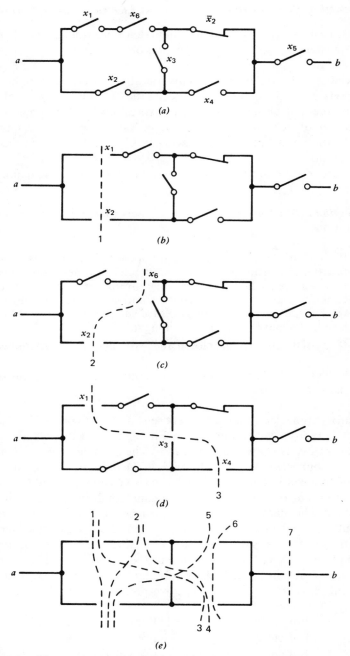

Fig. 2.2.3 Cut sets of the network in Fig. 2.2.2. (a) *Given network.* (b) *Cut set 1.* (c) *Cut set 2.* (d) *Cut set 3.* (e) *All cut sets.*

Analysis of Relay Contact Networks and Electrical Switch Networks 27

Procedure 2.2.2: Derivation of the Transmission of a Relay Contact Network by Cut Sets

Find all the cut sets of a network. Form the disjunction of all literals in each cut set. Then the product of all these disjunctions yields the transmission of the given network. □

On the basis of the cut sets in the network of Fig. 2.2.2 derived above, we get

$$f = (x_1 \vee x_2)(x_6 \vee x_2)\ (x_1 \vee x_3 \vee x_4)(x_6 \vee x_3 \vee x_4)$$
$$(\bar{x}_2 \vee x_3 \vee x_2)(\bar{x}_2 \vee x_4)(x_5). \quad (2.2.4)$$

This expression looks different from (2.2.3), but they are equivalent, since we can get identical truth tables for both expressions. Their equivalence can also be proved later by algebraic manipulation.

Procedure 2.2.2 yields the transmission of a relay contact network because all the cut sets correspond to all possible ways to disconnect two terminals, a and b, of a network, that is, all possibilities of making f equal 0. Any way to disconnect a and b which is not a cut set constitutes some cut set plus additional unnecessary open contacts, as can easily be seen.

The following property is worth noticing: each disjunction inside a pair of parentheses in (2.2.4) corresponds to a different cut set. A disjunction that contains the two different literals of some variable [e.g., $(\bar{x}_2 \vee x_3 \vee x_2)$ in (2.2.4) contains two literals, $\bar{x}_2$ and x_2, of the variable x_2] is identically equal to 1 and is insignificant in multiplying out f. Therefore every cut set that contains the two literals of some variable need not be considered in Procedure 2.2.2.

In this section only networks with two output terminals are analyzed. Networks with more output terminals can be analyzed in a similar way.

Combinational and Sequential Networks

The value of the transmission (or output) of any network discussed in this section is determined by the present values of the input variables of the network but is independent of the past values. Such a network is called a **combinational network** (the term "combinational" is used more often than "combinatorial," though some authors (see, e.g., p. 20 of [Clare 1973]) use the two interchangeably). Combinational networks are a major topic of Chapters 2 through 6 in this book.

In contrast, a network in which the output (or transmission) is dependent on the past values of inputs, as well as on the current values of inputs, is called a **sequential network**. Such a network must store the past values of inputs. As an example, consider the network in Fig. 2.2.4, where the magnet Y is part of a path between the output terminals a and b. Once contact x is

28 Implementation of Logic Operations

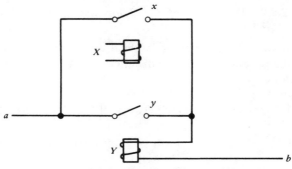

Fig. 2.2.4 A sequential network.

closed by energizing magnet X, magnet Y is energized and contact y is closed. Then, regardless of whether magnet X is later energized or not, contact y is kept closed since magnet Y remains energized by the current through y. Thus the transmission of this network becomes permanently 1. The path a–y–Y–b stores the information of whether or not X was energized in the past. (Note that in this example the output information goes through Y, and Y is fed back to the network input through y. This is essentially a loop along which the information repeatedly circulates.) This is a very simple example of a sequential network.

Malfunction of Networks

Relay contacts are opened or closed with a short time delay when currents in magnets change. Also, time delays for different contacts (contacts for the same magnet or those for different magnets) are not exactly equal. For example, when current X_2 is supplied to the corresponding magnet in Fig. 2.2.2, that is, when $X_2 = 1$, contact $\bar{x}_2$ is opened after a short delay. Contact x_2 also reacts after a short delay. The spring of contact x_2 is attracted to the magnet but must travel a short distance until the contact is closed. Hence the time until contact x_2 is closed is usually longer than the time until contact $\bar{x}_2$ is opened. This means that for a very short period contacts x_2 and $\bar{x}_2$ are both open. Since the behavior of these contacts is different from the ideal case, where contacts x_2 and $\bar{x}_2$ are instantly closed and opened, respectively, the network does not show a correct response during this short period. Also, in the case of the transfer relay shown in Fig. 2.1.6, both contacts, x and $\bar{x}$, are open during the short period when the central armature travels the gap between the two contacts. These deviations of contact behavior from the ideal case can be serious hazards to networks, particularly sequential networks as discussed in Chapter 7, unless the malfunction possibilities are taken into account when the network is designed.

Analysis of Relay Contact Networks and Electrical Switch Networks 29

For more detailed discussion of relay contact networks, see Appels and Geels [1966], Higonnet and Grea [1958], Caldwell [1958], Moisil [1969], Kohavi [1970], Harrison [1965] (in particular, Appendix 5), Oliver [1971], or National Association of Relay Manufacturers [1966].

Analysis of Electrical Switch Networks

Analysis of networks of electrical switches can be done in a similar manner.

*Exercises**

2.2.1 (R) For one of the relay contact networks in Fig. E2.2.1:

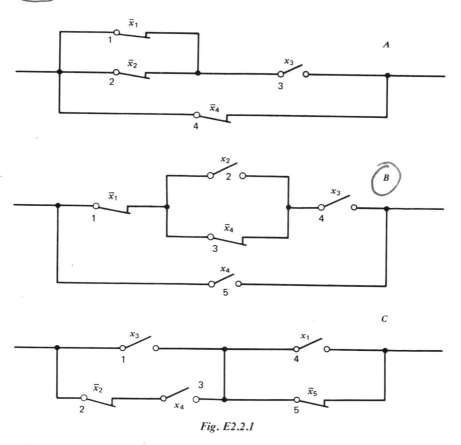

Fig. E2.2.1

* Since we have not yet discussed the simplification of switching expressions, these exercises need not be answered with the simplest switching expressions.

30   Implementation of Logic Operations

(i)   Derive the transmission by the series-parallel calculation method described at the beginning of Section 2.2.

(ii)   List tie sets, using the numbers attached to the contacts (e.g., $\{1, 4\}$ is a tie set for network C).

(iii)   Derive the transmission based on the tie sets.

2.2.2 (R)   For one of the relay contact networks in Fig. E2.2.2:

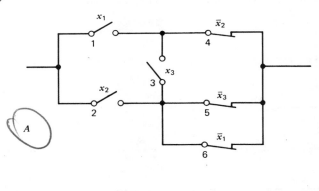

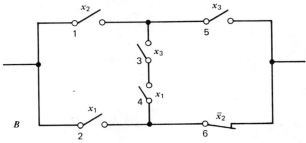

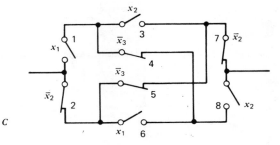

Fig. E2.2.2

Analysis of Relay Contact Networks and Electrical Switch Networks 31

(i) (a) List all tie sets, using the numbers attached to the contacts (e.g., {1, 3, 7} is a tie set for network C).
  (b) Derive the transmission, based on the tie sets obtained.

(ii) (a) List all cut sets, using the numbers attached to the contacts (e.g., {1, 2} is a cut set for network A).
  (b) Derive the transmission based on the cut sets obtained.

(iii) Show the equivalence of the two transmissions obtained, by writing the truth table.

(iv) Write the decimal specifications (the two decimal specifications are explained in Section 1.4).

2.2.3 (E) Derive the transmissions—f_{ab} between a and b, f_{ac} between a and c, and f_{bc} between b and c—of the relay contact network in Fig. E2.2.3.

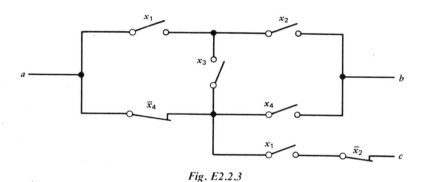

Fig. E2.2.3

2.2.4 (E) Synthesize a relay contact network which has one of the following transmissions:

  A. $x_1 x_2 \vee \bar{x}_1 x_3 \vee x_2 x_4$.   B. $x_1 x_2 \vee x_1 \bar{x}_3 \vee x_2 \bar{x}_3$.

2.2.5 (E) When "open" and "close" of a relay contact are interpreted to express 0 and 1, respectively, the series and parallel connections of two make-contacts x and y express the AND and OR functions of x and y, respectively. When "open" and "close" are interpreted to express 1 and 0, respectively, what functions* do they express?

2.2.6 (E) Derive a switching expression that is 1 if and only if the lamp is on for one of the relay contact networks in Fig. E2.2.6 when a battery is connected between terminals a and b (through a resistor). Assume that the resistance of the lamp is much higher than the resistance of the relay contacts.

---
*In this case some authors call these functions "hindrance" instead of "transmission."

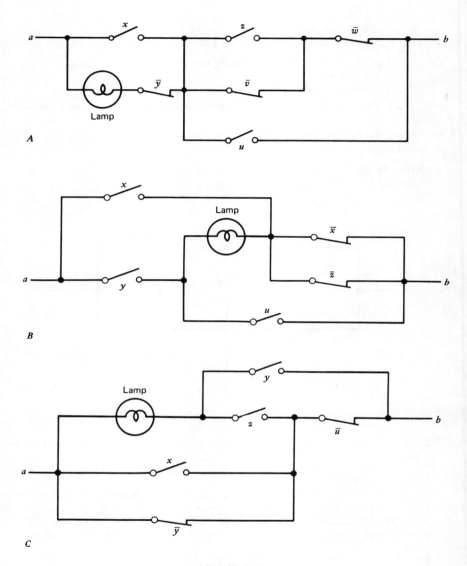

Fig. E2.2.6

Analysis of Relay Contact Networks and Electrical Switch Networks 33

2.2.7 (M) A lamp and relay contact networks whose transmissions are $p, q, r, s,$ and t are connected as shown in Fig. E2.2.7. Derive a switching expression that is 1 if and only if the lamp is on when a battery is connected between terminals a and b (through a resistor). Assume that the resistance of the lamp is much higher than the resistance of the relay contacts.

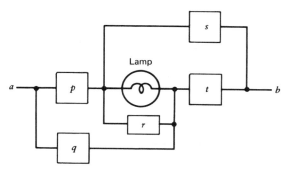

Fig. E2.2.7

2.2.8 (M+) Suppose that a relay contact network which has four terminals, $a, b, c,$ and d, is given as shown in Fig. E2.2.8a. An engineer wanted to derive a switching expression f, which is 1 if and only if the lamp is on, when the lamp is connected between terminals c and d, and a battery is connected through a resistor between terminals a and b. He calculated the switching expression by the following procedure. The resistance of the lamp is assumed to be much higher than the resistance of the relay contacts.

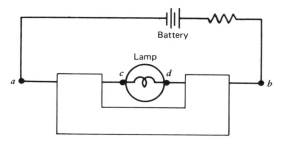

Fig. E2.2.8a   Relay contact network.

First he calculated the transmission g between terminals c and d of the relay contact network, after disconnecting the lamp, the battery, and the resistor, and short-circuiting terminals a and b (see Fig. E2.2.8b). Then he calculated the transmission h between terminals a and b after removing the external connection between a and b (see Fig. E2.2.8c). Then he concluded that $f = g \cdot \bar{h}$ is the desired switching expression f, which is 1 if and only if the lamp is on when the battery and resistor are connected between terminals a and b, as shown in Fig. E2.2.8a.

34 Implementation of Logic Operations

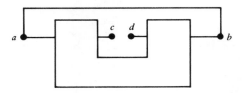

Fig. E2.2.8b *Calculation of transmission* **g** *between* **c** *and* **d**.

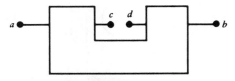

Fig. E2.2.8c *Calculation of transmission* **h** *between* **a** *and* **b**.

(i) Does he obtain a correct answer? If not, explain why and describe a correct calculation procedure.

(ii) Calculate the switching expression f of the network in Fig. E2.2.8d according to your own procedure.

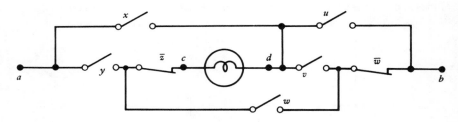

Fig. E2.2.8d *Relay contact network*.

2.2.9 (E) A student designed the relay contact network of Fig. E2.2.9, which was supposed to work in the following manner. After being pressed, switch S_1 immediately bounces back to the off position (i.e., the network is set to the initial state by S_1). Then, as S_2 is opened or closed, lamps L_1 and L_2 indicate whether switch S_2 is closed or open, respectively.

(i) Describe how this network works, assuming an ideal performance of relay contacts x and $\bar{x}$ (i.e., assume that when magnet X is energized, contacts x and $\bar{x}$ are instantly closed and open, respectively, and when X is deenergized, they are instantly open and closed, respectively).

Analysis of Relay Contact Networks and Electrical Switch Networks 35

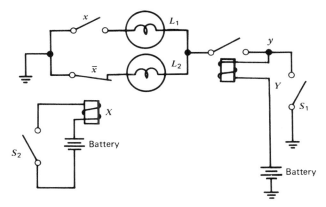

Fig. E2.2.9

(ii) Unfortunately, the network does not function as expected because contacts x and $\bar{x}$ do not work ideally. Explain why.

(iii) Redesign the network by a trial-and-error approach so that it works correctly. Assume that the resistances of the lamps are much higher than those of the relay contacts. (This is a malfunction problem of a relay contact network. As we will see later, electronic gate networks have similar problems.)

2.2.10 (M) (i) Using the switch shown in Fig. 2.1.9 at each of the two ends of a long hall, we want to turn the light on or off at the center of the hall. The light must be turned on and off by the switch at each end, independently of the switch position at the other end. Show a connection diagram that you believe is most economical. (Shorter total length of wire and switches with fewer terminals are less costly.)

(ii) We want to turn an electric light on or off at the center of a T-shaped hall by any of three switches at the ends of the hall, independently of switch positions at the other ends.

(a) Write a truth table, regarding the on-off state of the light as a switching function of the on-off positions of three switches.

(b) Design the wiring connection among the light and switches that you believe is most economical. Use appropriate switches of the types shown in Figs. 2.1.8 through 2.1.10 only. (Various types of switches in these three figures may be mixed.)

(c) Try to design a wiring connection of switches, using only switches shown in Fig. 2.1.9. (Use only one switch at each end of the hall.) If this is not possible, prove the infeasibility.

Remark: In house wiring one end of a light must be directly connected to one terminal (the ground) of a power supply, without going through switches, for the sake of safety. ∎

36 Implementation of Logic Operations

2.2.11 (M) Using an appropriate switch chosen from those shown in Figs. 2.1.8 through 2.1.10 at each of n different locations, we want to turn on or off one light. The light must be turned on and off by a switch at any location, independently of switch positions at other locations.

(i) Consider a truth table regarding the on-off state of the light as a switching function of the on-off positions of n switches. Then describe the characteristic nature of the table in a simple word statement, if you can.

(ii) Design a wiring connection for $n \geq 4$ that you believe is most economical. (Shorter total length of wire and switches with fewer terminals are less costly.)

2.3 Analysis of Networks of Electronic Gates

In this section we discuss transistor gates as typical electronic gates, though there are other electronic gates. We will try to minimize and simplify the discussion of the electronic aspects of these gates (this and the next section are the only ones that discuss electronics*), since in this book we are mainly interested in design procedures.

Transistors are classified into **metal oxide semiconductor field effect transistors** and **bipolar transistors**. The former term is usually abbreviated as MOSFETs. MOSFETs are now widely used and are classified further into many special types. Electronic gates that are implemented with MOSFETs are usually slow but inexpensive. Those implemented with bipolar transistors are usually fast but expensive. The latter usually consume more electrical power than MOSFETs. The technology, however, is changing, so this situation may not remain the same. The speed of some types of gates implemented with MOSFETs has been greatly improved. For simplicity, only one typical gate family for the MOSFET and one for the bipolar transistor are described, but important features of electronic gates from the viewpoint of logic design are well explained.

2.3.1. Analysis of MOS Networks

A metal oxide semiconductor field effect transistor, that is, a MOSFET, is an important type of transistor. A MOSFET is often further abbreviated as MOS. MOSFETs are extensively used in electronic desk calculators, microcomputers, and minicomputers, and also in memories in computers of all different scales. The MOS will be more extensively used in computers in the future.

The cross section of an n-channel MOSFET is shown in Fig. 2.3.1 (though it is not very important for us to completely understand the structure and

---

* Those who are not familiar with elementary electricity or electronics may consult, for example, pp. 3–65, 158–163, and 287–354 of R. J. Smith [1976].

Analysis of Networks of Electronic Gates 37

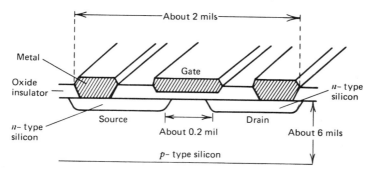

Fig. 2.3.1 MOSFET.* (The vertical dimension in this figure is greatly distorted for the sake of illustration; 1 mil = 0.001 in. = 25.4 µm.)

the working principle of a MOSFET for the rest of this book). An **n-channel MOSFET** is often called simply an **n-MOS**. In the lower part, two regions of *n*-type silicon are separated by *p*-type silicon. (These two regions are called **diffusion regions** since they are manufactured by a process termed diffusion.) One of the regions is called the **source** and the other the **drain**. (The source and the drain are usually interchangeable because of symmetric structure.† The "source" of a MOSFET must not be confused with an electrical power "source.") On top of these an oxide insulator layer and three metal pieces are attached. The metal piece in the central position is called the **gate**. (This must not be confused with "logic gate." Whenever we need to differentiate these two gates, a gate here will be called a MOSFET gate.) Note that the source and drain directly contact the metal pieces, while the gate is separated by the insulator from the *p*-type silicon (therefore no direct current flows between the gate and any of the source, the drain, or the *p*-type silicon, even if voltage is applied on the gate). A single MOSFET is very small, occupying only a few square mils, though its container (called a **package**) is much larger, 0.25 to 1 inch in dimension. A single *n*-channel MOSFET is denoted by the symbol shown in Fig. 2.3.2a. A single **p-channel MOSFET** (often abbreviated as ***p*-MOS**), which is obtained by interchanging *p* and *n* in Fig. 2.3.1, is denoted by the symbol in Fig. 2.3.2b. (Symbols other than those in Figs. 2.3.2a and 2.3.2b are also used by some authors.) **In this book, however, we**

---

* Precisely speaking, this is called an *n*-channel MOSFET of **enhancement type**, which is widely used. When *p* and *n* are interchanged in Fig. 2.3.1, we get another type of MOSFET, which is called a *p*-channel MOSFET of enhancement type.

† When a MOSFET is connected to a power supply, a diffusion region with a lower voltage (which supplies electrons) is called a source in the case of an *n*-MOS. (In the case of a *p*-MOS, a region with a higher voltage, which supplies holes, is called a source.)

38 Implementation of Logic Operations

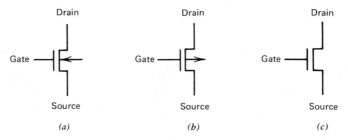

Fig. 2.3.2 Symbols for MOSFETs. (a) *Symbol for an* n-*channel MOSFET.* (b) *Symbol for a* p-*channel MOSFET.* (c) *Abbreviated symbol.*

will use an *n*-channel MOSFET denoted by the abbreviated symbol shown in Fig. 2.3.2c unless otherwise noted. (As a matter of fact, since we are concerned with logic operations of MOSFETs only, it does not matter whether a MOSFET is *n*-channel or *p*-channel, if positive or negative logic (to be discussed later in this section) is used, respectively.)

Logic Operations by MOSFETs

Assume that the source of an *n*-channel MOSFET is grounded and 5 volts from a power source is applied to the drain through the resistor, as shown in Fig. 2.3.3*a*.* When the voltage applied at the gate increases from 0 volt to 5 volts, the current from the power source to the ground increases as shown in Fig. 2.3.3*b*. Then, because of the voltage drop across the resistor, we get the relationship between gate voltage and output voltage shown in Fig. 2.3.3*c*. Let us use only two values of the gate voltage: 0.2 volt and 5 volts. If the gate voltage is 0.2 volt, there is essentially† no current from the power source to the ground and the output voltage is 5 volts. If the gate voltage is 5 volts, current flows and the output voltage is 0.2 volt. In other words, if the gate has low voltage (i.e., 0.2 volt), the output has high voltage (i.e., 5 volts); and if the gate has high voltage, the output has low voltage, as shown in Table 2.3.1*a*. if low and high voltages are denoted by 0 and 1, respectively, this table is reduced to the truth table of Table 2.3.1*b*. Thus, if the gate signal is denoted by x, the output may be denoted by $\bar{x}$ (as explained with Table 1.3.4). This means that the electronic circuit shown in Fig. 2.3.3*a* implements a logic gate for the NOT operation.

---

* The silicon substrate (i.e., the *p*-type silicon in Fig. 2.3.1) is usually connected to the ground or some bias voltage. But for the sake of simplicity, such connections of the substrates are not shown in any network throughout this book, since they are not directly related to logic operations.
† A negligibly small current flows.

Analysis of Networks of Electronic Gates 39

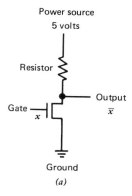

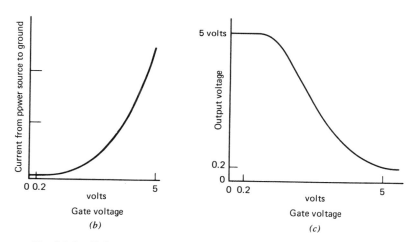

*Fig. 2.3.3 Voltage-current characteristics of an **n**-channel MOSFET.*

Table 2.3.1 Logic Operation of the Network in Fig. 2.3.3a

a. **Input-Output Voltage Relationship**

GATE	OUTPUT
Low voltage	High voltage
High voltage	Low voltage

b. **Truth table**

GATE (x)	OUTPUT ($\bar{x}$)
0	1
1	0

40 Implementation of Logic Operations

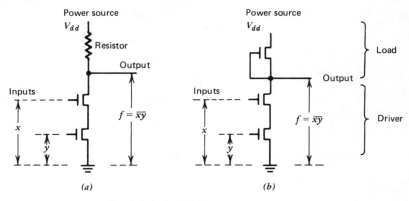

Fig. 2.3.4 MOSFETs in series.

Let us connect two MOSFETs in series, as in Fig. 2.3.4a, and then connect them to the power source of 5 volts, denoted by V_{dd}, through the resistor. From the voltage-current characteristic of the MOSFET in Fig. 2.3.3b, we can see that a large current flows from the power source to the ground only when both x and y terminals have 5 volts. (If at least one of them has 0.2 volt, the corresponding MOSFET blocks current. Thus no current flows through the two MOSFETs.) Then a voltage develops across the resistor, and the output voltage at terminal f becomes lower than before. This relationship between input and output voltages is shown in Table 2.3.2a. Thus, if low and high voltages at x, y, and f are interpreted as representing the binary signals 0 and 1, respectively, $f = 0$ only when $x = y = 1$, as shown in Table 2.3.2b. Thus function f is expressed as

$$f = \overline{xy}.$$

Table 2.3.2 Logic Operation of the Network in Fig. 2.3.4a

a. Input-Output Voltage Relationship			b. Truth Table of Table 2.3.2a in Positive Logic		
VOLTAGES AT					
x	y	f	x	y	f
Low	Low	High	0	0	1
Low	High	High	0	1	1
High	Low	High	1	0	1
High	High	Low	1	1	0

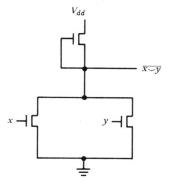

Fig. 2.3.5   MOSFETs in parallel.

Because a resistor requires a very large area on a chip, the resistor is usually replaced by another MOSFET whose gate is connected to its source, as shown in Fig. 2.3.4b.* This MOSFET works as a resistor and is called a **load MOSFET**, whereas other MOSFETs are collectively termed a **driver**.

Thus the series connection of two MOSFETs represents the complement of the AND function of x and y, that is, $(\overline{xy})$, as shown in Fig. 2.3.4b. The complement of AND is called **NAND**. By a similar observation, the parallel connection of two MOSFETs represents the complement of the OR function of x and y, that is, $\overline{x \vee y}$, as shown in Fig. 2.3.5. The complement of OR is called **NOR**. When Tables 2.3.1a and 2.3.2a are converted into Tables 2.3.1b and 2.3.2b, respectively, high and low voltages are interpreted as representing 1 and 0, respectively. This is called **positive logic**. **Throughout this book, positive logic will be used unless otherwise noted.** But interpretation of high and low voltages to represent 0 and 1, respectively, is also possible and is called **negative logic**.† (Exercise 2.1.) Both are used in practice. When the n-channel MOSFETs and the power sources of positive voltage in Fig. 2.3.3a and 2.3.4b are replaced by p-channel MOSFETs and power sources of negative voltage, Figs. 2.3.3a and 2.3.4b still represent the same functions, $\bar{x}$ and $\overline{xy}$, respectively, in negative logic.‡ (This is so because a p-channel

---

* A MOSFET of depletion type is widely used, though a MOSFET of enhancement type can also be used, with the gate connected to its drain.

† If negative logic is used instead of positive logic, then every disjunction and conjunction are replaced by conjunction and disjunction, respectively, in the logic operation in positive logic (Exercise 2.1).

‡ Generalizing this, we can say that an electronic circuit with p-channel MOSFETs in negative logic represents the same switching function as the circuit with the p-channel MOSFETs replaced by n-channel MOSFETs and interpreted in positive logic. Throughout this book only networks with n-channel MOSFETs in positive logic will be considered unless otherwise noted, but these networks may be interchangeably considered with p-channel MOSFETs in negative logic. (Thus the symbol in Fig. 2.3.2c is used throughout this book instead of the one in Fig. 2.3.2a or 2.3.2b.)

42 *Implementation of Logic Operations*

MOSFET is conductive or nonconductive when its gate voltage is low or high, respectively.)

MOS Gates (or MOS Cells)

Let us connect a number of MOSFETs in an arbitrary manner and then connect them to one load MOSFET, as shown in Fig. 2.3.6a (the connection is not restricted to the series or parallel connection). Such a set of interconnected MOSFETs connected to only one load MOSFET is called an **MOS gate**[*] or **MOS cell**. The output function of an MOS gate such as that in Fig. 2.3.6a can be calculated by using the calculation procedure for the transmission of a relay contact network. In other words, as illustrated in Fig. 2.3.6b, we first replace driver MOSFETs with the corresponding relay contacts and then **calculate the transmission according to the procedures in Section 2.2. Then we complement it**. The result is the output function f. (Readers should verify that f in Fig. 2.3.6a is $\overline{xy \vee vz \vee uyz \vee uvx}$.) Notice that all driver MOSFETs correspond to make-contacts only (not to break-contacts) and that only the load MOSFET corresponds to a break-contact relay. Since only a break-contact relay performs the NOT operation, an MOS gate contains only one NOT operation. Thus **the output of an MOS gate represents a switching expression that is the complement of a disjunction of products of input variables**, where these variables are the inputs to the driver MOSFETs

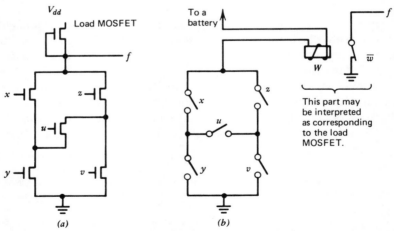

Fig. 2.3.6 Analogy between an MOS network and a relay contact network.

---

[*] If we want to differentiate from the MOSFET's gate, the gate here should be called an MOS logic gate, and the MOSFET's gate should be called a MOSFET gate, as mentioned already.

Analysis of Networks of Electronic Gates 43

(e.g., $f = \overline{xy \vee vz \vee uyz \vee uvx}$, represented by the output of the network in Fig. 2.3.6a). This type of switching expression (i.e., the complement of a disjunction of products of noncomplemented input variables; if a complemented variable, say $\bar{i}$, is given as an input variable, we replace it by a new variable, say x) is called a **negative function** of the input variables, as will be discussed in Chapter 5.

Remark 2.3.1: If there were available a gate whose switching expression did not contain NOT (e.g., a gate that represents AND or OR), it would be convenient for logic design. But it is physically difficult to implement such a gate with MOSFETs. ∎

Examples of networks of MOS (logic) gates are shown in Figs. 2.3.7 and 2.3.8. In Fig. 2.3.7 two MOS gates in the leftmost positions realize $\bar{x}$ and $\bar{y}$, respectively, at their outputs. Then $\bar{x}$ and $\bar{y}$ are fed as inputs to the last MOS gate. The entire network consists of three MOS gates. The network in Fig. 2.3.8 is more complicated and consists of four MOS gates. (Notice that each MOS logic gate consists of all MOSFETs whose drains and sources are connected by an unbroken line to the same load MOSFET. In Fig. 2.3.8, for example, MOSFETs D_1 through D_7 belong to the same logic gate, since the drains and sources of all these MOSFETs are connected by an unbroken line to load MOSFET L_1. But MOSFET D_8 does not belong to this logic gate, since the drain and source of D_8 are not connected to L_1.)

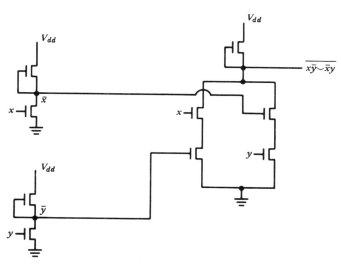

Fig. 2.3.7   An MOS network of three MOS gates.

44 *Implementation of Logic Operations*

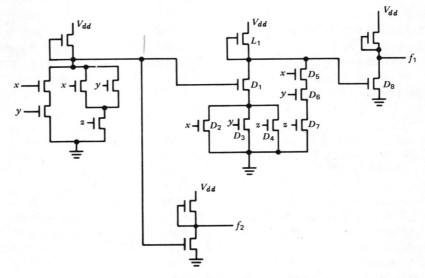

Fig. 2.3.8   An MOS network of four MOS gates.

As in the case of relay contacts, when a signal value at the (MOSFET) gate of an MOS (logic) gate in Fig. 2.3.4b changes, the output signal value changes after a short time delay (due to the internal physical mechanism). This time delay is called the **switching time** of the (logic) gate.

Implementation of MOS Networks

Next let us discuss how an MOS logic gate is implemented. Suppose that we want to implement the gate shown in Fig. 2.3.9a. When this MOS logic gate is to be implemented with discrete components, each of five MOSFETs is encased in a small container that has three leads, and these five containers (i.e., five discrete components) are connected by conductor wires as shown in Fig. 2.3.9b on a **pc board** (printed-circuit board). But when the logic gate of Fig. 2.3.9a is to be implemented with an integrated circuit, all the MOSFETs and connections are implemented on a single, tiny silicon substrate (whose size is about 10 mils times 10 mils), as shown in Fig. 2.3.9c. In this case, connections are realized not only by metal strips (*A*, *B*, *C*, etc., in Fig. 2.3.9c) deposited on the insulation layer but also by diffusion regions (the shaded areas in Fig. 2.3.9c), which are implemented by the diffusion process (this is the same process that implemented the source and drain regions in Fig. 2.3.1). The cutout view along the dotted line in Fig. 2.3.9c is shown in Fig. 2.3.9d. Then the substrate, that is, the IC chip, is encased in a container that has

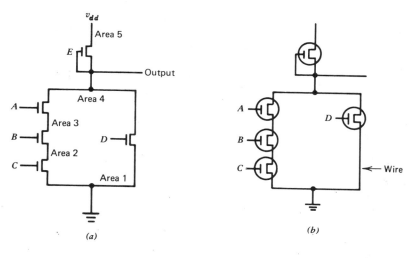
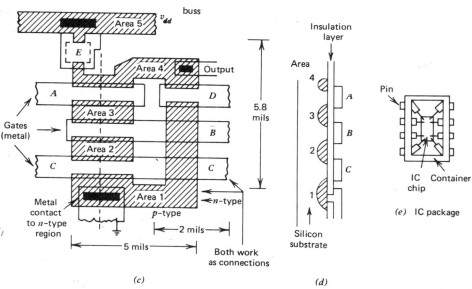

Fig. 2.3.9 Implementation of an MOS gate. (a) An MOS logic gate to be implemented. (*The area numbers correspond to those in Fig. 2.3.9c.*) (b) An implementation with discrete components. (*This occupies at least 1 square inch on a pc board. Circles denote discrete components, each of which is a MOSFET encased in a separate container.*) (c) Layout of Fig. 2.3.9a for IC implementation. (*The shaded areas denote diffusion regions, implementing the connections with the corresponding area numbers in 2.3.9a.*) (d) Cutout view along the dotted line in Fig. 2.3.9c. (e) IC package.

45

many pins used as input and output terminals (also terminals for the ground and power supply), as shown in Fig. 2.3.9e. The electrodes are connected to pins by gold wires, and a lid is sealed to the container. Thus we have completed an IC package.

Figure 2.3.9c is called a **layout** of the logic gate in Fig. 2.3.9a. In logic design it is vitally important, in terms of the speed and cost of a gate (also a network when a layout of a network is to be made on a single substrate), to have a good layout, as will be discussed later.

Each SSI package contains at most 20 MOS logic gates. Each MSI or LSI package contains a network (or networks) consisting of many logic gates.

Other MOS Logic Families

Currently MOS technology is advancing very rapidly. A number of MOS logic families,* in addition to the n-MOS and p-MOS, are available, such as CMOS, CMOS/SOS, VMOS, and DMOS. The complementary MOS, abbreviated as CMOS, is becoming very popular since it consumes little electric power, though it has a somewhat complex structure. The CMOS/SOS, a CMOS whose substrate is sapphire, has a higher speed than the CMOS. (See, e.g., Hittinger [1973].)

From the viewpoint of power sources, MOS gates are operated in basically two different ways. First, when direct current is used as the power source, the MOS is called a **static** MOS. The MOS discussed in this section is of the static type. Second, when clocked pulses (to be explained in Chapter 8) are used as the power source in order to synchronize the operations of all MOS gates, the MOS is called a **dynamic MOS**. This arrangement usually results in a reduction of power consumption.

The above analysis of logic performance is essentially identical for all MOS gates of different families (i.e., p-MOS, n-MOS, CMOS, etc.) or usage (i.e., static and dynamic MOS), though the VMOS can realize only the NOR operation because of its physical structure [Rodgers and Meindl 1974], unlike the p-MOS, n-MOS, and CMOS, which can implement negative functions. (Notice that NOR and NAND are special cases of a negative function.)

2.3.2 Analysis of Bipolar Transistor Networks

There are a few varieties of bipolar transistors. Let us choose an n–p–n transistor as a typical bipolar transistor.

Fig. 2.3.10 shows the cross section of an n–p–n transistor, which is a

---

* See, for example, Hnatek [1973, 1977b], Engineering Staff of AMI [1972], Carr and Mize [1972], or Cobbold [1970].

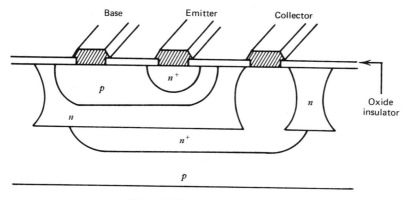

Fig. 2.3.10   An n-p-n *transistor.*

type of bipolar transistor (a *p–n–p* transistor is another important type of bipolar transistor). The transistor has three metal terminals called the base, the collector, and the emitter. In structure and material it is similar to a MOSFET but more complicated and somewhat larger. It occupies about 50 square mils, but its container is much larger (usually 0.25 to 0.5 inch in its dimension). The areas labeled as *p*, *n*, and n^+ in Fig. 2.3.10 represent *p*-type, *n*-type, and heavily doped *n*-type silicon, respectively. Fig. 2.3.11 shows the symbol for an *n–p–n* transistor (a *p–n–p* transistor is denoted by the same symbol except for the different direction of the arrow.)

Bipolar Transistor Gates

Gates for logic operations can be implemented with bipolar transistors in a manner different from that used for MOSFETs. Usually bipolar transistors are used for high-speed digital computers.

To get some notion of how such a gate is realized, let us connect an *n–p–n* transistor and resistors as shown by the solid line in Fig. 2.3.12*a*. The emitter of the transistor is grounded. When the voltage at the input increases, the current through the transistor also increases and the voltage at the output changes as shown in Fig. 2.3.12*b*. When the input voltage is 1 volt in Fig. 2.3.12*a*, the bipolar transistor conducts current between the collector and

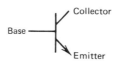

Fig. 2.3.11   Symbol for an n-p-n *transistor.*

48 *Implementation of Logic Operations*

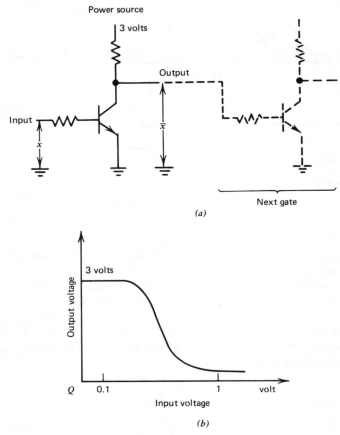

Fig. 2.3.12 RTL gate.

the emitter and the output voltage drops to about 0.1 volt, as can be seen from Fig. 2.3.12*b*. When the input voltage is 0.1 volt, the bipolar transistor does not conduct current and the output voltage is 3 volts. But in a network consisting of many gates, the output of this gate is connected to another gate, shown by the dotted line in Fig. 2.3.12*a*, and current flows through the output to the latter transistor (through its base toward the emitter). Because of this current, the output voltage will be lowered to about 1 volt from the 3 volts shown in Fig. 2.3.12*b* (the resistors are chosen so that 1 volt appears at the output). Thus, when high and low voltages at the input and output are interpreted to express the binary signal values 1 and 0, respectively (in other words, when positive logic is used), this transistor gate performs the NOT

Analysis of Networks of Electronic Gates 49

operation. In other words, if the input signal is denoted by x, the output represents $\bar{x}$.

Next, we connect one more transistor–resistor pair to this circuit, as shown in Fig. 2.3.13. Then, as we can easily see, only when at least one of x and y has high voltage, does the current flow through a transistor (or transistors) from the power source, a voltage develop across the resistor R, and the output voltage take the low voltage. Thus, if x and y denote the two signals at the inputs, the output of this gate expresses $f = \overline{x \vee y}$ (readers may check this by writing the truth table, as was done previously). In other words, this gate implements the NOR. If we connect one more input, z, the output expresses $f = \overline{x \vee y \vee z}$, and so on. (The NAND operation, e.g., $\overline{xyz}$, can be similarly implemented with an electronic circuit using p–n–p transistors instead of n–p–n transistors.)

The gates in Figs. 2.3.12a and 2.3.13 are examples of what is known as **resistor-transistor logic**, usually abbreviated as **RTL**. (Although RTL is currently not widely used—a similar version is preferred—it is discussed here because its simple structure is easy to understand.) Actually, RTL is another **IC logic family**, in addition to n-MOS or p-MOS, discussed previously. Other IC logic families, such as TTL or ECL gates (which are currently the most extensively used logic families in high-speed computers), have much more complicated electronic circuits and are somewhat harder to understand.

In the same manner as the IC implementation of an MOS network (or networks), all transistors, connections, and resistors of a bipolar transistor network (or networks) can be implemented on a single IC chip [Garrett 1970; Barna and Porat 1973; Taub and Schilling 1977].

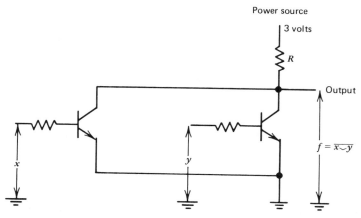

Fig. 2.3.13   RTL gate for NOR.

50 *Implementation of Logic Operations*

2.3.3 Features of Electronic Gate Networks

Let us discuss common features of the networks of electronic gates discussed in Sections 2.3.1 and 2.3.2.

Switching Time of a Gate

The output of a gate (an MOS gate in Section 2.3.1, or an RTL gate in Section 2.3.2) assumes a new value with a short time delay after the inputs assume new values. This time delay is called the **switching time** of the gate. Although switching times of gates are not all equal (indeed, are sometimes very different), **let us assume for simplicity that all gates in a network have equal delays, that is, identical switching times**, unless otherwise noted. Switching time differences among different gates sometimes have significant consequences, as discussed in later chapters.

> ***Remark 2.3.2:*** Strictly speaking, the gates in a network do not have exactly equal switching times, even if the network consists of all identical gates. Delays at gates closer to the network output are shorter than those at gates far from the network output, unless the gates are synchronized with clock pulses, as discussed in Chapter 8. This is so because gates closer to the network output start to change even before the switching time elapses. Therefore, when we cascade m copies of a gate that has time delay τ under individual measurement, the total time delay of this network is shorter than $m\tau$.
>
> Also, again strictly speaking, the time delay for the change of an input from low voltage to high voltage is usually different from that for the change from high voltage to low voltage. For simplicity, however, let us assume that the switching times are the same. ∎

Importance of NAND or NOR Gates

Notice the following important features of electronic gates. When an input signal has high voltage at the gate in Fig. 2.3.13, the gate output becomes low voltage; and when each input signal has low voltage, the gate output becomes high voltage. When transistors in a gate are connected so that this inverting action of signal level takes place, a **NOT** operation is always combined with some other logic operation such as **OR** or **AND**. It is difficult to separate the combination. If we want to implement **OR** only, for example, we need to use a complicated electronic circuit such as a circuit realized by cascading the **NOT** network in Fig. 2.3.12*a* to the output of the **NOR** network in Fig. 2.3.13, or we need to use electronic circuits which are not always desirable, as explained in the following paragraph. (Gates for **AND** or **OR** operations are commercially available as single gates, as well as **NOR** gates or **NAND** gates.) Thus a gate that performs a simpler logic operation is not always realized by a simpler electronic circuit. Also, for the sole purpose of performance improvement, such as speed increase or electric power consumption reduction, many transistors or other components (diodes or resistors) are often added to the electronic circuit that implements a gate. Thus, **in general, the complexity of the logic operation that a gate performs has no direct relationship to the complexity of the electronic circuit that implements the gate in question.**

We can connect transistors in a gate so that the inverting action of the signal level does **not** take place and the gate realizes a logic operation such as AND or OR without NOT, with a fairly simple electronic circuit (simpler than the cascade of Figs. 2.3.12*a* and 2.3.13). Such gates are commercially available.* But these gates do not necessarily have the best performance (e.g., they may show difficulty in maintaining signal voltage levels, instability of electronic operations, slow speed, excessively small signal voltages, or slight difficulty in manufacturing), though in many cases they are advantageous in terms of logic operations. Thus **gates that represent logic operations with NOT combined are most widely used in industry**. For this reason we will study logic design with NOR or NAND gates in detail in this book. We will also study logic design with AND and OR gates, because certain NOR (or NAND) gate networks can be obtained from networks consisting of a mixture of AND and OR gates [a typical example is a programmable logic array (PLA), to be discussed in Section 9.4], which are usually easier to treat theoretically than NOR or NAND gate networks. AND and OR operations are also important for networks implemented with TTL gates.

Symbols for Electronic Gates

Symbols for gates are illustrated in Fig. 2.3.14. (Often the MOS gate that realizes NOR or NAND is also denoted by these symbols. Of course, when an MOS gate represents a more complicated function than NOR or NAND, it cannot be denoted by any of these symbols, and we have to express it as in Fig. 2.3.7 or 2.3.8 [Skelley 1973, Dinsmore 1974].)

Analysis of Networks of Electronic Gates

A combinational network of electronic gates can be analyzed in the following manner. As an example, consider the combinational network of Fig. 2.3.15. Starting with the gates at the far left of this network, we write a switching expression for the output of each gate, moving toward the output gate. Then

---

* For example, CTμL9964 and others by Fairchild, and the emitter follower logic (EFL) family developed by Hewlett Packard [Skokan 1973, *Electronics*, January 1974].

52   Implementation of Logic Operations

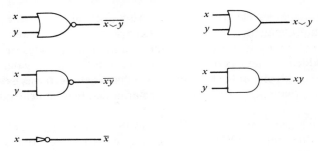

Fig. 2.3.14   Symbols for gates.

the expression for the output gate shows the switching expression for the entire network. Another approach is to compute the network output value for each combination of 0 and 1 for $x_1, x_2, \ldots, x_6$ by finding the output value at each gate, and then to form the truth table for the output gate. Another sample analysis of a combinational network of MOS gates is shown in Fig. 2.3.7. Thus analysis of a combinational network of electronic gates is simple and easy.

Since the output of a sequential network depends on the past values of input variables, as well as the current values, the network must remember the past values of inputs. This can be done by two approaches: first, loops of gates and connections along which the information circulates; and, second, memories. The analysis and synthesis of sequential networks based on loops will be discussed in Chapters 7 and 8. In general, the analysis of sequential networks is much more complicated than that of combinational networks.

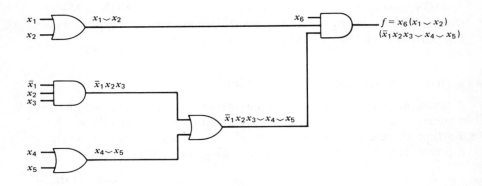

Fig. 2.3.15   Analysis of an electronic gate network.

2.4 Discrete Components and Integrated Circuits

In Section 2.3 we discussed electronic circuits such as MOS networks and bipolar transistor networks. These electronic circuits are realized by two different manufacturing approaches, that is, with discrete components and with integrated circuits, as illustrated with MOS in Fig. 2.3.9. It is very important for the logic designer to decide, before beginning the design, whether the network is to be realized with discrete components or with an integrated circuit, and then, in the latter case, to establish integration size, since the design criteria and the constraints to be considered in the design are different in each case.

As explained with respect to Figs. 2.3.9c, 2.3.9d, and 2.3.9e, when an electronic circuit for a gate such as the ones in Figs. 2.3.5, 2.3.9a, and 2.3.13, or an entire network consisting of many gates, such as that in Fig. 2.3.8, is manufactured simultaneously on a single, small, thin semiconductor sheet of a few to a few hundred mils on a side, called a **chip** (e.g., the silicon substrates in Figs. 2.3.1., 2.3.9d, and 2.3.10), and is encased in a single package (or a container), it is called an **integrated circuit*** (usually abbreviated as IC). In contrast, the conventional manufacturing process for an electronic circuit for an electronic gate, or for a network consisting of gates, is to separately manufacture individual components, such as transistors and resistors, and then encase them in separate packages, each component in one package. (For example, each of two transistors and three resistors in Fig. 2.3.13 is put into a separate package. Also, five MOSFETs are put into separate packages in Fig. 2.3.9b.) These packages are called **discrete components**. By connecting these discrete components by conductor wires, an electronic circuit is assembled on a pc board. Circuits and gates assembled in this manner are called **discrete circuits** and **discrete gates**, respectively.

Advantages and Disadvantages of Integrated Circuits

Integrated circuits have many advantages. **The speed of networks is improved by ICs.** On a pc board or along a twisted pair of wires, an electrical signal travels slowly because of a large parasitic capacity. On a pc board an electrical

---

* Strictly speaking, this is a **monolithic integrated circuit**. The other type, called a **hybrid integrated circuit**, is a package containing two or more monolithic integrated circuit chips. These chips are placed on a substrate inside the package and interconnected. In other words, a hybrid integrated circuit is simply another means to assemble monolithic integrated circuit chips more compactly than IC packages on a pc board. More recently, other means to assemble IC chips, replacing hybrid integrated circuits or pc boards, have been developed (see, e.g., Lyman [1977a, 1977b, 1978]). Since discussion of different assembling means is outside the scope of this book, from now on, for the sake of simplicity, only pc boards will be mentioned as a way of assembling chips.

signal travels about 6 inches in 1 nanosecond; it travels about 9 inches along a twisted pair of wires in a nanosecond. But an electrical signal travels connections on an IC chip in a much shorter time because of the short length and accordingly much smaller parasitic capacity. (A **parasitic capacitance** is a capacitance embodied without a component called a capacitor. A parasitic capacitance is usually embodied by terminals or connections against the ground and usually has a small value.) Also, the switching time of gates is improved because of reduced parasitic capacitance. **The reliability of ICs is superior to that of networks assembled with discrete components. Moreover, networks with ICs are more compact and weigh less than those with discrete components.** There are also some other advantages. But probably the most important is the **inexpensiveness of IC gates** when the same package is to be manufactured in large quantity. The cost is low because an entire network (i.e., all the gates and connections in a single IC chip) is manufactured simultaneously in large quantities, and labor for soldering and assembling is saved as well. The more gates a chip contains, the cheaper each gate is. When a chip contains a large number of gates, each gate becomes far less expensive than the package container that encases the gates (also, the cost is lower than the labor cost for soldering one connection).

On the other hand, **when the same package is manufactured in small quantities, each gate becomes very expensive** (much more than a gate in discrete components), because the high cost of initial design and manufacturing setups is distributed over small quantities and consequently constitutes a very large percentage of the cost of each package (i.e., labor and material costs for each package are less dominant). **Another important disadvantage of ICs is inflexibility in changing a network.** Even when we want to change the connections or gates in a network only slightly for some reason (e.g., if we find error in design, or if we need slightly different networks for new products), we must abandon the existing IC packages and must design and manufacture entirely new ones from scratch.

Discrete components will continue to be used because of their flexibility in realizing networks and their high electric power handling capability. Discrete components in a network can be easily (simply by desoldering components) and economically changed even if the design of the network needs to be altered during production. (This is convenient for new electronic products. See, e.g., p. 44 of McDermott [1972].) Also, when design and manufacture must be done in a short time, discrete components (also SSI) are preferred because those of LSI take a longer time.

Integration Size

Today IC chips of different integration sizes are used, depending on the application; these sizes are defined as follows:

Discrete Components and Integrated Circuits 55

SSI (small scale integration): the number of gates on a chip is less than 20.
MSI (medium scale integration): the number of gates on a chip is from 20 to 100.
LSI (large scale integration): the number of gates on a chip is from 100 to 1000.
VLSI (very large scale integration): the number of gates on a chip is more than 1000. [This is also called **GSI** (grand scale integration).]

In a broad sense, LSI includes VLSI. In particular, LSI means LSI and VLSI throughout this book, since the two are not essentially different from the logic design viewpoint.

The above definitions are not generally agreed upon. Some people define SSI as having fewer than 12 gates, and MSI as having 12 to 100 gates. Another definition* is based on the number of components: SSI with 1 through 99 components on a chip, MSI with 100 through 999 components, and LSI with 1000 or more components. Since we are concerned with gates in this book, classification of integration size by the number of gates is mentioned above. Actually, however, classification by the number of components (a gate consists of many components) is more realistic and common.

The chip size, or integration size, assuming fixed packing density of components per area, is limited by production yield and heat generation. Here **yield** (or production yield) is the ratio of the number of faultless IC packages to the total number of manufactured IC packages expressed in percentage. (Faulty packages among the manufactured packages must be dumped. The phrase "production yield" is commonly used for other products also.) As integration size increases, yield slowly decreases initially. Therefore, when integration size increases, cost per gate decreases. But **beyond a certain threshold integration size, cost per gate increases sharply** because of quick deterioration of production yield. Another important factor limiting integration size is heat generation, since maximum heat dissipation on one chip is a few watts (if more electric power is dissipated, the chip will be burned by excessive heat). For this reason the maximum integration size for gates of IC logic families based on bipolar transistors which generate great heat, such as ECL and TTL, is a few hundreds to about a thousand gates per chip (a very few hundred gates for ECL, and about a thousand gates for TTL; IIL, which will be discussed in Chapter 6, can have a large integration size comparable to that of MOS, though IIL is an IC logic family based on bipolar transistors). By contrast, MOS can easily have a few thousand or more gates per chip.

Integrated circuit packages that semiconductor manufacturers produce in large quantities for average users (such as SSI packages, many standard

---

* Some people define LSI as having 100 gates or more in the case of bipolar transistors, and as having 300 MOSFETs or more in the case of MOS.

56 Implementation of Logic Operations

networks in MSI, multipliers in LSI, and microprocessors in VLSI) are called **off-the-shelf packages**, in contrast to **custom-design IC packages**, which are designed according to the particular specifications of individual users.*

Design Problems with Different Integration Sizes

Integrated circuit packages of different integration sizes present diversified design problems, as follows.

Essentially all SSI packages are available as off-the-shelf packages. Examples of the simplest SSI packages, which contain a few separate gates, are shown in Fig. 2.4.1. Networks are formed by connecting these gates. In this sense, SSI packages are used in the same manner as discrete components —so to speak, a few discrete components in a block. In designing a network with discrete components or with SSI packages, minimization of the gate count yields the most economical network under appropriate assumptions such as equal cost for every gate (if the gate count is minimized, the number of SSI packages is correspondingly minimized).

An MSI chip usually contains a **standard network**, that is, a frequently used network such as an adder or a counter. For a large number of standard networks, MSI packages are commercially available as off-the-shelf packages. By interconnecting MSI and SSI packages, a complete digital system or a computer can be assembled. This can often be done using little switching algebra, because assembling such systems with these packages often requires knowledge only of the functional relationship between the inputs and outputs of the packages, which can often be more conveniently described in word statements than in switching algebra. Logic design becomes easy with MSI packages, if the situation allows the designer to use them.

An LSI chip (including the VLSI) contains many interconnected networks. It sometimes contains a subunit of a complete digital system (such as an arithmetic unit or a memory unit), or even a complete digital system. In designing a network in an LSI package, the design criteria for deriving an economical or fast network are very complex because of many complex constraints on network layout on a chip, as well as production considerations. By laying out networks appropriately on the chip (in particular, reducing parasitic capacitance), and thereby packing them into the chips as much as

---

* "Custom design" generally means a vendor's design that follows the specifications of an individual customer. But in the semiconductor industry, it means a design based on particular specifications, whether it is designed by a vendor or by a user in-house.

† Different symbols are sometimes used for power supplies. Usually V_{dd}, V_{ss}, V_{cc}, and V_{ee} mean the power supply terminals for the drain, source, collector, and emitter, respectively. In the CMOS, V_{dd} has been used as a convention, referring to the positive power supply terminal, though a power supply is actually connected to the source of a p-channel MOSFET in a CMOS cell.

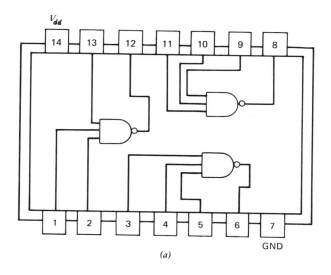

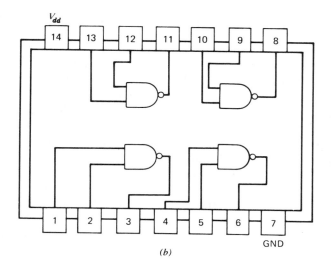

Fig. 2.4.1 SSI packages (V_{dd} and GND denote power source† and ground, respectively). (a) Triple three-input NAND gates. (b) Quadruple two-input NAND gates.

possible, cost is lowered and speed is raised. Logic design that is as error-free as possible is vitally important because correction in the case of LSI is costly and time-consuming. Thus design of LSI chips is the hardest.

When the designer is not attempting to attain lowest manufacturing cost and highest speed in LSI chip design, there are many approaches to alleviate the technical difficulty described above. For example, each different LSI chip can be designed simply by assembling, on the chip, layouts for subnetworks which were previously designed. This **library approach** (a library of layouts for networks or gates) can greatly reduce design errors and design time. Although the cost of design can be greatly reduced by the library approach, the manufacturing cost of the LSI chips is higher than that of chips made by the very careful design procedure mentioned above, because the chip size is not reduced to the full extent. (For further discussion, see Section 3.5.) The high manufacturing costs of LSI chips designed by this approach can be justified when the chips are parts of much more expensive products (e.g., sewing machines or expensive intelligent terminals), or when the shortest design time is very important for quick introduction of new products.

When the designer is attempting to attain lowest cost or highest speed in LSI chip design as much as possible, the design is very time-consuming and difficult, though it is technically very important and challenging. Often even standard networks must be redesigned from scratch* by adopting new layout rules or IC processing technology. Off-the-shelf LSI packages are usually designed in this manner.

In designing large networks, there is a choice between off-the-shelf LSI packages or custom-design LSI. When production volume is very high and no design change is expected for some time, custom LSI designed with the serious effort mentioned in the preceding paragraph (including custom design of microcomputers†) is usually preferable because of lower cost and better performance achieved by tailoring to their specific needs, though assembling off-the-shelf LSI packages has the unique advantage of short design time. When designers with nonsemiconductor manufacturers are concerned with the trade secrets of new products in which LSI is to be incorporated, they prefer custom LSI design, since competitors will then need more time to discover the internal mechanism of the new products.

A chip of smaller size is usually easy to manufacture and costs less (the

---

* When parasitic capacitance is very important, as in high-speed MOS networks, certain network configurations become very critical in terms of speed. For example, in the microprocessor 8080 of Intel, the carry-ripple adder is adopted rather than the look-ahead adder because the former is faster than the latter, in LSI implementation, though the latter is faster in discrete component implementation, as discussed in Section 9.2 (when new processing or layout rules which are different from those of the 8080 are developed, however, the situation may be different).

† See, for example, Gold [1976] or Puckett, Marley, and Gragg [1977].

Discrete Components and Integrated Circuits 59

cost depends, however, on other factors, such as production volume, in addition to chip size), so it is desirable to obtain networks which can be placed on the smallest chip possible. In LSI implementation, not only gates but also connections occupy a considerable area on a chip. Thus consideration of only the gate count appears to be insufficient (more so in the case of MOSFETs than of bipolar transistors). But even in LSI implementation, minimization of the gate count in a network, with minimization of the connection count as the second objective, usually appears to yield the most compact network, thus making LSI implementation cheaper [Muroga and Lai 1976]. If an entire system of networks cannot be packed into a single chip, and consequently must be distributed among many packages, serious attention must be paid to the fact that the maximum number of pins (pins are terminals of a package for outside connections) on each IC package is limited (usually at most 40 but can be as many as 150 [McDermott 1977]). If an entire network is inappropriately divided into subnetworks, some of them cannot be packed into single chips because of pin number limitations.

Semiconductor memories are also implemented in LSI. As these memories become cheaper and faster, they can sometimes be used to replace some gates in a network, in order to derive an economical network (though speed is usually lowered). This memory usage is different from the conventional use of memories to store programs, data, and intermediate computational results.

Logic design practice will be further discussed in Chapter 9.

Selection of IC Logic Families

Notice that the selection of IC logic families (RTL, ECL, *n*-MOS, CMOS, and others) and gate types (NOR gates, NAND gates, etc.) is not completely at the disposal of the logic designer. The designer usually must consider many aspects before choosing certain IC logic families and gate types. These aspects may be classified into **electronic performance**, **logic capability**, **economy**, and **availability**, as follows.

   1. Speed (i.e., switching time), electric power consumption, noise immunity (even if noise interferes, the circuit works correctly if noise amplitude is below a certain level), stability (even if the power source voltage fluctuates slightly, the circuit works correctly), and signal power amplification (since the electric signal must go through many gates, signal amplitude must be maintained; without amplification, the amplitude will decrease as the signal propagates through gates) are important aspects of **electronic performance**.

   2. When a certain (complete) set of gate types is used, networks may be designed with fewer gates or levels than are needed with some other set of gate types. This capability of gate types is called **logic capability**.

60   Implementation of Logic Operations

The number of inputs connected to a single gate is called the **fan-in** of the gate. Fan-in is limited for engineering reasons. If the number is too large, the operating condition of the electronic circuit that realizes the gate changes, and its operation may not be stable. The maximum number of inputs that can be connected to a single gate without damaging its stable operation is called the **maximum fan-in**. Maximum fan-in is usually 3 to 5 for a very fast gate. For a slow gate the number can be 10 or more, depending on the type of electronic circuit (usually it is not more than about 30).

The number of gates that are connected from the output of a single gate is called the **fan-out** of the gate. Fan-out is similarly limited for engineering reasons. The signal energy that a single gate can supply is finite. If the fan-out is too large, the following problems may develop. First, the signal amplitudes supplied to the succeeding gates may be lowered, resulting in malfunctions of these gates. Second, the condition for operation of the electronic circuit of the first gate may be changed because the succeeding gates lower the output impedance of the first gate, resulting in its faulty operation. The maximum number of gates that can be connected to a single gate output without damaging its stable operation is called the **maximum fan-out**. (This depends on the input impedances of gates in succeeding levels and the output impedance of the gate under consideration.) Maximum fan-out is usually from 3 to 5 for a fast gate and can be 10 to 50 for a slow gate [Millman and Halkias 1972, Taub and Schilling 1977].

Generally, networks can be synthesized with fewer gates, if each gate has a greater maximum fan-in or maximum fan-out. (The switching time of a gate usually tends to increase if the gate is designed so that the maximum fan-in or fan-out increases.) In this sense the maximum fan-in or fan-out restrictions are also important factors in logic capability.

An MOS gate has a greater logic capacity than a NOR or NAND gate, since it can express any negative function, including NOR and NAND. Also, an ECL gate has a greater logic capability than a NOR or an OR gate alone, since each ECL gate has double outputs (i.e., OR output and NOR output).

3.   The **cost** of a gate depends greatly on the choice of IC logic family (such as TTL, ECL, p-MOS, and CMOS), manufacturing processes, material, packages (in the case of the SSI, the packaging container costs more than the chip inside, so every gate costs more than gates in LSI, where the cost of each gate mainly depends on integration size), electronic implementation (wired logic, which will be discussed in Chapter 5, costs almost nothing), business practice, and so on. For example, a logic designer may find certain IC logic families or gate types cheaper than others because his or her firm buys them in large quantities (i.e., receives large-quantity discount). If a firm buys NOR gates in RTL very cheaply, the designer is obliged to use NOR gates.

Hence the situation is very complex. But throughout this book, let us assume, for the sake of simplicity, that **all gates cost the same**. This assump-

Discrete Components and Integrated Circuits 61

tion will be a reasonably good approximation of reality in most cases unless we are concerned with the detailed cost difference, which is sometimes vitally important, and the familiar statement "A network with fewer gates costs less than one with more gates" will usually be valid.

4. Availability is also an important aspect to consider. In the case of integrated circuits, users must be sure that like circuits are available from a second source, since one manufacturer may suddenly have trouble in production (production of integrated circuits is a very delicate process influenced by many factors). Some firms have gone into bankruptcy because ordered ICs were not delivered on time and these firms could not manufacture products without the ICs. Therefore firms usually do not order IC packages from a manufacturer unless they find a **second source**, that is, another manufacturer that can supply like IC packages. Often they use in-house IC production as a second source, even when they are not semiconductor manufacturers and in-house production is not economical.

In conclusion, logic designers do not have much freedom to choose gates of a certain IC logic family, even when they prefer it from a purely engineering viewpoint.

Exercises

2.1 (M−) When electronic circuits are given, it is the completely arbitrary choice of the designer to use positive logic or negative logic in expressing the logic performance of the circuits, since it does not change the nature of the analysis, that is, the electronic circuits involved are unaffected and the only difference in logic operations is that every AND and every OR is exchanged (i.e., each is the dual of the other logic operation, defined in Section 3.2).

(i) What function does the network of Fig. 2.3.4b (where n-channel MOSFETs are used) represent when we use negative logic?

(ii) What function does the network of Fig. 2.3.13 (where bipolar transistors are used) represent when we use negative logic?

(iii) What function does the network of Fig. 2.3.4b represent when the n-channel MOSFETs are replaced by p-channel MOSFETs, the power source with a positive voltage is replaced by a power source with a negative voltage, and negative logic is used? Is the function the same as the one shown in Fig. 2.3.4b?

2.2 (R) Translate the MOS network in Fig. E2.2 into a relay contact network. Then derive switching expressions for f and g.

62 Implementation of Logic Operations

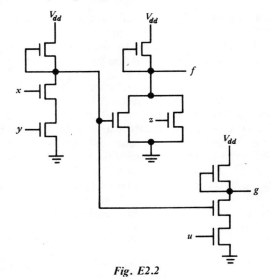

Fig. E2.2

2.3 (R) Write switching expressions for output functions (f, or f and g) in each MOS network in one of the sets of networks in Fig. E2.3.

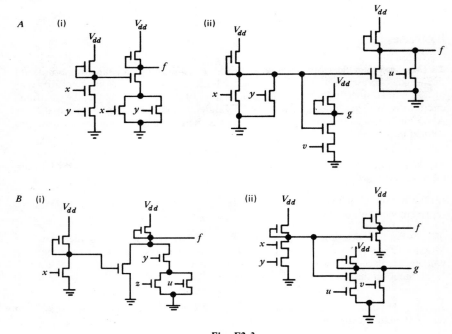

Fig. E2.3

2.4 (R) Write a switching expression for the output function f in one of the MOS networks in Fig. E2.4.

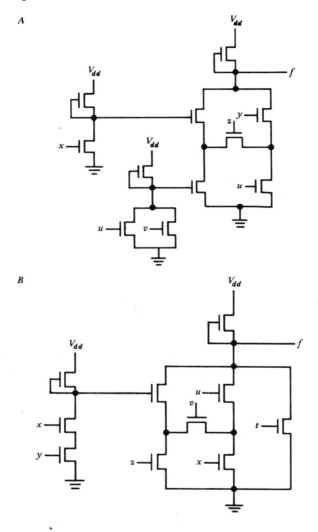

Fig. E2.4

2.5 (E) Write switching expressions for the two outputs, f_1 and f_2, of the MOS network in Fig. 2.3.8. If you can describe the logic performance of the network in a simple word statement (not only for each of f_1 and f_2 but also for the entire network), write it down.

64 Implementation of Logic Operations

2.6 (M) There is an integrated circuit package* consisting of four MOSFETs, as illustrated in Fig. E2.6a. One of these MOSFETs is used as a load MOSFET, and the terminal connected to it must always be connected to the power source.

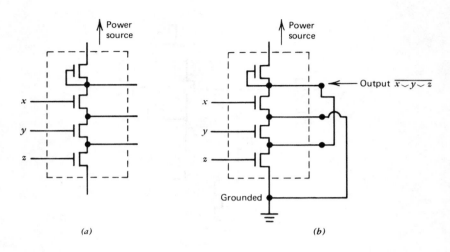

Fig. E2.6

By using and/or connecting the remaining seven terminals differently, this package can implement different functions. For example, the connection configuration shown in Fig. E2.6b implements the output function $\overline{x \vee y \vee z}$.

Show all the different functions that can be implemented by this package by different uses and/or connections of the seven terminals. At least one of these terminals must be grounded.

Show drawings also.

Notice that the package must have only one output, and since external variables x, y, z are interchangeable, you need to show only one of any set of equivalent functions ("equivalent" means "identical" by interchange of inputs; e.g., if you obtain $(\overline{xy})$, $(\overline{xz})$, or more, show one of them, say $(\overline{xy})$, only). Also, you are not allowed to connect each of the inputs x, y, z to any other terminal because of electronic performance problems.

2.7 (R) Write a switching expression for one of the gate networks in Fig. E2.7.

---

* Such a package is available from Fairchild (Fairchild 3102). By using these packages as building blocks, networks can be constructed. The gate in Fig. E2.6a is also used as a basic cell in an LSI chip, in which users can specify connections among a few hundred such cells [Nakano et al. 1978].

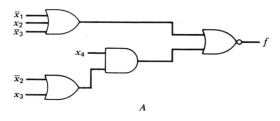

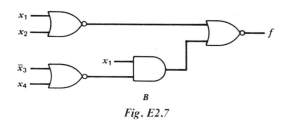

Fig. E2.7

2.8 (R) In each of the following cases, is it more appropriate to use discrete gates, SSI, MSI, LSI, or a combination of these? If a combination is necessary, discuss which integration size is appropriate in which part of the digital network.

(i) An automobile manufacturing company installs a simple digital network to control some aspects of the operation of an automobile (such as fuel injection and seat belt warning) in each car. The network must be manufactured in large quantities. But since new functions are added to the network every year, frequent design changes are expected.

(ii) A semiconductor manufacturing company makes a four-function calculator. The functions are fixed, and the company expects to manufacture the digital network inside the calculator without design changes, in large quantities for many years to come.

(iii) A minicomputer manufacturing company must change its designs almost every year because of the progress of IC technology. The production volume is moderate.

2.9 (R) Describe the important factors for a logic designer to consider in choosing a particular IC logic family or gate type.

2.10 (E) A network with multiple outputs is called a **decoder** (sometimes a demultiplexer) when exactly one output assumes 1 and all other outputs assume 0 for each combination of values of the input variables in such a manner that the output assuming 1 is different for each different combination of values of the input variables. A decoder can be implemented with a diode matrix as follows. (It can also be implemented with electronic gates.)

(i) The decoder with a diode matrix in Fig. E2.10 works as follows. A power source of $+10$ volts is applied between the terminal $+V$ and the ground. (The ground is not shown for simplicity.) Each of $x, \bar{x}, y, \bar{y}$ assumes 0 volt or 10 volts. Using positive logic, write a switching expression for each of the outputs $f_a, f_b, f_c,$ and f_d in terms of the input variables x and y. (This decoder with a diode matrix is often used for decoding a memory address when one of many memory address locations is to be selected.)

66 Implementation of Logic Operations

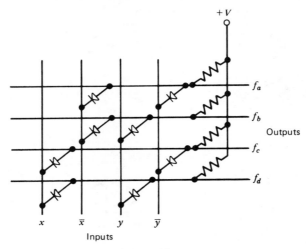

Fig. E2.10

(ii) Draw a diode matrix network for 3 variables, and write a switching expression for each output.

2.11 (E) Using AND gates only, design a decoder for two inputs, x and y. Assume that complemented and noncomplemented variables x, y, $\bar{x}$, and $\bar{y}$ are available as the decoder inputs. (For the definition of a decoder, see Exercise 2.10.) Show switching functions at the output terminals.

2.12 (M−) We want to design a network for a decimal number display of 8-4-2-1 binary-coded decimal code (or **8-4-2-1 BCD code**). Here each 8-4-2-1 BCD code word is represented by a four-bit binary number with bit weights 8, 4, 2, and 1 (e.g., 1 0 0 1 represents decimal number 9). Input variables x_1, x_2, x_3, x_4, which represent a 8-4-2-1 BCD code word (x_1 is the most significant bit), are fed into a decoder (see Exercise 2.10 for the definition of decoder), and then the outputs of the decoder are connected to the decimal number display, as shown in Fig. E2.12.

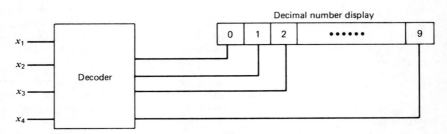

Fig. E2.12

Design the decoder with AND gates only, under the conditions stated below, assuming that both complemented and uncomplemented variables x_i and $\bar{x}_i$ ($i = 1, 2, 3, 4$) are available as inputs. Use as few gates as possible. Then show which output of the decoder indicates which decimal number.

The decoder is to be in two levels of gates (i.e., every path from an input of the network to an output must have exactly two levels of gates), and each gate has a maximum fan-in of 2.

2.13 (M) Design the decoder described in Exercise 2.12, replacing the condition there by the following:

The decoder need not be in two levels of gates, and each gate has a maximum fan-in of 2.

2.14 (M) A **half adder** is a network that adds two bits, whereas a full adder (Fig. 1.3.3) adds three bits (one of these three bits represents the carry from the previous bit position). It has two inputs, x and y, and two outputs, s and c, where c is the carry and s is the modulo-2 sum of x and y.

(i) Write decimal specifications and switching expressions for s and c.

(ii) Design a half adder with AND and OR gates only, assuming that complemented and noncomplemented variables x, y, $\bar{x}$, and $\bar{y}$ are available as the network inputs.

(iii) Design a full adder, using only two half adders and one OR gate.

2.15 (M) Design an MOS network for a half adder, using as few MOSFETs as possible. Assume that complemented and noncomplemented variables are available as network inputs (see Exercise 2.14 for the definition of a half adder).

CHAPTER 3

Fundamentals of Switching Algebra

To facilitate our analysis and synthesis of networks, let us further discuss switching algebra. If we can get simple switching expressions by algebraic manipulation, we can obtain simple networks, as we will see in later chapters. Tabular and map methods are conveniently used in logic design, as discussed in later chapters, but sometimes switching algebraic methods are more convenient.

Computer-aided design (CAD) is being widely used to facilitate logic design, layout on LSI chips, test, diagnosis, and manufacture. The switching algebraic methods are more suitable for CAD programs than the tabular or map methods. Error-free logic design is becoming vitally important as networks with ever-increasing size are being designed in LSI (or VLSI). Thus knowledge of switching algebra is becoming important for CAD program implementation.

3.1 Theorems of a Few Variables

Switching algebra is an algebraic system that consists of the set $\{0, 1\}$ (i.e., we deal with only two constants, 0 and 1) and operations of AND, OR, and NOT.

When we introduce switching variables x, y, z, each of which assumes the value of 0 or 1, the properties stated below hold. Although they are stated with only a few variables, their extension to many variables is easy.

(T1) $x \vee 1 = 1$ (T1') $x \cdot 0 = 0$
(T2) $x \vee 0 = x$ (T2') $x \cdot 1 = x$
(T3) $x \vee x = x$ (T3') $xx = x$ (Idempotency)
(T4) $x \vee y = y \vee x$ (T4') $xy = yx$ (Commutativity)

Theorems of a Few Variables 69

(T5) $(x \vee y) \vee z = x \vee (y \vee z)$ (T5') $(xy)z = x(yz)$ (Associativity)

(T6) $x \vee \bar{x} = 1$ (T6') $x\bar{x} = 0$ (Complementation)

(T7) $x(y \vee z) = xy \vee xz$ (T7') $x \vee yz = (x \vee y)(x \vee z)$ (Distributivity)

These properties can be proved by checking an equality for every combination of values of the variables. For example, (T6) is proved by showing that $x \vee \bar{x}$ in the left-hand side of (T6) is $0 \vee 1 = 1$ for $x = 0$ and $1 \vee 0 = 1$ for $x = 1$, always equaling 1 in the right-hand side of (T6). This method of proof is known as **perfect induction**. All the properties are no different from the ordinary algebra of real numbers with which we are familiar, except (T1), the idempotency (T3), and the complementation (T6) and (T6') [i.e., except for (T1), (T3), (T6), and (T6'), there are essentially no new properties which we need to memorize].

Notice that all the properties are grouped in pairs, and that one property in each pair can be obtained from the other by interchanging OR and AND operations and also 0 and 1. For example, (T2') is obtained from (T2), and vice versa. Any pair of properties with this nature is said to be **dual**. In each pair one property is called the dual of the other.

The following properties, which are also stated in dual pairs, can be proved by perfect induction. In each property the switching expression in the left-hand side of the identity is simplified into that in the right-hand side.

(T8) $x \vee xy = x$ (T8') $x(x \vee y) = x$ (Absorption)

(T9) $x \vee \bar{x}y = x \vee y$ (T9') $x(\bar{x} \vee y) = xy$ (Special case of consensus)

(T10) $xy \vee \bar{x}z \vee yz = xy \vee \bar{x}z$ (T10') $(x \vee y)(\bar{x} \vee z)(y \vee z) = (x \vee y)(\bar{x} \vee z)$ (Consensus)

(T11) $\overline{(\bar{x})} = x$ (Involution)

As a convention for the order of logic operations in any switching expression, **the calculation of conjunction precedes that of disjunction,** as mentioned in Section 1.3. If there are parentheses, the calculation inside the inner parentheses must be done first, and in each pair of parentheses the calculation of conjunction precedes that of disjunction.

The following property is useful in reducing the amount of work required to prove theorems.

70   Fundamentals of Switching Algebra

Theorem 3.1.1—Duality Theorem: If each $\vee$, $\cdot$, 1, and 0 in a switching identity or theorem is replaced by $\cdot$, $\vee$, 0, and 1, respectively, another identity or theorem is obtained.

[The new identity or theorem is called the **dual** of the original one. For example, from the identity $x \vee y \cdot (\bar{x} \vee \bar{y}) = x \vee y$, we get another identity, $x \cdot (y \vee \bar{x} \cdot \bar{y}) = x \cdot y$. Notice that the second identity is derived in the following manner. In the left-hand side in the first identity, the priority of calculation is in the order of the last $\vee$, the $\cdot$, and the first $\vee$. In the left-hand side in the second identity, the last $\cdot$ is correspondingly to be calculated first, the $\vee$ second, and then the first $\cdot$; thus parentheses are placed as shown. The right-hand side in the second identity can be similarly obtained.]

Proof: An identity or theorem can be proved by repeatedly applying properties (T1) through (T11) and (T1') through (T10'). Then, corresponding to this sequence of applications of properties, consider another sequence consisting of the duals of these properties. The new sequence is a proof for the second identity, for the following reason.

Every pair of properties in (T1) and (T1'), (T2) and (T2'), ..., (T10) and (T10'), and the property in (T11) are true. Therefore, if the sequence of properties applied to the original identity is correct, the dual sequence of properties on the dual identity is also correct.

The proof goes as follows for the above example:

For the first identity		dual	For the second identity	
$x \vee y \cdot (\bar{x} \vee \bar{y})$		$\longrightarrow$	$x \cdot (y \vee \bar{x} \cdot \bar{y})$	
$= x \vee y\bar{x} \vee y\bar{y}$	by (T7)	$\longrightarrow$	$= x \cdot (y \vee \bar{x}) \cdot (y \vee \bar{y})$	by (T7')
$= x \vee y \vee y\bar{y}$	by (T9)	$\longrightarrow$	$= x \cdot y \cdot (y \vee \bar{y})$	by (T9')
$= x \vee y \vee 0$	by (T6')	$\longrightarrow$	$= x \cdot y \cdot 1$	by (T6)
$= x \vee y$	by (T2)	$\longrightarrow$	$= x \cdot y$	by (T2')

Cases where switching expressions contain 0 or 1 can be shown in a similar way.   Q.E.D.

The above properties, (T1), ..., (T11), (T1'), ..., (T10'), will be used for simplification of switching expressions and also networks. For example, (T9) implies that the network in Fig. 3.1.1a can be simplified into the single gate shown in Fig. 3.1.1b. The duality theorem 3.1.1 implies that, if the network shown in Fig. 3.1.2a can be simplified into the gate in Fig. 3.1.2b, the network in Fig. 3.1.2c, which is obtained by replacing OR gates by AND

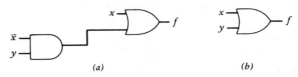

Fig. 3.1.1   *Application of* (T9). (a) $\bar{x}y \vee x$. (b) $x \vee y$.

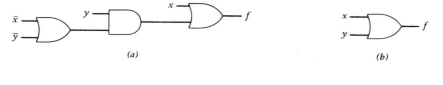

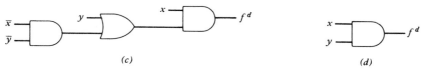

Fig. 3.1.2 Application of duality theorem. (a) $f = (\bar{x} \vee \bar{y})y \vee x$. (b) $f = x \vee y$. (c) $f^d = ((\bar{x} \cdot \bar{y}) \vee y)x$. (d) $f^d = xy$.

gates and AND gates by OR gates in Fig. 3.1.2a, can be simplified into the AND gate in Fig. 3.1.2d; in other words, the simplification from Fig. 3.1.2a to Fig. 3.1.2b holds in exactly the same way even if OR and AND gates are replaced by AND and OR gates, respectively.

Since we will discuss later in a more systematic way how to use the above properties, (T1), ..., (T11), (T1'), ..., (T10'), we need not memorize them at this point.

Remark 3.1.1: Among properties (T1), ..., (T11), (T1'), ..., (T10'), some can be obtained from others. For example, (T1'), (T2'), ..., can be obtained from (T1), (T2), ..., respectively, by the duality theorem, and vice versa [note that the dual identity of (T11) is (T11) itself]; (T3) is a special case of (T8) (with $y = 1$ inserted); (T6) results from (T9) (with $y = 1$ inserted) and (T1); (T9) results from (T10) and (T8). [If we replace y and z by 1 and y, respectively, in (T10), we get $x \vee \bar{x}y \vee y = x \vee \bar{x}y$. By $y \vee \bar{x}y = y$, that is, (T8) with x and y interchanged and with x replaced by $\bar{x}$, this becomes $x \vee y = x \vee \bar{x}y$, that is, (T9).] ∎

3.2 Theorems of *n* Variables

Let us denote the set of input variables by the vector expression $\mathbf{x} = (x_1, \ldots, x_n)$. There are 2^n different input vectors because each of these n variables assumes the value 1 or 0. Vectors with n components are often called **n-dimensional vectors** if we want to emphasize that there are n components. When the value of a switching function f is specified for each of the 2^n vectors (i.e., for every combination of the values of $x_1, \ldots, x_n$), f is said to be **completely specified**. Otherwise, f is said to be **incompletely specified**, that is, the value of f is specified for fewer than 2^n vectors. Input vectors for which the value of f is not specified are called **don't-care conditions**, as described in Section 1.4. Since these input vectors are never applied to a network whose output realizes f, the corresponding values of f need not be considered.

A switching function $f(x_1, \ldots, x_n)$ is sometimes denoted by $f(\mathbf{x})$, using the vector representation of variables, $\mathbf{x} = (x_1, \ldots, x_n)$.

An input vector $\mathbf{x}$ such that $f(\mathbf{x}) = 1$ is called a **true (input) vector** of f. An input vector $\mathbf{x}$ such that $f(\mathbf{x}) = 0$ is called a **false (input) vector** of f. For example, (0, 1), (1, 0), and (1, 1) are true vectors of f in Table 3.2.1.

Table 3.2.1 An Example of a Truth Table

x_1	x_2	f
0	0	0
0	1	1
1	0	1
1	1	1

If there exists a pair of input vectors, $(x_1, \ldots, x_{i-1}, 0, x_{i+1}, \ldots, x_n)$ and $(x_1, \ldots, x_{i-1}, 1, x_{i+1}, \ldots, x_n)$, which differ only in a particular variable x_i, such that f's values for these two vectors differ, the switching function $f(\mathbf{x})$ is said to be **dependent on** x_i. Otherwise it is termed **independent of** x_i. If f is independent of x_i, x_i is called a **dummy variable**. If $f(\mathbf{x})$ depends on all its variables, it is termed **nondegenerate**; otherwise, **degenerate**.

A particular switching function can be expressed in many different ways. For example, $x_1 \vee x_2$ can be expressed as $x_1 x_2 \vee x_1 \bar{x}_2 \vee \bar{x}_1 x_2, x_1 x_2 \vee x_1 \vee x_2, x_1 \vee \bar{x}_1 x_2, x_1 \vee x_2(\bar{x}_1 \vee \bar{x}_2)$, or $x_1 \vee x_2 x_3 \vee x_2 \bar{x}_3$. Although the last expression includes a new variable x_3, it is actually independent of x_3.

Switching Expressions and Expansions

As defined in Section 2.2, each variable x_i has two **literals**, x_i and $\bar{x}_i$.

Definition 3.2.1: (i) A conjunction (i.e., a logic product) of literals where a literal for each variable appears at most once is called a **term** (or a **product***). A term may consist of a single literal. A disjunction of

---

* "Product" may sometimes be ambiguous unless we add words, like "a product of literals," since $(x \vee y\bar{z})(u \vee v)$ is also a product.

terms is called a **disjunctive form*** (or a sum of products†). **(ii)** Similarly, a disjunction (i.e., a logic sum) of literals where a literal for each variable appears at most once is called an **alterm**. An alterm may consist of a single literal. A conjunction of alterms is called a **conjunctive form*** (or a product of sums†).

For example, $x_1 \bar{x}_2 x_3$ is a term, and $x_1 \lor x_2 \lor \bar{x}_3$ is an alterm. Also, $x_1 x_2 \lor x_1 \lor x_2$ and $x_1 \lor \bar{x}_1 x_2$ are disjunctive forms that are equivalent to the switching function $x_1 \lor x_2$, but $x_1 \lor x_2(\bar{x}_1 \lor \bar{x}_2)$ is not a disjunctive form, though it expresses the same function.

The following expressions of a switching function are important special cases of a disjunctive form and a conjunctive form.

Definition 3.2.2: Assume that n variables, $x_1, \ldots, x_n$, are under consideration. **(i)** A **minterm** (or a **fundamental product**, or a **canonical implicant**) is defined as the conjunction of exactly n literals, where exactly one literal for each variable (x_i and $\bar{x}_i$ are two literals of a variable x_i) appears. When a switching function f of n variables is expressed as a disjunction of minterms without repetition, it is called the **minterm expansion** of f (or the **canonical sum**, or the **canonical disjunctive form**). **(ii)** A **maxterm** (or a **fundamental sum**, or a **canonical implicate**) is defined as a disjunction of exactly n literals, where exactly one literal for each variable appears. When f is expressed as a conjunction of maxterms without repetition, it is called the **maxterm expansion** of f (or the **canonical product**, or the **canonical conjunctive form**).

For example, for $n = 3$, there exist $2^3 = 8$ minterms: $\bar{x}_1 \bar{x}_2 \bar{x}_3$, $\bar{x}_1 \bar{x}_2 x_3$, $\bar{x}_1 x_2 \bar{x}_3$, $\bar{x}_1 x_2 x_3$, $x_1 \bar{x}_2 \bar{x}_3$, $x_1 \bar{x}_2 x_3$, $x_1 x_2 \bar{x}_3$, and $x_1 x_2 x_3$. For the given function $x_1 \lor x_2 x_3$, the minterm expansion is $\bar{x}_1 x_2 x_3 \lor x_1 \bar{x}_2 \bar{x}_3 \lor x_1 \bar{x}_2 x_3 \lor x_1 x_2 \bar{x}_3 \lor x_1 x_2 x_3$ and the maxterm expansion is $(x_1 \lor x_2 \lor x_3)(x_1 \lor x_2 \lor \bar{x}_3)(x_1 \lor \bar{x}_2 \lor x_3)$, as explained in the following.

Any switching function can be uniquely expanded with minterms. If a function $f(x_1, x_2)$ of two variables is expanded as

$$f(x_1, x_2) = g_{00} \bar{x}_1 \bar{x}_2 \lor g_{01} \bar{x}_1 x_2 \lor g_{10} x_1 \bar{x}_2 \lor g_{11} x_1 x_2$$

with minterms $\bar{x}_1 \bar{x}_2, \bar{x}_1 x_2, x_1 \bar{x}_2, x_1 x_2$ and coefficients $g_{00}, g_{01}, g_{10}, g_{11}$, we obtain

$$g_{00} = f(0, 0), \qquad g_{01} = f(0, 1), \qquad g_{10} = f(1, 0), \qquad g_{11} = f(1, 1)$$

---

* For the sake of simplicity, throughout this book a disjunctive form is assumed not to contain terms that are identically 0 (e.g., $x_1 \bar{x}_2 x_2$) unless otherwise noted, since they are obviously redundant. Similarly, a conjunctive form is assumed not to contain alterms that are identically 1 (e.g., $x_1 \lor \bar{x}_2 \lor x_2 \lor x_3$).

† Since "sum" is used to mean "arithmetic sum" in later chapters, the term "sum of products" or "product of sums" is not used in this book.

74 Fundamentals of Switching Algebra

by inserting all combinations of 1 and 0 into x_1 and x_2 of $f(x_1, x_2)$. Thus $f(x_1, x_2)$ is expanded with minterms as

$$f(x_1, x_2) = f(0, 0)\bar{x}_1\bar{x}_2 \vee f(0, 1)\bar{x}_1 x_2 \vee f(1, 0)x_1\bar{x}_2 \vee f(1, 1)x_1 x_2. \quad (3.2.1)$$

Here the coefficients $f(1, 1), f(1, 0)$, and others are uniquely determined when the value of $f(x_1, x_2)$ is concretely given for each combination of values of x_1 and x_2. For example, if $f(x_1, x_2) = x_1 \vee x_2$, then $f(0, 0) = 0$ and $f(0, 1) = f(1, 0) = f(1, 1) = 1$. Thus this $f(x_1, x_2)$ is expanded with the minterms as

$$x_1 \vee x_2 = \bar{x}_1 x_2 \vee x_1 \bar{x}_2 \vee x_1 x_2. \quad (3.2.2)$$

Notice that **the minterm expansion 3.2.2, as a matter of fact, can be obtained directly from the truth table in Table 3.2.1 without using the general expression 3.2.1,** as follows. Consider only true vectors in Table 3.2.1. Then form a product of noncomplemented variables and complemented variables corresponding to 1's and 0's in each true vector, respectively. For true vector (0, 1), for example, we obtain the product $\bar{x}_1 x_2$. The disjunction of the products thus formed yields the minterm expansion 3.2.2.

Also, any switching function can be uniquely expressed with maxterms in a similar manner. For example, $f(x_1, x_2)$ can be expressed as follows, using maxterms $x_1 \vee x_2, x_1 \vee \bar{x}_2, \bar{x}_1 \vee x_2, \bar{x}_1 \vee \bar{x}_2$:

$$f(x_1, x_2) = (f(0, 0) \vee x_1 \vee x_2)(f(0, 1) \vee x_1 \vee \bar{x}_2)$$
$$(f(1, 0) \vee \bar{x}_1 \vee x_2)(f(1, 1) \vee \bar{x}_1 \vee \bar{x}_2). \quad (3.2.3)$$

Notice that in the minterm expansion the coefficient for $x_1\bar{x}_2$ is $f(1, 0)$, but in the maxterm expansion the coefficient for $\bar{x}_1 \vee x_2$ is $f(1, 0)$. In other words, in the former case 1 and 0 correspond to noncomplemented (e.g., $x_1, x_2, \ldots$) and complemented (e.g., $\bar{x}_1, \bar{x}_2, \ldots$) variables, respectively, whereas in the latter case 1 and 0 correspond to complemented and noncomplemented variables, respectively. Again, **the maxterm expansion can be obtained directly from the truth table for a given function**, as follows. Suppose that a truth table such as Table 3.2.2 is given. Consider only false vectors (not true vectors) in

Table 3.2.2 An Example of a Truth Table

x_1	x_2	$f = x_1 x_2$
0	0	0
0	1	0
1	0	0
1	1	1

Theorems of n Variables 75

Table 3.2.2, which is the truth table for $x_1 x_2$. Form an alterm of noncomplemented variables and complemented variables corresponding to 0's and 1's, respectively, in each **false** vector. Notice that in this case "alterm" is considered, and **0 and 1 are considered to correspond to noncomplemented and complemented variables, respectively**. For false input vector (0, 1), for example, we obtain the alterm $x_1 \vee \bar{x}_2$. Then the conjunction of these alterms yields the maxterm expansion

$$x_1 x_2 = (x_1 \vee x_2)(x_1 \vee \bar{x}_2)(\bar{x}_1 \vee x_2). \tag{3.2.4}$$

As can easily be seen, the expression that results from (3.2.3) for this function:

$$x_1 x_2 = (0 \vee x_1 \vee x_2)(0 \vee x_1 \vee \bar{x}_2)(0 \vee \bar{x}_1 \vee x_2)(1 \vee \bar{x}_1 \vee \bar{x}_2) \tag{3.2.5}$$

can be reduced to maxterm expansion 3.2.4 by deleting 0's and the last alterm containing 1 in this expression (3.2.5). If the minterm expansion $x_1 x_2$ for this function $x_1 x_2$ is compared with maxterm expansion 3.2.4 for the same function, we can see that minterm and maxterm expansions are very different expressions.

These expansions can be generalized to the case of n variables as follows.

Theorem 3.2.1: Let $x_i^{c_i}$ denote x_i for $c_i = 1$ and $\bar{x}_i$ for $c_i = 0$. Any switching function $f(x_1, x_2, \ldots, x_n)$ can be uniquely expanded in the minterm expansion:

$$f(x_1, x_2, \ldots, x_n) = \bigvee_{c_1, \ldots, c_n} f(c_1, c_2, \ldots, c_n) x_1^{c_1} x_2^{c_2} \ldots x_n^{c_n},$$

where the disjunction

$$\bigvee_{c_1, \ldots, c_n}$$

is formed over all possible combinations of 1 and 0 for $c_1, c_2, \ldots, c_n$.

Also $f(x_1, \ldots, x_n)$ can be uniquely expanded in the maxterm expansion:

$$f(x_1, x_2, \ldots, x_n) = \prod_{c_1, \ldots, c_n} \{f(\bar{c}_1, \bar{c}_2, \ldots, \bar{c}_n) \vee x_1^{c_1} \vee x_2^{c_2} \vee \cdots \vee x_n^{c_n}\},$$

where the conjunction

$$\prod_{c_1, \ldots, c_n}$$

is formed over all possible combinations of 1 and 0 for $c_1, c_2, \ldots, c_n$.

Proof: These two identities can be proved by comparing the left and right sides for all combinations of 1 and 0 for $x_1, x_2, \ldots, x_n$. As is easily seen, the coefficients are uniquely determined. Q.E.D.

76 Fundamentals of Switching Algebra

For example, although $x_1 \vee \bar{x}_1 x_2$, $x_1 \vee x_2$, and $x_1 x_2 \vee x_1 \vee x_2$ are different disjunctive forms for the same function, its minterm expansion is uniquely $x_1 \bar{x}_2 \vee \bar{x}_1 x_2 \vee x_1 x_2$. But notice that **the minterm expansion is unique only when the number of variables is specified, and the same function can be expressed by different minterm expansions when dummy variables are added**. For example, when a dummy variable x_3 is added, the minterm expansion of $x_1 \vee x_2$ in the above example with respect to three variables, x_1, x_2, and x_3, is

$$x_1 \vee x_2 = \bar{x}_1 x_2 \bar{x}_3 \vee \bar{x}_1 x_2 x_3 \vee x_1 \bar{x}_2 \bar{x}_3 \vee x_1 \bar{x}_2 x_3 \vee x_1 x_2 \bar{x}_3 \vee x_1 x_2 x_3.$$

In the above, minterm and maxterm expansions for the same function are derived separately. There is no simple way to derive one from the other directly.

Dualization of a Function

The following theorem is useful.

Theorem 3.2.2—De Morgan's Theorem:

$$\overline{(x_1 \vee x_2 \vee \ldots \vee x_n)} = \bar{x}_1 \bar{x}_2 \ldots \bar{x}_n$$

and

$$\overline{(x_1 x_2 \ldots x_n)} = \bar{x}_1 \vee \bar{x}_2 \vee \ldots \vee \bar{x}_n.$$

Proof: For the case of $n = 2$, it is easy to prove the identities by comparing the left and right sides for all combinations of 0 and 1 for two variables.

Assume that $\overline{(x_1 \vee x_2 \vee \cdots \vee x_n)} = \bar{x}_1 \bar{x}_2 \ldots \bar{x}_n$ holds. Then $\{\overline{(x_1 \vee x_2 \vee \cdots \vee x_n) \vee x_{n+1}}\}$ may be rewritten as $\overline{y \vee x_{n+1}}$ with $y = (x_1 \vee \cdots \vee x_n)$. Since the identity is true for two variables, we have $\overline{y \vee x_{n+1}} = \bar{y} \cdot \bar{x}_{n+1} = \overline{(x_1 \vee x_2 \vee \cdots \vee x_n)} \cdot \bar{x}_{n+1}$. This is equal to $\bar{x}_1 \bar{x}_2 \ldots \bar{x}_n \bar{x}_{n+1}$ by the assumption. The second identity can be proved in a similar way.   Q.E.D.

In the proof of Theorem 3.2.2 the statement is proved for a small value of n (i.e., $n = 2$ in this case). Then, after assuming that the statement is true for n, the case for $n + 1$ is proved. Thus, since the statement is true for $n = 2$, it is true for $n = 3$. Since it is true for $n = 3$, it is true for $n = 4$, and so on. This method of proving a theorem is called **finite induction** and is used to prove many other theorems.

De Morgan's theorem 3.2.2 has many applications. For example, it implies that a NOR gate with noncomplemented variable inputs $x_1, \ldots, x_n$ is interchangeable with an AND gate with complemented variable inputs $\bar{x}_1, \ldots, \bar{x}_n$, since the outputs of both gates express the same function. This is illustrated in Fig. 3.2.1 for $n = 2$.

Fig. 3.2.1 *Application of De Morgan's theorem.* (*Two gates are interchangeable*)

The dual of a function is an important concept.

Definition 3.2.3: The **dual** of a switching function $f(x_1, \ldots, x_n)$ is defined as $\bar{f}(\bar{x}_1, \bar{x}_2, \ldots, \bar{x}_n)$, where $\bar{f}$ denotes the inversion of the entire function $f(x_1, \ldots, x_n)$ [thus the notation $\overline{f(x_1, \ldots, x_n)}$ might be used instead of $\bar{f}$]. Let it be denoted by $f^d(x_1, \ldots, x_n)$. In particular, if $f(x_1, \ldots, x_n) = f^d(x_1, \ldots, x_n)$, then $f(x_1, \ldots, x_n)$ is called a **self-dual function**.

For example, when $f(x_1, x_2) = x_1 \vee x_2$ is given, we have $f^d(x_1, x_2) = \bar{f}(\bar{x}_1, \bar{x}_2) = \overline{\bar{x}_1 \vee \bar{x}_2}$. This is equal to $x_1 x_2$ by applying the first identity of Theorem 3.2.2. In other words, $f^d(x_1, x_2) = x_1 x_2$. The function $x_1 x_2 \vee x_2 x_3 \vee x_1 x_3$ is self-dual, as can be seen by applying the two identities of Theorem 3.2.2 (Exercise 3.2.11).

Notice that, if f^d is the dual of f, the dual of f^d is f.

As we will see later, the concept of a dual function has many important applications. For example, it is useful in the conversion of networks with different types of gates, as in Fig. 3.1.2, where the replacement of the AND and OR gates in Fig. 3.1.2a by OR and AND gates, respectively, yields the output function f^d in Fig. 3.1.2c, which is dual to the output f of Fig. 3.1.2a.

Another important example of an application is a CMOS gate (or CMOS cell), where **CMOS** stands for complementary MOS, a variation of the *n*-channel MOS explained in Section 2.3.1. The CMOS is extensively used in crystal watches because of its extremely low electric power consumption. A CMOS gate consists of two subcircuits, one with *p*-channel MOSFETS and the other with *n*-channel MOSFETS, as exemplified in Fig. 3.2.2. The *p*-channel MOSFET subcircuit works as a substitute for the load MOSFET in the *n*-channel MOS gate in Fig. 2.3.4b. The *p*-channel MOSFET subcircuit, however, does not work exactly as a load MOSFET in the sense of Section 2.3.1., since the same inputs, x and y, are connected to both *p*-channel and *n*-channel MOSFET subcircuits. In other words, the *p*-channel MOSFET subcircuit is, so to speak, a variable load MOSFET controlled by the inputs, and this makes the power consumption of a CMOS gate extremely low when inputs x and y do not change frequently.* If every MOSFET (no matter whether it is *p*-channel or *n*-channel) in a subcircuit in Fig. 3.2.2 is regarded as a make-contact of relay, the transmission g between the ground and the

---

* The power consumption increases linearly as the frequency of input change increases.

78 Fundamentals of Switching Algebra

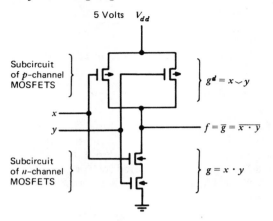

Fig. 3.2.2   CMOS gate.

output terminal for f is xy (consequently, the output function f becomes its complement $\bar{g} = \overline{xy}$), and the transmission between power supply V_{dd} and the output terminal for f becomes the dual of g, i.e., $g^d = x \vee y$. Although Fig. 3.2.2 shows a CMOS for a particular function $\overline{xy}$, **a CMOS gate* for a general function f consists of an n-channel MOSFET subcircuit whose transmission is $g = \bar{f}$ and a p-channel MOSFET subcircuit whose transmission is its dual g^d.** In other words, the output function f is related to the transmission g of the n-channel MOSFET subcircuit only and has nothing to do with the p-channel MOSFET subcircuit, which serves simply to reduce electric power consumption. (For the rest of this book, it is not important to understand how a CMOS gate works electronically, but if you want to understand, see Exercise 3.2.21.)

The following theorem shows a more convenient method of computing the dual of a function than direct use of Definition 3.2.3 (i.e., inversion of the function's value with its variables inverted).

Theorem 3.2.3—Generalized De Morgan's Theorem: Let $f(\mathbf{x})$ be a function expressed by $\vee$, $\cdot$, and $^-$ (and possibly also by parentheses and the constants 0 and 1). Let $g(\mathbf{x})$ be a function that is obtained by replacing every $\vee$ and $\cdot$ by $\cdot$ and $\vee$, respectively, throughout the switching expression of $f(\mathbf{x})$ (and also, if 0 or 1 is contained in the original expression, by replacing 0 and 1 by 1 and 0, respectively). Then

$$f^d(\mathbf{x}) = g(\mathbf{x}).$$

* There are variations of the CMOS explained here. For example, see the CMOS with diodes added [Capell, Knoblock, Mather, and Lopp 1977; Mei 1976].

Theorems of n Variables 79

(For example, when $f = x_1 \vee x_2 \cdot \bar{x}_3$ is given, $f^d = x_1 \cdot (x_2 \vee \bar{x}_3)$ is obtained by this theorem. Here notice that in f the calculation of · precedes that of $\vee$, and in f^d the $\vee$ must correspondingly be calculated first [thus the parentheses are placed as $(x_2 \vee \bar{x}_3)$]. When $f = x_1 \vee 0 \cdot x_2 \cdot \bar{x}_3$ is given, $f^d = x_1 \cdot (1 \vee x_2 \vee \bar{x}_3)$ results by this theorem. When $f = x_1 \vee 1 \cdot x_2 \cdot \bar{x}_3$ is given, $f^d = x_1 \cdot (0 \vee x_2 \vee \bar{x}_3)$ results. When $f = x_1 \vee A \cdot x_2 \cdot \bar{x}_3$ is given, where A is a constant coefficient 0 or 1, $f^d = x_1 \cdot (\bar{A} \vee x_2 \vee \bar{x}_3)$ results, summarizing the above two cases.)

Proof: First we can easily prove the following identities, where u and v are functions of variables x_i's $(i = 1, 2, \ldots, n)$:

(i) $(u \vee v)^d = u^d v^d$.
(ii) $(u \cdot v)^d = u^d \vee v^d$.
(iii) $(\bar{u})^d = \overline{(u^d)}$.
(iv) $(0)^d = 1$.
(v) $(1)^d = 0$.
(vi) $(x_i)^d = x_i$ for $i = 1, 2, \ldots, n$.

For example, the left-hand side of (i) is $\{\overline{u(\bar{x}_1, \ldots, \bar{x}_n) \vee v(\bar{x}_1, \ldots, \bar{x}_n)}\}$ according to Definition 3.2.3. Then, by De Morgan's theorem, this becomes $\bar{u}(\bar{x}_1, \ldots, \bar{x}_n)\bar{v}(\bar{x}_1, \ldots, \bar{x}_n)$, which is the right-hand side of (i).
When an expression for a function $f(x_1, \ldots, x_n)$ is given, the expression for f^d can be converted into an expression without dual operations d by repeatedly applying the above identities. Whenever $\vee, \cdot, 0$, or 1 appears during this conversion, it is replaced by $\cdot, \vee, 1$, or 0, respectively, as we can see in (i), (ii), (iv), or (v), respectively. Therefore the theorem statement is true.
For example, let us derive the dual of $f = x_1 \vee x_2 \cdot (x_3 \vee \bar{x}_1 x_4)$. Here

$$\begin{aligned}
f^d &= \{x_1 \vee x_2 \cdot (x_3 \vee \bar{x}_1 \cdot x_4)\}^d, \\
&= (x_1)^d \cdot \{x_2 \cdot (x_3 \vee \bar{x}_1 \cdot x_4)\}^d && \text{by (i)}, \\
&= x_1 \cdot \{x_2^d \vee (x_3 \vee \bar{x}_1 \cdot x_4)^d\} && \text{by (vi) and (ii)}, \\
&= x_1 \cdot (x_2 \vee x_3^d \cdot (\bar{x}_1 \cdot x_4)^d) && \text{by (vi) and (i)}, \\
&= x_1 \cdot (x_2 \vee x_3 \cdot ((\bar{x}_1)^d \vee x_4^d)) && \text{by (vi) and (ii)}, \\
&= x_1 \cdot (x_2 \vee x_3 \cdot (\overline{x_1^d} \vee x_4)) && \text{by (iii) and (vi)}, \\
&= x_1 \cdot (x_2 \vee x_3 \cdot (\bar{x}_1 \vee x_4)) && \text{by (vi)}.
\end{aligned}$$

If the last expression is compared with the original expression for f, it is seen that $\vee$ and $\cdot$ have been interchanged (with appropriate use of parentheses).
Even if 0 or 1 is included in the original expression, it can be replaced by 1 or 0, respectively.   Q.E.D.

For example, the dual of $x_1 \vee x_2 x_3$ is $\overline{\bar{x}_1 \vee \bar{x}_2 \bar{x}_3}$ according to Definition 3.2.3, which is a somewhat complicated expression. But by using the generalized De Morgan's theorem we can immediately obtain the expression without bars, $x_1 \cdot (x_2 \vee x_3) = x_1 x_2 \vee x_1 x_3$.

80 Fundamentals of Switching Algebra

Possible Pitfalls in Algebraic Manipulation

Some of the theorems and properties of switching algebra are not different from the corresponding ones of ordinary algebra of real numbers. For example, the conjunction $x \cdot y$ is not different from ordinary multiplication. But the disjunction $x \vee y$ is different from ordinary addition. It is easy to overlook such subtle differences in manipulation of expressions in switching algebra from those in ordinary algebra, as the following example shows. In ordinary algebra, if $x + y = x + z$, then $y = z$ must follow. But in switching algebra, even if $x \vee y = x \vee z$, $y = z$ does not necessarily follow. Even if $z = \bar{x}y$, which is obviously not identical to y, the above equality $x \vee y = x \vee z$ holds because of (T9). (All solutions of z can be found by the approach stated in Exercises 3.2.23 and 3.2.24. For a general solution of Boolean equations, see Rudeanu [1974].) A similar situation occurs with conjunction: even if $xy = xz$, $y = z$ does not necessarily follow and $z = \bar{x} \vee y$ can be a solution. During manipulation of switching expressions, readers are cautioned against forgetting these subtleties.

Symmetric Functions

Symmetric functions are an important concept.

When a function $f(x_1, \ldots, x_n)$ remains unchanged for every permutation of its variables $x_1, x_2, \ldots, x_n$ (i.e., for interchange of variables in all possible ways), f is called a **totally symmetric function of $x_1, x_2, \ldots, x_n$**, or f is said to be **totally symmetric in $x_1, x_2, \ldots, x_n$**. Examples are $x_1 x_2 \vee x_2 x_3 \vee x_1 x_3$ and $x_1 \bar{x}_2 \vee \bar{x}_1 x_2$. (No matter how we permute x_1, x_2, and x_3, $x_1 x_2 \vee x_2 x_3 \vee x_1 x_3$ becomes itself. Also, $x_1 \bar{x}_2 \vee \bar{x}_1 x_2$ becomes itself after permutation of x_1 and x_2.) But $x_1 \vee x_2 x_3$ is not totally symmetric, since permutation of x_1 and x_2 converts $x_1 \vee x_2 x_3$ into $x_2 \vee x_1 x_3$, which is different from the original $x_1 \vee x_2 x_3$. Notice that, **if we consider the complement of some variables, say $\bar{x}_i, \ldots, \bar{x}_n$, some functions become totally symmetric in $x_1, x_2, \ldots, \bar{x}_i, \ldots, \bar{x}_n$**. For example, $\bar{x}_1 x_2 \bar{x}_3 \vee x_1 \bar{x}_2 \bar{x}_3 \vee x_1 x_2 x_3$ is totally symmetric in x_1, x_2, and $\bar{x}_3$, though not totally symmetric in x_1, x_2, and x_3. (This can be seen from the fact that the function becomes $\bar{x}_1 x_2 y \vee x_1 \bar{x}_2 y \vee x_1 x_2 \bar{y}$, which is totally symmetric in x_1, x_2, and y, when $\bar{x}_3$ is expressed as y.)

In contrast to totally symmetric functions, when function $f(x_1, x_2, \ldots, x_n)$ is unchanged for every permutation of a subset of variables, f is said to be **partially symmetric**. For example, $x_1 \vee x_2 x_3$ is partially symmetric in x_2 and x_3, since $x_1 \vee x_2 x_3$ becomes itself after permutation of x_2 and x_3.

The variables in which a function is symmetric are called **symmetric variables**.

Since complementation of variables in all possible ways must be considered, there is no simple way to find symmetric variables when functions are complex, though we can do better than the brute-force method which literally executes the above definitions, based on some properties such as Exercises 3.2.27 and 3.2.28. [Harrison 65] [Kohavi 70]. The following theorem may sometimes be helpful.

Theorem 3.2.4 [Shannon 1938]: A function $f(x_1, x_2, \ldots, x_n)$ is totally symmetric in $x_1, x_2, \ldots, x_n$ if and only if there exist integers $I_1, I_2, \ldots, I_t$, where $0 \le I_i \le n$ for $i = 1, 2, \ldots, t$, such that the value of f is 1 when and only when I_i of the n variables assumes the value 1. (For example, the parity function $x \oplus y \oplus z$ in Table 1.4.1 is 1 when and only when the number of 1's in x, y, and z is 1 or 3—in other words, $I_1 = 1$ and $I_2 = 3$.)

Proof: Suppose that f is totally symmetric. Then, for any permutation of the variables, the value of f is unchanged and also the number of variables that assume the value 1 is unchanged. Hence f is still 1 if I_i of the n variables assume the value 1, and 0 otherwise. Conversely, suppose that f is 1 if I_i of the n variables assume the value 1, and 0 otherwise. Then this means that the value of f is unchanged for any permutation of the variables.   Q.E.D.

If we want to find the possibility of symmetry in some complemented and other noncomplemented variables, say $\bar{x}_1, \ldots, \bar{x}_{p-1}, x_p, \ldots, x_n$, the number of 1's in $\bar{x}_1, \ldots, \bar{x}_{p-1}, x_p, \ldots, x_n$ must be counted.

Many functions that are important in logic design practice are totally symmetric functions. Examples are the sum and carry outputs of an adder (e.g., Fig. 1.3.3) and the parity function in Table 1.4.1. Notice that networks for totally symmetric functions do not necessarily have symmetric configurations of gates and connections (e.g., the adder network in Fig. 9.2.2, which has a minimum number of gates).

Simple Questions for Self-Study
___

Before proceeding further, try the following simple questions. (Answers are given after the following exercises.)

1. Is the function f shown in Table 3.2.3 independent of some variable (or variables)?

Table 3.2.3

x	y	z	f
0	0	0	0
0	0	1	1
0	1	0	0
0	1	1	1
1	0	0	1
1	0	1	0
1	1	0	1
1	1	1	0

82 Fundamentals of Switching Algebra

2. Which of the following switching expressions is (a) an alterm, (b) a disjunctive form, (c) a conjunctive form, (d) a minterm expansion, or (e) a maxterm expansion? [Some switching expressions may be combinations of some of (a), (b), ..., (e).]

(i) $x \vee yz$.
(ii) $(x \vee y)(x \vee \bar{z})$.
(iii) $(x \vee yz)(x \vee y)$.
(iv) $xy \vee \bar{x}\bar{y}$.
(v) $x \vee \bar{y} \vee z$.
(vi) $(x \vee \bar{y} \vee z)(\bar{x} \vee y \vee z)$.

3. Derive the minterm and maxterm expansions of $f = x \vee y\bar{z}$ with respect to $x, y,$ and z.

4. Derive the dual of $x \vee \bar{y}z$.

5. Apply De Morgan's theorem 3.2.2 to the following functions:

   (i) $\overline{x \vee (\bar{y} \vee \bar{z})}$.   (ii) $\overline{xy \vee z \vee \bar{x}(u \vee v)}$.

6. In what variables is the following function symmetric?

$$x_1 \vee x_2 x_3 \vee \bar{x}_2 \bar{x}_3.$$

Exercises

3.2.1 (R) Prove one of the following sets of identities:
A. (T5) and (T10) in Section 3.1.
B. (T5′) and (T10′) in Section 3.1.

3.2.2 (R) Derive an identity by replacing every $\vee$, $\cdot$, 1, and 0 by $\cdot$, $\vee$, 0, and 1, respectively, in the left-hand and right-hand sides of the following identity:

$$x \vee \bar{x}(y \vee 0) = x \vee y.$$

3.2.3 (R) Find whether the following function f is dependent on each variable:

$$f = xy \vee yz \vee \bar{z}.$$

3.2.4 (E) Prove that, if a function f is independent of a particular variable x_i, there exists a switching expression for f that does not contain any literals of x_i.

3.2.5 (M−) Derive a necessary and sufficient condition for $f(x_1, \ldots, x_i, \ldots, x_n) \oplus f(x_1, \ldots, \bar{x}_i, \ldots, x_n)$ to be identically 0, where only one variable x_i of function f is changed to $\bar{x}_i$ in the second term $f(x_1, \ldots, \bar{x}_i, \ldots, x_n)$.

   Remark: This expression is known as the **Boolean difference** and is used in a method to diagnose a network for f (Section 9.6). ■

3.2.6 (R) (i) Obtain the minterm expansion and the maxterm expansion for $x_1 x_2 \vee x_2 \bar{x}_3 \vee \bar{x}_2 x_3$ with respect to x_1, x_2, x_3.
(ii) Obtain the maxterm expansion for $x_1 x_2 \vee \bar{x}_1 \bar{x}_2 \vee x_1 x_3$ with respect to x_1, x_2, x_3.
(iii) Obtain the minterm expansion for $(x_1 \vee x_2)(\bar{x}_1 \vee x_3)(\bar{x}_1 \vee \bar{x}_2)$ with respect to x_1, x_2, x_3. Also obtain it with respect to x_1, x_2, x_3, x_4.

Theorems of n Variables 83

3.2.7 (E) Prove the expansions of a function $f(x_1, \ldots, x_n)$:

$$f(x_1, \ldots, x_n) = x_1 f(1, x_2, \ldots, x_n) \vee \bar{x}_1 f(0, x_2, \ldots, x_n),$$
$$f(x_1, \ldots, x_n) = x_1 x_2 f(1, 1, x_3, \ldots, x_n) \vee x_1 \bar{x}_2 f(1, 0, x_3, \ldots, x_n)$$
$$\vee \bar{x}_1 x_2 f(0, 1, x_3, \ldots, x_n) \vee \bar{x}_1 \bar{x}_2 f(0, 0, x_3, \ldots, x_n),$$
. . . .

Also prove the expansions of $f(x_1, \ldots, x_n)$:

$$f(x_1, \ldots, x_n) = (x_1 \vee f(0, x_2, \ldots, x_n))(\bar{x}_1 \vee f(1, x_2, \ldots, x_n)),$$
$$f(x_1, \ldots, x_n) = (x_1 \vee x_2 \vee f(0, 0, x_3, \ldots, x_n))(x_1 \vee \bar{x}_2 \vee f(0, 1, x_3, \ldots, x_n))$$
$$(\bar{x}_1 \vee x_2 \vee f(1, 0, x_3, \ldots, x_n))(\bar{x}_1 \vee \bar{x}_2 \vee f(1, 1, x_3, \ldots, x_n)),$$
. . . .

Remark: These expansions can be extended to the case with m variables factored out, where $m \leq n$, though the expansion only for $m = 1$ and 2 are shown above. These are called **Shannon's expansions**. Of course, when $m = n$, the expansions become the minterm and maxterm expansions. ∎

3.2.8 (R) Convert the following functions into disjunctive forms, using De Morgan's theorem 3.2.2 repeatedly:

(i) $\overline{(\bar{x} \vee y)(y \vee \bar{x}z)}$. (ii) $\overline{(xy \vee \bar{x}\bar{y})(u \vee v)(w \vee \overline{u\bar{v}})}$.

3.2.9 (E) Prove that, if $x \vee yz = u$, then $\bar{x}u \vee y\bar{u} = \bar{x}y$.

3.2.10 (R) Obtain, in disjunctive forms, the duals of the following functions:

(i) $x_1 x_2 \vee x_3 x_4 \vee \bar{x}_1 x_3$. (ii) $x_1(\bar{x}_2 x_3 \vee x_2 \bar{x}_3) \vee x_1 \bar{x}_3$.

3.2.11 (R) Verify that $x_1 x_2 \vee x_2 x_3 \vee x_1 x_3$ is self-dual.

3.2.12 (E) Prove that, if f^d is the dual of a function f, the dual of $\bar{f}$ is $\overline{f^d}$.

3.2.13 (E) Assume that $f(\mathbf{x})$ is a self-dual function. In comparing the number of true vectors and the number of false vectors, can you say whether or not one is smaller than the other? Explain why.

3.2.14 (M) When a vector $\mathbf{B} = (b_1, \ldots, b_n)$ is given, $\bar{\mathbf{B}}$ denotes the vector where all components of $\mathbf{B}$ are complemented, that is, $\bar{\mathbf{B}} = (\bar{b}_1, \ldots, \bar{b}_n)$. Prove that f is self-dual if and only if $f(\mathbf{A}) \neq f(\mathbf{B})$ for every pair of vectors $\mathbf{A}$ and $\mathbf{B}$ such that $\mathbf{A} = \bar{\mathbf{B}}$.

3.2.15 (E) Derive the dual function of $x_1 \vee x_2 g(x_3, x_4)$, where g is a function of x_3 and x_4.

3.2.16 (M−) Suppose that we have a series-parallel network of relay contacts. When every parallel connection and every series connection of contacts are replaced by a series and a parallel connection, respectively, what switching function does the transmission of the resultant network represent? Explain why.

3.2.17 (E) Design an MOS gate that represents the dual of the function represented by the MOS gate in Fig. E3.2.17, using as few MOSFETs as possible. Assume that x, $\bar{x}$, y, and $\bar{y}$ are available as inputs.

84   *Fundamentals of Switching Algebra*

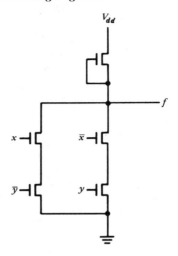

Fig. E3.2.17

3.2.18 (M)   Parity function $x_1 \oplus x_2 \oplus \cdots \oplus x_n$ is 1 if the number of variables that assume the value 1 is odd, and 0 if the number is even. Derive the minterm expansion for each of the following functions, and then determine whether or not each function is self-dual:

(i) $x_1 \oplus x_2$.   (ii) $x_1 \oplus x_2 \oplus x_3$.

3.2.19 (M)   A self-dual function $f(x_1, x_2, \ldots, x_n) = x_1 x_2 \vee x_1 f_1 \vee x_2 f_2$ is given, where f_1 and f_2 are functions of $x_3, \ldots, x_n$ but independent of x_1 and x_2. A student tried to find a relationship between f_1 and f_2 and concluded that $f_1 = f_2$. Prove or disprove this conclusion (i.e., if the student is correct, give a reason; if the student is incorrect, give a correct relation between f_1 and f_2 with a reason).

3.2.20 (R)   Design a CMOS network for function $f = xy \vee \bar{x}\bar{y}$, assuming that only x and y, but not $\bar{x}$ and $\bar{y}$, are available as network inputs. Use as few MOSFETs as possible.

3.2.21 (M)   A CMOS gate (or cell) consists of two subcircuits N_1 and N_2, as illustrated in Fig. E3.2.21, where a and b are connected to the power source of a positive voltage and the ground, respectively, as shown in Fig. 3.2.2. Subcircuit N_1 consists of p-channel MOSFETs, each of which becomes conductive when the voltage applied to its MOSFET gate is low and almost nonconductive* when the voltage is high. Subcircuit N_2 consists of n-channel MOSFETs, each of which becomes conductive when the voltage applied to its MOSFET gate is high and almost nonconductive* when the voltage is low, as discussed in Section 2.3. The same voltages are simultaneously applied to N_1 and N_2 as

---

* Leakage current (i.e., negligibly small current) flows. Thus only leakage current flows between the power supply and the ground all the time except for the short transition periods due to input changes, making the electric power consumption of the CMOS extremely small.

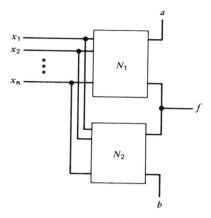

Fig. E3.2.21

the inputs $x_1, \ldots, x_n$. Subcircuits N_1 and N_2 are constructed such that, when the path between a and f is conductive, the path between b and f is almost nonconductive,* and vice versa.

(i) If N_1 and N_2 are in a general connection configuration (i.e., not necessarily a series-parallel connection configuration), what relationship exists between the transmission between a and f and the transmission between b and f? Also explain why. (*Hint*: As explained in the text, consider the transmission by regarding MOSFETs as make-contacts of relays. Then by replacing the make-contacts back to the original MOSFETs, consider the conductivity of the subcircuits.

(ii) When the network configuration of N_1 in a series-parallel connection with p-channel MOSFETs is known, we can construct N_2 with n-channel MOSFETs, based on the network configuration of N_1, so that the entire network satisfies the above requirement. Describe how to do it.

3.2.22 (M) Solve the following set of simultaneous equations with respect to $x, y, z, u,$ and v:

$$x \vee \bar{x}\bar{y} = 0, \qquad xy = yz(u \vee v), \qquad \text{and} \qquad (\bar{u} \vee \bar{v})y \vee x\bar{y} = z.$$

3.2.23 (M) We want to find all solutions for f that satisfy $x_1 f^d = x_1 x_2$, assuming that f is a function of only x_1 and x_2. To solve this, first express f in its minterm expansion:

$$f = A\bar{x}_1\bar{x}_2 \vee B\bar{x}_1 x_2 \vee C x_1\bar{x}_2 \vee D x_1 x_2,$$

where each of A, B, C, D assumes 1 or 0, and then determine the coefficients $A, B, C,$ and D in such a way that $x_1 f^d = x_1 x_2$ is satisfied for every combination of values of x_1 and x_2. (This is a general approach to algebraically finding all solutions that satisfy given constraints.)

---

* See the preceding footnote.

3.2.24 (M) **A.** When $x \vee y = x \vee z$ holds, find all solutions for z that satisfy this equality, as a function of x, y, or both. (*Hint*: You might use the following approach. Express z in its minterm expansion $z = A\bar{x}\bar{y} \vee B\bar{x}y \vee Cx\bar{y} \vee Dxy$ as we did in Exercise 3.2.23. Then determine coefficients A, B, C, D in such a way that $x \vee y = x \vee z$ is satisfied. You may also use any other approach.)

B. Let $f_1 = f(x, y)$ and $f_2 = f(x, z)$ be two functions obtained from the same function f of two variables (e.g., $f_1 = x\bar{y} \vee \bar{x}y$ and $f_2 = x\bar{z} \vee \bar{x}z$). Find all solutions for f that satisfy $zf_1 = yf_2$. (*Hint*: You might use the following approach. Express $f(x, y)$ in its minterm expansion $f(x, y) = A\bar{x}\bar{y} \vee B\bar{x}y \vee Cx\bar{y} \vee Dxy$ in order to let $f(x, y)$ express an arbitrary function. Then, replacing y in this expression by z, obtain $f(x, z) = A\bar{x}\bar{z} \vee B\bar{x}z \vee Cx\bar{z} \vee Dxz$. Then determine A, B, C, D in such a way that $zf_1 = yf_2$ holds.)

3.2.25 (R) For each of the following functions, find the largest set of variables in which it is symmetric:

(i) $xyz \vee \bar{x}y \vee x\bar{y}$.

(ii) $xyz \vee \bar{x}yz \vee x\bar{y}z \vee xy\bar{z}$.

(iii) $x\bar{y}z \vee \bar{x}yz \vee \bar{x}\bar{y}\bar{z}$.

3.2.26 (R) Derive a switching expression for each of the following functions of four variables, expressed with the symbols in Theorem 3.2.4:

(i) The function totally symmetric in x_1, x_2, x_3, and x_4, with $I_1 = 1$, $I_2 = 2$, and $I_3 = 4$ only.

(ii) The function totally symmetric in $\bar{x}_1, \bar{x}_2, x_3$, and x_4, with $I_1 = 0$, $I_2 = 1$, and $I_3 = 4$ only.

3.2.27 (M) Prove the following statements:

(i) The disjunction of two totally symmetric functions is also totally symmetric.

(ii) A totally symmetric function of n variables, $f(x_1, x_2, \ldots, x_n)$, can be expanded with two other symmetric functions of $n-1$ variables, g and h, as follows:

$$f(x_1, x_2, \ldots, x_n) = \bar{x}_1 g(x_2, \ldots, x_n) \vee x_1 h(x_2, \ldots, x_n).$$

3.2.28 (E) It is known that a function $f(x_1, x_2, \ldots, x_n)$ is totally symmetric in $x_1, x_2, \ldots, x_n$ if and only if f is unchanged for the following two permutations: **(i)** the permutation of x_1 and x_2 only, and **(ii)** the cyclic permutation, that is, replacement of x_i by x_{i+1} for $i = 1, 2, \ldots, n$, where $x_{n+1} = x_1$.

Verify this with the function

$$f(x_1, x_2, x_3, \bar{x}_4) = x_1 x_2 x_3 \bar{x}_4 \vee \bar{x}_1 x_2 x_3 \bar{x}_4 \vee x_1 \bar{x}_2 x_3 \bar{x}_4 \vee x_1 x_2 \bar{x}_3 \bar{x}_4 \vee x_1 x_2 x_3 x_4.$$

Answers for the simple questions at the end of Section 3.2: **1.** f is independent of y. **2.** (i) b; (ii) c; (iii) none; (iv) b and d; (v) a, b, and special cases of c and e; (vi) c and e. **3.** Minterm expansion $f = \bar{x}y\bar{z} \vee x\bar{y}\bar{z} \vee x\bar{y}z \vee xy\bar{z} \vee xyz$ and maxterm expansion $f = (x \vee y \vee z)(x \vee y \vee \bar{z})(x \vee \bar{y} \vee \bar{z})$. **4.** $f^d = \bar{x} \vee y\bar{z}$ by Definition 3.2.3, or $f^d = x(\bar{y} \vee z)$ by Theorem 3.2.4. (Notice that both expressions represent the same function.) **5.** (i) $\overline{x \vee (\bar{y} \vee \bar{z})} = \bar{x}(\overline{\bar{y} \vee \bar{z}}) = \bar{x}(y \vee z)$; (ii) $\overline{xy \vee z \vee \bar{x}(u \vee v)} = (\overline{xy})\bar{z}(\overline{\bar{x} \cdot (u \vee v)}) = (\bar{x} \vee \bar{y})\bar{z}(x \vee u \vee v)$. **6.** Symmetric in x_2 and x_3.

3.3 Karnaugh Maps

A Karnaugh map is simply a different way of representing a truth table. Its more concise representation saves time and space for writing 0's and 1's. For the case of four variables, for example, a Karnaugh map consists of 16 cells, that is, small squares as shown in Fig. 3.3.1a. (In a truth table we need to write 0 and 1 for x_1, x_2, x_3, x_4 64 times instead of the 16 bits in Fig. 3.3.1a.) Here each two-bit number along the horizontal line shows the values of x_1 and x_2, and each two-bit number along the vertical line shows the values of x_3 and x_4. The binary numbers are arranged in such a way that adjacent numbers differ in only one bit position. Also, the two numbers in the first and last columns differ in only one bit position and are interpreted to be adjacent. Also, the two numbers in the top and bottom rows are

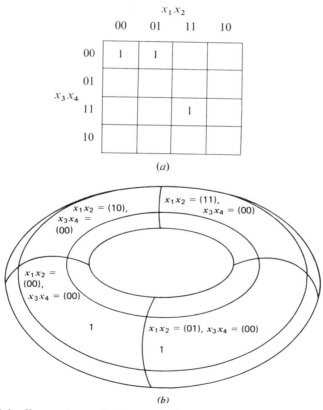

Fig. 3.3.1 Karnaugh map for four variables. (a) Karnaugh map. (b) Torus expression of Fig. 3.3.1a.

88 *Fundamentals of Switching Algebra*

similarly interpreted to be adjacent. Thus **the four cells in the top row are interpreted to be adjacent to the four cells in the bottom row. The four cells in the first column are interpreted to be adjacent to the four cells in the last column.** With this arrangement of cells and this interpretation,* a Karnaugh map is more than a concise representation of a truth table; it can express many important algebraic concepts, as we will see later. A Karnaugh map is a two-dimensional representation of the 16 cells on the surface of a torus, as shown in Fig. 3.3.1b, when the map extends endlessly in vertical and horizontal directions. Notice that each pair of cells positioned in diagonal corners is not adjacent in Fig. 3.3.1a.

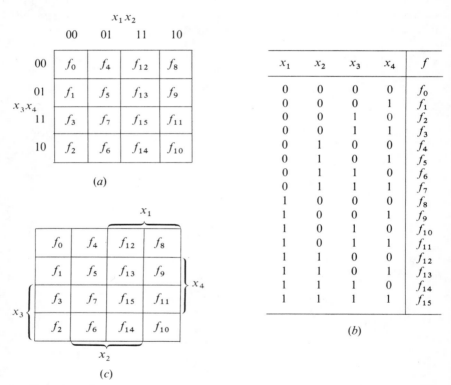

Fig. 3.3.2 *Correspondence between the cells in a Karnaugh map and the rows in a truth table.* (a) *Karnaugh map labeled with binary numbers.* (b) *Truth table.* (c) *Karnaugh map labeled with letters.*

---

* Karnaugh maps are sometimes mistakenly called Veitch diagrams. But Veitch diagrams do not have this adjacency of rows or columns [Veitch 1952]. (Veitch diagrams was rediscovery of the diagrams in [Marquand 1881].)

Fig. 3.3.2 shows the correspondence between the cells in the map (Fig. 3.3.2a) and the rows in the truth table (Fig. 3.3.2b). Notice that the rows in the truth table are not distributed in a consecutive order in the Karnaugh map. The Karnaugh map labeled with variable letters instead of with binary numbers, shown in Fig. 3.3.2c, is also often used. A 1 or 0 (0 is often not shown in each cell) shows the function's value corresponding to a particular cell. In other words, it implies the existence or absence, respectively, of the minterm corresponding to that cell in the minterm expansion of the given function: for example, Fig. 3.3.1a corresponds to the minterms of the function $\bar{x}_1 \bar{x}_2 \bar{x}_3 \bar{x}_4$ ∨ $\bar{x}_1 x_2 \bar{x}_3 \bar{x}_4$ ∨ $x_1 x_2 x_3 x_4$. (For example, the cell with coordinates $(x_1\ x_2, x_3, x_4) = (0\ 1\ 0\ 0)$ represents the minterm $\bar{x}_1 x_2 \bar{x}_3 \bar{x}_4$. Since the cell contains 1, the minterm expansion of the function contains the minterm $\bar{x}_1 x_2 \bar{x}_3 \bar{x}_4$.)

If we "don't-care" some values of the function, we enter the symbol d in the corresponding cells.

Patterns of Karnaugh maps for two and three variables are shown in Figs. 3.3.3a and 3.3.3b, respectively. As will be shown later, we can extend this treatment to the cases of more variables, but the maps become increasingly complicated.

Let us illustrate some algebraic concepts defined so far on Karnaugh maps. Cells that contain 1's are called **1-cells** (similarly, **0-cells**). Let us consider two 1-cells that are adjacent rowwise or columnwise (recall that the two 1-cells at the ends of the same column or row are also adjacent). Then there is exactly one variable, say y, such that literal y is contained in the minterm of one of the 1-cells, and the other literal, $\bar{y}$, is contained in that of the other 1-cell. Thus these two minterms can be combined, by identity (T6) in Section 3.1, into the product of all the literals in these minterms other than the literals of y. For example, Fig. 3.3.4a has two adjacent 1-cells that express minterms $x_1 x_2 x_3 x_4$ and $x_1 \bar{x}_2 x_3 x_4$, shown in the dotted-line loops. They differ only in the literals x_2 and $\bar{x}_2$ of variable x_2. The disjunction of them, $x_1 x_2 x_3 x_4$ ∨ $x_1 \bar{x}_2 x_3 x_4 = (x_2 \vee \bar{x}_2) x_1 x_3 x_4$, is equal to $x_1 x_3 x_4$ by identity (T6) in Section 3.1. If we encircle the two 1-cells in Fig. 3.3.4a by a solid-line loop, the loop

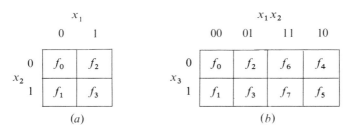

Fig. 3.3.3 Karnaugh maps for two and three variables. (a) *Two variables.* (b) *Three variables.*

90 *Fundamentals of Switching Algebra*

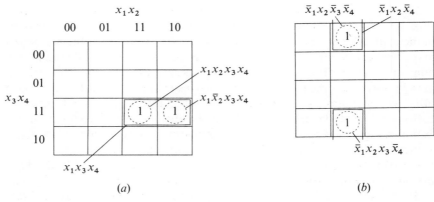

Fig. 3.3.4 Products expressed by combined 1-cells.

expresses the product (or term) $x_1 x_3 x_4$. Similarly, the loop that contains the two adjacent 1-cells in Fig. 3.3.4b expresses the product $\bar{x}_1 x_2 \bar{x}_4$. Thus **a loop that encircles two adjacent 1-cells expresses a product with one less literal than each of the minterms that the original 1-cells represent.** In this case, a variable whose two literals appear noncomplemented in one loop and complemented in the other is deleted (e.g., x_2 in Fig. 3.3.4a).

Next consider Fig. 3.3.5a, which contains two pairs of 1-cells such that the 1-cells in each pair are adjacent and then the two pairs are adjacent, in the same rows or columns. When the two pairs of 1-cells are combined, separately, by dotted-line loops, these loops express the two products $\bar{x}_1 \bar{x}_2 x_4$ and $\bar{x}_1 x_2 x_4$, respectively, by the above argument. These two dotted-line loops are rowwise adjacent, so let us combine them by a solid-line loop. Then the loop expresses the product with one less literal $\bar{x}_1 x_4$, because $\bar{x}_1 \bar{x}_2 x_4 \vee \bar{x}_1 x_2 x_4 = \bar{x}_1 x_4$ holds by identity (T6) in Section 3.1. Similarly, the solid-line loop that combines two rowwise adjacent dotted-line loops expresses product $\bar{x}_3 \bar{x}_4$ in Fig. 3.3.5b.

Next we combine two adjacent dotted-line loops that are adjacent rowwise or columnwise, as shown in Fig. 3.3.5c or 3.3.5d, where each dotted-line loop is rectangular and now contains four 1-cells. Then we get the product x_2 with one less literal.

By the above observation, we can conclude the following. **Suppose that two rectangular loops (or square loops, as special cases), each consisting of 2^i 1-cells, are adjacent rowwise or columnwise on a Karnaugh map, where i is a nonnegative integer. If these loops express products, say Py and $P\bar{y}$, we get a new rectangular loop consisting of 2^{i+1} 1-cells by combining the two loops, and the new loop expresses product P** because $Py \vee P\bar{y} = P$ holds by (T6) in Section 3.1.

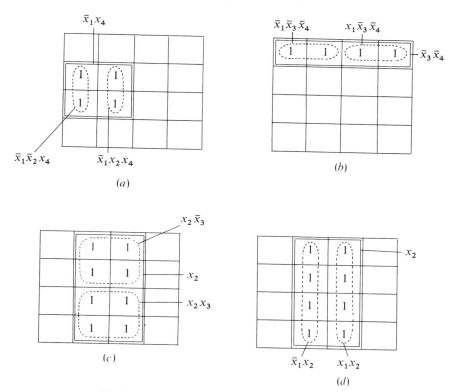

Fig. 3.3.5 *Products expressed by combined 1-cells.*

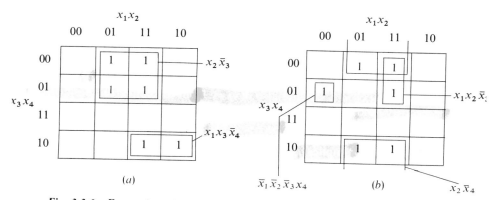

Fig. 3.3.6 *Expression of disjunctive forms on Karnaugh maps.* (a) $f = x_2\bar{x}_3 \vee x_1 x_3 \bar{x}_4$. (b) $f = \bar{x}_1 \bar{x}_2 \bar{x}_3 x_4 \vee x_1 x_2 \bar{x}_3 \vee x_2 \bar{x}_4$.

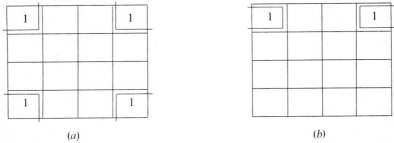

Fig. 3.3.7 Products expressed by combined 1-cells in adjacent corners. (a) $\bar{x}_2 \bar{x}_4$. (b) $\bar{x}_2 \bar{x}_3 \bar{x}_4$.

Therefore an arbitrary disjunctive form of a function f can be expressed with rectangular loops, corresponding to the terms in this disjunctive form, such that each loop consists of 2^i 1-cells. For example, the disjunctive forms $x_2 \bar{x}_3 \vee x_1 x_3 \bar{x}_4$ and $\bar{x}_1 \bar{x}_2 \bar{x}_3 x_4 \vee x_1 x_2 \bar{x}_3 \vee x_2 \bar{x}_4$ are expressed by the Karnaugh maps in Figs. 3.3.6a and 3.3.6b, respectively.

It is important to notice that the four 1-cells in the corners, as shown in Fig. 3.3.7a, can be also combined as representing product $\bar{x}_2 \bar{x}_4$. The two 1-cells in adjacent corners in Fig. 3.3.7b represent $\bar{x}_2 \bar{x}_3 \bar{x}_4$.

3.4 Implication Relations and Prime Implicants

In this section we discuss the algebraic manipulation of switching expressions, that is, how to convert a given switching expression into others. This is very useful for simplification of a switching expression. Although simplification of a switching expression based on a Karnaugh map, which will be discussed later, is convenient in many cases, algebraic manipulation is more convenient in other situations.

First let us introduce important terminology used often in switching algebra.

Definition 3.4.1: Let two switching functions be $f(\mathbf{x})$ and $g(\mathbf{x})$. If every $\mathbf{x}$ satisfying $f(\mathbf{x}) = 1$ satisfies also $g(\mathbf{x}) = 1$ but the converse does not necessarily hold, we write

$$f(\mathbf{x}) \subseteq g(\mathbf{x}), \qquad (3.4.1)*$$

and we say that f **implies** g. In addition, if there exists a certain $\mathbf{x}$ satisfying simultaneously $f(\mathbf{x}) = 0$ and $g(\mathbf{x}) = 1$, we write

$$f(\mathbf{x}) \subset g(\mathbf{x}), \qquad (3.4.2)†$$

---

* Some authors use $\subset$, $\rightarrow$, or $\leftarrow$ instead of $\subseteq$. The adoption of $\subseteq$ here is based on the conceptual analogy to $\leq$ in the sense that $\subseteq$ means $<$ or $=$.
† Some authors use $\subsetneq$ instead of $\subset$.

Implication Relations and Prime Implicants   93

and we say that f **strictly implies** g. Therefore (3.4.1) means $f \subset g$ or $f = g$. These relations are called **implication relations**. The left-hand and right-hand sides of (3.4.1) or (3.4.2) are called **antecedent** and **consequent**, respectively. If an implication relation holds between f and g, that is, if $f \subseteq g$ or $f \supseteq g$ holds, f and g are said to be **comparable** (more precisely, "$\subseteq$ comparable" or "implication comparable"). Otherwise, they are **incomparable**.

When two functions, f and g, are given, we can find by the following methods at least whether or not there exists an implication relation between f and g.

The first method is to use a truth table for f and g, directly based on Definition 3.4.1. If and only if there is no row in which $f = 1$ and $g = 0$, the implication relation

$$f \subseteq g$$

holds. Furthermore, if there is at least one row in which $f = 0$ and $g = 1$, the relation is tightened to

$$f \subset g.$$

Table 3.4.1. shows the truth table for

$$f = x_1 \vee x_2 x_3$$

and

$$g = x_1 \vee x_2.$$

There is no row with $f = 1$ and $g = 0$, so $f \subseteq g$ holds. Furthermore, there is a row with $f = 0$ and $g = 1$, so the relation is actually $f \subset g$.

The second method is to use a Karnaugh map, as illustrated in Fig. 3.4.1. When the 1-cells that represent f are contained among those that represent g, we have implication relation $f \subseteq g$. For example, $f \subseteq g$ (actually $f \subset g$)

Table 3.4.1 $f \subset g$

x_1	x_2	x_3	f	g
0	0	0	0	0
0	0	1	0	0
0	1	0	0	1
0	1	1	1	1
1	0	0	1	1
1	0	1	1	1
1	1	0	1	1
1	1	1	1	1

94   **Fundamentals of Switching Algebra**

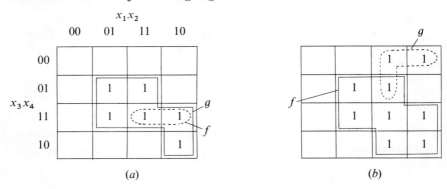

Fig. 3.4.1 Implication relation on a Karnaugh map. (a) *If* f ⊆ g, *the loop for* f *must be inside that for* g. (b) *If* f *and* g *are incomparable, neither the solid loop for* f *nor the dotted loop for* g *is inside the other.*

holds between $g = x_2 x_4 \lor x_1 \bar{x}_2 x_3$ and $f = x_1 x_3 x_4$ in Fig. 3.4.1a. But f and g in Fig. 3.4.1b are incomparable, since some 1-cells of g are not 1-cells of f and some 1-cells of f are not 1-cells of g.

These methods are simple. But if there are many variables, writing a large truth table or map for given functions is cumbersome. When switching expressions for f and g are simple, the following algebraic method based on Definition 3.4.1 is convenient even when the number of variables is large. (For a more indirect algebraic method, which is not necessarily simpler than the following method, see Exercise 3.5.2)

The third method is to determine the values of variables which make f equal 1. Then, if and only if g becomes 1 for these values, the implication relation

$$f \subseteq g$$

holds. For example, when $f = x_1 \lor x_2 x_3$ and $g = x_1 \lor x_2$ are given, $x_1 \lor x_2 x_3 = 1$ holds only for $x_1 = 1$ or $x_2 x_3 = 1$, that is, for the following two cases:

(i) $x_1 = 1$ and   (ii) $x_2 = x_3 = 1$.

For case (i), $g = x_1 \lor x_2 = 1$. For case (ii) also, $g = 1$. Thus $f \subseteq g$. Here notice that the values of x_2 and x_3 in case (i) and the value of x_1 in case (ii) need not be specified. Let us try to be more specific than ⊆, that is, let us try to find whether this ⊆ can be replaced by ⊂ or =. Since $f = x_1 \lor x_2 x_3 = 0$ and $g = x_1 \lor x_2 = 1$ for $x_1 = x_3 = 0$ and $x_2 = 1$, we may rewrite the implication relation as $f \subset g$.

In the following, we further illustrate the third method.

Implication Relations and Prime Implicants

Example 3.4.1: Function $x_1x_2 \vee \bar{x}_1\bar{x}_2$ does not imply $x_1x_3 \vee x_2x_3$ as follows. We can see that $x_1x_2 \vee \bar{x}_1\bar{x}_2 = 1$ means $x_1x_2 = 1$ or $\bar{x}_1\bar{x}_2 = 1$. When $x_1x_2 = 1$, that is, $x_1 = x_2 = 1$, we have $x_1x_3 \vee x_2x_3 = x_3$, which can be 0. (Similarly, when $\bar{x}_1\bar{x}_2 = 1$, i.e., $x_1 = x_2 = 0$, we have $x_1x_3 \vee x_2x_3 = 0$.)

Example 3.4.2: An implication relation is sometimes not obvious. An example is $x_2x_3 \subseteq x_1x_2 \vee \bar{x}_1x_3$, where x_2x_3 does not contain all literals in x_1x_2 or $\bar{x}_1x_3$. If $x_2x_3 = 1$, we have $x_2 = x_3 = 1$. Hence $x_1x_2 \vee \bar{x}_1x_3$ becomes $x_1 \vee \bar{x}_1 = 1$. Thus x_2x_3 certainly implies $x_1x_2 \vee \bar{x}_1x_3$.

Although "g implies f" means "if $g = 1$, then $f = 1$," notice that "g does not imply f" does not necessarily mean "f implies g" but does mean either "f implies g" or "g and f are incomparable." In other words, it does mean "if $g = 1$, then f becomes a function other than the constant function, which is identically equal to 1." (As a special case, f could be identically equal to 0.) Notice that "$\bar{g}$ does not imply f" does not necessarily mean "if $g = 0$, then $f = 0$."

The implication itself is a useful concept in simplifying networks (e.g., Exercise 3.5.7), as well as switching expressions (as seen later).

As already defined, a product of literals is called a **term** if a literal for each variable appears at most once.

Definition 3.4.2: An **implicant** of a switching function f is a term that implies f.

For example, $x_1, x_2, x_1x_2, x_1\bar{x}_2$, and x_1x_3 are examples of implicants of the function $x_1 \vee x_2$. But $\bar{x}_1\bar{x}_2$ is not. Notice that x_1x_3 is an implicant of $x_1 \vee x_2$ even though $x_1 \vee x_2$ is independent of x_3. (Notice that every product of an implicant of f with any dummy variables is also an implicant of f. Thus f has an infinite number of implicants.) But x_1x_2 is not an implicant of $f = x_1x_3 \vee \bar{x}_2$ because x_1x_2 does not imply f. (When $x_1x_2 = 1$, we have $f = x_3$, which can be 0. Therefore, even if $x_1x_2 = 1$, f may become 0.) Some implicants are not obvious from a given expression of a function. For example, $x_1x_2 \vee \bar{x}_1x_3$ has implicants x_2x_3 and $x_2x_3x_4$. Also, $x_1\bar{x}_2 \vee x_2x_3 \vee \bar{x}_1x_3$ has an implicant x_3 because, if $x_3 = 1$, $x_1\bar{x}_2 \vee x_2x_3 \vee \bar{x}_1x_3$ becomes $x_1\bar{x}_2 \vee x_2 \vee \bar{x}_1 = x_1 \vee x_2 \vee \bar{x}_1$ [by (T9) in Section 3.1], which is equal to 1 [by (T6) in Section 3.1].

Definition 3.4.3: A term P is said to **subsume** another term Q if all the literals of Q are contained among those of P. If a term P which subsumes another term Q contains literals that Q does not have, P is said to **strictly subsume** Q.

For example, term $x_1\bar{x}_2 x_3 \bar{x}_4$ subsumes $x_1 x_3 \bar{x}_4$ and also itself. More precisely speaking, $x_1\bar{x}_2 x_3 \bar{x}_4$ strictly subsumes $x_1 x_3 \bar{x}_4$. Notice that Definition 3.4.3 can be equivalently stated as follows: "A term P is said to subsume another term Q if P implies Q, that is, $P \subseteq Q$. Term P strictly subsumes another term Q if $P \subset Q$."

Notice that, when we have terms P and Q, we can say, "P implies Q" or, equivalently, "P subsumes Q." But the word "subsume" is ordinarily not used in other cases, except for comparing two alterms (as we will see in Definition 4.3.2). For example, when we have functions f and g which are not in single terms, we usually do not say "f subsumes g."

On a Karnaugh map, if the loop representing a term P (always a single rectangular loop consisting of 2^i 1-cells because P is a product of literals) is part of the 1-cells representing function f, or is contained in the loop representing a term Q, P implies f or subsumes Q, respectively. Fig. 3.4.2 illustrates this. Conversely, it is easy to see that, if a term P, which does not contain any dummy variables of f, implies f, the loop for P must consist of some 1-cells of f, and if a term P, which does not contain any dummy variables of another term Q, implies Q, the loop for P must be inside the loop for Q.

The concept of "prime implicant" in the following paragraphs is useful for deriving the simplest disjunctive form for a given function f (recall that switching expressions for f are not unique) and consequently for deriving the simplest network realizing f.

Definition 3.4.4: A **prime implicant*** of a given function f is defined as an implicant of f such that no other term subsumed by it is an implicant of f.

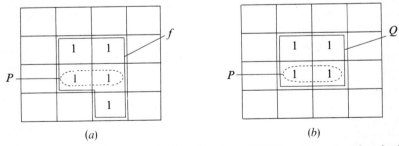

Fig. 3.4.2 Comparison of "imply" and "subsume." (a) Term **P** implies a function **f**. (b) Term **P** subsumes a term **Q**.

---

* An implicate and a prime implicate, which are concepts dual to an implicant and a prime implicant, respectively, will be discussed in Chapter 4 for ease of explanation.

For example, when $f = x_1x_2 \lor \bar{x}_1x_3 \lor x_1x_2x_3 \lor \bar{x}_1x_2x_3$ is given, x_1x_2, $\bar{x}_1x_3$, and x_2x_3 are prime implicants. But $x_1x_2x_3$ and $\bar{x}_1x_2x_3$ are not prime implicants, though they are implicants (i.e., if any of them is 1, then $f = 1$). Prime implicants of a function f can be obtained from other implicants of f by stripping off unnecessary literals until further stripping makes the remainder no longer imply f. Thus x_2x_3 is a prime implicant of $x_1x_2 \lor \bar{x}_1x_3$, and $x_2x_3x_4$ is an implicant of this function but not a prime implicant. As seen from this example, some implicants and accordingly some prime implicants are not obvious from a given expression of a function. Notice that, unlike implicants, **a prime implicant cannot contain a literal of any dummy variable of a function** (Exercise 3.5.12).

On a Karnaugh map all prime implicants of a given function f of up to four variables can be easily found. As is readily seen from Definition 3.4.4, **each rectangular loop that consists of 2^i 1-cells, with i chosen as large as possible, is a prime implicant of f. If we find all such loops, we will have found all prime implicants of f.** Suppose that a function f is given as shown in Fig. 3.4.3a. Then the prime implicants are shown in Figs. 3.4.3b, 3.4.3c, and 3.4.3d. In each figure we cannot make the size of the loop any bigger. (If we increase the size of any one of these loops, the new loop will contain a number of 1-cells that is not 2^i for any i, or will include one or more 0-cells.)

Consensus

Next, let us systematically find all prime implicants, including those not obvious, for a given switching function.

To facilitate our discussion, let us define a consensus.

Definition 3.4.5: Assume that two terms, P and Q, are given. If there is exactly one variable, say x, appearing noncomplemented in one term and complemented in the other—in other words, if $P = xP'$ and $Q = \bar{x}Q'$ (no other variables appear complemented in either P' or Q', and noncomplemented in the other)—then the product of all literals except the literals of x, that is, $P'Q'$ with duplicates of literals deleted, is called the **consensus*** of P and Q.

For example, if we have two terms, $x_1\bar{x}_2x_3$ and $\bar{x}_1\bar{x}_2x_4\bar{x}_5$, the consensus is $\bar{x}_2x_3x_4\bar{x}_5$. But $x_1\bar{x}_2x_3$ and $\bar{x}_1x_2x_4\bar{x}_5$ do not have a consensus because two variables, x_1 and x_2, appear noncomplemented and complemented in these two terms.

---

* "Consensus" is defined somewhat differently by various authors (see, e.g., p. 288 of Dietmeyer [1971].)

98 *Fundamentals of Switching Algebra*

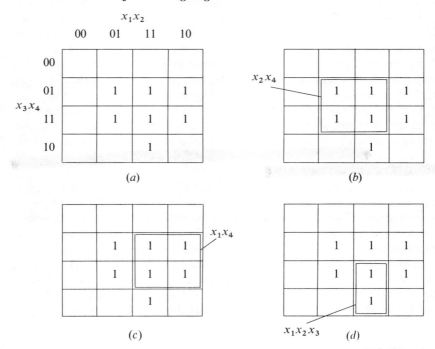

Fig. 3.4.3 Expression of prime implicants on Karnaugh maps. **(a)** *A function* **f**. **(b)** *A prime implicant.* **(c)** *A prime implicant.* **(d)** *A prime implicant.*

A consensus can easily be shown on a Karnaugh map. For example, Fig. 3.4.4 shows a function $f = x_1 x_2 \vee \bar{x}_1 x_4$. In addition to the two loops shown in Fig. 3.4.4a, which corresponds to the two prime implicants $x_1 x_2$ and $\bar{x}_1 x_4$ of f, this f can have another rectangular loop, which consists of 2^i 1-cells with i chosen as large as possible, as shown in Fig. 3.4.4b. This third loop, which represents $x_2 x_4$, the consensus of $x_1 x_2$ and $\bar{x}_1 x_4$, intersects the two loops in Fig. 3.4.4a and is contained within the 1-cells that represent $x_1 x_2$ and $\bar{x}_1 x_4$. This is an important characteristic of a loop representing a consensus. Notice that these three terms, $x_2 x_4$, $x_1 x_2$, and $\bar{x}_1 x_4$, are prime implicants of f. Another example is $\bar{x}_1 x_4$, which is the consensus of two terms, $\bar{x}_1 \bar{x}_2 x_4$ and $\bar{x}_1 x_2 x_4$, in Fig. 3.3.5a. In this case the consensus $\bar{x}_1 x_4$ is subsumed by the original terms, $\bar{x}_1 \bar{x}_2 x_4$ and $\bar{x}_1 x_2 x_4$. **When rectangular loops of 2^i 1-cells are adjacent (not necessarily exactly in the same row or column), the consensus is a rectangular loop of 2^i 1-cells, with i chosen as large as possible, that intersects and is contained within these two loops. Therefore, if we obtain all largest possible rectangular loops of 2^i 1-cells, we can obtain all prime**

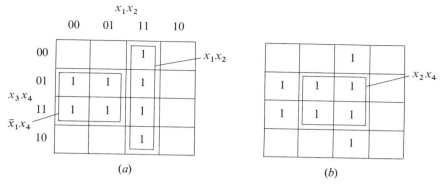

Fig. 3.4.4 *Expression of a consensus on a Karnaugh map.* (a) *A function* f. (b) *Consensus as a prime implicant.*

implicants, including consensuses, which intersect and are contained within other pairs of loops. Sometimes a consensus term can be obtained from a pair consisting of another consensus and a term, or a pair of other consensuses that do not appear in a given expression, like the example $x_1\bar{x}_2 \vee x_2 x_3 \vee \bar{x}_1 x_3$, where $x_1 \bar{x}_2$ and $x_2 x_3$ yield $x_1 x_3$, which in turn yields consensus x_3 with $\bar{x}_1 x_3$. Each such consensus is also obtained among the above largest possible rectangular loops.

In the paragraph that immediately follows Definition 3.4.4, $x_2 x_3$ is mentioned as a prime implicant of $f = x_1 x_2 \vee \bar{x}_1 x_3$, a fact that is not obvious from the expression of f; but $x_2 x_3$ is, as a matter of fact, the consensus of the two terms $x_1 x_2$ and $\bar{x}_1 x_3$ in f. This can be generalized as follows.

Theorem 3.4.1: If two terms P and Q have a consensus, their consensus implies $P \vee Q$.

Proof: Since P and Q have a consensus, they can be expressed as $P = P'x$ and $Q = Q'\bar{x}$, where P' and Q' are products of literals other than x and $\bar{x}$ in P and Q, respectively. Then $P'Q'$ is the consensus. When $P'Q' = 1$, $P' = Q' = 1$ results and $P \vee Q$ becomes $P'x \vee Q'\bar{x} = x \vee \bar{x} = 1$. Thus $P'Q'$ implies $P \vee Q$.
Q.E.D

Generalizing this theorem, we can easily see that **every consensus that is obtained from terms of a given function f implies f.** In other words, every consensus generated is an implicant of f, though not necessarily a prime implicant.

Simple Questions for Self-Study

Before proceeding further, try the following simple questions. (Answers are given after the next exercises.) In this and the next chapters, many algebraic words are introduced, but readers will eventually become accustomed to them by working on the exercises in these and later chapters.

1. Choose a proper word (or words) in each of the following statements.

    (a) $x \vee yz$ $\begin{pmatrix} \text{implies} \\ \text{subsumes} \\ \text{is implied by} \\ \text{is subsumed by} \end{pmatrix}$ $x \vee yzu$.

    (b) $xy\bar{z}$ $\begin{pmatrix} \text{implies} \\ \text{subsumes} \\ \text{is implied by} \\ \text{is subsumed by} \end{pmatrix}$ xy.

2. Is each of the following terms an implicant of the function $f = xy \vee \bar{x}\bar{y} \vee xz$?

    (a) xy.   (b) x.   (c) xzu.   (d) $\bar{y}z$.

3. Which of (a), (b), (c), and (d) are prime implicants of f in Question 2?

4. Derive a consensus for each of the following pairs of products:

    (a) $xyz, x\bar{y}u$.   (b) $\bar{x}yz, x\bar{y}u$.

Derivation of Prime Implicants from a Disjunctive Form

The derivation of all prime implicants of a given function f is easy, using a Karnaugh map. If, however, the function has five or more variables, the derivation becomes complicated on a Karnaugh map, as we will see in Chapter 4. Therefore let us discuss an algebraic method, which is convenient for implementation in a computer program.

The algebraic method to find all prime implicants of a given function, called the **iterated-consensus method**, was proposed for the first time by A. Blake in 1937 (according to Brown [1968]).

Procedure 3.4.1—Iterated-Consensus Method: Derivation of All Prime Implicants of a Given Function

When a switching function f is given in a disjunctive form, we can derive all prime implicants by repeated applications of the following two operations in any order:

1. Delete any term that subsumes another in the current expression for f.

Implication Relations and Prime Implicants   101

2. If any two terms in the current expression for f have a consensus, the disjunction of this consensus and the current expression is used as the next expression for f. (As will be shown in Corollary 3.4.2, the next expression with the new consensus added still represents the original f.)

Notice that, if the consensus formed in operation 2 subsumes any existing term, it is immediately eliminated by operation 1, so such a consensus need not be considered in operation 2.

The application of operations 1 and 2 is stopped when the following conditions simultaneously hold: (a) no term subsumes another, and (b) each pair of terms has either no consensus or a consensus that subsumes some term in the current expression. Then all prime implicants of f are terms in the last expression. (For a proof of this statement, see, e.g., Quine [1955] or p. 262 of Nelson [1968].) The last expression is called the **complete sum** or the **all-prime-implicant disjunction**.   □

The complete sum is obviously unique because all prime implicants of f are present.

For example, suppose that the switching function $f = \bar{x}_1 \bar{x}_2 \vee x_1 x_2 x_3 \vee \bar{x}_1 x_2 x_3 \vee x_1 \bar{x}_3$ is given. First, $\bar{x}_1 \bar{x}_2$ and $x_1 \bar{x}_3$ have the consensus $\bar{x}_2 \bar{x}_3$, which does not subsume any existing term. Therefore make the disjunction of $\bar{x}_2 \bar{x}_3$ and the original f (operation 2) as

$$f = \bar{x}_1 \bar{x}_2 \vee x_1 x_2 x_3 \vee \bar{x}_1 x_2 x_3 \vee x_1 \bar{x}_3 \vee \bar{x}_2 \bar{x}_3. \qquad (3.4.3)$$

Next, $x_1 x_2 x_3$ and $\bar{x}_1 x_2 x_3$ of f of (3.4.3) have the consensus $x_2 x_3$, which does not subsume any term of (3.4.3). Therefore make the disjunction of $x_2 x_3$ and (3.4.3) (operation 2) as

$$f = \bar{x}_1 \bar{x}_2 \vee x_1 x_2 x_3 \vee \bar{x}_1 x_2 x_3 \vee x_1 \bar{x}_3 \vee \bar{x}_2 \bar{x}_3 \vee x_2 x_3. \qquad (3.4.4)$$

However, $x_1 x_2 x_3$ subsumes $x_2 x_3$. Therefore delete $x_1 x_2 x_3$ (operation 1):

$$f = \bar{x}_1 \bar{x}_2 \vee \bar{x}_1 x_2 x_3 \vee x_1 \bar{x}_3 \vee \bar{x}_2 \bar{x}_3 \vee x_2 x_3. \qquad (3.4.5)$$

Then $\bar{x}_1 x_2 x_3$ subsumes $x_2 x_3$. Therefore delete $\bar{x}_1 x_2 x_3$ (operation 1). Thus

$$f = \bar{x}_1 \bar{x}_2 \vee x_1 \bar{x}_3 \vee \bar{x}_2 \bar{x}_3 \vee x_2 x_3. \qquad (3.4.6)$$

Then $\bar{x}_1 \bar{x}_2$ and $x_2 x_3$ have the consensus $\bar{x}_1 x_3$. Therefore make the disjunction of this with (3.4.6) (operation 2) as

$$f = \bar{x}_1 \bar{x}_2 \vee x_1 \bar{x}_3 \vee \bar{x}_2 \bar{x}_3 \vee x_2 x_3 \vee \bar{x}_1 x_3. \qquad (3.4.7)$$

Then $x_1 \bar{x}_3$ and $x_2 x_3$ have the consensus $x_1 x_2$. Therefore make the disjunction as follows:

$$f = \bar{x}_1 \bar{x}_2 \vee x_1 \bar{x}_3 \vee \bar{x}_2 \bar{x}_3 \vee x_2 x_3 \vee \bar{x}_1 x_3 \vee x_1 x_2. \qquad (3.4.8)$$

102 *Fundamentals of Switching Algebra*

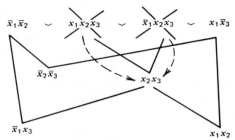

Fig. 3.4.5 An example for the iterated-consensus method.

This is the complete sum for f, because no pair of terms has a new consensus and no term subsumes another. The above procedure is illustrated in Fig. 3.4.5, where the dotted lines show a subsuming relationship.

Notice that the complete sum for f does not contain any dummy variables of f even if the initial expression for f contains some (Exercise 3.5.12).

An important property of a consensus is obtained from the following theorem.

Theorem 3.4.2: A function g implies another function f if and only if $f = f \vee g$ holds.

Proof: If $f \supseteq g$, we add f to both sides; then $f = f \vee f \supseteq f \vee g$ holds. Therefore $f \supseteq f \vee g$. Also, $f \subseteq f \vee g$ obviously holds. Combining the above two implication relations, we get $f = f \vee g$.

If $f \supseteq g$ does not hold, $f = 0$ and $g = 1$ hold for some combination of values of variables. Then $f = f \vee g$ does not hold, since its left-hand and right-hand sides are 0 and 1, respectively, for this combination of values of variables. Q.E.D.

This theorem is useful in simplifying switching functions as well as networks (for an example of simplification of a network, see Exercise 3.5.6).

Corollary 3.4.1: A term Q subsumes another term P if and only if $P = P \vee Q$ holds.

Proof: Q subsumes P if and only if Q implies P. Thus this is a special case of Theorem 3.4.2. Q.E.D.

For example, when $Q = xyz$ and $P = xy$ are given, $xyz \vee xy$ can be reduced to xy according to this corollary. Corollary 3.4.1 is a generalization of the absorption law (T8) in Section 3.1.

Corollary 3.4.2: Let P and Q be a pair of terms that has consensus $P'Q'$, where $P = P'x$ and $Q = Q'\bar{x}$, and where P' and Q' are free of x or $\bar{x}$. Then
$$P \vee Q = P \vee Q \vee P'Q'.$$

Proof: This results from Theorems 3.4.1 and 3.4.2.   Q.E.D.

Each application of operation 1 or 2 of Procedure 3.4.1 produces an expression equivalent to the original because of Corollary 3.4.1 or 3.4.2, respectively. In other words, **any intermediate expression obtained during the iterated-consensus method is also equivalent to f.** (This property itself is often useful for simplification of a switching expression.) Thus the complete sum, that is, disjunction of all prime implicants, is equivalent to f.

An obvious consequence of the iterated-consensus method is that **consensus formation generates new terms that are candidates for prime implicants not present in the given expression of a function.** (Note that some consensus terms generated by operation 2 may be deleted by operation 1 later, so not every consensus generated will necessarily be a prime implicant.)

Let us introduce the following definition in order to facilitate our discussion.

Definition 3.4.6: Suppose that p products, $A_1, \ldots, A_p$, are given. A variable such that only one of its literals appears in $A_1, \ldots, A_p$ is called a **monoform variable**. A variable such that both of its literals appear in $A_1, \ldots, A_p$ is called a **biform variable**.

For example, when $x_1\bar{x}_2$, $\bar{x}_1 x_3$, $x_3\bar{x}_4$, and $\bar{x}_2 x_4$ are given, x_1 and x_4 are biform variables, since x_1 and x_4 as well as their complements appear, and x_2 and x_3 are monoform variables.

Tison showed that the iterated-consensus method can be modified so that all prime implicants can be found more efficiently by generating consensuses in a more systematic manner [Tison 1965, 1967]. In the iterated-consensus method, comparison of every pair of terms is time-consuming when there are many prime implicants (e.g., some functions of 10 variables have a few hundred prime implicants, and a large number of comparisons are required). In the following procedure the number of comparisons is usually greatly reduced.

Procedure 3.4.2—The Tison Method: Derivation of All Prime Implicants of a Given Function

Assume that a function f is given in a disjunctive form $f = P \vee Q \vee \cdots \vee T$, where $P, Q, \ldots, T$ are terms, and that we want to find all prime implicants of f. Let S denote the set $\{P, Q, \ldots, T\}$.

1. Among $P, Q, \ldots, T$ in set S, first delete every term subsuming any other term. Among all biform variables, choose one of them.

104   Fundamentals of Switching Algebra

For example, when $f = x_1x_2x_4 \vee x_1x_3 \vee x_2\bar{x}_3 \vee x_2x_4 \vee \bar{x}_3\bar{x}_4$ is given, delete $x_1x_2x_4$, which subsumes x_2x_4. The biform variables are x_3 and x_4. Let us choose x_3 first.

2. For each pair of terms, that is, one with the complemented literal of the chosen variable and the other with the noncomplemented literal of that variable, generate the consensus. Then add the generated consensus to S. From S, delete every term that subsumes another.

For our example, we choose x_3 as the first biform variable. Thus we get consensus x_1x_2 from pair x_1x_3 and $x_2\bar{x}_3$, and consensus $x_1\bar{x}_4$ from pair x_1x_3 and $\bar{x}_3\bar{x}_4$.

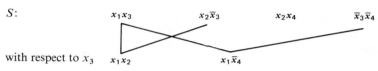

3. Choose another biform variable* in the current S. (When a subsuming term is deleted, a variable that was biform may become monoform. If so, choose one of the current biform variables.) Then go to Step 2. If all biform variables are tried, go to Step 4.

For the above example, for the second iteration of Step 2, we choose x_4 as the second biform variable and generate two new consensuses as follows:

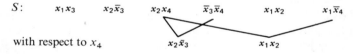

But when they are added to S, each one subsumes some term contained in S. Therefore they are eliminated.

4. The procedure terminates, and all the products in S are desired prime implicants.

For the above example, prime implicants are x_1x_3, $x_2\bar{x}_3$, x_2x_4, $\bar{x}_3\bar{x}_4$, x_1x_2, $x_1\bar{x}_4$. □

Usually the Tison method is more efficient than the iterated consensus method, because once each biform variable is tried in Steps 1 and 3, there is no need to repeat it (in this sense the Tison method is a one-pass procedure). The

---

* It is generally hard to find a best order in choosing biform variables. Starting with a biform variable whose literals appear less frequently in the given set S seems to be a better selection to finish Procedure 3.4.2 in a shorter time, though for some functions this may not be the case.

Tison method may be considered to be a streamlined version of the iterated-consensus method, since both methods are based on the same idea. When the number of prime implicants increases, the Tison method becomes increasingly efficient, whether processed by hand or by computer. But one advantage of the iterated-consensus method is the simplicity of the rules, which are somewhat easier to remember.

Another well-known method for the derivation of prime implicants is the Quine–McCluskey method, which will be discussed with related subjects in Chapter 4. This method starts with the minterm expansion of a given function, whereas the iterated-consensus method or the Tison method starts with any disjunctive form, a more general concept than the minterm expansion. Though the Quine–McCluskey method may be convenient when a function is given in minterm expansion, the iterated-consensus method or the Tison method tends, in general, to be more convenient because of the flexibility of starting with any disjunctive form for a given function.

The complete sum, or the disjunction of all prime implicants, is the first important step in deriving the most economical network for a given function under certain conditions, as discussed in later chapters. Also, it is useful for simplifying switching expressions or proving the equivalence of two switching expressions. (Equivalence can also be proved by deriving the minterm or maxterm expansions, or by drawing Karnaugh maps for the two functions under comparison.) The complete sum has other applications, such as the detection of a function's independence of some variables.

▲ Derivation of Prime Implicants from a Conjunctive Form

Let us derive all prime implicants of a function f when f is given in an arbitrary conjunctive form. This problem is radically different from the preceding one —to derive all prime implicants from a disjunctive form—because, as the following theorem asserts, all prime implicants can be derived from a conjunctive form for f without consensus formation (which is essential for the preceding problem), and yet the derivation is generally more time-consuming, despite the simplicity of the derivation procedure.

Theorem 3.4.3 [Nelson 1954]: Suppose that an arbitrary conjunctive form is given for a function f. When the conjunctive form is expanded into a disjunctive form by multiplying out the alterms [i.e., repeating distributivity (T7) in Section 3.1] and deleting terms that subsume others, the disjunctive form thus obtained is the complete sum for f, that is, the disjunction of all prime implicants of f.

(A proof is left to readers as Exercise 3.5.19.)

For example, function $f = (x_1 \lor x_2)(x_2 \lor \bar{x}_3)(\bar{x}_1 \lor \bar{x}_2 \lor x_3)$ is expanded into a disjunctive form, $x_1 x_2 x_3 \lor x_1 \bar{x}_2 \bar{x}_3 \lor \bar{x}_1 x_2 \lor x_2 x_3 \lor \bar{x}_1 x_2 \bar{x}_3$, where duplicated literals in each term (e.g., $x_2 x_2 x_3$) and terms that become identically equal to 0 (e.g., $x_1 \bar{x}_1 x_2$) are of course deleted. Upon deleting subsuming terms, this becomes $x_1 \bar{x}_2 \bar{x}_3 \lor \bar{x}_1 x_2 \lor x_2 x_3$, the complete sum for f.

Notice that no consensus formation is involved in the procedure of Theorem 3.4.3, and yet all prime implicants are generated. Unfortunately, the expansion of a conjunctive form into a disjunctive form is usually very time-consuming. All prime implicants from a conjunctive form can be derived much more quickly if the conjunctive form is not expanded into a disjunctive form, as, for example, by the semantic tree method [Slagle, Chang, and Lee 1970; Bredeson and Hulina 1971b; Das 1971].

Once in a while the expansion of a conjunctive form for a function f into a disjunctive form is necessary in logic design. The computational techniques to be discussed in Sections 4.6 and 5.2 may sometimes be helpful, though a drastic reduction in computation time should not always be expected.

3.5 Algebraic Design Methods and Computer-Aided Design

"Design automation" was a magic term in the early 1960s. During this period substantial efforts were made throughout the computer industry to automate the design of digital systems (such as logic design, electronic design, mechanical design, and manufacturing design)—in other words, to completely replace human designers by computer programs. Though the efforts were highly successful in some areas, complete automation was found to be extremely difficult when design requires complex procedures. Logic design is a notable example. Logic networks designed by design automation computer programs required too many gates. (The number of gates was typically doubled or more, compared with human design. See, e.g., Friedman and Yang [1970].) Since then, design automation with human intervention has been emphasized, rather than complete automation; this is called **computer-aided design (CAD)**.

Advantages and Limitation of Computer-Aided Design

Since reduction of errors and time is becoming vitally important in logic design as LSI or VLSI is widely used with ever-increasing integration size, serious efforts to develop computer programs for CAD are being made and will continue. For CAD computer program development, algebraic methods are more suitable than the tabular or map methods conventionally used for

Algebraic Design Methods and Computer-Aided Design 107

hand processing. (Algebraic methods are sometimes more useful even for hand processing. An example will be presented in Section 4.6.)

Computer-aided design is being applied wherever applicable. An example is the layout of networks on an LSI chip by the library approach, mentioned in Section 2.4. By this CAD approach, design time and errors can be greatly reduced. But it is difficult to reduce the chip size, a major cost factor, to the extent possible with hand processing, though simple tasks are done by CAD even in the case of hand processing. (If all possibilities necessary for chip size minimization are tried by a CAD program, processing time will be astronomically long.) Thus designers of cost-competitive LSI chips make their layouts without using the library approach. The difference between chip size by the library approach and that by hand processing is said to be 15 percent or more, depending on the sophistication of the CAD and also the design problems. With the 15 percent difference, the cost of the chip by the library approach is more than doubled when the chip size is critical. (Even with hand processing, a 10 percent difference can easily result if the processing is not carefully done.) Thus the library approach is not appropriate when the design is to be technologically very advanced or complex. Another example is CAD for compact PLAs (programmable logic arrays), which are a special type of read-only memories (Section 9.4). As PLAs are widely used, the development of CAD for compact PLAs is becoming very important. Another example is CAD for the simulation of logic networks (Section 9.6).

Problems with Computer-Aided Design Implementation

The CAD approach is being applied wherever applicable, not only in the computer industry but also in other types of manufacturing, as digital systems and technology are used elsewhere than in computers. However, most CAD computer programs are proprietary, so designers must develop their own from scratch, since they cannot buy good CAD programs from other firms [Schindler 1977]. Furthermore, in most cases CAD programs require continuous maintenance, since they must be modified constantly because of technological progress or the introduction of new products. It is impossible to develop CAD computer programs that incorporate all future technological progress. (Even if this could be done, omnibus computer programs incorporating every possibility would be extremely inefficient and difficult to use.)

In many cases, particularly in logic design, CAD computer programs have the nature of combinatorial mathematics and are extremely sensitive to the details of computer program implementation. Even when the same design procedures are used, devising simple speed-up techniques in details of the procedures or in programming often improves computational efficiency drastically (a few times, hundreds of times, thousands of times, or

108 Fundamentals of Switching Algebra

more), depending on the types of procedures, because of their combinatorial nature. (The improvements are greater when better whole procedures are chosen.) When the scale of design problems is greater, the improvements also tend to be greater. Thus this book discusses algebraic methods in some detail, seeking to develop the ability to devise appropriate techniques to speed up processing, whether by computer or by hand. (Theorem-proving exercises are very useful for this purpose.)

This does not mean, however, that logic design by hand is not important. If a design is simple enough to be processed in a short time (say, several minutes) by hand, the designer need not bother with CAD programs, which are usually cumbersome in setting up of input data. Also, without CAD programs, the designer can understand better the nature of the problems and procedures and can make flexible judgments. Hence algebraic methods and tabular or map methods are complementary in usefulness, because of their different features.

In any case, designers often can do their work more cleverly and efficiently if they think through the algebraic or engineering properties of given design problems, and understand them, before starting the design.

Computer-aided design is discussed in many papers,* though many key aspects are omitted because of their proprietary nature.

Exercises†

3.5.1 (R) Find whether an implication relation holds between f and g in each case:

(i) $f = x_1 x_2 \vee \bar{x}_2 x_3, \quad g = x_1 x_2 \vee x_1 x_3 \vee \bar{x}_1 \bar{x}_2$.

(ii) $f = x_1 \bar{x}_2 \vee \bar{x}_1 x_2 \vee x_1 x_3, \quad g = x_1 x_2 \vee x_2 x_3$.

3.5.2 (M−) Prove that $f\bar{g}$ is identically 0 if and only if $f \subseteq g$. On the basis of this property, solve Exercise 3.5.1.

3.5.3 (R) Prove or disprove that, if $g(\mathbf{x}) \subseteq h(\mathbf{x})$, then $\bar{g}(\mathbf{x}) \subseteq \bar{h}(\mathbf{x})$. If you disprove, show a correct relation with a proof.

---

* For example, Breuer [1972b, 1975], van Cleemput [1976a, 1976b, 1977], Lewin [1977], and the issues of *Proceedings of Design Automation Workshop*, published by the Institute of Electrical and Electronics Engineers.

† Theorem-proving exercises are intended to help students understand algebraic concepts so that logic design or CAD program implementation can be done cleverly and efficiently. Unless students understand the material in the text completely, they cannot prove these simple results. If on the other hand, they understand the definitions and procedures given, these exercises can be easily proved. Also, students should regard study of the answers to be distributed or discussed as an important part of their overall effort, since the answers are usually much easier than might be expected from the abstract appearance of the exercise statements.

Algebraic Design Methods and Computer-Aided Design 109

3.5.4 (E) Prove or disprove that, if $g(x) \subseteq h(x)$, then $g^d(x) \subseteq h^d(x)$ holds between their dual functions $g^d(x)$ and $h^d(x)$. If you disprove, show a correct relation with a proof.

3.5.5 (E) Prove one of the following sets of relations:

A. (i) If $f(x) \supseteq g(x)$ and $g(x) \supseteq h(x)$, then $f(x) \supseteq h(x)$.
    (ii) $f(x) \vee g(x) \supseteq f(x)$.

B. (i) $f(x) \supseteq f(x)g(x)$.
    (ii) If $f(x) \supseteq g(x)$, then $f(x) \cdot g(x) = g(x)$.

3.5.6 (E) Suppose that a network contains an OR gate that has two sets of inputs such that the disjunction of the first set of inputs represents a function $g(x)$, and the disjunction of the second set of inputs a function $h(x)$, as shown in Fig. E3.5.6a. Prove that, if $h(x) \subseteq g(x)$, the second set of inputs can be eliminated without changing the output f of the OR gate, as shown in Fig. 3.5.6b.

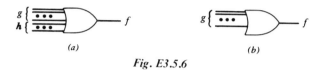

Fig. E3.5.6

3.5.7 (M−) Suppose that we have a network which consists of a subnetwork and two NOR gates, as shown in Fig. E3.5.7, where the subnetwork may consist of a mixture of different gate types (e.g., NOR gates may be included, or there may be a mixture of NAND and OR gates). Gate 1 has one input which represents a function $g(x)$ and other inputs whose disjunction represents a function $h(x)$. Gate 2 has inputs from the subnetwork and other inputs whose disjunction represents a function $i(x)$. Prove that, if $h(x) \vee i(x) \supseteq g(x)$, we can eliminate the input $g(x)$ without changing the output function f of the entire network [Lai, Nakagawa, and Muroga 1974]. (Application of this property will be discussed in Section 6.3.)

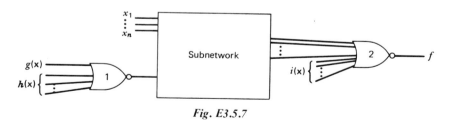

Fig. E3.5.7

3.5.8 (M−) There exists a pair of input vectors $(a_1, \ldots, a_n)$ and $(b_1, \ldots, b_n)$ such that $a_i = b_i$ for every i except a particular k, $a_k = 1$ and $b_k = 0$, and $f(a_1, \ldots, a_n) = 0$ and $f(b_1, \ldots, b_n) = 1$.
A student concluded that $\bar{x}_k$ implies function f. Prove or disprove this conclusion.

3.5.9 (M−) Prove or disprove that, if $\bar{x}_1$ does not imply f and $\bar{x}_2$ does not imply f, then $\bar{x}_1\bar{x}_2$ does not imply f.

3.5.10 (R) (i) Check whether each of the following is an implicant of $x_1 x_2 x_3 \vee x_1 \bar{x}_2 \vee \bar{x}_2 x_3 \vee x_2 \bar{x}_3$, by the third (algebraic) method discussed after Definition 3.4.1: $x_1 \bar{x}_2 x_3$, $\bar{x}_1 x_2 x_3$, $x_1 x_3$, x_1.
(ii) Check whether each of the following is a prime implicant of $x_1 \bar{x}_2 \vee \bar{x}_1 x_3 \vee x_2 x_4$, by the third (algebraic) method discussed after Definition 3.4.1: $x_2 x_4$, $x_1 \bar{x}_2 x_3$, $x_3 x_4$, $x_2 x_3$.

3.5.11 (R) Solve Exercise 3.5.10 by using Karnaugh maps.

3.5.12 (E) Prove that every prime implicant of a switching function f cannot contain a literal of any dummy variable of f.

3.5.13 (R) Illustrate the derivation of all prime implicants of the following function by the iterated-consensus method on Karnaugh maps (i.e., draw a Karnaugh map for each step of the method, showing changes of loops):

$$f = \bar{x}_1 x_2 \bar{x}_3 \vee x_1 x_2 \bar{x}_3 x_4 \vee x_1 \bar{x}_2 x_4 \vee x_1 x_2 x_3 x_4 \vee \bar{x}_1 x_2 x_3 \bar{x}_4.$$

3.5.14 (R) Find all prime implicants of the following functions, using Karnaugh maps:

(i) $f = \bar{x}_1 x_2 x_4 \vee x_2 x_3 \bar{x}_4 \vee x_1 x_2 x_4 \vee \bar{x}_1 \bar{x}_2 \bar{x}_3$.
(ii) $f = x_1 \bar{x}_3 \bar{x}_4 \vee x_1 x_3 \bar{x}_4 \vee \bar{x}_1 \bar{x}_2 \bar{x}_3 \bar{x}_4 \vee \bar{x}_1 \bar{x}_2 x_3 \bar{x}_4 \vee \bar{x}_1 x_2 x_3 x_4$.

3.5.15 (R) Find all prime implicants for one of the following functions by the Tison method (Procedure 3.4.2):

A. $f = x_1 x_2 x_5 \vee x_3 x_4 x_5 \vee \bar{x}_2 x_3 x_4 \vee \bar{x}_1 \bar{x}_3 x_5 \vee x_3 x_6 x_7 \vee x_2 \bar{x}_3 x_4 x_5 \vee x_2 \bar{x}_3 x_5 x_6$.
B. $f = \bar{x}_1 x_2 \vee \bar{x}_2 x_3 \vee x_3 x_7 \vee x_1 \bar{x}_2 \vee x_3 x_8 \vee x_4 x_8 \vee x_1 x_5 x_6 \vee x_2 x_5 x_6 \bar{x}_7 \vee x_3 x_5 x_6 \bar{x}_8$.
C. $f = x_1 x_2 x_3 \vee \bar{x}_1 \bar{x}_2 x_4 x_7 \vee \bar{x}_1 x_2 x_3 x_5 \vee x_1 \bar{x}_3 x_6$.

(Karnaugh maps are difficult to use for these functions because of many variables.)

3.5.16 (E) The network in Fig. E3.5.16 can be simplified by deriving the complete sum for its output function f. On the basis of the complete sum, design a new network with AND or OR gates, using as few gates as possible, under a maximum fan-in of 3 for each gate. Variables for each i, x_i and $\bar{x}_i$ are available as network inputs.

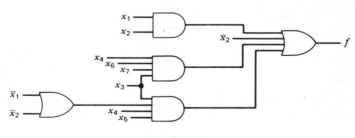

Fig. E3.5.16

3.5.17 (M −) Given products of literals P, Q, R, S, and T, prove that, if R is the consensus of P and Q, and P is the consensus of S and T, then R implies $Q \vee S \vee T$. [Mott 1960]

3.5.18 (M) Let A, B, and C each represent a product of literals. If
(i) C implies $A \vee B$,
(ii) C does not imply A, and
(iii) C does not imply B,
then prove that the consensus of A and B exists and is implied by C.

▲ **3.5.19 (M)** Prove Theorem 3.4.3.

▲ **3.5.20 (M)** Prove that we can get all prime implicants of a given function f by the following procedure (without forming any consensus):

(i) Expand $\bar{f}$, the complement of f, into a disjunctive form. Then delete any terms that subsume others. Denote the resultant expression by Ψ. (For example, for $f = x_1 x_2 \vee \bar{x}_2 x_3 x_4 \vee x_3 \bar{x}_4$, we get $\Psi = \bar{x}_1 x_2 x_4 \vee \bar{x}_1 \bar{x}_3 \vee \bar{x}_2 \bar{x}_3$.)

(ii) Expand $\bar{\Psi}$, the complement of Ψ, into a disjunctive form. Then delete any terms that subsume others. The resultant expression is the complete sum for f. (For the above example, we get $\bar{\Psi} = x_1 x_2 \vee x_1 x_3 \vee \bar{x}_2 x_3 \vee x_3 \bar{x}_4$.) [Hulme and Worrell 1975]

3.5.21 (M −) Suppose that a function f can be expressed with some prime implicants $P_1, P_2, \ldots, P_m$ of f, as $f = P_1 \vee P_2 \vee \cdots \vee P_m$. Then prove that, if $P_1 \subseteq g$ holds, where g is a disjunction of some of $P_2, P_3, \ldots, P_m$ (without P_1), then P_1 is a prime implicant of g. [Mott 1960]

3.5.22 (R) Determine whether or not the following identity is valid, by deriving the complete sums for both sides:

$$x_1 x_2 \vee x_1 \bar{x}_2 x_4 \vee \bar{x}_1 \bar{x}_2 x_3 = (x_1 \vee x_3)(x_2 \vee x_3) \vee (x_2 \vee x_4)(\bar{x}_3 \vee \bar{x}_4).$$

3.5.23 (R) Determine which of all variables the following function f is dependent on or independent of, by deriving the complete sum for f:

$$f = \bar{x}_1 x_3 \vee x_1 x_2 \bar{x}_4 \vee x_3 x_4 \vee x_2 x_3 x_5 x_6.$$

(If a function f is independent of a certain variable x_i, no prime implicant of f contains any literal of x_i because of the property in Exercise 3.5.12.)

Answers for the simple questions in Section 3.4: **1.** (a) "Is implied by"; (b) "subsumes" or "implies." **2.** (a) Implicant; (b) not implicant; (c) implicant; (d) implicant. **3.** (a) and (d). **4.** (a) xzu; (b) no consensus.

▲ 3.6 Boolean Algebra

The switching algebra discussed so far is a special case of Boolean algebra as defined in this section. In switching algebra we dealt with only two elements, 1 and 0, and we defined binary operations between them, such as AND. In Boolean algebra, we deal with a set of elements, whose number may be more than two, and define operations among them. A binary operation on an ordered pair of elements in a set specifies a unique element. If the specified element also belongs to the given set, the binary operation is said to satisfy the **closure property**.

Definition 3.6.1: An algebraic system consisting of a set B of elements $\{a, b, \ldots\}$ and two binary operations, $\vee$ and $\cdot$, that satisfy the closure property is called a **Boolean algebra** if the following postulates are satisfied:

(P1) The operations $\vee$ and $\cdot$ are commutative; for each $a, b \in B$,
$$a \vee b = b \vee a \quad \text{and} \quad a \cdot b = b \cdot a.$$

(P2) Each operation is distributive over the other; for each $a, b, c \in B$,
$$a \cdot (b \vee c) = a \cdot b \vee a \cdot c \quad \text{and} \quad a \vee b \cdot c = (a \vee b) \cdot (a \vee c).$$

(P3) In B there exist identity elements denoted by 0 and 1, relative to the operations $\vee$ and $\cdot$, respectively, that is,
$$a \vee 0 = a \quad \text{and} \quad a \cdot 1 = a \quad \text{for every } a \in B.$$

(The notations 0 and 1 are used here, but other arbitrary notations may be substituted.)

(P4) For each element $a \in B$ there exists an element $\bar{a} \in B$ such that
$$a \vee \bar{a} = 1 \quad \text{and} \quad a \cdot \bar{a} = 0.$$

The set of postulates presented here is one of several sets that can describe a Boolean algebra. For other equivalent definitions see Huntington [1904, 1933].

From this set of postulates, it is possible to prove the following theorems:

$a \vee a = a$	$a \cdot a = a$	(Idempotent)
$a \vee 1 = 1$	$a \cdot 0 = 0$	(Null elements)
$\overline{(\bar{a})} = a$		(Involution)
$a \vee a \cdot b = a$	$a \cdot (a \vee b) = a$	(Absorption)
$a \vee (b \vee c) = (a \vee b) \vee c$	$a \cdot (b \cdot c) = (a \cdot b) \cdot c$	(Associative laws)
$\overline{(a \vee b)} = \bar{a} \cdot \bar{b}$	$\overline{(a \cdot b)} = \bar{a} \vee \bar{b}$	(De Morgan's laws)

For example, the first statement of the idempotent theorem can be proved in the following manner:

$$
\begin{aligned}
a \vee a &= (a \vee a) \cdot 1 & &\text{by (P3)} \\
&= (a \vee a) \cdot (a \vee \bar{a}) & &\text{by (P4)} \\
&= a \vee (a \cdot \bar{a}) & &\text{by (P2)} \\
&= a \vee 0 & &\text{by (P4)} \\
&= a & &\text{by (P3)}
\end{aligned}
$$

Boolean Algebra

For proofs of other theorems, see Whitesitt [1961]. Also, we can derive the duality theorem: Every theorem or identity is transformed into a second theorem or identity if $\vee$ and $\cdot$, and 0 and 1, are interchanged throughout.

The above discussion shows that **switching algebra is simply a Boolean algebra for which the set B consists of only the two identity elements, 0 and 1**. An example of a Boolean algebra with more than two elements is $B = \{0, a, b, 1\}$ with $\vee$ and $\cdot$ defined in Table 3.6.1.

Table 3.6.1   A Boolean Algebra: $B = \{0, a, b, 1\}$

$\vee$	0	a	b	1		$\cdot$	0	a	b	1
0	0	a	b	1		0	0	0	0	0
a	a	a	1	1		a	0	a	0	a
b	b	1	b	1		b	0	0	b	b
1	1	1	1	1		1	0	a	b	1

Notice that **in the literature switching algebra is very often called Boolean algebra, without strictly differentiating the two**.

Boolean algebra itself is an interesting branch of mathematics [Dwinger 1971, Halmos 1963, Sikorski 1964, Rudeanu 1974].

Since knowledge of Boolean algebra is becoming common,* many applications of it will be developed, though most of them may require only elementary knowledge of Boolean algebra (e.g., a biology student tried to apply switching algebra to a biological problem).

Logic networks for essentially all digital systems or computers are currently based on binary values, as mentioned in Section 1.3. Logic networks based on many values (e.g., those based on Boolean algebra) have been explored for many years.†

▲ Exercises

3.6.1 (M)   Prove the null element theorem, $a \vee 1 = 1$, for every $a \in B$, using the postulates of Definition 3.6.1.

3.6.2 (M)   Prove $a \cdot 0 = 0$ for every $a \in B$, using the postulates of Definition 3.6.1.

---

* Recently, some eighth grade students have learned Boolean algebra, though not in detail (see, e.g., Pearson and Allen [1970]).

† See, for example, Dao, Russell, Preedy, and McCluskey [1977], Dao [1977], and the issues of *Proceedings of the International Symposium on Multivalued Logic*, published by the IEEE Computer Society.

114 *Fundamentals of Switching Algebra*

3.6.3 (M) Prove that element $\bar{a}$ is uniquely determined for any element a in a Boolean algebra, using the postulates in Definition 3.6.1.

3.6.4 (M−) Prove the idempotent theorem, $a \cdot a = a$, for $a \in B$, using the postulates of Definition 3.6.1.

3.6.5 (E) Prove $\bar{a} \vee ab = \bar{a} \vee b$ for $a, b \in B$, using the postulates of Definition 3.6.1.

3.6.6 (M+) Using the postulates of Definition 3.6.1, prove that a Boolean algebra that consists of only three elements, $B = \{0, 1, a\}$, does not exist.

3.6.7 (DT) Prove De Morgan's law, $\overline{(a \vee b)} = \bar{a} \cdot \bar{b}$, for every $a, b \in B$, using the postulates of Definition 3.6.1.

▲ 3.7 Propositional Logic

In addition to switching algebra, another special case of Boolean algebra is propositional logic. A **proposition** in propositional logic is a declarative sentence that is either true or false but not both. A statement that is ambiguous or self-contradictory is not a proposition. For example, the following statements are propositions:

"Silicon is a semiconductor."
"10 is smaller than 3."
"Creatures live on some planets."
"The sun is shining."

The first is always true, whereas the second is always false. The third is either true or false but not both, though we cannot decide at present because of the lack of scientific evidence. The fourth is sometimes true but is false at other times. But "All students are mediocre" is not a proposition because "mediocre" is an ambiguous word. (The truth or falsity of the sentence may not be ambiguous sometimes, because "mediocre" may not be ambiguous for people in some community.) "This statement you are reading is false" is not a proposition because it is contradictory. In other words, if we assume that the statement is true, then from its content we conclude that the sentence is false. If we assume that the statement is false, then from its content we conclude that the sentence is true.

Arbitrary or unspecified propositions are denoted by the letters p, q, r, ..., which are called **propositional variables**. When a proposition p is true, we write $p = T$; when it is false, we write $p = F$. Two propositions, p and q, are said to be **equal** if, whenever one is true, the other is true, and vice versa. For example, when

p: x is an odd integer

and

q: x can be expressed as $2y + 1$ with y an integer

are two given propositions, $p = q$ results.

Table 3.7.1

p	$\bar{p}$
T	F
F	T

Table 3.7.2

p	q	$p \wedge q$
F	F	F
F	T	F
T	F	F
T	T	T

When a proposition p is negated, the new proposition is denoted by $\sim p$ or $\bar{p}$ and is called the **denial** of p. If p is true, $\bar{p}$ is false, and vice versa, as shown in the truth table in Table 3.7.1.

Two propositions can be connected by a logical connective to form a new proposition. Suppose that two propositions, p and q, are given. The **conjunction** of p and q is denoted by $p \wedge q$ (or $p \cdot q$ or $p \& q$) and is true only when both p and q are true, and is false otherwise, as shown in Table 3.7.2. The conjunction corresponds to the common usage of "and" to combine two statements. For example, if

p: It rains

and

q: The outdoor temperature is above 30°C

are given, then

$p \wedge q$: It rains, and the outdoor temperature is above 30°C.

A logical connective corresponding to the common usage of "or" might be considered. But this case is not as simple as that of "and." Consider the following two propositions:

1. A thief must use the door or window to get out of the room.
2. A thief used a knife or a gun.

In the first sentence, a thief must use one of the two, door or window, but not both. In the second sentence, he might use both a knife and a gun. Apparently the two usages of "or" are different, and we must differentiate them. In propositional logic, "or" in the first sentence is called **EXCLUSIVE-OR** and the other "or" is called **INCLUSIVE-OR**. The latter is also called **disjunction**, denoted by $\vee$. The truth table for disjunction is shown in Table 3.7.3. Thus, with the preceding example for p and q, a new proposition with EXCLUSIVE-OR is "It rains, or the outdoor temperature is above 30°C, but not both," and a new proposition with INCLUSIVE-OR is "It rains, or the outdoor temperature is above 30°C, or both."

Table 3.7.3

p	q	$p \vee q$
F	F	F
F	T	T
T	F	T
T	T	T

Table 3.7.4

p	q	$p \to q$
F	F	T
F	T	T
T	F	F
T	T	T

When the propositional constants T and F are replaced by 1 and 0, respectively, "disjunction," "conjunction," and "denial" have the same meanings as "disjunction," "conjunction," and "negation" in switching algebra, respectively, since the following four postulates are satisfied. (They can be proved on the basis of truth tables.)

(P1) $\quad p \vee q = q \vee p, \quad p \wedge q = q \wedge p.$

(P2) $\quad p \wedge (q \vee r) = (p \wedge q) \vee (p \wedge r),$
$\quad\quad\; p \vee (q \wedge r) = (p \vee q) \wedge (p \vee r).$

(P3) $\quad F \vee p = p, \quad T \wedge p = p.$

(P4) $\quad p \vee \bar{p} = T, \quad p \wedge \bar{p} = F.$

Thus **propositional logic is equivalent to switching algebra.**

A logical connective called **material implication** (or simply **implication**) is often used in propositional logic because of convenience. This is denoted by $p \to q$ and is false only when p is true and q is false, as shown in Table 3.7.4. (Thus, if p is false, $p \to q$ is true, regardless of whether q is true or false.) Material implication can be expressed with other connections such as $p \to q = \bar{p} \vee q$. For the preceding example for p and q, $p \to q$ is "It does not rain, or the outdoor temperature is above 30°C, or both."

For further study of propositional logic, see, for example, Whitesitt [1961] or Hilbert and Ackermann [1950].

Applications

So far, propositions are defined as declarative sentences that can be determined to be true or false. But it is convenient for some applications (e.g., Exercise 3.7.6) to use interrogative sentences that can be answered "yes" or "no." For example, the proposition "It rains," denoted by a propositional variable p, can be replaced by "Does it rain?" Then we can have $p = 1$ or 0 (or 0 or 1), corresponding to "yes" or "no," instead of p being T or F.

In the switching theory which we have discussed so far, we have already applied propositional logic, as shown in the following:

Propositional variable	Proposition
X:	Magnet of a relay X is energized.
x:	A make-contact of a relay X is closed.
x_1:	Voltage at input 1 of a gate is high.
x_2:	Voltage at input 2 of a gate is high.
f:	Voltage at the output of a gate is high.

If the gate is an AND gate, "Voltage at the output of the gate is high" is true only when both "Voltage at input 1 of the gate is high" and "Voltage at input 2 of the gate is high" are true. Thus, using x_1, x_2, and f above, we have

$$f = x_1 \cdot x_2.$$

Propositional logic has been used in computer-aid design (CAD) including computer performance requirement description.*

In the software area, propositional logic has important applications. **Data base management systems** represent the most important computer application in industry, and many firms are using it on a grandiose scale. Propositional logic is its indispensable tool.†

Another important application in software is the **decision table**, which is extensively used in business transaction processing by computers. (An example of a decision table is shown in Exercise 3.7.6.) In the mid-1950s, General Electrical Co. began experimental work with decision tables. Since the usefulness of the method has been recognized, many software packages for decision tables (such as DETAB-X, CENTAB, TAB 40, and FORTAB) are

---

* See, for example, the special issues on "Hardware description languages," December 1974, and "Hardware description language applications," June 1977, of the *IEEE Computer*, and also Hill and Peterson [1973].

† See, for example, Fagin [1977], Delobel and Casey [1973], and Date [1977]. More specifically, propositional logic is also important for information retrieval (see, e.g., Brandhorst [1966], Belzer [1971], Sommar and Dennis [1969], or Iker [1967]).

118 *Fundamentals of Switching Algebra*

available. These packages can be conveniently used by simply feeding in appropriate decision tables. There are numerous publications about decision tables.*

Propositional logic has other applications in computer programming.†

▲ Exercises

3.7.1 (R) Which of the following sentences or phrases represent propositions? (Assume that all words and grammar used are unambiguous.)

(i) Beautiful yellow cars.
(ii) All mathematics is difficult, and some mathematics is impossible.
(iii) Give me a glass of water.
(iv) The earth rotates.
(v) Is he a foreigner?
(vi) If all cars have brakes, no owner of a car needs to worry about a car accident.

3.7.2 (R) Let p and q denote the propositions "The sky is blue" and "It rains", respectively. Write the following propositions in reasonable English:

(i) pq.
(ii) $p \vee q$.
(iii) $p\bar{q} \vee \bar{p}q$.
(iv) $\overline{(p \vee q)}$.
(v) $\overline{(pq)}$.

3.7.3 (R) Let p and q denote these propositions about numbers: "a is greater than b" and "a is smaller than c," respectively. Translate the following propositions into propositional expressions in p and q.

(i) a is not greater than b, and a is smaller than c.
(ii) Either a is greater than b, or a is smaller than c.
(iii) a is neither greater than b nor smaller than c.
(iv) Either a is greater than b and smaller than c, or a is not greater than b and is not smaller than c.
(v) a is greater than b, or a is smaller than c, but a is not simultaneously greater than b and smaller than c.

3.7.4 (E) A certain company had the following rule for hiring, in the days when discrimination was still legal. Applicants had to be

>male, married, and under 40, or
>male, unmarried, and under 40, or
>female, married, and under 40, or
>female, unmarried, and over 40.

---

* See, for example, McDaniel [1968, 1970], Gildersleeve [1970], Harrison [1971], London [1972], Humby [1973], Montalbano [1962], or Pooch [1974].
† Arden and Graham [1959], Arden, Galler, and Graham [1962], Huskey and Wattenburg [1961], Prather and Casstevens [1978], Kantorovich [1957], Rosin [1960].

Propositional Logic   119

Translate the hiring rule into a propositional expression, using the following symbols:

   male p,   married q,   under 40 r.

Then simplify the expression, if possible.

3.7.5 (M −) Suppose that two three-bit binary numbers, $A = (a_2 a_1 a_0)$ and $B = (b_2 b_1 b_0)$, are given, where a_2 and b_2 are the most significant bits. Then the proposition "$A > B$" can be expressed as

$$(a_2 a_1 a_0 > b_2 b_1 b_0) = (a_2 > b_2) \vee (a_2 = b_2) \wedge \{(a_1 > b_1) \vee (a_1 = b_1) \wedge (a_0 > b_0)\},$$

where propositions such as $a_2 > b_2$ are expressed with pairs of parentheses.

(i) Prove that this expression for $(a_2 a_1 a_0 > b_2 b_1 b_0)$ can be rewritten as the following switching expression:

$$a_2 \bar{b}_2 \vee \overline{(a_2 \oplus b_2)} \{a_1 \bar{b}_1 \vee \overline{(a_1 \oplus b_1)} a_0 \bar{b}_0\}.$$

(The symbol $\oplus$ is defined in Section 1.4.)

(ii) Derive two expressions, corresponding to the two above, for "$A > B$," where A and B are two four-bit binary numbers.

3.7.6 (E) An airline company has a counter for reservations and ticket sales, with the following instructions for the attendant. If a customer requests a first-class seat and it is available, issue a first-class ticket and subtract one from the total number of available seats. If the first class is not available, place the customer's name on a waiting list for the first class. The same procedure is taken for tourist class.

For this situation a **decision table** is prepared in Table E3.7.6. The table is divided into four areas. The upper left area and lower left area show conditions and actions, respectively. The upper right area and lower right area show different rules and the corresponding actions to be taken, respectively; a cross denotes an action to be taken. For example, rule 3 is "If the request is for the first class and first class is not available, then place the request on the first-class waiting list." In the upper right area, Y and N denote "yes" and "no", respectively, and the blank means that a corresponding condition is irrelevant.

Can you further simplify this table? If so, show a simplified table.

Table E3.7.6   Decision Table

					RULES			
	CONDITIONS AND ACTIONS	1	2	3	4	5	6	7
Condition	Is request for first-class?	Y	Y	Y	N	N	N	N
	Is first-class available?	Y	Y	N	Y	Y	N	N
	Tourist class available?	Y	N		Y	N	Y	N
Action	Issue a first-class ticket, and subtract one from total.	×	×					
	Place on first-class waiting list.			×				
	Issue tourist ticket, and subtract one from total.				×		×	
	Place a tourist-class waiting list.					×		×

Remark: If the situation expressed in these rules is expressed in a flow chart, it will be complex. But once a computer program is formed in decision table form, it is easy to code, debug, and also update whenever changes are necessary, since only the decision table needs to be changed. ■

Exercises

3.1 (E) Write the minterm expansion and the maxterm expansion for each of the following functions:
- A. (i) $f(x, y, z) = \Sigma(1, 4, 7)$.
  - (ii) $f(x, y, z) = \Pi(1, 4, 7)$.
- B. (i) $f(x, y, z) = \Sigma(0, 2, 5)$.
  - (ii) $f(x, y, z) = \Pi(0, 2, 5)$.
- C. (i) $f(x, y, z) = \Sigma(1, 2, 6)$.
  - (ii) $f(x, y, z) = \Pi(1, 2, 6)$.

3.2 (M−) Derive the complete sum for the output function f of the MOS network in Fig. E3.2.

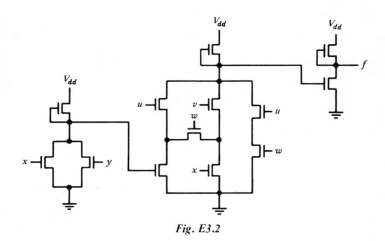

Fig. E3.2

3.3 (M−) Derive a disjunctive form for the output function f of the MOS network in Fig. E3.3, in the following manner. **(i)** Derive the transmission of the output gate between the output terminal for f and the ground, in a conjunctive form by Procedure 2.2.2, and then **(ii)** derive a disjunctive form for f, by applying De Morgan's theorem to the complement of the conjunctive form obtained. (Derivation of a disjunctive form for f in this manner is easier for this particular network than conversion of f expressed in the complement of a disjunctive form.)

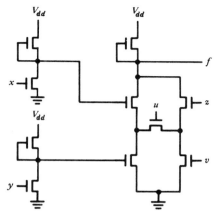

Fig. E3.3

3.4 (M) Derive a MOS network for the following function f, using as few MOSFETs as possible:

$$f = (\bar{u} \vee \bar{v})(\bar{u} \vee \bar{w} \vee y)(x \vee y)(x \vee \bar{w} \vee \bar{v}).$$

(*Hint*: If a bridge connection of MOSFETs like the one in Fig. 2.3.6a is used, the number of MOSFETs can be reduced in this case.)

Remark: Generally it is not easy to find appropriate bridge connections such that the number of MOSFETs is reduced. Therefore, in practice, series-parallel connections of MOSFETs are usually used in MOS gates to save design time, though bridge connections are sometimes used (e.g., in some commercially available microprocessors). ∎

3.5 (M) **(i)** Let P be the consensus of two terms, A and B. Prove that A and P do not have a consensus.

(ii) Consider the operation to derive the consensus from a pair of terms. Is this operation commutative? Associative? Is it distributive with respect to conjunction and disjunction? (For each question, explain why.)

3.6 (M) A patient must pass five tests, x_1, x_2, x_3, x_4, and x_5, in order to be released from a hospital.

> Dr. A can conduct tests x_1, x_4, and x_5.
> Dr. B can conduct tests x_2 and x_4.
> Dr. C can conduct tests x_1 and x_3.
> Dr. D can conduct tests x_1 and x_4.
> Dr. E can conduct tests x_3 and x_5.

By deriving a switching expression for the requirements, answer the following:

(i) Find the minimum number of doctors the patient must see in order to have all five tests.

(ii) Who is the essential doctor (or doctors) without whom the patient cannot have all five tests?

(iii) Find all the minimal combinations of doctors such that each combination performs all five tests and elimination of any doctor in each combination results in failure to perform all five tests.

3.7 (M+) Students x_1, x_2, x_3, x_4, and x_5 are planning an automobile trip for which the following conditions must be met:

(1) If x_4 goes, x_5 must go.
(2) If x_2 goes, x_1 and x_4 also must go.
(3) Either x_1 or x_2 or both must go.
(4) Either x_3 or x_5 but not both must go.
(5) Either both x_1 and x_3 go or do not go.

Derive a switching expression $f(x_1, x_2, x_3, x_4, x_5)$ that shows the combinations of students who can go on the trip together, where x_1, x_2, x_3, x_4, x_5 mean that the corresponding students are selected to go, when not complemented. Simplify f as much as possible.

3.8 (M) Prove that, if $\bar{x}\bar{y} \vee zu = 0$, then

$$\bar{y}v \vee x\bar{y} \vee \bar{z}(\bar{x} \vee u) = \bar{y}(x \vee \bar{u}) \vee yu \vee \bar{x}\bar{z}\bar{u}.$$

3.9 (M) Prove that $xf \vee \bar{x}f^d$ is self-dual, where f is a function and x is a variable that may or may not be a variable of f.

3.10 (M) Given the identity $xy\bar{z} \vee \bar{x}\bar{y}z = 0$, find all solutions for function f such that $z = xf \vee \bar{y}\bar{f}$ is satisfied, where f is a function of x, y, z only. (*Hint*: considering that constant $xy\bar{z} \vee \bar{x}\bar{y}z = 0$ relates z to x and y, derive z as a function of x and y. Then derive all solutions for f as a function of x and y only. See Exercise 3.2.23 or the hint in Exercise 3.2.24.A.)

3.11 (M) Fig. E3.11 shows a network to turn on or off a lamp. The network consists of a diode and some relay contacts. The magnet for make-contact y is denoted by Y. (The magnets X_i for contacts x_i and $\bar{x}_i$, $i = 1, 2, 3, 4$, are not shown.)

(i) When does the lamp light? (Derive a switching expression that assumes 1 only when the lamp lights.)

(ii) Redraw the network, replacing the diode by one of make-contacts x_1, x_2, x_3, x_4 or break-contacts $\bar{x}_1, \bar{x}_2, \bar{x}_3, \bar{x}_4$, so that the operation of contact y is unchanged. Find all solutions.

3.12 (M−) A technical paper contained the following statement: "There is a function f of x_1 and x_2 only, satisfying the following two conditions:

(1) $\bar{x}_i$ does not imply f for each i.
(2) $f = 1$ for $x_1 = x_2 = 1$."

A student concluded that f must be $f = x_1 x_2$ by considering the truth table in Table E3.12.

Prove or disprove the student's conclusion. If there are other solutions for f, show all of them.

Table E3.12

x_1	x_2	f
0	0	0
0	1	0
1	0	0
1	1	1

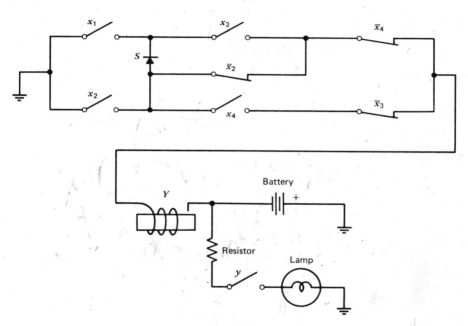

Fig. E3.11

CHAPTER **4**

Simplification of Switching Expressions

Simplification of switching expressions is very important for designing economical and fast networks—more specifically, two-level networks of AND and OR gates, two-level networks of NOR gates (to be discussed in Chapter 6), two-level networks in programmable logic arrays (usually abbreviated as PLA, to be discussed in Chapter 9), which are widely used as a convenient LSI implementation of networks, and networks with complex IC gates such as ECL (emitter coupled logic) or TTL (transistor–transistor logic) gates.

Karnaugh maps are convenient for the simplification of switching expressions of a few variables, since in many cases the pictorial nature of the map makes it easy for us to grasp the entire procedure. Thus Karnaugh maps are an indispensable tool for logic design.

After the use of the Karnaugh map is discussed, tabular and algebraic methods for the simplification of a switching expression are described. When we need to deal with five or more variables, these methods are useful, since Karnaugh maps become increasingly complex. In particular, the algebraic method discussed at the end of this chapter is very powerful for hand processing.

4.1 Design Objectives and Minimal Sums

When we design logic networks, we usually try to reduce the cost, space, response time, or electric power consumption of the networks as much as possible, under restrictions such as maximum fan-in. If one of these factors becomes unreasonably poor when we reduce others, we may need to compromise among them. Since networks are designed under very diversified situations (to be discussed in Section 9.1), the objectives and restrictions under which networks must be designed are correspondingly very diversified. Some restrictions are simple, such as maximum fan-in or fan-out, and others

Design Objectives and Minimal Sums 125

are complex, such as layout rules, pin number limitation of IC packages, requirements from architectural considerations, or diagnosability. It is practically impossible to design a network with one of these diversified design objectives under complex restrictions. Design Objectives 1 and 2 given below, are usually considered, however, because of their simplicity. **When restrictions are complex, networks can be designed without much difficulty under these design objectives first, without considering the restrictions, and then the designed networks can be modified so that the restrictions are satisfied. By this approach we can usually obtain networks that are satisfactory for practical use.** (Probably there is no better means.) Of course, **if restrictions are simple, they can be taken into account during the design of the networks**, as we will see in some procedures in this book. But it is usually practical to modify after designing networks without considering part or all of the restrictions, when the restrictions are complex.

Design Objective 1: Minimization of the Number of Gates, as the Primary Objective, and of the Number of Connections, as the Secondary Objective

When networks are to be designed with discrete gates, or SSI packages (each with a few separate gates like those in Fig. 2.4.1), in most cases each gate can be assumed to have approximately the same cost, area, and electric power consumption. Thus let us make this assumption* for the sake of simplicity. Each connection can usually be assumed to have much less cost and area than does a gate. (How much less depends on the electronic implementation of the networks.†) Then, by designing a network with discrete gates or SSI packages under Design Objective 1, we can minimize the costs (if SSI packages are used, the number of packages is minimized correspondingly by Design Objective 1), areas on pc boards, and power consumption‡ of the network. (See Remark 4.1.1 for the area covered by connections.) Since each connection ends at an input of a gate, the number of connections (the output connection of an output gate is usually not counted) is equal to the number of inputs to all gates. Thus reduction of the number of connections is equivalent

---

* The cases in which gates have different costs, areas, or power consumptions can be treated by slight modifications of our design procedures, as we will mention later, though the procedures will then become more complex.

† For example, a logic operation in some logic families can be realized by tying down connections without using a gate; this is referred to as "wired logic" in Section 5.5. In this case the logic operation may be regarded as having a cost and an area comparable to those of connections. Also, in the case of PLAs (to be discussed later), the cost of connections is fixed, so only the cost of gates can be minimized.

‡ Reduction of power consumption means a reduction of temperature rise, which usually results in a reduction of failure rate. This also means a reduction of cost of a power supply.

126 Simplification of Switching Expressions

to reduction of the number of inputs to all gates. (Throughout this book, differentiate "inputs to gates" from "inputs to a network" or "network inputs.")

When, in the case of LSI (or VLSI) or MSI, entire networks are to be designed on IC chips, we need to carefully examine what Design Objective 1 means, depending on the IC logic family. The most economical networks can usually be obtained by minimizing the IC chip size. (Other cost reduction factors, like production volume, are not directly related to logic design. Production yield generally increases as chip size decreases.) Gates and connections both cover significant areas on the chip, and the layout of a network on the chip requires consideration of many complex rules. Hence the design of the most compact networks (i.e., the smallest chip size) and consequently the most economical networks is very complex. But when ECL or TTL is used, gates occupy much larger areas than connections, and following Design Objective 1 usually yields the networks with the smallest chip size, cost, and power consumption. However, when MOS gates are used, connections can cover an area comparable to that covered by gates, depending on the situation. (Connections inside a gate are regarded as part of the gate.) Unfortunately, no procedure of logic design is known to minimize the connection area as the primary objective, before the actual layout of a designed logic network on a chip. (Since the layout is cumbersome and time-consuming, it is strongly desirable to have a logic network that is more or less guaranteed to lead to the most compact network.)

Design Objective 1, however, appears to yield the smallest chip size and consequently the most economical networks, as a rule of thumb, at least in the case of small networks, judging from the following computational experiment. NOR gate networks obtained under Design Objective 1 were compared with those designed by reversing the two objectives in Design Objective 1, that is, by minimizing the number of connections as the primary objective, and the number of gates as the secondary objective.* The result of this experiment with a few hundred functions of a few variables was that the networks in the two cases were completely identical in the majority of the functions,† and when they were not identical, differences in gate or connection counts were not great. From this experiment and some theoretical considerations [Muroga and Lai 1976], we can probably say that the most compact network size (or chip size) coincides exactly with the minimum value of the gate count in practically all cases; and even if this is not true, the two lie in the same neighborhood, when the networks are of processable size. (At values that are distant from this minimum one, the network size

---

* This minimization was done by a complex design procedure called the integer programming logic design method, to be mentioned in Sections 5.4 and 6.6.
† A preliminary computational experiment with AND and OR gates also showed the same tendency.

Design Objectives and Minimal Sums

definitely increases.) However, if the networks are too large, generally no design procedures are known,* regardless of whether we want to minimize the number of gates or of connections (though only required functions can be realized,† as will be discussed in Section 9.2). Of course, if large networks are designed by decomposing them into small networks, as we do in practice, Design Objective 1 is a reasonable choice within this framework.

Design Objective 1 is used because we have simple design procedures based on it and we have no simple procedures for other design objectives, such as minimization of the number of connections as the primary objective (except the integer programming logic design method to be mentioned in Sections 5.4 and 6.6, which is complex). Therefore the above observation based on experiment is convenient. When there are complex restrictions such as layout rules, the common practice in logic design is to design networks with Design Objective 1, taking into account some of the restrictions that can easily be incorporated, and then to modify the designed networks according to the remaining restrictions.

Remark 4.1.1 — The area covered by connections on a pc board or an IC chip: Strictly speaking, the number of connections is different from the total cost or area of the connections. (The relationship among them is complex. For example, the cost of connections depends partly on the number of connections, because of costs incurred for making holes on a pc board and soldering them, and partly on the total length, because of the areas covered by the connections on a pc board.) But we can reduce the total cost or area of connections by reducing the number of connections, since we can approximately express the total cost or area as a monotonically increasing function of the number of connections [Muroga and Lai 1976]. Notice that the length of a connection is highly dependent on the placement of gates. ■

Whereas Design Objective 1 is used primarily for the minimization of cost, area, or power consumption, the following objective is concerned with the maximization of network speed. Design Objective 1 is also indirectly related to network speed, as discussed a few paragraphs later.

Design Objective 2: Minimization of the Number of Levels

Consider all chains of gates connected from each input to the output of a network. Then the number of gates in the longest chain is called the **number of levels (or gate delay) of the network**. Roughly speaking, if the number of levels is minimized, the network gives the fastest response at its output, assuming that all gates have equal switching times. No matter which—

---
* Except in a few cases such as the minimal adders discussed in Section 9.2.
† Such as the design of multiplier networks.

128 Simplification of Switching Expressions

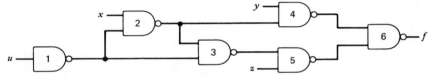

Fig. 4.1.1 Network whose effective gate delay is smaller than the number of levels.

discrete gates, SSI packages, or an LSI package—we use, this design objective yields the fastest network under the assumption that every gate has the same switching time.

The number of levels of a network, however, does not show a real delay time in the network response against the input changes, as the example in Fig. 4.1.1 shows. The real delay time measured as a multiple of the switching time τ of each gate is called the **effective gate delay** (or **effective number of levels**) of the network. The number of levels in the network in Fig. 4.1.1 is 5, since the number of gates in the longest chain, u–1–2–3–5–6, is 5. But the effective gate delay is only 4, as can be seen by determining when f changes its value for a change in u's value (from 0 to 1 or from 1 to 0), against every possible combination of values of the other inputs, x, y, and z. Thus the number of levels is simply an upper bound on the real delay, the effective gate delay. Since determination of the latter, despite its importance, is much more time-consuming than finding the former, we will consider only the number of levels until the effective gate delay is discussed again, along with related problems, such as the problem of faulty output values of combinational networks during the transition period, in Section 6.7.

In the preceding discussion of Design Objective 2, we assumed that every gate has the same switching time. If gates have different switching times, however, Design Objective 2 does not necessarily lead to the fastest network, since the network response time to input changes is not proportional to the effective gate delay. This takes place with gates of some logic families when they have large fan-out connections. When MOS gates on an LSI chip for high speed have large fan-out connections, they have very large switching times,* unlike gates of logic families such as TTL, whose switching times are much less affected by parasitic capacitance. Also, in this case, Design

---

* This is so because they have limited power output, and consequently their switching time becomes sensitive to parasitic capacitance. As explained in Section 9.2, the carry-ripple adder is used in Intel's microcomputer 8080 because it has smaller parasitic capacitance in MOS implementation due to the compactness of the network, and consequently it is faster than the look-ahead adder, which is faster if all gates have equal switching times. Whenever new processing or layout rules are developed, logic designers usually redesign even standard networks to reduce parasitic capacitance, and accordingly to increase network speed.

Objective 1 is indirectly related to the response time of the network for the following reason. If the network is designed with excessively many gates without considering Design Objective 1, the entire network will have large parasitic capacitances and consequently a slow response time. Also, in the case of gates of some other logic families, such as ECL gates, if the power consumption of the network becomes excessively large, without consideration of Design Objective 1, then the network cannot be placed on a single chip because of excessive heat, but must be split into more than one chip. Since connections among these chips must be placed on a pc board, and these connections on a pc-board have much greater parasitic capacitances than those on a chip, a network split on different chips has slower speed [Van Tuyl and Liechti 1977]. The situation is very complex.

As discussed above, the relationship among Design Objectives 1 and 2 and more direct design objectives, such as minimization of cost, area, power consumption and speed, is very complicated. Nevertheless, Design Objectives 1 and 2 are very important as approximate objectives (though the extent of approximation differs case by case), since there are no better, simpler objectives, and networks derived by Design Objective 1 or 2 can usually be modified to meet the original design objectives or restrictions to a satisfactory extent. When we design a network, either Design Objective 1 or 2 is usually chosen as the primary one. As mentioned above, when more than one network meets this primary objective, we usually choose a network that meets the other, as the secondary objective.

Minimal Sum

In this chapter, using only AND and OR gates, we shall synthesize the fastest network, that is, a two-level network, since the number of levels cannot be reduced further unless a given function can be realized with a single AND or OR gate. If there is more than one such network, we will derive the most economical network. In other words, Design Objective 2 is chosen with Design Objective 1 as the secondary objective. This network we shall derive has a close relation to important concepts in switching algebra, that is, irredundant disjunctive forms and minimal disjunctive forms (or minimal sums), as we shall now discuss.

For many functions, some terms in their complete sums are redundant. In other words, even if we eliminate some terms from a complete sum, the remaining expression may still represent the original function for which the complete sum was obtained. Thus we have the following concept.

Definition 4.1.1: An **irredundant disjunctive form** for f (sometimes called an irredundant sum-of-products expression or an irredundant

130 *Simplification of Switching Expressions*

sum) is a disjunction of prime implicants such that removal of any of the prime implicants makes the remaining expression not express the original f.

An irredundant disjunctive form for a function is not necessarily unique.

Definition 4.1.2: Prime implicants that appear in every irredundant disjunctive form for f are called **essential prime implicants** (or core prime implicants) of f. Prime implicants that do not appear in any irredundant disjunctive form for f are called **absolutely eliminable prime implicants** of f. Prime implicants that appear in some irredundant disjunctive forms for f but not in all are called **conditionally eliminable prime implicants** of f.

For example, when $f = x_1 x_3 \bar{x}_5 \lor x_2 \bar{x}_3 x_5 \lor \bar{x}_1 \bar{x}_3 x_4 \bar{x}_5 \lor x_2 x_3 \bar{x}_5 \lor x_3 \bar{x}_4 \bar{x}_5 \lor \bar{x}_1 x_2 x_4 \bar{x}_5 \lor \bar{x}_1 \bar{x}_3 \bar{x}_4 x_5 \lor \bar{x}_1 \bar{x}_2 \bar{x}_4 x_5 \lor \bar{x}_1 \bar{x}_2 x_3 \bar{x}_4 \lor \bar{x}_1 x_2 \bar{x}_3 x_4$ is given, all terms in this expression, which is the complete sum for f, are prime implicants of f. But this expression is not an irredundant disjunctive form for f, because even if the last two terms, $\bar{x}_1 \bar{x}_2 x_3 \bar{x}_4$ and $\bar{x}_1 x_2 \bar{x}_3 x_4$, for example, are removed, the remaining expression assumes the same value as does f for every combination of values of x_1, x_2, x_3, x_4, and x_5. This function has the following four irredundant disjunctive forms (we will discuss later how to derive all irredundant disjunctive forms for f):

(a) $x_1 x_3 \bar{x}_5 \lor x_2 \bar{x}_3 x_5 \lor \bar{x}_1 \bar{x}_3 x_4 \bar{x}_5 \lor x_2 x_3 \bar{x}_5 \lor x_3 \bar{x}_4 \bar{x}_5 \lor \bar{x}_1 \bar{x}_2 \bar{x}_4 x_5$
(b) $x_1 x_3 \bar{x}_5 \lor x_2 \bar{x}_3 x_5 \lor \bar{x}_1 \bar{x}_3 x_4 \bar{x}_5 \lor x_3 \bar{x}_4 \bar{x}_5 \lor \bar{x}_1 \bar{x}_2 \bar{x}_4 x_5 \lor \bar{x}_1 x_2 x_4 \bar{x}_5$
(c) $x_1 x_3 \bar{x}_5 \lor x_2 \bar{x}_3 x_5 \lor \bar{x}_1 \bar{x}_3 x_4 \bar{x}_5 \lor x_2 x_3 \bar{x}_5 \lor \bar{x}_1 \bar{x}_3 \bar{x}_4 x_5 \lor \bar{x}_1 \bar{x}_2 x_3 \bar{x}_4$
(d) $x_1 x_3 \bar{x}_5 \lor x_2 \bar{x}_3 x_5 \lor \bar{x}_1 \bar{x}_3 x_4 \bar{x}_5 \lor x_3 \bar{x}_4 \bar{x}_5 \lor \bar{x}_1 x_2 x_4 \bar{x}_5 \lor \bar{x}_1 \bar{x}_3 \bar{x}_4 x_5$
$\lor \bar{x}_1 \bar{x}_2 x_3 \bar{x}_4$

Each of them is an irredundant disjunctive form, since deletion of any term makes the remaining expression no longer represent f. Among the 10 prime implicants, $x_1 x_3 \bar{x}_5$, $x_2 \bar{x}_3 x_5$, and $\bar{x}_1 \bar{x}_3 x_4 \bar{x}_5$ are essential prime implicants, since they appear in every irredundant disjunctive form for f, and $\bar{x}_1 x_2 \bar{x}_3 x_4$ is an absolutely eliminable prime implicant, since it does not appear in any irredundant disjunctive form. Other prime implicants, that is, $x_2 x_3 \bar{x}_5$, $x_3 \bar{x}_4 \bar{x}_5$, $\bar{x}_1 x_2 x_4 \bar{x}_5$, $\bar{x}_1 \bar{x}_3 \bar{x}_4 x_5$, $\bar{x}_1 \bar{x}_2 \bar{x}_4 x_5$, and $\bar{x}_1 \bar{x}_2 x_3 \bar{x}_4$, are conditionally eliminable prime implicants, since each of them appears in some irredundant disjunctive forms but not in others.

Any function f of n variables has at least one irredundant disjunctive form consisting of at most 2^{n-1} prime implicants (Exercise 4.14).

Irredundant disjunctive forms, and particularly the concepts defined in the following, play an important role in switching theory.

Design Objectives and Minimal Sums   131

Definition 4.1.3: Among all irredundant disjunctive forms of f, those with a minimum number of prime implicants are called **sloppy minimal sums**.* Among sloppy minimal sums, those with a minimum number of literals are called **minimal sums** (or minimal disjunctive forms) for f. [If we need to differentiate minimal sums explicitly from sloppy minimal sums, minimal sums may be called real (or absolute) minimal sums.]

Among the five irredundant disjunctive forms for the function in the above example, (a), (b), and (c) are sloppy minimal sums but (d) is not, since (a), (b), and (c) have six terms each and (d) consists of seven terms. Among these sloppy minimal sums, (a) is a minimal sum but (b) and (c) are not, since (a) consists of 20 literals and (b) and (c) consist of 21 literals each.

Irredundant disjunctive forms for a given function f can be obtained by deleting prime implicants one by one from the complete sum in all possible ways, after obtaining the complete sum by the iterated-consensus method or the Tison method discussed in Section 3.4. Then the minimal sums can be found among the irredundant disjunctive forms. Usually, however, this approach is excessively time-consuming because a function has many prime implicants, on the average, as can be seen in the statistics on the number of prime implicants shown in Figs. 4.1.2 and 4.1.3 [Cobham, Fridshal and North 1962]. The statistics shown in these figures are based on a randomly chosen function (notice that it is impossible to solve all 2^{2^n} functions for a large n, because of the excessive computation time required) and in this sense may not give a very accurate picture, but they show roughly how big the problem may become, on the average. When the number of variables is too large, derivation of a minimal sum is usually impossible. For example, when the number of variables n is 9, a function consisting of $\frac{1}{2} \times 2^9$ minterms has, on the average, roughly 260 prime implicants, according to Fig. 4.1.2. But according to Fig. 4.1.3, the average number of prime implicants in a minimal sum of the same function is roughly 90. Therefore we must find an appropriate set of 90 among the 260 prime implicants, a task which is excessively time-consuming. Some functions of n variables are known to have prime implicants whose number increases in the order of 3^n [Dunham and Fridshal 1959].

In this and the following chapter, we will discuss efficient methods to derive minimal sums within reasonable computation time when the number of variables is few, or when the number of variables is large but the relationship among prime implicants (the relationship will be discussed in the next chapter) is not complex.

---

* Some authors define "sloppy minimal sum" as "minimal sum."

132 *Simplification of Switching Expressions*

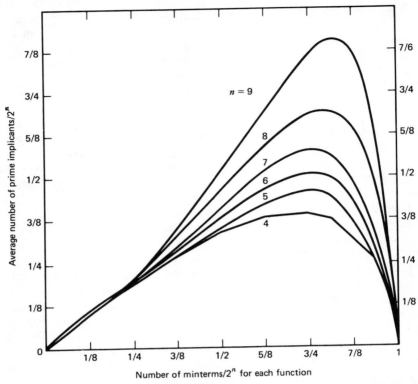

Fig. 4.1.2   Average number of prime implicants for a function of n *variables.*

Why are we interested in irredundant disjunctive forms or minimal sums? Assume that we have a disjunctive form for f. For example,

$$f = x_5 \lor x_1 x_3 x_4 \lor \bar{x}_1 \bar{x}_3 x_4 \lor \bar{x}_1 x_2 x_3 \lor x_1 x_2 x_3 \lor x_1 \bar{x}_2 x_3 \lor \bar{x}_1 x_3 \bar{x}_4$$

is given. This function can be implemented on the basis of this disjunctive form, as shown in Fig. 4.1.4a, using AND and OR gates. In this figure, the output of each AND gate represents a term consisting of two or more literals in this disjunctive form, while a term consisting of a single literal such as x_5 can be directly connected to the OR gate. This network consists of six AND gates and one OR gate, with 25 connections. Notice that the required number of AND gates is equal to the number of terms consisting of two or more literals in this disjunctive form, and the number of inputs from variables to all gates (including the OR gate) is equal to the number of literals contained in this disjunctive form.

Design Objectives and Minimal Sums   133

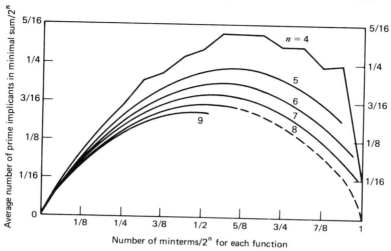

Fig. 4.1.3 Average number of prime implicants for a minimal sum representing a function of n variables.

As seen in this example, based on a disjunctive form for a given function f,

$$\Phi = Q_1 \vee Q_2 \vee \cdots \vee Q_t,$$

where terms $Q_1, Q_2, \ldots, Q_t$ are not necessarily prime implicants of f, we can derive a network consisting of t AND gates and one OR gate. As we will see later, if a disjunctive form has the smallest number of terms and then the smallest number of literals, the number of AND gates and then the number of connections required for a network are minimized, even if some terms consist of single literals (such as x_5 in the above example, which can be directly connected to the OR gate). Therefore let us show that, if a disjunctive form for f has the minimum number of terms t and then the minimum number of literals, each term must be a prime implicant of f and the form must be a minimal sum.

Theorem 4.1.1: Suppose that we want to find a disjunctive form for a given function which has the minimal number of literals among all disjunctive forms that have the minimal number of terms. Then minimal sums are the only solutions for this problem (notice that each term in a minimal sum must be a prime implicant).

Proof: Suppose that we have a disjunctive form for a given function f,

$$\Phi = Q_1 \vee Q_2 \vee \cdots \vee Q_t,$$

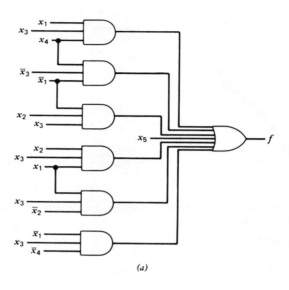

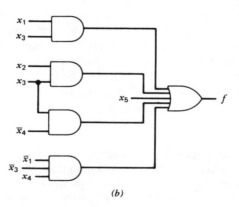

Fig. 4.1.4 Networks based on different switching expressions for a function f.
(a) *Network based on disjunctive form*, $f = x_5 \vee x_1 x_3 x_4 \vee \bar{x}_1 \bar{x}_3 x_4 \vee \bar{x}_1 x_2 x_3 \vee x_1 x_2 x_3 \vee x_1 \bar{x}_2 x_3 \vee \bar{x}_1 x_3 \bar{x}_4$. (b) *Network based on minimal sum*, $f = x_5 \vee x_1 x_3 \vee x_2 x_3 \vee x_3 \bar{x}_4 \vee \bar{x}_1 \bar{x}_3 x_4$.

where terms $Q_1, \ldots, Q_t$ are not necessarily prime implicants of f. We replace each Q_i in the above Φ by a prime implicant P_i which Q_i subsumes, that is, $Q_i \subseteq P_i$. (If Q_i is already a prime implicant, it is rewritten as P_i.) Thus we get

$$\Phi' = P_1 \vee P_2 \vee \cdots \vee P_t.$$

Since $Q_i \subseteq P_i$ for every i, $\Phi \subseteq \Phi' = P_1 \vee \cdots \vee P_t$ holds. Since Φ represents f by the assumption, this means that

$$f \subseteq P_1 \vee \cdots \vee P_t.$$

Each P_i is a prime implicant of f, so whenever any P_i becomes 1, f becomes 1. Thus

$$f \supseteq P_1 \vee \cdots \vee P_t.$$

Combining these two implication relations, we obtain

$$f = P_1 \vee \cdots \vee P_t.$$

We then delete prime implicants (including duplicate ones) from this expression until further deletion of any prime implicant makes the remaining expression no longer represent f.

Therefore, whenever we have a disjunctive form with t terms, we can obtain an irredundant disjunctive form consisting of t or fewer prime implicants by this procedure. Hence, if we repeat this procedure on every possible disjunctive form Φ for f, there must be irredundant disjunctive forms with a minimal number of terms. Then those with a minimal number of literals among them are minimal sums for f.
Q.E.D.

The following theorem is often useful.

Theorem 4.1.2: If a prime implicant of a function* f consists of a single literal (x_i or $\bar{x}_i$), then it is an essential prime implicant of f. Furthermore, no prime implicant of f that consists of two or more literals contains x_i or $\bar{x}_i$, when x_i or $\bar{x}_i$ is an essential prime implicant of f.

Proof: Let a prime implicant of f that consists of a single literal be x. (The case of $\bar{x}$ can be similarly proved.) Suppose that f has another prime implicant xQ or $\bar{x}Q$, consisting of two or more literals, where Q is a product of literals of variables other than x.

But xQ cannot be a prime implicant of f, since it subsumes x. Also, $\bar{x}Q$ cannot be a prime implicant of f, since it subsumes the consensus Q of $\bar{x}Q$ and x (i.e., $\bar{x}Q$ is not in the complete sum of f if the iterated-consensus method is applied, starting with a disjunctive form that contains $\bar{x}Q$ and x).

Therefore no prime implicant consisting of two or more literals contains x or $\bar{x}$. Also, it is obvious that no other single literal prime implicant can be $\bar{x}$. Thus every prime implicant other than x is independent of x. Therefore, unless every irredundant disjunctive form contains x, we have the contradiction of having an irredundant disjunctive form that is independent of x.
Q.E.D.

---

* The function f is assumed not to be identically 1.

136 *Simplification of Switching Expressions*

Let us group the terms of an irredundant disjunctive form for f into those of single literals and those of two or more literals, as

$$f = (a \vee b \vee \cdots) \vee (P \vee Q \vee \cdots),$$

where $a, b, \ldots$ are single literals (e.g., $a = x_i$, $b = \bar{x}_j, \ldots$) and $P, Q, \ldots$ are terms consisting of two or more literals. When the total number of terms in f is minimized, the number of terms in $(P \vee Q \vee \cdots)$ is also minimized, because $(a \vee b \vee \cdots)$ is present in every irredundant disjunctive form, by Theorem 4.1.2. Therefore, whether a minimal sum contains single-literal terms or not, it yields a desired network that has the minimum number of AND gates in the first level. Also, since literals that appear in these single-literal terms do not appear in other terms by Theorem 4.1.2, these literals need to be connected only to the OR gate. Then the total number of connections, including those from the AND gates to the OR gate, is minimized, since the number of AND gates is minimized. Summarizing the above, we get the following Theorem.

Theorem 4.1.3: Whether a minimal sum contains single-literal terms or not, the minimal sum for a given function f yields a two-level AND-OR network for f that has the minimal number of AND gates and then, as the secondary objective, the minimal number of connections.

Returning to the above example, we see that function f shown in Fig. 4.1.4a has two irredundant disjunctive forms, $x_5 \vee x_1 x_3 \vee x_3 \bar{x}_4 \vee \bar{x}_1 \bar{x}_3 x_4 \vee x_2 x_3$ and $x_5 \vee x_1 x_3 \vee x_3 \bar{x}_4 \vee \bar{x}_1 \bar{x}_3 x_4 \vee \bar{x}_1 x_2 x_4$. Of the two, the first one is the minimum sum and the network based on this is shown in Fig. 4.1.4b. Obviously, this network has the fewest AND gates and then the fewest connections. Thus, if the cost of every gate is the same and is much higher than the costs of connections, the network based on the minimal sum is the most economical (one OR gate is always necessary in the second level), because it has the fewest gates and then the fewest connections.

Assumption Regarding Two-Level AND-OR Networks

The above derivation of networks that have the minimum costs, areas, or power consumptions is a motivation for our discussion of irredundant disjunctive forms or minimal sums, assuming that the networks are realized with AND and OR gates in two levels, and that both x_i and $\bar{x}_i$ are available as

Design Objectives and Minimal Sums 137

inputs for each i. If networks are to be realized with other types of gates such as NOR gates, our situation may be different. Thus for a while let us consider the realization of networks with AND and OR gates. The synthesized networks have other problems. If a minimal sum for a given function contains many terms, the OR gate must have many inputs. Also, for a large n, some AND gates may have many inputs. As mentioned in Section 2.4, however, each gate cannot have a large number of inputs (if it does, the gate may not work correctly).

Thus **the above design procedure of a minimal network based on a minimal sum is meaningful under the following assumption**. Let us keep this assumption for a while, unless otherwise noted.

Assumption 4.1.1

1. The number of levels is at most two.
2. Only AND gates and an OR gate are used in the first and second (output) levels, respectively. (Networks with OR gates in the first level and an AND gate in the second level will be discussed in Section 4.4.)
3. Complemented variables $\bar{x}_i$'s as well as noncomplemented x_i's for each i are available as the network inputs.
4. No maximum fan-in restriction is imposed on any gate in a network to be designed.
5. Among networks realizable in two levels, we will choose networks that have a minimum number of gates. Then, from those with the minimum number of gates, we will choose a network that has a minimum number of connections.

As discussed above, a network based on a minimal sum can be realized in at most two levels (i.e., in one level if a given function is simply the AND or OR function, and in exactly two levels otherwise, when there is no maximum fan-in restriction). In other words, under Assumption 4.1.1, networks designed based on a minimal sum, or those designed based on a minimal product, to be discussed later,* are the most economical among the fastest networks, if each gate is assumed to have the same cost (much higher than the costs of connections) and switching time. Thus Design Objective 2 is met first, and Design Objective 1 secondarily.

---

* Networks based on a minimal sum have AND gates in the first level and one OR gate in the second level. Those based on a minimal product have OR gates in the first level and one AND gate in the second level, as will be discussed in Section 4.4.

138 *Simplification of Switching Expressions*

If we do not have the restriction "at most two levels," we can generally have a network of fewer gates. For example, for the function $f = x_1(\bar{x}_2 \vee \bar{x}_3 \vee \bar{x}_4 \vee \bar{x}_5 \vee \bar{x}_6 \vee \bar{x}_7) \vee \bar{x}_1 x_2 x_3 x_4 x_5 x_6 x_7$, we have the network of eight gates in two levels in Fig. 4.1.5a, based on the minimal sum expression (also eight gates based on the minimal product). But f can be realized with the network of only four gates in three levels shown in Fig.

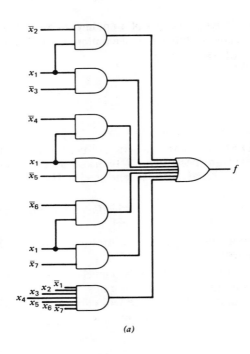

(a)

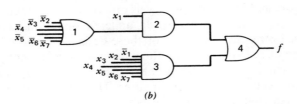

(b)

Fig. 4.1.5 *Networks for a function* $f = x_1(\bar{x}_2 \vee \bar{x}_3 \vee \bar{x}_4 \vee \bar{x}_5 \vee \bar{x}_6 \vee \bar{x}_7) \vee \bar{x}_1 x_2 x_3 x_4 x_5 x_6 x_7$. (a) *Two-level network of eight gates.* (b) *Multilevel network of four gates.*

Derivation of Minimal Sums by Karnaugh Map   139

4.1.5b. (Design of multilevel networks will be discussed in Section 5.4, though no simple systematic approach is known, unlike the design of two-level networks.) Some functions of n variables are known to have two-level networks, whose gate count is at least kn times the gate count of multilevel networks, where k is a constant coefficient [Muroga, to be published]. Thus multilevel network implementation of some functions is greatly advantageous in reducing the gate count, compared with two-level network implementation (though it tends to have the spurious output signal problem during a transient period, as discussed in Section 6.7).

Other Network Implementations Based on Minimal Sums

Although the importance of a minimal sum is discussed in relation to a two-level network of AND gates and an OR gate, a minimal sum or a minimal product is important also in relation to networks with other types of gates (e.g., NOR or NAND gates in Chapter 6), and networks with complex gates of some logic families such as ECL and TTL. Also, a minimal sum or a sloppy minimal sum is important in implementing a network in a PLA, which is widely used. As will be discussed in Section 9.4, a PLA is essentially a two-level network of AND and OR gates, where the gates are prearranged in matrix form on an integrated circuit chip, and the only connections are custom-made to each particular design specification. This is economical because making only connections for each specific design renders IC fabrication inexpensive, while maintaining the advantage of the reasonably high speeds of integrated circuits. Derivation of a minimal sum or a sloppy minimal sum is important in making the size of PLAs small (accordingly, economical).

Remark 4.1.2: If a TTL network is designed based on a minimal sum, the numbers of expanders and emitters in input transistors can be reduced. ∎

4.2   Derivation of Minimal Sums by Karnaugh Map

Because of its pictorial nature, a Karnaugh map is a very powerful tool for deriving by hand complete sums, irredundant disjunctive forms, and also minimal sums. Concepts such as prime implicants and consensuses, which are algebraically defined in the preceding chapter, can be better understood on a map, as illustrated in Chapter 3.

140   Simplification of Switching Expressions

One can derive all prime implicants, irredundant disjunctive forms, and minimal sums by the following procedure on a Karnaugh map, when the number of variables is small enough for the map to be manageable.

Procedure 4.2.1: Derivation of Minimal Sums on a Karnaugh Map

There are three steps in Procedure 4.2.1:

1. On a Karnaugh map, encircle all the 1-cells with rectangles (also squares as a special case), each of which consists of 2^i 1-cells, choosing i as large as possible, where i is a nonnegative integer. Let us call these loops **prime-implicant loops**, since they correspond to prime implicants in the case of the Karnaugh map for four or fewer variables. (In the cases of a five- or six-variable map, the correspondence is more complex, as will be explained later.) Examples are shown in Fig. 4.2.1a.

2. Cover all the 1-cells with prime-implicant loops so that removal of any loops leave some 1-cells uncovered. These sets of loops represent **irredundant disjunctive forms**. Figs. 4.2.1b through 4.2.1e represent four irredundant disjunctive forms, obtained by choosing the loops in Fig. 4.2.1a in four different ways. For example, if the prime-implicant loop $x_1 \bar{x}_3 x_4$ is omitted in Fig. 4.2.1b, the 1-cells for $(x_1, x_2, x_3, x_4) =$ (1 1 0 1) and (1 0 0 1) are not covered by any loops.

3. From the sets of prime-implicant loops formed in Step 2 for irredundant disjunctive forms, choose the sets with a minimum number of loops. Among these sets, the sets that contain as many of the largest loops as possible (a larger loop represents a product of fewer literals) represent minimal sums. Fig. 4.2.1c expresses the unique minimal sum of this function, since it contains one less loop than Fig. 4.2.1b, 4.2.1d, or 4.2.1e.

□

It can be seen, from the definitions of prime implicants, irredundant disjunctive forms, and minimal sums, why Procedure 4.2.1 works. (From the illustration of a prime implicant on a Karnaugh map, it is easy to see that all prime implicants of a given function f can be found in Step 1. Then by the definitions of "irredundant disjunctive form" and "minimal sum," Steps 2 and 3, respectively, result. Also notice that only a rectangular loop which consists of 2^i 1-cells can denote a product of literals, as explained in Section 3.3.)

When we derive irredundant disjunctive forms or minimal sums by Procedure 4.2.1, the following property is useful. When we find all prime-implicant loops by Step 1, some 1-cells may be contained in only one loop. Such 1-cells are called **distinguished 1-cells** and are labeled with asterisks. (The 1-cells shown with asterisks in Fig. 4.2.1a are distinguished 1-cells.) A prime-implicant loop that contains distinguished 1-cells is called an **essential**

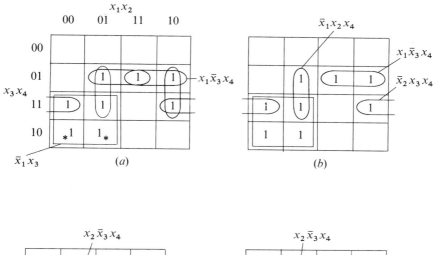
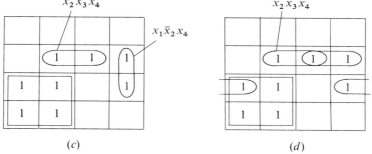
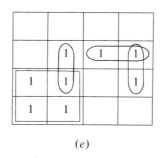

Fig. 4.2.1 The function in Fig. 4.2.1a has four sets of prime-implicant loops representing irredundant disjunctive forms in Fig. 4.2.1b through 4.2.1e.

142 *Simplification of Switching Expressions*

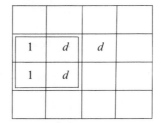

Fig. 4.2.2   Map with don't-care conditions.

prime-implicant loop. The corresponding prime implicant is an essential prime implicant, as defined in Section 4.1. In every irredundant disjunctive form and every minimal sum to be found in Steps 2 and 3, respectively, essential prime implicants must be included, since each 1-cell on the map must be contained in at least one prime-implicant loop and distinguished 1-cells can be contained only in essential prime-implicant loops. Hence, if essential prime-implicant loops are first identified and chosen, Procedure 4.2.1 is quickly processed. In Figs. 4.2.1b through 4.2.1e, the prime-implicant loop that contains the asterisks in Fig. 4.2.1a is included.

Even if the don't-care condition d is contained in some cells, prime implicants can be formed in the same manner, by simply regarding d as being 1 or 0 whenever necessary to draw a greater prime-implicant loop. For example, in Fig. 4.2.2 we can draw a greater rectangular loop by regarding two d's as being 1. One d is left outside and is regarded as being 0. We need not consider loops consisting of d's only.

Maps for Five and Six Variables

The Karnaugh map is most useful for functions of four or fewer variables, but it often serves also for functions of five or six variables. A map for five variables consists of two four-variable maps, as shown in Fig. 4.2.3, one for each value of the first variable. A map for six variables consists of four four-variable maps, as shown in Fig. 4.2.4, one for each combination of values of the first two variables. Note that **the four maps in Fig. 4.2.4 are arranged so that binary numbers represented by x_1 and x_2 differ in only one bit horizontally and vertically** (the map for $x_1 = x_2 = 1$ goes to the bottom right-hand side).

In a five-variable map, 1-cells are combined in the same way as in the four-variable case, with the additional feature that rectangular loops of 2^i 1-cells which are on different four-variable maps can be combined to form a greater loop replacing the two original loops only if they occupy the same relative position on their respective four-variable maps. Notice that these

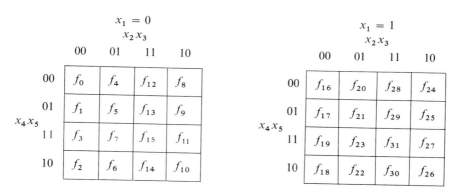

Fig. 4.2.3 Karnaugh map for five variables.

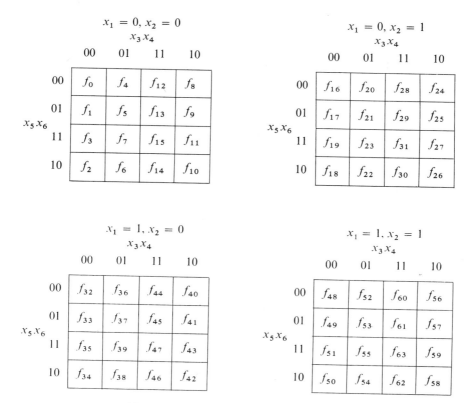

Fig. 4.2.4 Karnaugh map for six variables.

144 *Simplification of Switching Expressions*

loops may be inside other loops in each four-variable map. For example, if f_{15} and f_{31} are 1, they can be combined, but even if $f_{15} = f_{29} = 1$, f_{15} and f_{29} cannot. In a six-variable map, only 1-cells in two maps which are horizontally or vertically adjacent can be combined. In Fig. 4.2.4, for example, if f_5 and f_{37} are 1, they can be combined; but even if f_5 and f_{53} are 1, f_5 and f_{53} cannot. Also, four 1-cells that occupy the same relative position in all four four-variable maps can be combined as representing a single product. For example, $f_5, f_{21}, f_{37}, f_{53}$ can be combined if they are 1.

In the case of a five-variable map, we can find prime-implicant loops as follows.

Procedure 4.2.2: Derivation of Minimal Sums on a Five-Variable Map

1. Unlike Procedure 4.2.1 for a function of four variables, this step requires the following two substeps to form prime-implicant loops:

   (a) On each four-variable map, encircle all the 1-cells with rectangles, each of which consists of 2^i 1-cells, choosing the number of 1-cells contained in each rectangle as large as possible. Unlike the case of Procedure 4.2.1, **these loops are not necessarily prime-implicant loops** because they may not represent prime implicants, depending on the outcome of Substep (b).

   In Figs. 4.2.5 and 4.2.6, for example, loops formed in this manner are shown with solid lines.

   (b) On each four-variable map, encircle all the 1-cells with rectangles, each of which consists of 2^i 1-cells in exactly the same relative position on the two maps, choosing i as great as possible. Then connect each pair of the corresponding loops with an arc. [On each four-variable map, some of these loops may be inside some loops formed in Substep (a).]

   In Fig. 4.2.5 one pair of loops is newly formed. One member of the pair, shown in a dotted line, is contained inside a loop formed in Substep (a). The pair is connected with an arc. The other loop coincides with a loop formed in Substep (a). In Fig. 4.2.6 two pairs of loops are formed; one pair is newly formed, as shown in dotted lines, and the second pair is the connection of loops formed in Substep (a).

   The loops formed in Substep (b) and also those formed in Substep (a) but not contained in any loop formed in Substep (b) are **prime-implicant loops** since they correspond to prime implicants.

Derivation of Minimal Sums by Karnaugh Map 145

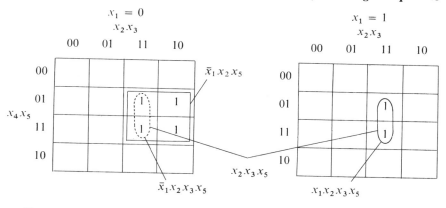

Fig. 4.2.5 *Prime-implicant loops on a map for five variables. (Prime implicants of this function are $\bar{x}_1 x_2 x_5$ and $x_2 x_3 x_5$.)*

In Fig. 4.2.5 the loop formed in Substep (a), which represents $x_1 x_2 x_3 x_5$, is contained in the prime-implicant loop formed in Substep (b), which represents $x_2 x_3 x_5$. Thus that loop is not a prime-implicant loop, and consequently $x_1 x_2 x_3 x_5$ is not a prime implicant.

2. Processes for deriving irredundant disjunctive forms and minimal sums are the same as Steps 2 and 3 of Procedure 4.2.1. □

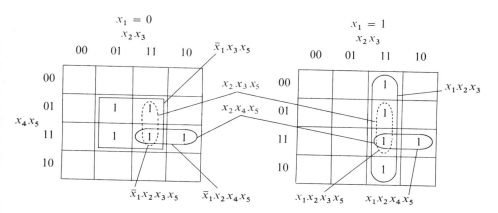

Fig. 4.2.6 *Prime-implicant loops on a map for five variables. (Prime implicants of this function are $\bar{x}_1 x_3 x_5$, $x_2 x_3 x_5$, $x_2 x_4 x_5$, and $x_1 x_2 x_3$.)*

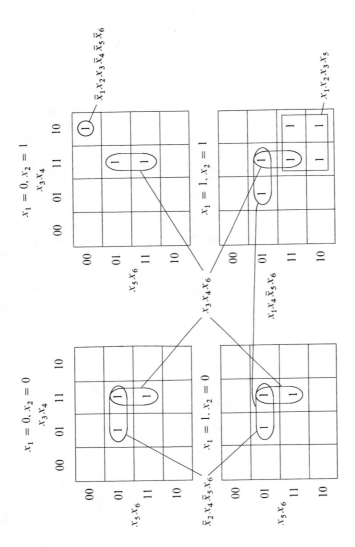

Fig. 4.2.7 *Prime-implicant loops on a map for six variables (Prime implicants of this function are $\bar{x}_2 x_4 \bar{x}_5 x_6$, $\bar{x}_1 x_2 x_3 \bar{x}_4 \bar{x}_5 \bar{x}_6$, $x_1 x_4 \bar{x}_5 x_6$, $x_1 x_2 x_3 x_5$, and $x_3 x_4 x_6$.)*

Prime Implicates, Irredundant Conjunctive Forms 147

In the case of a six-variable map, the derivation of prime-implicant loops requires three substeps, as follows.

(a) On each four-variable map, encircle all the 1-cells with rectangles, each of which consists of 2^i 1-cells, choosing i as great as possible.
(b) Find the rectangles (each of which consists of 2^i 1-cells) occupying the same relative position on every two adjacent four-variable maps, choosing i as great as possible. (Two maps in diagonal positions are not adjacent.) Thus we need four comparisons of two maps (i.e., upper two maps, lower two maps, left two maps, and right two maps).
(c) Then find the rectangles (each of which consists of 2^i 1-cells) occupying the same relative position on all four-variable maps, choosing i as great as possible. **Prime-implicant loops** are loops formed at substeps (c), loops formed at (b) but not contained inside those at (c), and loops formed at (a) but not contained inside those formed at (b) or (c).

An example is shown in Fig. 4.2.7.

Irredundant disjunctive forms and minimal sums are derived in the same manner as in the case of four variables.

Procedures 4.2.1 and 4.2.2 can be extended to the cases of seven or more variables with increasing complexity. It is usually hard to find a minimal sum, however, since each prime-implicant loop consists of 1-cells scattered in many loops.

4.3 Prime Implicates, Irredundant Conjunctive Forms, and Minimal Products

So far, we have discussed the concepts of implicants, prime implicants, irredundant disjunctive forms, and minimal sums, all based on the concept of the disjunctive form, and have derived a minimal network in two levels that has AND gates in the first level and one OR gate in the second level. In this section, let us define "implicates," "irredundant conjunctive forms," and "minimal products," all based on the concept of conjunctive form, and then derive a minimal network that has OR gates in the first level and one AND gate in the second level.

First, let us represent the maxterm expansion of a given function f on a Karnaugh map. Unlike the map representation of the minterm expansion of f, where each minterm contained in the expansion is represented by a 1-cell on a Karnaugh map, each maxterm is represented by a 0-cell. Suppose that a function f is given as shown by the 1-cells in Fig. 4.3.1a. This function can be expressed in the maxterm expansion:

$$f = (\bar{x}_1 \vee x_2 \vee x_3 \vee x_4)(\bar{x}_1 \vee x_2 \vee x_3 \vee \bar{x}_4)(\bar{x}_1 \vee \bar{x}_2 \vee \bar{x}_3 \vee x_4)$$
$$(x_1 \vee \bar{x}_2 \vee x_3 \vee x_4)(x_1 \vee \bar{x}_2 \vee x_3 \vee \bar{x}_4)(x_1 \vee \bar{x}_2 \vee \bar{x}_3 \vee \bar{x}_4)$$
$$(x_1 \vee x_2 \vee x_3 \vee \bar{x}_4)(x_1 \vee x_2 \vee \bar{x}_3 \vee \bar{x}_4).$$

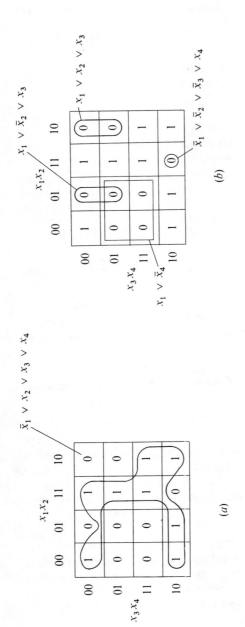

Fig. 4.3.1 *Representation of a function f with alterms on a map.* (a) f. (b) *Alterms.*

Prime Implicates, Irredundant Conjunctive Forms 149

The first maxterm, $(\bar{x}_1 \vee x_2 \vee x_3 \vee x_4)$, for example, is represented by the 0-cell that has coordinates $(x_1, x_2, x_3, x_4) = (1\,0\,0\,0)$ in the Karnaugh map in Fig. 4.3.1a. Notice that each literal in the maxterm is complemented or noncomplemented, corresponding to 1 or 0 in the corresponding coordinate, respectively, instead of corresponding to 0 or 1. All other maxterms are similarly represented by 0-cells. It may look somewhat strange that these 0-cells represent the f expressed by 1-cells on the Karnaugh map. But the first maxterm, $(\bar{x}_1 \vee x_2 \vee x_3 \vee x_4)$, for example, assumes the value 0 only for $(x_1, x_2, x_3, x_4) = (1\,0\,0\,0)$ and assumes the value 1 for all other combinations of the coordinates. The situation is similar with other maxterms. Therefore the conjunction of these maxterms becomes 0 only when any of the maxterms becomes 0. Thus these 0-cells represent f through the corresponding maxterms. This is what we discussed about the maxterm expansion in Section 3.2 or the Π-decimal specification of f in Section 1.4, based on the false vectors of f.

Like the representation of a product on a Karnaugh map in Figs. 3.3.4 and 3.3.5, any alterm can be represented by a rectangular loop of 2^i 0-cells by repeatedly combining a pair of 0-cell loops that are horizontally or vertically adjacent in the same rows or columns, where i is a nonnegative integer. For example, the two adjacent 0-cells in the same column representing maxterms $(\bar{x}_1 \vee x_2 \vee x_3 \vee x_4)$ and $(\bar{x}_1 \vee x_2 \vee x_3 \vee \bar{x}_4)$ can be combined to form alterm $\bar{x}_1 \vee x_2 \vee x_3$, by deleting literals x_4 and $\bar{x}_4$, as shown in a solid-line loop in Fig. 4.3.1b.

The function f in Fig. 4.3.1 can be expressed in the conjunctive form $f = (\bar{x}_1 \vee x_2 \vee x_3)(x_1 \vee \bar{x}_2 \vee x_3)(x_1 \vee \bar{x}_4)(\bar{x}_1 \vee \bar{x}_2 \vee \bar{x}_3 \vee x_4)$, using a minimum number of such loops. The alterms in this expansion are represented by the loops shown in Fig. 4.3.1b.

Definition 4.3.1: **An implicate** of f is an alterm implied by a function f.

Notice that an implicate's relationship with f is opposite to that of an implicant that implies f.

For example, $(x_1 \vee x_2 \vee x_3)$ and $(x_1 \vee x_2 \vee \bar{x}_3)$ are implicates of function $x_1 \vee x_2$, since whenever $x_1 \vee x_2 = 1$, both $x_1 \vee x_2 \vee x_3$ and $x_1 \vee x_2 \vee \bar{x}_3$ become 1. The implication relationship between an alterm P and a function f can sometimes be more conveniently found, however, by using the property "f implies P if $f = 0$ whenever $P = 0$," which is a restatement of the property "f implies P if $P = 1$ whenever $f = 1$" in Definition 3.4.1. Thus $(x_1 \vee x_3), (x_1 \vee \bar{x}_2)$, and $(x_1 \vee x_2 \vee x_3)$ are implicates of $f = (x_1 \vee \bar{x}_2)(x_2 \vee x_3)$, because when $(x_1 \vee x_3), (x_1 \vee \bar{x}_2)$, or $(x_1 \vee x_2 \vee x_3)$ is $0, f$ is 0. [For example, when $(x_1 \vee x_3)$ is $0, x_1 = x_3 = 0$ must hold. Thus $f = (\bar{x}_2)(x_2) = 0$. Consequently, $(x_1 \vee x_3)$ is an alterm implied by f,

150 *Simplification of Switching Expressions*

though this is not obvious from the given expressions of f and $(x_1 \vee x_3)$.] Also, $(x_1 \vee \bar{x}_2 \vee x_4)$ and $(x_1 \vee x_2 \vee x_3 \vee \bar{x}_4)$, which contain the literals of a dummy vairable x_4 of this f, are implicates of f.

Definition 4.3.2: An alterm P is said to **subsume** another alterm Q if all the literals in Q are among the literals of P.

The alterm $(x_1 \vee \bar{x}_2 \vee x_3 \vee \bar{x}_4)$ subsumes $\bar{x}_2 \vee x_3$. Summarizing the two definitions of "subsume" for "term" in Definition 3.4.3 and "alterm" in Definition 4.3.2, we have that **"P subsumes Q" simply means "P contains all the literals in Q," regardless of whether P and Q are terms or alterms**. But, as illustrated in Table 4.3.1, the relationships between "subsume" and "imply" in the two cases are opposite. If an alterm P subsumes another alterm Q, then $Q \subseteq P$ holds, whereas if a term P subsumes another term Q, then $P \subseteq Q$ holds.

Definition 4.3.3: A **prime implicate** of a function f is defined as an implicate of f such that no other alterm subsumed by it is an implicate of f.

In other words, if deletion of any literal from an implicate of f makes the remainder not an implicate of f, the implicate is a prime implicate. For example, $x_2 \vee x_3$ and $x_1 \vee x_3$ are prime implicates of $f = (x_1 \vee \bar{x}_2)(x_2 \vee x_3)(\bar{x}_1 \vee x_2 \vee x_3)$, but $x_1 \vee x_3 \vee x_4$ is not a prime implicate of f, though it is still an implicate of this f. As seen from these examples, some implicates are not obvious from a given conjunctive form of f. [Such an example is $x_1 \vee x_3$ in the above example. As a matter of fact, $x_1 \vee x_3$ can be obtained as the consensus (of the following Definition 4.3.4) of two alterms,

Table 4.3.1 Relationship between "Subsume" and "Imply"

TERMS	ALTERMS
P subsumes Q	P subsumes Q
$\Updownarrow$	$\Updownarrow$
$P \subseteq Q$ (i.e., P implies Q) e.g., $x_1 \bar{x}_2 \bar{x}_3$ $\begin{Bmatrix}\text{implies} \\ \text{subsumes}\end{Bmatrix}$ $x_1 \bar{x}_3$	$Q \subseteq P$ (i.e., Q implies P) e.g., $(x_1 \vee \bar{x}_2 \vee \bar{x}_3)$ subsumes $(x_1 \vee \bar{x}_3)$ and $(x_1 \vee \bar{x}_3)$ implies $(x_1 \vee \bar{x}_2 \vee \bar{x}_3)$

$(x_1 \vee \bar{x}_2)$ and $(x_2 \vee x_3)$, of f.] Also notice that, unlike implicates, prime implicates cannot contain a literal of any dummy variable of f.

On a Karnaugh map **a loop for an alterm P that subsumes another alterm Q is contained in the loop for Q.** For example, in Fig. 4.3.2 the dotted-line loop for alterm $(x_1 \vee \bar{x}_3 \vee x_4)$, which subsumes alterm $(x_1 \vee \bar{x}_3)$, is contained in the solid-line loop for $x_1 \vee \bar{x}_3$. Thus **a rectangular loop that consists of 2^i 0-cells, with i as large as possible, represents a prime implicate of f.**

The consensus of two **alterms**, V and W, is defined in a manner similar to Definition 3.4.5.

Definition 4.3.4: If there is exactly one variable, say x, appearing noncomplemented in one alterm and complemented in the other (i.e., if two alterms, V and W, can be written as $V = x \vee V'$ and $W = \bar{x} \vee W'$), where V' and W' are alterms free of literals of x, the disjunction of all literals except those of x (i.e., $V' \vee W'$ with duplicate literals deleted) is called the **consensus** of the two alterms V and W.

For example, when $V = x \vee y \vee \bar{z} \vee u$ and $W = \bar{x} \vee y \vee u \vee \bar{v}$ are given, their consensus is $y \vee \bar{z} \vee u \vee \bar{v}$.

It would be desirable to call the consensus of two alterms by some other name, in order to differentiate it from the consensus of two products, given in Definition 3.4.5. Unfortunately, both are called "consensus." From the context we must decide which definition—3.4.5 or 4.3.4—the term "concensus" refers to.

On a Karnaugh map a consensus is represented by the largest rectangular loop of 2^i 0-cells that intersects, and is contained within, two adjacent loops of 0-cells that represent two alterms. For example, in Fig.

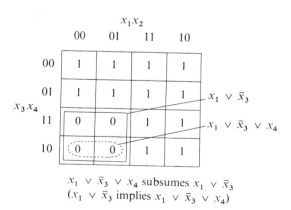

Fig. 4.3.2 *An alterm that subsumes another alterm.*

152 Simplification of Switching Expressions

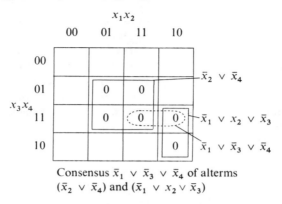

Consensus $\bar{x}_1 \vee \bar{x}_3 \vee \bar{x}_4$ of alterms
$(\bar{x}_2 \vee \bar{x}_4)$ and $(\bar{x}_1 \vee x_2 \vee \bar{x}_3)$

Fig. 4.3.3 *Consensus of two adjacent alterms.*

4.3.3 the dotted-line loop represents the consensus of two alterms, $(\bar{x}_2 \vee \bar{x}_4)$ and $(\bar{x}_1 \vee x_2 \vee \bar{x}_3)$, which are represented by the two adjacent loops.

All prime implicates of a function f can be algebraically obtained from a conjunctive form for f by modifying the iterated-consensus method (Procedure 3.4.1) or the Tison method (Procedure 3.4.2), that is, by using dual operations in these methods. (This is left to readers as Exercise 4.4.13.)

We can define the following concepts, which are dual to irredundant disjunctive forms, sloppy minimal sums, minimal sums, essential prime implicants, complete sums, and others.

Definition 4.3.5: **An irredundant conjunctive form** for a function f is a conjunction of prime implicates such that removal of any of them makes the remainder not express f. The **sloppy minimal products** are irredundant conjunctive forms for f with a minimum number of prime implicates. The **minimal products** (or **minimal conjunctive forms**) are sloppy minimal products with a minimum number of literals. Prime implicates that appear in every irredundant conjunctive form for f are called **essential prime implicates** of f. **Conditionally eliminable prime implicates** are prime implicates that appear in some irredundant conjunctive forms for f, but not in others. **Absolutely eliminable prime implicates** are prime implicates that do not appear in any irredundant conjunctive form for f. The **complete product** for a function f is the product of all prime implicates of f.

Derivation of Minimal Products by Karnaugh Map

Minimal products can be derived by the following procedure, based on a Karnaugh map.

Procedure 4.3.1: Derivation of Minimal Products on a Karnaugh Map

Consider the case of a map for four or fewer variables (cases for five or more variables are similar, using more than one four-variable map).

1. Encircle **0-cells**, instead of 1-cells, with rectangular loops, each of which consists of 2^i 0-cells, choosing i as large as possible. These loops are called **prime-implicate loops** because they represent prime implicates (not prime implicants). Examples of prime-implicate loops are shown in Fig. 4.3.4 (including the dotted-line loop). The **prime implicate** corresponding to a loop is formed by making a **disjunction** of literals, instead of a conjunction, using a noncomplemented variable corresponding to 0 of a coordinate of the map and a complemented variable* corresponding to 1. Thus, corresponding to the loops of Fig. 4.3.4, we get the prime implicates

   $(x_1 \vee \bar{x}_4)$, $(\bar{x}_1 \vee \bar{x}_2 \vee \bar{x}_3)$,   and $(\bar{x}_2 \vee \bar{x}_3 \vee \bar{x}_4)$.

2. Each **irredundant conjunctive form** is derived by the conjunction of prime implicates corresponding to a set of loops, so that removal of any loop leaves some 0-cells uncovered by loops. An example is the set of two solid-line loops in Fig. 4.3.4, from which the irredundant conjunctive form $(x_1 \vee \bar{x}_4)(\bar{x}_1 \vee \bar{x}_2 \vee \bar{x}_3)$ is derived.

3. Among sets of a minimum number of prime-implicate loops, the sets that contain as many of the largest loops as possible yield **minimal products**.

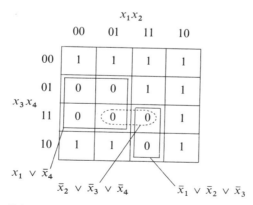

Fig. 4.3.4 Prime-implicate loops and the corresponding prime implicates.

---

* Recall that, in forming a prime implicant, variables corresponding to 0's were complemented and those corresponding to 1's were noncomplemented.

154 *Simplification of Switching Expressions*

The don't-care conditions are dealt with in the same manner as in the case of minimal sums. In other words, whenever possible, we can form a larger prime-implicate loop by interpreting some d's as 0-cells. Any prime-implicate loops consisting of only d-cells need not be formed. □

From the above illustration of algebraic concepts such as prime implicates on a Karnaugh map, it is obvious why Procedure 4.3.1 yields all prime implicates of a given function f.

Procedure 4.3.1 can be extended to five or more variables in the same manner as Procedure 4.2.1 was extended in Section 4.2, though the procedure will be increasingly complex.

4.4 Design of Two-Level Minimal Networks with AND and OR Gates

On the basis of a minimal product for a given network, we can design a network that has OR gates in the first level and one AND gate in the second level, with a minimum number of gates as the primary objective and a minimum number of connections as the secondary objective. (This can be proved in the same manner as Theorems 4.1.1, 4.1.2, and 4.1.3.) This design procedure is meaningful under the following assumption.

> *Assumption 4.4.1:* This is the same as Assumption 4.1.1 except for OR gates in the first level and one AND gate in the output level. (In Assumption 4.1.1 we have AND gates in the first level and one OR gate in the second level.)

Suppose that we want to obtain a two-level network with a minimum number of gates and then, as a secondary objective, a minimum number of connections under Assumption 4.1.1 or 4.4.1, regardless of whether we have AND and OR gates, respectively, in the first and second levels, or in the second and first levels. In this case we have to design a network based on the minimal sum and another based on the minimal product, and then choose the better network. Suppose that we want to design a two-level AND/OR network for the function shown in Fig. 4.4.1a. This function has only one minimal sum, as shown with loops in Fig. 4.4.1a. Also, it has only one minimal product, as shown in Fig. 4.4.2a. The network in Fig. 4.4.2b, based on this minimal product, requires one less gate, despite more loops, than the network based on the minimal sum in Fig. 4.4.1b, and consequently the network in Fig. 4.4.2b is preferable.

Design of Two-Level Minimal Networks with AND and OR Gates 155

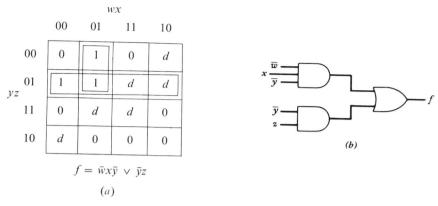

Fig. 4.4.1 *Minimal sum and the corresponding network.* **(a)** *Minimal sum.* **(b)** *Network based on the minimal sum in Fig. 4.4.1a.*

Karnaugh maps have been widely used, often with some modifications (Remark 4.4.1), because of convenience.

Remark 4.4.1: With slight modifications, Karnaugh maps can be conveniently used for cases other than those discussed so far. One such example is a variable-entered Karnaugh map, as explained in Exercise 4.4.20. This map can express functions of more variables than those we have discussed. Another scheme for representing a five- or six-variable function on a four-variable Karnaugh map is described in Vikas [1975].

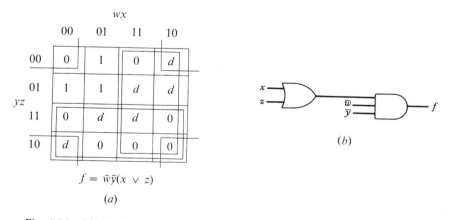

Fig. 4.4.2 *Minimal product and the corresponding network.* **(a)** *Minimal product.* **(b)** *Network based on the minimal product in Fig. 4.4.2a.*

156 Simplification of Switching Expressions

Other examples of modification of the Karnaugh map are as follows. Minimization of the number of inputs to gates, regardless of the number of gates or the number of levels, based on a Karnaugh map is discussed in Levine [1967]. Derivation of the conjunction of disjunctive forms such as $(\bar{x}_1\bar{x}_2 \vee x_1x_2)$ $(\bar{x}_3x_4 \vee x_3\bar{x}_4)$ on a map is discussed in Karnaugh [1953] (notice that $(\bar{x}_1\bar{x}_2 \vee x_1x_2)$ is not an alterm). Functions up through nine variables can be handled on a map [Jesse 1972]. Also, a network with threshold gates (to be discussed in Chapter 5), each of which has three inputs with equal weights, can be conveniently designed on Karnaugh maps [Miiler and Winder 1962].

When there are many prime-implicant loops on a map (or maps), it is easy to overlook redundant loops. A simple technique to prevent this by entering letters for prime implicants into cells is discussed in Marcovitz and Shub [1969].

Karnaugh maps are conveniently used to find raceless state assignments for sequential networks in Chapter 7. ∎

Exercises

4.4.1 (M+) Prove that a prime implicant P of a function f is an essential prime implicant if and only if P cannot be derived when the iterated consensus method (Procedure 3.4.1) is applied to any disjunction of other prime implicants.

4.4.2 (R) For each function in one of the following pairs of functions in A, B, and C, use the Karnaugh map method:

(i) Show the complete sum, underlining the essential prime implicants.

(ii) Identify at least two prime implicants each of which is the consensus of other prime implicants, if possible.

(iii) Show one minimal sum on a map.

(iv) Determine the number of different minimal sums.

(v) If there are any irredundant disjunctive forms that are not minimal sums, show one of them on a map.

A. $f_1(w, x, y, z) = \Sigma(0, 1, 4, 5, 6, 7, 9, 10, 13, 14, 15)$,
$f_2(w, x, y, z) = \Sigma(0, 2, 3, 4, 5, 6, 7, 8, 9, 10, 11, 13, 14, 15)$.

B. $f_3(w, x, y, z) = \Sigma(3, 5, 6, 7, 9, 10, 11, 12, 13, 14, 15)$,
$f_4(w, x, y, z) = \Sigma(0, 1, 2, 3, 5, 6, 7, 8, 9, 10, 12, 13, 14, 15)$.

C. $f_5(w, x, y, z) = \Sigma(0, 1, 2, 3, 5, 6, 8, 9, 10, 12, 13, 14, 15)$.
$f_6(w, x, y, z) = \Sigma(2, 4, 5, 6, 8, 9, 10, 13, 14)$.

4.4.3 (R) For one of the following functions in A and B, do the following:

(i) Express the function on a Karnaugh map, encircling the prime implicants.

(ii) List all prime implicants, underlining essential prime implicants.

A. $f(v, w, x, y, z) = \Pi(3, 5, 8, 9, 12, 13, 14, 15, 16, 17, 20, 21, 22, 23, 24, 25, 26, 28, 29, 30, 31)$.

B. $f(v, w, x, y, z) = \Pi(2, 5, 6, 7, 13, 15, 16, 18, 20, 21, 22, 24, 26, 28, 29)$.

4.4.4 (R) For each function in one of the following pairs of functions in A, B, and C, use the Karnaugh map method:

Design of Two-Level Minimal Networks with AND and OR Gates   157

(i) Show the complete product, underlining the essential prime implicates.

(ii) Identify at least two prime implicates each of which is the consensus of other prime implicates, if possible.

(iii) Show one minimal product on a map.

(iv) Determine the number of different minimal products.

(v) If there are any irredundant conjunctive forms that are not minimal products, show one of them on a map.

A. $f_1(w, x, y, z) = \Sigma(2, 3, 8, 10, 11, 12)$,
$f_2(w, x, y, z) = \Sigma(2, 3, 8, 12, 14)$.

B. $f_3(w, x, y, z) = \Sigma(0, 1, 2, 4, 8)$,
$f_4(w, x, y, z) = \Sigma(2, 3, 8, 11, 12)$.

C. $f_5(w, x, y, z) = \Sigma(3, 4, 7, 11)$,
$f_6(w, x, y, z) = \Sigma(6, 11)$.

4.4.5 (E)   For one of the following functions in A and B, do the following:

(i) Express the function on a Karnaugh map, encircling the prime implicates.

(ii) List all prime implicates, underlining essential prime implicates.

A. $f(v, w, x, y, z) = \Pi(3, 5, 8, 9, 12, 13, 14, 15, 16, 17, 20, 21, 22, 23, 24, 25, 26, 28, 29, 30, 31)$.

B. $f(v, w, x, y, z) = \Pi(2, 5, 6, 7, 13, 15, 16, 18, 20, 21, 22, 24, 26, 28, 29)$.

4.4.6 (E)   Identify all absolutely eliminable prime implicants and all absolutely eliminable prime implicates for each of the following functions, using a Karnaugh map:

(i) $f_1(w, x, y, z) = \Sigma(1, 5, 6, 7, 8, 9, 12, 13, 14, 15)$.

(ii) $f_2(w, x, y, z) = \Sigma(1, 3, 4, 12)$.

4.4.7 (E)   Is there a Karnaugh map which contains don't-care conditions and in which at least one loop of a minimal sum intersects at least one loop of a minimal product? If so, show an example.

4.4.8 (M−)   Find a completely specified function of no more than four variables that has more prime implicants than minterms.

4.4.9 (E)   When a function f is given in an arbirtary form with parentheses other than a disjunctive or conjunctive form, the following approach to the preparation of a Karnaugh map may sometimes be more convenient than rewriting f into the minterm expansion or calculating the value of f for each combination of values of the variables.

Suppose that f is given as

$$f = g_1 \cdot g_2 \vee g_3,$$

for example, using subfunctions $g_1, g_2,$ and g_3. Then prepare a Karnaugh map for each of $g_1, g_2,$ and g_3, as shown in Fig. E4.4.9a.

Draw another Karnaugh map. In each cell of this map, calculate the value of f according to the given expression $f = g_1 \cdot g_2 \vee g_3$. You have now obtained the Karnaugh map for f.

158 Simplification of Switching Expressions

g_1

	1		
	1		
		1	

g_2

	1	1	
	1		
		1	

g_3

		1	
	1		1

Fig. E4.4.9a.

In practice, the above procedure can be performed on only one map, as shown in Fig. E4.4.9b, skipping the maps for subfunctions [Smother 1970].

f

	$1 \cdot 1 \vee 0 = 1$	$0 \cdot 1 \vee 0 = 0$	
	$1 \cdot 1 \vee 1 = 1$		
	$0 \cdot 0 \vee 1 = 1$		$1 \cdot 0 \vee 1 = 1$
		$0 \cdot 1 \vee 0 = 0$	

Fig. E4.4.9b.

When g_1, g_2, g_3, and f are given, as shown in Fig. E4.4.9c, derive g_4 in a minimal sum, where

$$f = g_1 \cdot g_2 \vee g_3 \vee g_4.$$

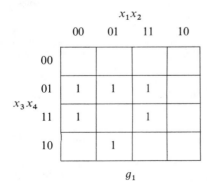
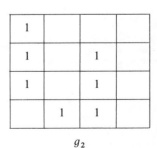

Fig. E4.4.9c.

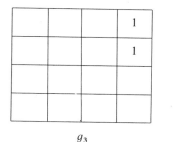

Fig. E4.4.9c. *(Continued)*

4.4.10 (R) **(i)** Check algebraically (without the use of a Karnaugh map) whether each of the following:

$$(x_1 \vee x_4), (\bar{x}_2 \vee x_3), (\bar{x}_1 \vee x_2), (\bar{x}_1 \vee x_3), (x_2 \vee x_3)$$

is a prime implicate (not implicant) of

$$f = (x_2 \vee x_4)(x_1 \vee \bar{x}_2)(\bar{x}_1 \vee x_3).$$

(ii) Show five implicates (not implicants) of the above f that are not prime implicates.

4.4.11 (E) Suppose that two alterms, U and V, are given, and U is free of the variable x. When U does not subsume V but the alterm $x \vee U$ subsumes V, which of the following is true?

(i) V has no literal of x.
(ii) V has the literal x.
(iii) V has the literal $\bar{x}$.

4.4.12 (M−) Prove that, if two alterms, U and V, have a consensus R, the conjunction UV implies R.

4.4.13 (M−) **(i)** Develop an algebraic procedure to obtain all prime implicates from a given function f in a conjunctive form, by modifying Procedure 3.4.1 or 3.4.2. Explain why your procedure works. (Avoid the obvious approach of first converting f into a disjunctive form, and applying Procedure 3.4.1 or 3.4.2 without its modification.)
(ii) Derive all prime implicates for

$$f = (\bar{x}_1 \vee \bar{x}_2)(x_1 \vee x_2 \vee x_3)(\bar{x}_1 \vee x_2 \vee x_3)(x_1 \vee \bar{x}_3)$$

by your procedure. (Notice that this function is dual to that in the example used to illustrate Procedure 3.4.1. This can be seen by comparing your derivation of all prime implicates with the derivation of all prime implicants in that example.)

4.4.14 (M) **(i)** A student tried to derive all the prime implicates for a function f, which is given in a disjunctive form. He developed the following procedure. Prove or disprove it. If it is disproved, develop a correct procedure.

1. Convert the complement $\bar{f}$ of f into a disjunctive form, using De Morgan's theorem.
2. Derive all the prime implicants of $\bar{f}$ by Procedure 3.4.1 or 3.4.2.

160   *Simplification of Switching Expressions*

3. When all prime implicants of $\bar{f}$ are denoted by $P, Q, \ldots, T$, derive alterms by replacing the AND operations by OR operations and complementing all the literals in each of these prime implicants. (For example, if $P = x_1 \bar{x}_2 x_3$, the corresponding alterm is $\bar{x}_1 \vee x_2 \vee \bar{x}_3$.) These alterms are all prime implicates of f.

(ii) Derive all the prime implicates for the following function f, by the above procedure if you proved it, or by your own procedure if you disproved it.

$$f = \bar{x}_1 x_2 \bar{x}_3 \vee x_1 \bar{x}_2 x_3.$$

4.4.15 (R) Design a two-level network with a minimum number of gates for one of the following functions. Use only AND and OR gates, and assume that $\bar{w}, \bar{x}, \bar{y}, \bar{z}$, as well as w, x, y, z, are available at the inputs of the network.

A. $f(w, x, y, z) = \Sigma(3, 5, 6, 9, 12, 13, 14, 15) + d(0, 1, 7)$.
B. $f(w, x, y, z) = \Sigma(1, 3, 4, 7, 11, 13, 14) + d(2, 5, 12, 15)$.
C. $f(w, x, y, z) = \Sigma(3, 6) + d(1, 4, 7, 13)$.

4.4.16 (M−) Fig. E4.4.16 shows an MOS gate that a student designed. Derive an MOS gate realizing the same output function f with fewer MOSFETs. Assume that complemented and noncomplemented variables are available as inputs.

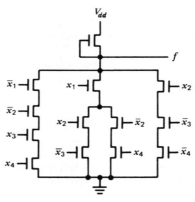

Fig. E4.4.16

4.4.17 (M−) Derive all functions $f(x, y, z)$ that satisfy

$$f = (xy \vee yz)f,$$

using a Karnaugh map, where f is a function of variables x, y, and z only. Explain why they are solutions.

4.4.18 (M) Design a two-level network with a minimum number of AND and OR gates for a function $g(w, x, y, z)$ that makes the function

$$f = \bar{w} x \bar{y} \bar{z} \vee \bar{w} \bar{x} z \vee \bar{w} y z \vee w x \bar{y} z \vee w \bar{x} y \bar{z} \vee g(w, x, y, z)$$

Design of Two-Level Minimal Networks with AND and OR Gates

self-dual, determining g on a Karnaugh map. Assume that complemented and non-complemented variables are available as network inputs. Explain concisely why your approach can yield all solutions.

4.4.19 (M) Suppose that we have a two-level network with AND and OR gates based on a minimal sum for a self-dual function f. Prove that, if every AND gate and OR gate are replaced by an OR gate and AND gate, respectively, this network is a two-level network corresponding to a minimal product for f.

4.4.20 (M−) A **variable-entered Karnaugh map*** is a modification of a Karnaugh map such that the map can express a function of more variables than it can in the ordinary manner discussed in the text. Disjunctive forms of extra variables are entered in some cells in a Karnaugh map, as illustrated in Fig. E4.4.20a. These extra variables are called **entered variables**. (These disjunctive forms should not be complex; otherwise the map becomes too messy.)

$x_3 x_4$ \ $x_1 x_2$	00	01	11	10
00		1	$\bar{x}_7$	1
01	x_8	1	x_8	d
11	1	d	$x_5 x_6 \vee x_8$	1
10	1	1		1

(a)

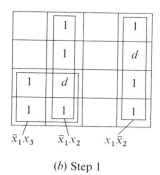

(b) Step 1

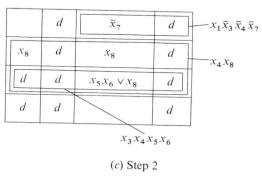

(c) Step 2

Fig. E4.4.20

---

* The idea described here is a slight extension, by L. Dornhoff, of the one in Burgoon [1972] and Clare [1973].

162 Simplification of Switching Expressions

$x_1 x_2$

$x_3 x_4$		00	01	11	10
	00	d	1	1	1
	01		d		1
	11		1		$x_5 x_6$
	10	1	1	$x_5 x_6 \vee x_7$	x_7

(d)

$x_1 x_2$

$x_3 x_4$		00	01	11	10
	00		1	1	1
	01		$\bar{x}_5 x_6 \vee x_7 x_8$	x_5	d
	11	1		x_6	$x_6 \vee x_7 x_8$
	10	1		1	$x_7 x_8$

(e)

Fig. E4.4.20 (Continued)

The switching expression that a variable-entered Karnaugh map expresses can be obtained by the following procedure:

Step 1. Form prime-implicant loops, using only 1-cells and d-cells. In this case, regard all the cells with entered variables as 0-cells.

Choose the minimal number of prime-implicant loops that cover all these 1-cells, using as large loops as possible. Then write the terms corresponding to these loops.

For the variable-entered map in Fig. E4.4.20a, for example, Step 1 is illustrated in Fig. E4.4.20b. Three terms, $\bar{x}_1 x_3$, $\bar{x}_1 x_2$, and $x_1 \bar{x}_2$, have been obtained.

Step 2. For each of the different products of entered variables, form prime-implicant loops by regarding 1-cells and d-cells all as d-cells, and the cells with that product of entered variables as 1-cells. (All other cells are regarded as 0-cells.)

Choose the minimal number of prime-implicant loops that cover all these cells, using as large loops as possible. Then write the terms corresponding to these loops, multiplying each term by that product of entered variables.

Step 2 is illustrated in Fig. E4.4.20c for the map of Fig. E4.4.20a. For the product of entered variables, $x_5 x_6$, we obtain the prime-implicant loop representing $x_3 x_4$.

By multiplying $x_5 x_6$, we obtain $x_3 x_4 x_5 x_6$. For two other products of entered variables we similarly obtain products. Thus, in all, three terms, $x_1 \bar{x}_3 \bar{x}_4 \bar{x}_7$, $x_4 x_8$, and $x_3 x_4 x_5 x_6$, are obtained.

Step 3. The disjunction of the terms obtained in Steps 1 and 2 is the function that the given variable-entered Karnaugh map expresses. Thus Fig. E4.4.20a expresses $\bar{x}_1 x_3 \vee \bar{x}_1 x_2 \vee x_1 \bar{x}_2 \vee x_1 \bar{x}_3 \bar{x}_4 \bar{x}_7 \vee x_4 x_8 \vee x_3 x_4 x_5 x_6$.

With this procedure a variable-entered map can concisely express a function that may require more maps if it is expressed on the ordinary Karnaugh map. (As in the ordinary Karnaugh map, there may be more than one way of forming loops. But when the variable-entered map contains no don't-cares, these different sets of loops express the same function.)

(i) Derive the disjunctive forms that the variable-entered maps in Figs. E4.4.20d and E4.4.20e express.

(ii) Express each of the following functions on a single variable-entered map that has x_1, x_2, x_3, x_4 as its coordinates (like those in Fig. E4.4.20). Treat x_5 through x_8 as entered variables. (Do not enter d in any cell of the map.)

a. $f = \bar{x}_2 x_4 \vee \bar{x}_2 x_5 \vee x_1 x_3 x_6 x_7 \vee x_3 x_4 x_8$.
b. $f = \bar{x}_1 x_3 \vee \bar{x}_2 x_3 x_4 \vee x_5 \vee x_1 \bar{x}_2 x_4 x_6 \bar{x}_7 \vee x_3 x_4 x_8$.

(Note that, if we use the ordinary Karnaugh maps, (a) and (b) in (ii) require 16 maps each.)

4.5 Tabular Method to Derive a Minimal Sum

When we want to derive minimal sums for functions of five or more variables, the Karnaugh map becomes increasingly complicated. Therefore different methods are desirable, such as the tabular methods discussed in this section.

Procedure 4.5.2 is a tabular method to derive a minimal sum when all prime implicants of a given function f are known. The prime implicants of f that are required for Procedure 4.5.2 can be obtained by algebraic Procedure 3.4.1 or 3.4.2; but when f is given in its minterm expansion, tabular Procedure 4.5.1 may be convenient.

Unlike the Karnaugh map, whose complexity suddenly increases beyond four variables, Procedure 4.5.2 has a uniform procedural complexity. Procedures 4.5.1 and 4.5.2 together are often termed the **Quine–McCluskey method** and are easy to understand, explicitly illustrating the basic ideas of derivation of a minimal sum.

Tabular Derivation of All Prime Implicants

Procedure 4.5.1: Derivation of all Prime Implicants

1. Partition all true input vectors of a given function f into groups, such that all input vectors in the same group have the same number of 1's;

164   Simplification of Switching Expressions

place the group with the least number of 1's at the top of the list, followed by groups of increasing numbers of 1's. Then separate adjacent groups by a horizontal line.

The list for $f = \Sigma(0, 2, 5, 6, 7, 8, 9, 10, 13, 15)$, for example, is shown in Table 4.5.1a.

2. Compare each input vector in the first group with the input vectors in the second group, in order to find the input vectors in the second group that differ from the input vector in the first group in only one bit position. Whenever you find such a pair of input vectors, place a check mark after these vectors, and enter the combined vector in the second list, shown in Table 4.5.1b, in the following format: the decimal numbers for the pair of vectors, followed by the common bits of these vectors, along with a dash (—) to replace the different bit. Even though each input vector in the first group may be combined with more than one vector in the second group, only a single check mark needs to be entered for that vector.

For example, vectors 0 and 2 can be combined, since they differ only in their third bit position. Therefore the combined vector is entered in the first row in Table 4.5.1b.

When all vectors in the first group (only one vector in the above example) are compared, draw a horizontal line under the first group in the second list in Table 4.5.1b.

3. Go down to the next group in the first list. In the same manner, compare this group with the next lower group in the first list and enter new combined vectors, if discovered, in the second list. Repeat this process until the group second from the bottom is compared with the bottom group.

Thus the second list is prepared in Table 4.5.1b.

4. Let us work on the second list. In the same manner, compare each group with the next lower group in the list in order to prepare the third list. But the combination rule is somewhat different. Only when the two input vectors in a pair have dashes in the same position, and differ in only one bit position which has no dash, can they be combined; enter the combined vector in the third list, entering a dash in the differing bit position.

For example, the first vector, 0, 2 (0 0 — 0), in the second list (Table 4.5.1b) differs from the sixth vector, 8, 10 (1 0 — 0), only in the first bit position. Therefore 0, 2, 8, 10 (— 0 — 0) is entered in the third list (Table 4.5.1c), entering a dash in the first bit position.

Table 4.5.1 Derivation of All Prime Implicants

a. **First List**

DECIMAL REPRESENTATION OF INPUT VECTOR	INPUT VECTOR				
	w	x	y	z	
0	0	0	0	0	✓
2	0	0	1	0	✓
8	1	0	0	0	✓
5	0	1	0	1	✓
6	0	1	1	0	✓
9	1	0	0	1	✓
10	1	0	1	0	✓
7	0	1	1	1	✓
13	1	1	0	1	✓
15	1	1	1	1	✓

b. **Second List**

	w	x	y	z	
0, 2	0	0	—	0	✓
0, 8	—	0	0	0	✓
2, 6	0	—	1	0	
2, 10	—	0	1	0	✓
8, 9	1	0	0	—	
8, 10	1	0	—	0	✓
5, 7	0	1	—	1	✓
5, 13	—	1	0	1	✓
6, 7	0	1	1	—	
9, 13	1	—	0	1	
7, 15	—	1	1	1	✓
13, 15	1	1	—	1	✓

c. **Third List**

	w	x	y	z
0, 2, 8, 10	—	0	—	0
5, 7, 13, 15	—	1	—	1

d. **Prime Implicants**

2, 6:	(0 — 1 0)	$\bar{w}y\bar{z}$
8, 9:	(1 0 0 —)	$w\bar{x}\bar{y}$
6, 7:	(0 1 1 —)	$\bar{w}xy$
9, 13:	(1 — 0 1)	$w\bar{y}z$
0, 2, 8, 10:	(— 0 — 0)	$\bar{x}\bar{z}$
5, 7, 13, 15:	(— 1 — 1)	xz

5. Continue to prepare new lists until you cannot generate a new list, that is, no pair of vectors can be combined in the last list. (In each new list we need not enter the same combined vectors generated by sets of the same decimal numbers ordered differently. For example, (0, 2, 8, 10) generates the combined vector (— 0 — 0) in Table 4.5.1c. The set (0, 8, 2, 10) also generates this vector but need not be entered. We need to list the combined vector only once, though check marks are placed on the input vectors corresponding to different sets of decimal numbers for the combined vector.)

The unchecked vectors throughout the lists represent the prime implicants of f. In other words, if we form the product corresponding to each vector in the following manner, ignoring dashes, we have a prime implicant for each vector: we form the product of complemented variables corresponding to the 0's and noncomplemented variables corresponding to the 1's.

For the above example, the unchecked vectors are (0 — 1 0), (1 0 0 —), (0 1 1 —), (1 — 0 1), (— 0 — 0) and (— 1 — 1). Accordingly, the prime implicants of f are $\bar{w}y\bar{z}$, $w\bar{x}\bar{y}$, $\bar{w}xy$, $w\bar{y}z$, $\bar{x}\bar{z}$ and xz, as shown in Table 4.5.1d.

Furthermore, **the unchecked vectors show the implication relationships among the minterms and the prime implicants.** For example, the unchecked vector 0, 2, 8, 10 (— 0 — 0) means that each of the minterms corresponding to decimal numbers 0, 2, 8, and 10 implies the prime implicant $\bar{x}\bar{z}$. Thus, by Procedure 4.5.1, we can find not only all prime implicants but also the implication relationships between prime implicants and minterms that are necessary for Procedure 4.5.2. □

As can easily be seen, Procedure 4.5.1 is essentially the iterated-consensus method (Procedure 3.4.1), starting with the minterm expansion of a function (i.e., the combination of vectors is essentially the generation of a consensus from a pair of products, and entering check marks means the elimination of terms that subsume the consensus), though the computational efficiency is improved by partitioning true vectors into groups according to the number of 1's in the vectors.

Procedure 4.5.1 can easily be implemented in computer programs. It is a convenient means of finding all prime implicants when a function is given in a minterm expansion, truth table, or Karnaugh map. But when a function is given in a disjunctive form, it must be converted into its minterm expansion, which usually contains many more terms than the given disjunctive form, if it is to be processed by Procedure 4.5.1. (For other approaches, see Remarks 4.5.1 and 4.5.2.)

Remark 4.5.1: A different procedure, based on true vectors, which is more complex but more efficient than Procedure 4.5.1 is discussed by Morrealle [Morrealle 1967,

Tabular Method to Derive a Minimal Sum 167

1970a, 1970b, 1970c; Ikran and Roy 1976]. It appears also to be more efficient than Procedure 3.4.2 when we need to start with a minterm expansion, particularly if a function is symmetric. ■

Remark 4.5.2: The Tison method (Procedure 3.4.2) can also be processed by using vector expressions in the same manner as Procedure 4.5.1 [e.g., $w\bar{x}z$ is expressed as (1 0 — 1)]. The way of processing these vectors will be different from that of Procedure 4.5.1. ■

Tabular Derivation of Minimal Sums

Compared with the derivation of prime implicants, derivation of minimal sums is usually much more time-consuming.

The following procedure yields only one minimal sum but can be modified so that all irredundant disjunctive forms, all sloppy minimal sums, or all minimal sums can be derived, although usually at the expense of a significant increase in processing time.

Procedure 4.5.2: Derivation of a Minimal Sum

1. **Prepare a prime-implicant table** that has as its coordinates all prime implicants vertically and minterms horizontally, as shown in Table 4.5.2. When we want a minimal sum, we arrange prime implicants into groups such that the prime implicants in the same group have the same number of literals; we place the group with the fewest literals at the top, followed by groups with increasing number of literals, and draw horizontal lines between groups. (This grouping* of the prime implicants will be used only in Step 3.)
   Minterms are expressed by corresponding decimal numbers (e.g., $\bar{x}_1\bar{x}_2\bar{x}_3 x_4 x_5$ is expressed by 3). A cross (×) is entered at the intersection of a row and a column if the prime implicant corresponding to the row is implied by the minterm corresponding to the column (e.g., $A = \bar{x}_2 x_4 x_5$ is implied by minterm 3, i.e., $\bar{x}_1\bar{x}_2\bar{x}_3 x_4 x_5$). In this case the row is said to **cover** the column. Notice that the implication relationship between each prime implicant and a minterm is given by Procedure 4.5.1 as a set of decimal numbers. [For example, 0, 2, 8, 10 (— 0 — 0) in Table 4.5.1d means that prime implicant $\bar{x}\bar{z}$ is implied by each of the minterms with decimal numbers 0, 2, 8, and 10.] Also notice that the implication relationship between prime implicants and minterms can easily be derived even if the prime implicants are obtained by algebraic Procedure 3.4.1 or 3.4.2. [For example, when a prime implicant $w\bar{y}$ of a four-variable function $f(w, x, y, z)$ is obtained, we rewrite $w\bar{y}$ in vector form (1 — 0 —), and derive all the vectors by assigning

---
* If we are not interested in "real" minimal sums but want sloppy minimal sums, this grouping is not necessary. As mentioned in the second footnote in Section 4.1, sloppy minimal sums may be sufficient in some electronic implementations.

168 Simplification of Switching Expressions

0 or 1 to each dash, that is, (1 0 0 0), (1 0 0 1), (1 1 0 0), and (1 1 0 1). The decimal numbers that represent these vectors denote the decimal numbers of all minterms that imply the prime implicant $w\bar{y}$. Repeating this process for every prime implicant, we can derive the implication relationship.]

An irredundant disjunctive form corresponds to a set of rows chosen so that every column has at least one × in the chosen rows and so that, if any row is removed from the set, some column does not have × in the set. In other words, every column must be covered by a set of rows such that removal of any row from the set makes some columns not covered. The problem of choosing such rows, generally known as a **covering problem** (or a set-covering problem or a minimum-covering problem), is often encountered not only in logic design but also in other areas. (The table size reduction techniques to be discussed here are also useful in solving covering problems encountered in areas other than logic design. This is another important motivation for discussing the subject here. See, e.g., Lemke, Salkin, and Spielberg [1971] or Salkin [1975].)

Among such sets of rows, those that have a minimum number of rows and then have as many rows as possible located in higher groups of rows represent minimal sums.

Let us try to derive only one minimal sum, instead of deriving all irredundant disjunctive forms or all minimal sums. If we are satisfied with finding only one minimal sum, the procedure can usually be efficiently processed, as we will see.

First let us reduce the size of the table by the following steps.

Table 4.5.2 Prime Implicant Table

	3	4	5	7	⑩	11	13	15	17	19	20	22	23	㉔	25	28	29
$A = \bar{x}_2 x_4 x_5$	×		×						×			×					
$B = \bar{x}_1 x_4 x_5$	×		×			×		×									
$C = \bar{x}_1 x_3 x_5$				×	×				×	×							
*$D = x_1 x_2 \bar{x}_4$														×	×	×	×
$E = \bar{x}_1 \bar{x}_2 x_3 \bar{x}_4$		×	×														
$F = \bar{x}_2 x_3 \bar{x}_4 \bar{x}_5$	×										×						
$G = x_2 x_3 \bar{x}_4 x_5$							×										×
*$H = \bar{x}_1 x_2 \bar{x}_3 x_4$				×	×												
$I = x_1 x_3 \bar{x}_4 \bar{x}_5$											×					×	
$J = x_1 \bar{x}_2 x_3 \bar{x}_5$											×	×					
$K = x_1 \bar{x}_3 \bar{x}_4 x_5$									×						×		
$L = x_1 \bar{x}_2 \bar{x}_3 x_5$									×	×							
$M = x_1 \bar{x}_2 x_3 x_4$													×	×			

Tabular Method to Derive a Minimal Sum

2. Selection of essential rows: As explained with respect to Fig. 4.2.1a, any column which contains only a single × represents a distinguished minterm, and the row which contains that × represents an essential prime implicant. Such a column and row will be called a **distinguished column** and an **essential row**, respectively. In Table 4.5.2 distinguished columns are encircled. Essential rows are labeled with asterisks as a reminder that these rows must be selected in the final solution. Remove essential rows from the table. Also eliminate all columns that contain ×'s in essential rows, since these columns are covered by the essential rows and henceforth we need not consider them.

In Table 4.5.2, rows D and H are essential rows, so label them with an *. Then delete columns 10, 11, 24, 25, 28, and 29, since they contain ×'s in these rows. Thus Table 4.5.2 is reduced to Table 4.5.3, and henceforth we need to consider only this table.

Table 4.5.3 Table Reduced by Removing Essential Rows and Columns Covered by These Rows

	3	4	5	7	13	15	17	19	20	22	23
A	×		×				×				×
B	×		×		×						
C		×		×	×	×					
E	×	×									
F	×							×			
G				×							
I								×			
J								×	×		
K					×						
L					×	×					
M										×	×

3. Elimination of dominated rows

Definition 4.5.1: A row P is said to **dominate** another row Q in a reduced prime-implicant table if row P has ×'s in all the columns in which row Q has ×'s. (This includes a special case in which two rows have ×'s in exactly the same columns, i.e., the two rows are **equal**.)

If a row R in a reduced table is dominated by another row S of the table and the prime implicant corresponding to R consists of no fewer* literals than the

---

* If we need not find a minimal sum, but want a sloppy minimal sum, dominated rows can be deleted without comparing the number of literals.

prime implicant corresponding to S, row R can be removed without losing the possibility of deriving at least one minimal sum from the remaining table. In other words, if a row R is dominated by another row S that belongs to a group in the same or a higher position in a table, R can be removed.

In Table 4.5.3, for example, row G is dominated by row C, as shown with an arrow. Furthermore, the prime implicant corresponding to row G, that is, $x_2 x_3 \bar{x}_4 x_5$, has no fewer literals than that corresponding to C, that is, $\bar{x}_1 x_3 x_5$ (i.e., C is in a higher group of rows than G, as seen in Table 4.5.3). Thus row G can be eliminated, and we can still derive at least one minimal sum from the remaining table. The reason is that, if a set of rows that contains row G represents a minimal sum, the same set of rows with G replaced by C also represents a minimal sum, since the set still covers all the columns, and the prime implicant corresponding to row C has no more literals than that corresponding to row G. But if the prime implicant corresponding to row C has more literals than that corresponding to row G, this is not true.

Similarly, rows K and I can be eliminated, since they are dominated by L and J, respectively, and all these four rows belong to the same group (i.e., they have an equal number of literals). Thus we have obtained Table 4.5.4.

Table 4.5.4 Table Reduced by Eliminating Dominated Rows

	3	4	5	7	13	15	17	19	20	22	23
A	×			×			×				×
B	×			×		×					
C			×	×	×	×					
E		×	×								
F		×							×		
J									×	×	
L					×		×				
M										×	×

4. Elimination of dominating columns

Definition 4.5.2: A column is said to **dominate** another column if the former has ×'s in all the rows in which the latter has ×'s. (This includes a special case in which two columns have ×'s in exactly the same rows, i.e., the two columns are **equal**.)

In Table 4.5.4 column 7 dominates column 3, as shown with an arrow. Column 3 requires that row A or row B must be selected, while column 7 requires that

Tabular Method to Derive a Minimal Sum   171

row A, row B, or row C must be selected. Then the removal of column 7 has no influence on the final selection of a minimal sum, since, whenever column 3 is covered by some row, column 7 is also covered by the same row.

As seen in this example, **dominating columns can be eliminated from a reduced table without affecting the final selection of a minimal sum.***
Notice that dominating columns can be eliminated here, whereas dominated rows can be eliminated in Step 3.

Similarly, columns 5, 15, and 19 can be eliminated by comparisons with 13, 13, and 17, respectively. Thus we have obtained Table 4.5.5.

Table 4.5.5   Table Reduced by Eliminating Dominating Columns

	3	4	⑬	⑰	20	22	23
A	×						×
B	×						
**C				×			
E		×					
F		×			×		
J					×	×	
**L			×				
M						×	×

5. *Selection of secondary essential rows*: In Table 4.5.5 there are two columns encircled, each of which has only a single ×. The rows that contain these ×'s are called **secondary essential rows** and labeled with double asterisks. These rows become essential ones only after some columns and rows have been eliminated in the preceding steps, but they may not become essential if other columns and rows† are eliminated. Thus they are called "secondary" essential rows. A minimal sum derived from this reduced table must contain prime implicants corresponding to these secondary essential rows, but some other minimal sums may not.

---

* Precisely speaking, even when we try to derive all irredundant disjunctive forms, all sloppy minimal sums, or all minimal sums with Procedure 4.5.2 by skipping Step 3, the elimination of dominating columns does not lose any irredundant disjunctive forms, sloppy minimal sums, or minimal sums.

† When we repeat Steps 3 through 5 in Step 6, we have freedom in choosing a different sequence of these steps, and we may obtain a different solution, depending on what sequence we choose. (Also see Remark 4.5.5.)

Simplification of Switching Expressions

Secondary essential rows must be selected. Remove them from the table. Also eliminate all columns that contain ×'s in these rows.
Thus we have obtained Table 4.5.6.

Table 4.5.6 Table Reduced by Removing Secondary Essential Rows

	3	4	20	22	23
A	×				×
B	×				
E		×			
F		×	×		
J			×	×	
M				×	×

6. *Repetition of Steps 3 through 5 in any order.*

First, apply Step 3 to Table 4.5.6 and eliminate rows B and E, which are dominated by rows A and F for prime implicants consisting of no more literals, respectively. Table 4.5.7 results. From Table 4.5.7, remove secondary essential rows and all the columns which contain ×'s in these rows, to obtain Table 4.5.8. Applying Step 3, eliminate J. Since only row M is left now, it must be selected to cover the remaining column, 22. The procedure terminates, having covered all the columns in the initial Table 4.5.2. During the procedure we selected $D = x_1 x_2 \bar{x}_4$, $H = \bar{x}_1 x_2 \bar{x}_3 x_4$, $C = \bar{x}_1 x_3 x_5$, $L = x_1 \bar{x}_2 \bar{x}_3 x_5$, $A = \bar{x}_2 x_4 x_5$, $F = \bar{x}_2 x_3 \bar{x}_4 \bar{x}_5$, and $M = x_1 \bar{x}_2 x_3 x_4$ by picking up only rows labeled with asterisks. Thus the minimal sum that we have obtained is $f = x_1 x_2 \bar{x}_4 \vee \bar{x}_1 x_2 \bar{x}_3 x_4 \vee \bar{x}_1 x_3 x_5 \vee x_1 \bar{x}_2 \bar{x}_3 x_5 \vee \bar{x}_2 x_4 x_5 \vee \bar{x}_2 x_3 \bar{x}_4 \bar{x}_5 \vee x_1 \bar{x}_2 x_3 x_4$.

Table 4.5.7 Table Reduced by Eliminating Dominated Rows

	③	④	20	22	23
**A	×				×
**F		×	×		
J			×	×	
M				×	×

Table 4.5.8 Last Table Reduced by Removing Secondary Essential Rows

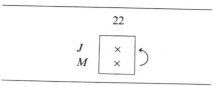

To derive a minimal sum, the steps of elimination and selection described above may be applied in any order. By applying them in an appropriate order, however, we might get the result faster.

7. There are cases in which the above procedure does not eliminate all columns completely. In other words, no more rows or columns can be eliminated from the table by the procedure described. The remaining table, which contains at least two ×'s in each column, is called **cyclic**. As an example, consider Table 4.5.9.

Table 4.5.9 Cyclic Table

	3	5	7	9	11	13
$A = \bar{x}_2 x_3 x_4$	×				×	
$B = \bar{x}_1 x_3 x_4$	×		×			
$C = \bar{x}_1 x_2 x_4$		×	×			
$D = x_2 \bar{x}_3 x_4$		×				×
$E = x_1 \bar{x}_3 x_4$				×		×
$F = x_1 \bar{x}_2 x_4$				×	×	

One method to derive a minimal sum from a cyclic table is the **Petrick method**. For simplicity, assume that we are given a cyclic table in which it is impossible to eliminate rows or columns by Steps 2 through 6 (we will later discuss the case where a cyclic table is encountered after eliminating rows or columns). The Petrick method will generate all irredundant disjunctive forms, and we will choose a minimal sum among them [Petrick 1959]. First, we associate to each row of the prime-implicant table a new switching variable, which is 1 if the associated row is selected in a possible minimal sum, and is 0 otherwise. On the basis of these variables, we want to construct a new switching function, called the **Petrick function** or **the prime-implicant function**, which is 1 only when each column has at least one × in the selected rows (i.e., only when the selected rows cover all the columns, and consequently the disjunction of the selected prime implicants represents f).

174 *Simplification of Switching Expressions*

As an example, let us construct a Petrick function for Table 4.5.9 and denote it by p. Let us use $A, B, \ldots$ as switching variables associated with the rows in this table. Since column 3 of Table 4.5.9 contains ×'s in rows A and B, p must be 0 when rows A and B are not selected, that is, when $A = B = 0$. In other words, $A = 1$ or $B = 1$ is necessary for p to become 1. This condition is realized by setting $p = (A \lor B)q$, where q is a function of the variables of the rows to be determined from consideration of the other columns. Since column 5 has ×'s in rows C and D, p must be written as $p = (A \lor B)(C \lor D)r$ for a similar reason, where r is a function of the variables of the rows to be determined from consideration of the remaining columns. Repeating this process, we obtain $p = (A \lor B)(C \lor D)(B \lor C)(E \lor F)(A \lor F)(D \lor E)$. Another way to see why this p is a Petrick function for Table 4.5.9 is this: only when every column is covered by a selected set of rows, at least one of the literals in each alterm in this p is 1, and, accordingly, the conjunction of these alterms becomes 1.

Next let us **derive the complete sum of the Petrick function** p. This can be done by multiplying out the alterms of p and deleting terms that subsume others:

$$p = (A \lor B)(C \lor D)(B \lor C)(E \lor F)(A \lor F)(D \lor E)$$

(If appropriate pairs of alterms are multiplied out as indicated by the arcs, the computation is simpler, since some terms that subsume others in each product can be deleted.)

$= (A \lor AB \lor AF \lor BF)(C \lor BC \lor CD \lor BD)(E \lor DE \lor EF \lor DF)$

$= (A \lor BF)(C \lor BD)(E \lor DF)$

$= (AC \lor BCF \lor ABD \lor BDF)(E \lor DF)$

$= ACE \lor ACDF \lor BCEF \lor BCDF \lor ABDE \lor ABDF \lor BDEF$

$\lor BDF$

$= ACE \lor BDF$

$\lor ACDF \lor BCEF \lor ABDE.$

Multiplying out the Petrick function expressed in a conjunctive form into the complete sum is generally time-consuming, though the above example is very simple. When any term becomes 1, $p = 1$. This means that each term expresses a set of rows such that omission of any row does not cover all columns, since each term is a prime implicant of p (i.e., if any literal is removed from a prime implicant, the remainder does not imply function p) and p is a function that is 1 only when all columns are covered. In other words, each term corresponds to an irredundant disjunctive form of the function f given in Table 4.5.9. Since we are interested in a minimal sum, **we find only** terms with the fewest literals in the complete sum for p. Then we derive irredundant disjunctive forms for f, from these terms of p. Among them, the irredundant disjunctive forms for f which have the minimum number of literals are minimal sums of f. For the above example, terms with the fewest literals in the complete sum for p are ACE and BDF. The corresponding irredundant disjunctive forms for f are $\bar{x}_2 x_3 x_4 \vee \bar{x}_1 x_2 x_4 \vee x_1 \bar{x}_3 x_4$ and $\bar{x}_1 x_3 x_4 \vee x_2 \bar{x}_3 x_4 \vee x_1 \bar{x}_2 x_4$. Since each of them contains nine literals, they are minimal sums of f.   □

A Petrick function may be derived for an initially given prime-implicant table before going into Steps 2 through 6, or a reduced table, which results after eliminating or selecting rows and columns according to Steps 2 through 6. In the latter case, switching variables corresponding to selected rows must be attached as multiplicands to the Petrick function derived from the reduced table; then at least one of the terms in the complete sum for the resultant Petrick function corresponds to a minimal sum, though possibly not all irredundant disjunctive forms are expressed by the terms in this complete sum. Usually, the latter case yields a minimal sum more quickly if we want to derive only one minimal sum.

Branch-and-Bound Method to Solve a Cyclic Table

A cyclic table can also be solved by the **branch-and-bound method**. Let us work on the cyclic table shown in Table 4.5.10, where each prime implicant consists of the same number of literals. First, consider a row, say A in Table 4.5.10. Here we have the following two cases in which the problem is said to be **branched** based on A:

(a) Suppose that A is chosen in a minimal sum.
Then, by applying Steps 3 through 6 of Procedure 4.5.2, derive a minimal sum.
(b) Suppose that A is not chosen in a minimal sum.
Hence row A is deleted. Then, by applying Steps 3 through 6, derive a minimal sum.

(Notice that these minimal sums may not really be minimal sums for f, since they are "minimal" under the assumption that A is chosen or not chosen, though they are all disjunctive forms for f. Thus later we will examine whether they are really minimal sums for f.)

If we encounter another cyclic table during the execution of (a) or (b), we repeat (a) and (b) by branching based on a new row, instead of A, in the new cyclic table. We repeat this whenever we encounter a cyclic table.

In each case we derive a minimal sum. [In some cases we may be able to detect that any minimal sum which we are trying to derive is going to have

176 *Simplification of Switching Expressions*

Table 4.5.10 Cyclic Table

PRIME IMPLICANT	MINTERM										
	0	2	6	7	10	11	12	13	15	16	20	28
$A = x_1\bar{x}_2 x_3 \bar{x}_5$					×	×						
$B = x_1 x_3 x_4 \bar{x}_5$						×			×			
$C = \bar{x}_2 x_3 \bar{x}_4 \bar{x}_5$		×			×							
$D = \bar{x}_1 \bar{x}_2 \bar{x}_4 \bar{x}_5$	×	×										
$E = \bar{x}_1 x_3 \bar{x}_4 \bar{x}_5$		×	×									
$F = \bar{x}_1 x_2 x_3 \bar{x}_5$			×	×								
$G = x_2 x_3 x_4 \bar{x}_5$				×					×			
$H = x_1 x_2 x_4 \bar{x}_5$								×	×			
$I = x_1 x_2 \bar{x}_3 \bar{x}_5$							×	×				
$J = x_1 x_2 \bar{x}_3 \bar{x}_4$							×					×
$K = x_2 \bar{x}_3 \bar{x}_4 x_5$											×	×
$L = \bar{x}_1 \bar{x}_3 \bar{x}_4 x_5$										×	×	
$M = \bar{x}_1 \bar{x}_2 \bar{x}_3 \bar{x}_4$	×									×		

more terms than some minimal sum obtained in previous cases, for example, by counting the number of (secondary) essential rows. If so, we need not complete the execution to derive that particular minimal sum. If a **bounding operation** like this (other bounding operations are conceivable) can be applied, we can reduce the computation time.] The best results among these minimal sums are some of the minimal sums for f. (The other minimal sums turn out not to be minimal sums of f.)

Table 4.5.11 Reduced Table at Node 2 (A is chosen)

	0	2	6	7	12	13	15	16	20	28
D	×	×								
E		×	×							
F			×	×						
G				×			×			
H						×	×			
I					×	×				
J					×					×
K									×	×
L								×	×	
M	×						×			

Tabular Method to Derive a Minimal Sum 177

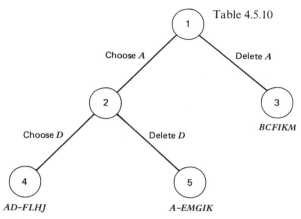

Fig. 4.5.1 *Tree to illustrate the branch-and-bound method.*

This branch-and-bound method is illustrated in Fig. 4.5.1. Table 4.5.10, shown at node 1, is branched to nodes 2 and 3 based on row A. At node 2, after choosing A, we can reduce the table at node 1 to the table shown in Table 4.5.11 by Steps 3 through 6. (Since A is chosen in Table 4.5.10, columns 10 and 11 are covered, so we delete columns 10 and 11. Then B and C can be deleted since they are dominated by G and D, respectively, yielding Table 4.5.11.) Since Table 4.5.11 is cyclic, let us branch based on D to nodes 4 and 5. At node 4 we can delete rows E and M, dominated by F and L, respectively, in Table 4.5.12a, which is obtained by deleting columns 0 and 2, covered by

Table 4.5.12 **Table to be Reduced at Node 4**

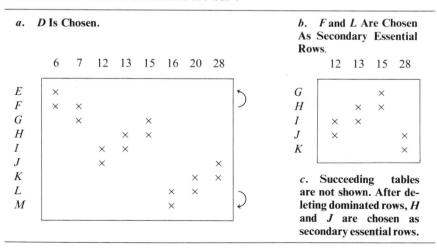

178 *Simplification of Switching Expressions*

the chosen row D in Table 4.5.11. Since rows F and L become secondary essential rows, F and L are chosen, deleting covered columns 6, 7, 16, and 20, and consequently yielding Table 4.5.12b. Continuing to apply Steps 3 through 6 (the corresponding intermediate tables are not shown), we choose rows H and J as secondary essential rows, because of the deletion of the dominated rows G and K. Thus, at node 4, we have obtained the minimal sum AD–$FLHJ$, as shown in Fig. 4.5.1. At node 5, rows E and M are chosen as secondary essential rows in Table 4.5.13a, which is obtained by deleting row D from Table 4.5.11 at node 2. Deleting columns 0, 2, 6, and 16, which are covered by E and M, we obtain Table 4.5.13b. Continuing to apply Steps 3 through 6 (the corresponding intermediate tables are not shown), we choose rows G, I, and K as secondary essential rows. Thus, at node 5, we

Table 4.5.13   Table to Be Reduced at Node 5

a. **D is Deleted.**

	0	2	6	7	12	13	15	16	20	28
E	×	×								
F			×	×						
G			×				×			
H						×	×			
I					×	×				
J					×					×
K									×	×
L								×	×	
M	×						×			

b. **E and M Are Chosen As Secondary Essential Rows.**

	7	12	13	15	20	28
F	×					
G	×			×		
H			×	×		
I		×	×			
J		×				×
K					×	×
L				×		

c. Succeeding tables are not shown. Deleting dominated rows, G, I, and K are chosen as secondary essential rows.

have obtained the minimal sum A–$EMGIK$, as shown in Fig. 4.5.1. At node 3 we can similarly obtain the minimal sum $BCFIKM$, as shown also in Fig. 4.5.1.

Consequently, at these three nodes, we have obtained three sets of rows: $ADFLHJ$, $AEMGIK$, and $BCFIKM$. If we obtain irredundant disjunctive forms corresponding to these, each of them consists of 6 prime implicants and 24 literals (e.g., we get $x_1\bar{x}_2 x_3 \bar{x}_5 \vee \bar{x}_1 \bar{x}_2 \bar{x}_4 \bar{x}_5 \vee \bar{x}_1 x_2 x_3 \bar{x}_5 \vee x_1 x_2 x_4 \bar{x}_5 \vee x_1 x_2 \bar{x}_3 \bar{x}_4 \vee \bar{x}_1 \bar{x}_3 \bar{x}_4 x_5$ corresponding to $ADFLHJ$). Thus all of them are really minimal sums of f.

Network Implementation Based on Procedure 4.5.2

Notice that, in implementing a two-level network of AND and OR gates based on the results of Procedure 4.5.2, prime implicants consisting of single literals can be directly connected to the OR gate in the second level without the use of AND gates. (This is sometimes not permitted, e.g., when certain electronic implementations such as PLAs are used, or when we want to have the same time delay from every variable input to the network output.)

Advantages and Disadvantages of Procedure 4.5.2

Procedure 4.5.2 clearly illustrates important concepts involved in deriving a minimal sum in logic design (and also in solving the covering problem in other areas such as operations research). But it presents two problems. The initial prime-implicant table is often too large to be conveniently handled, because all minterms and prime implicants of a given function must be shown, even when the table can eventually be reduced to a small size. Also, the expansion of a Petrick function into its complete sum is usually very time-consuming.

Modifications of Procedure 4.5.2

If we want to find all minimal sums, we can do so by eliminating dominated rows in Step 3 of Procedure 4.5.2 only when the prime implicants corresponding to the dominated rows have more literals than those corresponding to the dominating rows. If we want to find all irredundant disjunctive forms, we can do so by skipping only Step 3. Usually, however, it is very time-consuming to execute Procedure 4.5.2 without Step 3. If we are interested only in one minimal sum, computation time is greatly saved by the elimination of rows and columns in all the steps, as seen in Procedure 4.5.2. If we are not concerned about the number of connections in a network, the processing can be further accelerated, since the numbers of literals in the prime implicants corresponding to two rows under comparison need not be compared in Step 3 and we can usually eliminate more dominated rows in this step.

180 Simplification of Switching Expressions

Remark 4.5.3: The branch-and-bound method is powerful in solving certain types of problems by computers (not only in logic design but also in other areas). Unlike the situation for many other methods and procedures, the details of the method must be specified, being tailored to the properties of the individual problems to be solved, and in this sense the "branch-and-bound method" is a set of principles for procedures. The computational speed is highly dependent on how the details are implemented. (See, for example, Salkin [1975].) ∎

Remark 4.5.4: There are other methods to derive irredundant disjunctive forms or minimal sums.

J. P. Roth has made many significant contributions to algebraic procedures, including the extraction procedure and the selection of essential prime implicants without generating all prime implicants [Roth 1958, 1959].

Irredundant disjunctive forms or minimal sums can also be obtained from a prime-implicant table by an integer programming approach [Cobham, Fridshal, and North 1961; Grasselli and Luccio 1968].

The semantic tree method [Slagle, Chang, and Lee 1970] can be applied to calculate the Petrick function. If we are interested in deriving only one minimal sum, we halt the execution of the semantic tree method whenever we obtain the first success node. ∎

Remark 4.5.5: The following properties are known:

1. The elimination of rows and/or columns by Steps 2 through 6 of Procedure 4.5.2 yields a unique reduced table, except for the difference in the order and labels of columns and rows, regardless of the order in which these steps are applied [Cobham, Fridshal, and North 1961]. (Here "labels" designates the decimal numbers labeled to the columns and the prime implicants labeled to the rows.)

2. Let A be a matrix, where each element is either 0 or 1. If each column of A has at least one 1 and no two rows of A are identical, there exists a function f whose prime-implicant table can be reduced to A only by Step 2 of Procedure 4.5.2, that is, only by removal of essential rows [Cobham and North 1963]. ∎

4.6 Algebraic Method to Derive Minimal Sums

In this section we describe an algebraic method to efficiently derive minimal sums of a given function f in a concise manner, unlike the previous methods based on a table or map that is cumbersome to draw, and whose size rapidly increases as the number of variables increases. The following algebraic method not only eliminates the large spaces and time-consuming efforts required for tables or maps in Procedure 4.5.2 (though Procedure 4.5.2 is very useful for illustrating basic concepts), but also can be processed by hand in a much shorter time than is needed to derive the complete sum for a Petrick function and choose a minimal sum in Procedure 4.5.2 (as described in textbooks). Some functions that require several hours for the calculation of a

Petrick function by Procedure 4.5.2 (possibly longer, depending on how often errors are made and corrected) can be processed by hand in several minutes by Procedure 4.6.1, to find a minimal sum.

Procedure 4.6.1: Algebraic Derivation of a Minimal Sum

Suppose that all the prime implicants of a given function f have been obtained. When we try to derive a minimal sum, these prime implicants are represented by letters in such a way that letters for prime implicants with fewer literals (i.e., with lower costs) appear before those with more literals in the alphabetical order. (This order of letters* will be used only in Step 3.)

1. **Form the Petrick function for f.** If the prime implicants are obtained by Procedure 4.5.1, we can form the Petrick function directly, without drawing the prime implicant table in Step 1 of Procedure 4.5.2. For example, if the prime implicants in Table 4.5.1d are denoted as $A = \{0, 2, 8, 10\}$, $B = \{5, 7, 13, 15\}$, $C = \{2, 6\}$, $D = \{8, 9\}$, $E = \{6, 7\}$, $F = \{9, 13\}$, we get the Petrick function

$$p = \overset{0}{A}(A \vee \overset{2}{C})B(C \vee \overset{5}{E})(B \vee \overset{6}{E})(A \vee \overset{7}{D})(D \vee \overset{8}{F})A(B \vee \overset{9}{F})B$$

by forming alterms corresponding to minterms 0, 2, 5, 6, 7, 8, 9, 10, 13, and 15 in this order. (When prime implicants are derived on Karnaugh maps, we can obtain a Petrick function in a similar manner, without using a prime-implicant table.)

If the prime implicants are obtained by Procedure 3.4.1 or 3.4.2, we can similarly form the Petrick function directly. For example, suppose that a four-variable function $f(w, x, y, z)$ has a prime implicant $\bar{w}y$. We rewrite it in vector form, that is, $(0 - 1 -)$, where the dash denotes a missing variable in this prime implicant. By assigning 0 or 1 to each dash, we derive all the vectors: (0 0 1 0), (0 0 1 1), (0 1 1 0), (0 1 1 1). All the decimal numbers that these vectors represent, that is, 2, 3, 6, and 7, express the minterms, each of which implies the prime implicant $\bar{w}y$. Thus, by the approach in the preceding paragraph, we can write the Petrick function.

To illustrate our procedure, suppose that the Petrick function

$$p = (G \vee H)(F \vee M)(L \vee M \vee O)(G \vee I)(I \vee J)(K \vee L)(D \vee H)$$
$$(D \vee E)(B \vee F)(A \vee B \vee D)(A \vee D \vee E)(E \vee J)(B \vee C)$$
$$(A \vee B \vee C)(C \vee K)(A \vee C \vee E) N(N \vee O)$$

---

* If we are not interested in a minimal sum only, but want to derive all irredundant disjunctive forms or sloppy minimal sums, letters can be assigned to the prime implicants in any order. As mentioned in the second footnote in Section 4.1, minimization of the number of connections that results from minimization of the number of literals is not important in some electronic implementations.

has been obtained for some function f, where prime implicants A through E consist of 4 literals each and the remaining prime implicants consist of 5 literals each.

2. **In each product of alterms** (if Step 4 is not applied yet, p consists of a single product of alterms), **delete alterms that subsume other single- or multiletter alterms in that product** (i.e., delete multiletter alterms that subsume other single- or multiletter alterms, and also delete single-letter alterms that are identical to other single-letter alterms).

   For our example, which consists of a single product of alterms at this stage, the multiletter alterm $(N \vee O)$ can be deleted, since it subsumes the single-letter alterm N. Also, multiletter alterms $(A \vee D \vee E)$ and $(A \vee B \vee C)$ can be deleted, since they subsume other multiletter alterms, $(D \vee E)$ and $(B \vee C)$, respectively. Thus we get

   $$p = (G \vee H)(F \vee M)(L \vee M \vee O)(G \vee I)(I \vee J)(K \vee L)(D \vee H)$$
   $$(D \vee E)(B \vee F)(A \vee B \vee D)(E \vee J)(B \vee C)(C \vee K)$$
   $$(A \vee C \vee E)N.$$

3. **In each product of alterms, consider a pair of letters, say L_i and L_j. If one letter, L_j, appears in every alterm where the second letter, L_i, appears** (but L_j may appear in alterms where L_i does not appear), **delete the second letter, L_i throughout this product of alterms. In this case, when we are trying to derive a single minimal sum, L_i cannot be deleted if it has a lower cost*** (i.e., the prime implicant corresponding to L_i consists of fewer literals) **than L_j.** (In this cost comparison, the alphabetical order adopted in Step 1 is convenient.)

   For our example, compare the letters O and M. Letter O can be deleted, since M appears in every alterm where O appears [O appears only in the alterm $(L \vee M \vee O)$ with M], and the prime implicants corresponding to these letters consist of the same number of literasl, five. Thus p is reduced to the following expression:†

   $$p \Rightarrow (G \vee H)(F \vee M)(L \vee M)(G \vee I)(I \vee J)(K \vee L)(\underline{D \vee H})(D \vee E)$$
   $$(B \vee F)(A \vee B \vee D)(E \vee J)(B \vee C)(C \vee K)(A \vee \underline{C \vee E})N.$$

4. **In each product of alterms, choose two or possibly more multiletter alterms** such that Steps 2 and 3 will be applicable as often as possible after this step (4) is performed (e.g., choose alterms that share common letters, perhaps as many common letters as possible). **Then multiply out these**

---

* If we are not interested in a minimal sum, but want to derive a sloppy minimal sum, this cost comparison is not necessary, and the letters can be assigned to the prime implicants in any order in Step 1.

† The symbol "$\Rightarrow$" is used to show that p is reduced to a new expression, but the new expression is not necessarily equal to p.

Algebraic Method to Derive Minimal Sums 183

alterms into a single disjunctive form. [If more than two multiletter alterms are to be multiplied out, **never** multiply them out into two or more disjunctive forms. For example, if $(D \vee H)$, $(D \vee E)$, $(B \vee C)$, and $(C \vee K)$ in the above reduced expression for Petrick function p are to be multiplied out, multiply them out into a **single** disjunctive form, $(DC \vee DBK \vee HEC \vee HEBK)$, but not into two disjunctive forms, $(D \vee HE)$ and $(C \vee BK)$. If you derive two or more disjunctive forms, the calculation will usually become very messy, time-consuming, and prone to errors.]

In this case, if too many alterms are multiplied out, the multiplication may become too time-consuming. If so, multiply out only a few alterms.

Then distribute the obtained disjunctive form over the remaining expression in each product of alterms, without changing any other alterms.

For our example, which still consists of a single product of alterms, multiply out two multiletter alterms, $(D \vee H)$ and $(D \vee E)$ (shown underlined in the reduced expression for p in Step 3) into a single disjunctive form, $(D \vee HE)$. Then distribute $(D \vee HE)$ over the remaining expression, as follows:

$p \Rightarrow (D \vee HE)(G \vee H)(F \vee M)(L \vee M)(G \vee I)(I \vee J)(K \vee L)(B \vee F)$
$\quad (A \vee B \vee D)(E \vee J)(B \vee C)(C \vee K)(A \vee C \vee E)N$
$\Rightarrow D(G \vee H)(F \vee M)(L \vee M)(G \vee I)(I \vee J)(K \vee L)(B \vee F)$
$\quad (A \vee B \vee D)(E \vee J)(B \vee C)(C \vee K)(A \vee C \vee E)N$
$\vee HE(G \vee H)(F \vee M)(L \vee M)(G \vee I)(I \vee J)(K \vee L)(B \vee F)$
$\quad (A \vee B \vee D)(E \vee J)(B \vee C)(C \vee K)(A \vee C \vee E)N.$

5. **Repeat Steps 2, 3, and 4 in any order.** (Usually we had better apply Step 4 only when we cannot apply Step 2 or 3.)

    For our example, Steps 2 and 3 can be applied as follows:*

    $p \Rightarrow D(G \vee \cancel{H})(F \vee M)(\cancel{L \vee M})(\cancel{G \vee I})(I \vee J)(K \vee L)(\cancel{B \vee F})$
    $\quad (\cancel{A \vee B \vee D})(\cancel{E \vee J})(B \vee C)(C \vee K)(\cancel{A} \vee C \vee E)N$
    $\vee HE(\cancel{G \vee H})(F \vee M)(\cancel{L \vee M})(G \vee I)(\cancel{I \vee J})(K \vee L)(\cancel{B \vee F})$
    $\quad (A \vee B \vee D)(\cancel{E \vee J})(\cancel{B \vee C})(\cancel{C \vee K})(\cancel{A \vee C \vee E})N.$

    Here $\times$ and $/$ mean deletion by Steps 2 and 3, respectively. First let us process the first product of alterms. By Step 2, $(A \vee B \vee D)$ can be deleted, since it subsumes D. Letter H in $(G \vee H)$ can be deleted by Step 3, since G appears in every alterm where H appears throughout the first product of alterms, and the prime implicants corresponding to H and G consist of the same number of literals. Then $(G \vee I)$ can be deleted by Step 2, since it subsumes G, which is now a single-letter alterm. Letter I in $(I \vee J)$ can be deleted by Step 3. Then $(E \vee J)$ can be deleted by Step 2, since it subsumes J, which is now a single-letter alterm. Letter A, and then E, can be deleted in $(A \vee C \vee E)$ by Step 3. Then $(B \vee C)$ and $(C \vee K)$ can be deleted in Step 2, since they subsume C,

---

* If we want to check our processing of p later, we label letters and alterms with numbers in the order of deleting them.

which is now a single-letter alterm. Letter K in $(K \vee L)$ can be deleted by Step 3. Then $(L \vee M)$ can be deleted by Step 2, since it subsumes L. Letter M in $(F \vee M)$ can be deleted by Step 3. Then $(B \vee F)$ can be deleted by Step 2, since it subsumes F. Thus the first product of alterms becomes simply $DGFJLCN$.

Now let us process the second product of alterms in a similar manner. First, $(G \vee H)$ can be deleted, since it subsumes H. Then $(E \vee J)$ and $(A \vee C \vee E)$ can be deleted, since they subsume E. After deleting G in $(G \vee I)$ by Step 3, $(I \vee J)$ can be deleted, since it subsumes I. The letters A and D in $(A \vee B \vee D)$ can be deleted by Step 3. Then $(B \vee F)$ and $(B \vee C)$, which subsume B, can be deleted. Letter F in $(F \vee M)$ can be deleted by Step 3, and $(L \vee M)$ can be deleted, since it subsumes M. Letter L in $(K \vee L)$ can be deleted by Step 3. Then $(C \vee K)$ can be deleted, since it subsumes K. Thus the second term becomes $HEMIKBN$.

Hence, p is reduced to the following two terms:

$$p \Rightarrow DGFJLCN \vee HEMIKBN.$$

6. **When you obtain a disjunctiver form for p, delete terms that contain more letters than others.**

7. **From the remaining terms, derive irredundant disjunctive forms for f. These forms are irredundant disjunctive forms for f which have the minimum number of terms. Among them, irredundant disjunctive forms that have the minimum number of literals are minimal sums for f.**

For our example, Step 6 cannot delete any terms from the disjunctive form for p, since the two terms consist of the same number of letters. Then, in Step 7, we have two irredundant disjunctive forms for f corresponding to these two terms, $CDFGJLN$ and $BEHIKMN$. Since both have the same total number of literals, they are minimal sums for f, though by this procedure (as in the case of Procedure 4.5.2) we are not sure whether f has more minimal sums.

As a second example, let us apply Procedure 4.6.1 to the function given in Table 4.5.2 for Procedure 4.5.2. We get the Petrick function:

$$p = (A \vee B)(E \vee F)(\cancel{C \vee E})(\cancel{A \vee B \vee C})H(\cancel{B \vee H})(C \vee \cancel{G})(\cancel{B \vee C})$$
$$(K \vee L)(\cancel{A \vee L})(\cancel{F \vee I \vee J})(J \vee M)(\cancel{A \vee M})D(\cancel{D \vee K})(\cancel{D \vee I})$$
$$(\cancel{D \vee G}) \Rightarrow AFHCLMD$$

Thus the same minimal sum as that in Step 6 of Procedure 4.5.2 has been obtained by working on only one switching expression, instead of tables as in Procedure 4.5.2. □

Comparison of Procedure 4.6.1 and Other Procedures

If the Petrick function p in this example is processed by Procedure 4.5.2, a much longer time and more space for hand processing will be required.

At first look, comparison of alterms or letters for deletion in a switching expression appears more cumbersome than that of rows or columns in Procedure 4.5.2. But our experience reveals the contrary. Comparison of rows or columns in a table in Procedure 4.5.2 is more time-consuming and prone

Algebraic Method to Derive Minimal Sums 185

to errors, particularly in large prime-implicant tables. (Try Exercise 4.11 if you want to compare Procedures 4.6.1 and 4.5.2.)

In Procedure 4.5.2 new prime-implicant tables need to be drawn after deletion of rows or columns (or selection of rows) for a while, in order to avoid mistakes in further table reduction, although, in principle, a single prime-implicant table is sufficient for reduction until we get a cyclic table. Drawing a prime-implicant table itself in Procedure 4.5.2 is much more time-consuming than writing a switching expression in Procedure 4.6.1. In Procedure 4.6.1, only when we multiply out the Petrick function in Step 4, do we need to write a new switching expression. Otherwise, we can apply Steps 2 and 3 repeatedly on the same switching expression, deleting multi-letter alterms and letters with $\times$ and $/$, respectively. In this sense Procedure 4.6.1 requires very little space compared with the previous methods, which call for tables or maps.

If we multiply out the Petrick function given in Step 1 in a straightforward manner, as we did in Procedure 4.5.2, Step 7, the multiplication is usually very time-consuming. For example, the Petrick function given at the end of Step 1 of Procedure 4.6.1 has potentially $2^{12}3^5$ (roughly a million) terms. Even though some terms that subsume others are deleted during the multiplication, we still need to handle too many terms, and consequently the straightforward multiplication of the Petrick function is much more time-consuming than Procedure 4.6.1.

Unlike the Karnaugh map, Procedure 4.6.1 is relatively independent of the number of variables, in addition to the advantage of space saving. Of course, processing time increases as the numbers of prime implicants and minterms increase.

If the number of variables is four or less, the Karnaugh map is usually easier to use than Procedure 4.6.1, but even in this case some functions are complex on a four-variable map. Fig. 4.6.1 is such an example. Since this function has many prime implicants, it is not easy to derive a minimal sum on a map. In other words, one can derive an irredundant disjunctive form that looks like a minimal sum. But if one is not sure whether the form is really minimal, it may be desirable to use Procedure 4.5.2 to find a solution that is guaranteed to be a minimal sum. If a Petrick function is calculated as stated in Procedure 4.5.2 (i.e., deriving its complete sum and then finding a minimal sum), the process is very time-consuming. Using Procedure 4.6.1, we can derive a minimal sum in a much shorter time (Exercise 4.6).

Why Procedure 4.6.1 Works

It is easy to see why Procedure 4.6.1 works, once it is understood why Procedure 4.5.2 works. Notice that Step 2 of Procedure 4.6.1 corresponds

186 *Simplification of Switching Expressions*

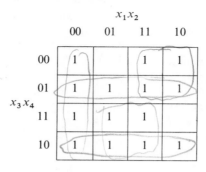

Fig. 4.6.1

to the elimination of columns covered by (secondary) essential rows, and also the elimination of any dominating columns in Procedure 4.5.2. Also, Step 3 of Procedure 4.6.1 corresponds to the elimination of any dominated rows in Procedure 4.5.2. Furthermore, cyclic tables are handled by Procedure 4.6.1 in a simple manner, by distributing disjunctive forms in Step 4, whereas cyclic tables require a complex treatment in Step 7 of Procedure 4.5.2, which is different from the other steps in the same procedure. In other words, the branching operations in Step 7 of Procedure 4.5.2 are done simply by writing new switching expressions in Procedure 4.6.1. Procedure 4.6.1, because of its Steps 2 and 3, looks like an algebraic version of Procedure 4.5.2. But, precisely speaking, Step 4 of Procedure 4.6.1 is somewhat different from the binary branching in the branch-and-bound method in Step 7 of Procedure 4.5.2. In Step 4 of Procedure 4.6.1, we may branch to more than two nodes, followed by removal of many secondary essential rows. [In contrast, the binary branching in Procedure 4.5.2 may be interpreted as the distribution of each product of alterms in Step 4 of Procedure 4.6.1 based on Shannon's expansion (described in Exercise 3.2.7), as described in Remark 4.6.1. This algebraic version of Procedure 4.5.2 is also much more efficient than Procedure 4.5.2. Its efficiency may be comparable to that of Procedure 4.6.1.]

The choice of appropriate multiletter alterms (to be multiplied out into a disjunctive form) in Step 4 of Procedure 4.6.1 is important for the speed-up in executing Procedure 4.6.1.

Modifications of Procedure 4.6.1

Notice that all minimal sums of a given function f may not be derived in Step 7, though one minimal sum can always be derived. If we want to find all minimal sums of f instead of a single minimal sum, we delete L_i in Step 3 only when the prime implicant corresponding to L_i consists of more literals than that corresponding to L_j. Also, if we want to find all irredundant disjunctive forms of f, we must skip Steps 3 and 6. But skipping Step 3 usually makes processing much more time-consuming.

Remark 4.6.1: A function $g(A, B, C, \ldots)$ of variables $A, B, C, \ldots$ can be expanded as follows, based on Shannon's expansion, described in Exercise 3.2.7:

$$g(A, B, C, \ldots) = A \cdot g(1, B, C, \ldots) \vee \bar{A} \cdot g(0, B, C, \ldots).$$

When g is a product of alterms consisting of noncomplemented literals, the above expression can be rewritten as follows:

$$g(A, B, C, \ldots) = A \cdot g(1, B, C, \ldots) \vee g(0, B, C, \ldots),$$

since, for $A = 1$, the right-hand side of this equality becomes $g(1, B, C, \ldots) \vee g(0, B, C, \ldots) = g(1, B, C, \ldots)$ because $g(0, B, C, \ldots)$ subsumes $g(1, B, C, \ldots)$.

Thus, if Step 4 of Procedure 4.6.1 is replaced by the above expansion—in other words, if each product of alterms for a Petrick function is expanded as

$$g(A, B, C, \ldots) = A \cdot g(1, B, C, \ldots) \vee g(0, B, C, \ldots) \qquad (4.6.1)$$

with respect to an appropriately chosen letter A—we have obtained an algebraic version of the branch-and-bound method in Step 7 of Procedure 4.5.2.

For example, suppose that we have a Petrick function

$$p = (A \vee B)(A \vee C \vee D)(E \vee F).$$

By expansion 4.6.1 we get

$$p = A(1 \vee B)(1 \vee C \vee D)(E \vee F) \vee (0 \vee B)(0 \vee C \vee D)(E \vee F)$$
$$= A(E \vee F) \vee B(C \vee D)(E \vee F). \qquad (4.6.2)$$

The above algebraic version of Procedure 4.5.2 can derive a minimal sum in a much shorter time than is possible with Procedure 4.5.2. Moreover, this version probably has processing efficiency comparable to that of Procedure 4.6.1 in most cases, though Procedure 4.6.1 sometimes appears to be slightly better. (For some functions this version may need extra intermediate switching expressions before reaching an expression that Procedure 4.6.1 can yield by a single application of Step 4. But for some other functions, when all multiletter alterms that contain a certain letter are not multiplied out in Step 4 of Procedure 4.6.1, duplicate terms may be generated later, whereas this cannot occur in the above algebraic version of Procedure 4.5.2.)

In contrast to expression 4.6.2, by Step 4 of Procedure 4.6.1, we get

$$p = (A \vee BC \vee BD)(E \vee F) \qquad \text{[by multiplying } (A \vee B) \text{ and } (A \vee C \vee D)\text{]}$$
$$= A(E \vee F) \vee BC(E \vee F) \vee BD(E \vee F). \blacksquare$$

Remark 4.6.2: For hand processing, Procedure 4.6.1 and also the algebraic version of Procedure 4.5.2 described in Remark 4.6.1 are much easier to use, and can derive a minimal sum in a much shorter time than is possible with Procedure 4.5.2. For computer processing, however, we need to incorporate many more techniques for speed-up than straightforward implementation of Procedure 4.5.2 or 4.6.1, if we want to have efficient computer programs. In particular, appropriate

188 Simplification of Switching Expressions

bounding methods, the storage of tables used in Procedure 4.5.2, the algebraic expressions used in Procedure 4.6.1, and other techniques are essential for speed-up. (Computer programs to derive a sloppy minimal sum have been improved by techniques devised by M. H. Young and S. Muroga at least 30 to 100 times in computation efficiency over straightforward implementation of Procedure 4.5.2 or 4.6.1. These programs can process all seven-variable problems each in a reasonable amount of computer time [M. H. Young and S. Muroga, to be published].) ∎

EXERCISES

4.1 (RT) Derive a minimal sum for one of the following functions, using Procedures 4.5.1 and 4.5.2:
  A. $f(v, w, x, y, z) = \Sigma(2, 3, 6, 12, 13, 14, 19, 24, 25, 26, 27, 28, 29, 30)$.
  B. $f(v, w, x, y, z) = \Sigma(2, 3, 6, 7, 12, 13, 15, 18, 19, 20, 22, 24, 25, 26, 27, 28, 29)$.

4.2 (RT) By Procedure 4.5.2, derive a minimal sum for the following six-variable function, all prime implicants of which are given in terms of minterms expressed in decimal numbers:

$A = \{24, 25, 28, 29, 56, 57, 60, 61\}$, $B = \{8, 12, 24, 28\}$, $C = \{0, 4, 8, 12\}$,
$D = \{0, 2, 4, 6\}$, $E = \{2, 6, 34, 38\}$, $F = \{34, 38, 42, 46\}$,
$G = \{42, 43, 46, 47\}$, $H = \{38, 39, 46, 47\}$, $I = \{49, 51, 57, 59\}$,
$J = \{33, 49\}$, $K = \{34, 50\}$, $L = \{33, 37\}$, $M = \{43, 59\}$,
$N = \{37, 39\}$, $O = \{50, 51\}$.

4.3 (E) On the basis of a Karnaugh map, explain with a simple example why Procedure 4.5.1 yields all prime implicants.

4.4 (E) With a simple example, explain how the elimination of rows or columns in Procedure 4.5.2 can be interpreted on a Karnaugh map.

4.5 (R) For one of the functions given in Exercise 4.1, derive all the prime implicants by Procedure 4.5.1. Then derive a minimal sum by Procedure 4.6.1.

4.6 (R) The function f shown in Fig. 4.6.1 has the following Petrick function p. Derive a minimal sum for f by Procedure 4.6.1.

$$p = (A \vee J \vee K)(A \vee C \vee E \vee J)(A \vee D \vee F \vee K)(A \vee E \vee F)$$
$$(C \vee E \vee G)(D \vee F \vee H)(E \vee F \vee G \vee H)(I \vee J \vee K \vee L)$$
$$(C \vee I \vee J)(D \vee K \vee L)(B \vee I \vee L)(B \vee C \vee G \vee I)$$
$$(B \vee D \vee H \vee L)(B \vee G \vee H),$$

where

$A = \bar{x}_1 \bar{x}_2$, $B = x_1 x_2$, $C = \bar{x}_3 x_4$, $D = x_3 \bar{x}_4$, $E = \bar{x}_1 x_4$,
$F = \bar{x}_1 x_3$, $G = x_2 x_4$, $H = x_2 x_3$, $I = x_1 \bar{x}_3$, $J = \bar{x}_2 \bar{x}_3$,
$K = \bar{x}_2 \bar{x}_4$, $L = x_1 \bar{x}_4$.

4.7 (R) Discuss how we should modify Procedure 4.6.1 when we want to find only one sloppy minimal sum. Then derive a sloppy minimal sum for the six-variable function, all

Algebraic Method to Derive Minimal Sums   189

prime implicants of which are given in terms of min terms expressed in decimal numbers as follows:

$A = \{4, 20\}$, $B = \{10, 11, 42, 43\}$, $C = \{5, 7, 13, 15\}$,
$D = \{24, 25, 28, 29, 56, 57, 60, 61\}$, $E = \{4, 5\}$, $F = \{3, 7, 19, 23\}$,
$G = \{13, 29\}$, $H = \{3, 7, 11, 15\}$, $I = \{20, 28, 52, 60\}$,
$J = \{20, 22, 52, 54\}$, $K = \{17, 25\}$, $L = \{17, 19\}$,
$M = \{22, 23, 54, 55\}$.

4.8 (M−) Discuss how we should modify Procedure 4.6.1 when we want to find all sloppy minimal sums of a function f.

4.9 (RT) **(i)** When we want to find all irredundant disjunctive forms of a given function instead of a single minimal sum, discuss how we should modify Procedure 4.5.2 and also Procedure 4.6.1.

When we want to find all minimal sums, how should we modify these procedures?

(ii) Derive all irredundant disjunctive forms for the following function by Procedures 4.5.1 and 4.6.1:

$$f(w, x, y, z) = \Sigma(0, 4, 5, 8, 10, 11, 13, 15).$$

Then, among these, identify minimal sums and nonminimal irredundant disjunctive forms.

4.10 (M−) **(i)** When we want to find a minimal sum for an incompletely specified function, how should we modify Procedures 4.5.1 and 4.5.2?

(ii) Derive a minimal sum for the following function:

$$f(w, x, y, z) = \Sigma(2, 3, 6, 7, 13) + d(5, 8, 9, 15).$$

4.11 (M − T) Using Procedure 4.5.2, derive a minimal sum for the following six-variable function, all prime implicants of which are given in terms of minterms expressed in decimal numbers. When multiplying out straightforwardly the Petrick function formed for a cyclic table is found to be too time-consuming, the Petrick function may be processed by switching to Procedure 4.6.1. But apply Procedure 4.6.1 to the Petrick function obtained after table reduction by Procedure 4.5.2.

$A = \{8, 12, 24, 28\}$, $B = \{34, 38, 42, 46\}$, $C = \{38, 39, 46, 47\}$,
$D = \{42, 43, 46, 47\}$, $E = \{11, 43\}$, $F = \{42, 58\}$, $G = \{24, 56\}$,
$H = \{56, 58\}$, $I = \{37, 39\}$, $J = \{33, 37\}$, $K = \{1, 33\}$,
$L = \{3, 11\}$, $M = \{2, 34\}$, $N = \{1, 3\}$, $O = \{2, 3\}$.

▲ **4.12 (M−)** **(i)** By the algebraic version of Procedure 4.5.2 described in Remark 4.6.1, derive a minimal sum for each of the example functions used to illustrate Procedure 4.6.1 and the function shown in Exercise 4.6.

(ii) Discuss what type of Petrick function makes the use of the algebraic version of Procedure 4.5.2 more advantageous than applying Procedure 4.6.1, or vice versa. (A precise discussion may not be easy.)

4.13 (E) Using the statistics in Fig. 4.1.1, estimate the largest average size of initial prime-implicant tables (such as Table 4.5.2) for a function f of n variables for $n = 5, 7$, and 9. Write the approximate size (the product of the number of rows and the number of columns) in decimal numbers.

190 Simplification of Switching Expressions

4.14 (M) **(i)** Show that there exists a function of n variables that has a minimal sum consisting of exactly 2^{n-1} prime implicants.

(ii) Prove that any function of n variables has at least one irredundant disjunctive form consisting of no more than 2^{n-1} prime implicants.

> **Remark:** Some functions have more than 2^{n-1} prime implicants, but at least one of the irredundant disjunctive forms for each of these functions does not contain more than 2^{n-1} prime implicants. ∎

4.15 (M) Let $f_k(x_1, \ldots, x_n)$ be a function that is equal to 1 if and only if exactly k of the n variables are 1.

(i) Prove or disprove that any such function f_k has a unique minimal sum.

(ii) Show the number of prime implicants that f_k has. What value of k maximizes this number? Then determine whether this maximum number is greater than 2^{n-1} or not.

(iii) The disjunction of f_k for all odd values of k, that is, $f_1 \vee f_3 \vee \cdots \vee f_n$ (if n is odd) or $f_1 \vee f_3 \vee \cdots \vee f_{n-1}$ (if n is even), is the parity function. Determine whether each f_k can have more prime implicants than the parity function. Explain why.

4.16 (M) Prove or disprove one of the following sets of statements:

A. **(i)** If a function f that is not identically 0 does not have any essential prime implicants, f has two or more irredundant disjunctive forms.

(ii) If every conjunction and disjunction are replaced by disjunction and conjunction, respectively, in a minimal sum for a function f, we obtain a minimal product for f.

B. **(i)** If a function f has a unique minimal sum, all the prime implicants of f are essential.

(ii) If a function f has a unique minimal sum, f has a unique minimal product.

4.17 (E) A **multiplexer** is a network with 2^k data inputs, k control inputs, and two outputs* (an example for $k = 2$ is shown in Fig. E4.17a), where, corresponding to each combination of the values of control inputs $c_1, \ldots, c_k$, one of the data inputs d_0,

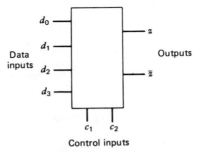

Fig. E4.17a A multiplexer.

---
* Some multiplexers have only one output (either z or $\bar{z}$).

$d_1, \ldots, d_{2^k-1}$ is connected to output z and the complement of this appears at output $\bar{z}$. For example, when $(c_1, c_2) = (0\ 0)$, $z = d_0$ and $\bar{z} = \bar{d}_0$; when $(c_1, c_2) = (0\ 1)$, $z = d_1$ and $\bar{z} = \bar{d}_1$; and so on.

A multiplexer with 2^k data inputs is usually called a **2^k-input multiplexer**.

In addition to multiplexing data channels (like a single-pole multiposition switch), a multiplexer can be used to implement a function of $k+1$ variables as follows.

Suppose that a function f of three variables, $x_1, x_2,$ and x_3, is given as shown in the Karnaugh map in Fig. E4.17b. Let us connect variables x_1 and x_2 to the control inputs

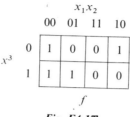

Fig. E4.17b

as shown in Fig. E4.17c. When $(x_1, x_2) = (0\ 0)$, the value of f is 1 regardless of the value of x_3, as seen in the first column of Fig. E4.17b. Hence we connect the constant value 1 to data input d_0 in Fig. E4.17c. When $(x_1, x_2) = (0\ 1)$, f is 0 and 1, corresponding to

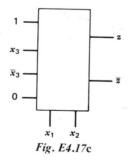

Fig. E4.17c

0 and 1 of x_3, respectively (i.e., $f = x_3$). Hence we connect x_3 to data input d_1 in Fig. E4.17c. When $(x_1, x_2) = (1\ 0)$, the value of f is 1 and 0, corresponding to 0 and 1 of x_3, respectively, as seen in the fourth column of Fig. E4.17b. Hence we connect $\bar{x}_3$ to data input d_2 in Fig. E4.17c. Similarly, we connect the value 0 (corresponding to the third column in Fig. E4.17b) to data input d_3 in Fig. E4.17c. Then output z of the multiplexer in Fig. E.17c expresses the function f given in Fig. E4.17b.

(i) Determine the input connections of a multiplexer (Fig. E4.17a) for the function given in the Karnaugh map in Fig. E4.17d.

(ii) Determine the input connections of a multiplexer (Fig. E4.17a) for the function given in the Karnaugh map in Fig. E4.17e. In this case maximize the number of constant inputs 0, since the constant input 0 can be easily implemented.* (Try to connect an appropriate pair of variables other than the pair x_1 and x_2 to the control inputs.)

	$x_1 x_2$			
x_3	00	01	11	10
0	0	1	0	1
1	0	0	1	1

Fig. E4.17d

	$x_1 x_2$			
x_3	00	01	11	10
0	1	0	0	1
1	0	1	0	0

Fig. E4.17e

Remark: Multiplexers are commercially available in MSI packages with 2, 4, 8, or 16 data inputs. When a function of three or four variables is implemented with discrete gates or SSI packages, more than one package may be necessary. But if we use a multiplexer, only one MSI package is always sufficient, so a multiplexer is a convenient means to implement an arbitrary function of three or four variables. For a function of five or more variables we need two or more packages, as shown in Exercise 4.20. (For a general discussion of multiplexers, see Yau and Tang [1968], Anderson [1969], or Van Holten [1974].)

4.18 (M−) Determine the input connections of a multiplexer† that has eight data inputs and three control inputs, for the functions shown in Table E4.18. In this case maximize the number of constant inputs 0 by choosing an appropriate set of three

Table E4.18

		$x_1 x_2$			
		00	01	11	10
	00	1	1	1	0
$x_3 x_4$	01	0	0	1	0
	11	0	0	0	1
	10	1	0	1	0

variables as the control inputs. (It is assumed that the constant input 0 can easily be implemented.) (*Hint*: Notice that modification of the four-variable Karnaugh map to a map that has a single variable coordinate vertically and three variable coordinates horizontally facilitates the calculation.)

4.19 (E) Design the multiplexer in Fig. E4.17a of Exercise 4.17, using AND, OR, and NOT gates. Use as few gates as possible.

---

* Constant input 1 instead of 0 may sometimes be desirable, depending on electronic implementation of the constant 1 or 0.
† For the definition of "multiplexer," see Exercise 4.17.

Algebraic Method to Derive Minimal Sums 193

4.20 (M − T) When there are five or more variables, f can be implemented by a network of multiplexers* in two levels, as shown in Fig. E4.20. (If f is implemented

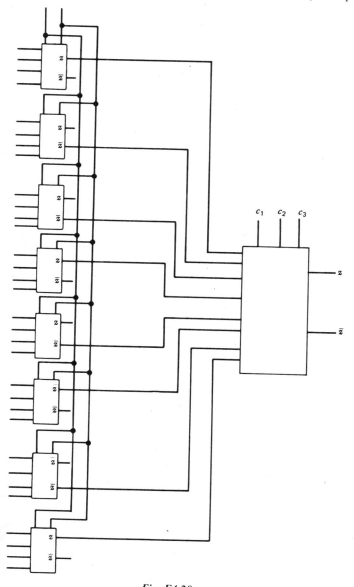

Fig. E4.20

* For the definition of "multiplexer," see Exercise 4.17.

194 *Simplification of Switching Expressions*

with a single multiplexer, the package requires too many pins, making implementation of the multiplexer with a single MSI package impractical.) The multiplexers in the first level have four data inputs each, and the multiplexer in the second level has eight data inputs.

Determine the input connections of these multiplexers for the following function:

$f(x_1, x_2, x_3, x_4, x_5, x_6)$
$= \Sigma(2, 4, 5, 6, 8, 9, 10, 11, 12, 13, 14, 15, 24, 25, 26, 28, 29, 30, 31, 32, 33, 37,$
$\qquad 38, 43, 44, 48, 51, 55, 58, 59, 63).$

In this case, connect x_4 and x_5 to the control inputs of every multiplexer in the first level, and x_6 to appropriate data inputs of these multipliers. Then connect x_1, x_2, and x_3 to the control inputs of the multiplexer in the output (i.e., second) level.

An economical network can generally be obtained by the following considerations:

1. By using an appropriate output, either z or $\bar{z}$, of each multiplexer in the first level, the number of constant inputs 0 is maximized. This is desirable, since constant 0 can easily be implemented.

2. If there is a pair of multiplexers in the first level such that one has an output function g, and the other its complement $\bar{g}$, one of these multiplexers can be eliminated, since both g and $\bar{g}$ can be obtained as the outputs of one of these multiplexers.

(*Hint*: Draw eight three-variable Karnaugh maps, each of which has the coordinates x_4, x_5, and x_6, though four four-variable maps may be thought of for this six-variable function. In other words, use Karnaugh maps with a modification.)

Remark: For further general implementation techniques, see Barna and Porat [1973] or Tabloski and Mowle [1976]. ∎

4.21 (M) In the MOS network consisting of two gates shown in Fig. E4.21, one gate with output g is faulty. Using a Karnaugh map, find an MOS gate with a minimal number of MOSFETs to replace this faulty gate. Any variables x, y, z and/or their complements are available as inputs to this gate. Explain concisely why your approach yields a solution.

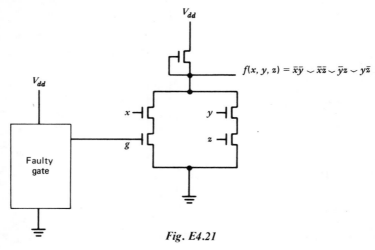

Fig. E4.21

CHAPTER 5

Advanced Simplification Techniques and Basic Properties of Gates

Irredundant disjunctive forms have unique algebraic properties in terms of consensuses. On the basis of these algebraic properties, we can sometimes derive minimal sums or irredundant disjunctive forms very quickly, without going through the derivation of the implication relationship between prime implicants and minterms.

Multiple-output networks are important for logic design practice, but are not easy to design. In this chapter some design methods, based on tables and algebraic manipulations, are presented.

The design of networks under arbitrary restrictions is generally difficult. Certain heuristic approaches are shown with examples.

Lastly, important algebraic properties of different gate types are discussed. Then the limitation of the gate concept is explained from the viewpoint of logic design practice. Although the concept of gates is convenient, good logic design requires a knowledge of electronic circuits, since complex logic operations can sometimes be realized by simple modifications of the electronic circuits that implement gates.

5.1 Derivation of Irredundant Disjunctive Forms without Use of Minterms

In the methods in Chapter 4, the derivation of minimal sums or irredundant disjunctive forms was based on minterms in map, table, or algebraic form. But when the number of variables is large, the number of minterms rapidly increases, and to deal with minterms is cumbersome and time-consuming.

Irredundant disjunctive forms of a function f have a unique algebraic relationship to consensuses. From this algebraic relationship, minimal sums or irredundant disjunctive forms can, in some cases, be obtained very quickly from the prime implicants of f, without using minterms, even if the number of variables of f is large. Thus this approach is very convenient in some cases for hand processing, although the approach in general is usually much more time-consuming, even for computer processing.

This completely algebraic approach to finding all irredundant disjunctive forms for a given function f was developed mostly by Ghazala [1957], Mott [1960], Chang and Mott [1965], Tison [1965, 1967], and Cutler and Muroga [to be published].

To facilitate our discussion, let us introduce the concept of the generalized consensus, which, as the term implies, is a generalization of the consensus defined in Section 3.4.

Definition 5.1.1: Suppose that the implication relation

$$A \subseteq A_1 \vee A_2 \vee \cdots \vee A_t$$

holds among $t + 1$ products of literals, $A, A_1, A_2, \ldots, A_t$. If deletion of any literal from A or any product in the consequent (i.e., the right-hand side) invalidates this implication relation, the implication relation is said to be **irredundant** and is denoted as

$$A \subseteq A_1 \vee A_2 \vee \cdots \vee A_t \text{ (irr.)},$$

adding "(irr.)."

Definition 5.1.2: If the irredundant implication relation

$$A \subseteq A_1 \vee A_2 \vee \cdots \vee A_t \text{ (irr.)}$$

holds among $t + 1$ products, $A, A_1, A_2, \ldots, A_t$, then A is called the **generalized consensus** of $A_1, A_2, \ldots, A_t$.

For given products $A_1, A_2, \ldots, A_t$, the generalized consensus is uniquely determined if the irredundant implication relation holds, as Tison proved (Exercise 5.1.11). Of course, the consensus of two products defined in Definition 3.4.5 is a special case of the generalized consensus of the above definition for $t = 2$.

In Fig. 5.1.1a, $x_1 x_3$ at the bottom of the tree is the generalized consensus of the products in the top level, that is, $x_1 \bar{x}_2$, $x_2 x_3 \bar{x}_4$, and $x_1 x_4$, because

$$x_1 x_3 \subseteq x_1 \bar{x}_2 \vee x_2 x_3 \bar{x}_4 \vee x_1 x_4 \text{ (irr.)}$$

holds, as can easily be seen, and this relation does not hold if any literal in the antecedent (i.e., the left-hand side) or any product in the consequent is removed. In Fig. 5.1.1a the products in the top level are indexed with A, B, and C, as $x_1 \bar{x}_2 A$, $x_2 x_3 \bar{x}_4 B$, and $x_1 x_4 C$, and then their generalized consensus $x_1 x_3$ is indexed with ABC, in order to show the generalized-consensus relationship among them. As illustrated in Fig. 5.1.1a, **the generalized consensus of any given products can be obtained by repeatedly deriving the**

Derivation of Irredundant Disjunctive Forms 197

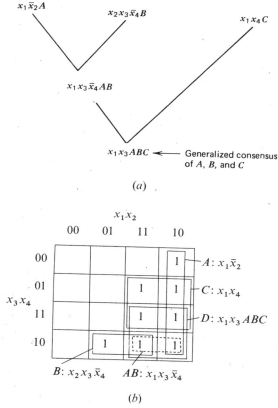

Fig. 5.1.1 *Generalized consensus.* (a) *Formation tree.* (b) *Illustration of Fig. 5.1.1a on map.* (D *is the generalized consensus of* A, B, *and* C.)

consensus of exactly two products (one of them or both may be consensuses generated) in the sense of Definition 3.4.5, as Tison proved. In other words, the consensus $x_1 x_3 \bar{x}_4 AB$ is generated from the two products $x_1 \bar{x}_2 A$ and $x_2 x_3 \bar{x}_4 B$; then the generalized consensus $x_1 x_3 ABC$ is generated as the consensus of this and the remaining product $x_1 x_4 C$. This is illustrated on the Karnaugh map in Fig. 5.1.1b. Product D is the generalized consensus of A, B, and C. The consensus of two products, A and B, is first generated as the dotted-line loop. Then D is generated as the consensus of this loop and product C.

Let us call the tree in Fig. 5.1.1a a **generalized-consensus formation tree**, whose **root** is the generalized consensus $x_1 x_3 ABC$. (In the following, this may be called a **formation tree** for simplicity.) Notice here that a generalized-consensus formation tree may or may not contain, as part of it, other generalized-consensus formation trees. For example, the tree in Fig. 5.1.1a includes the subtree whose root is $x_1 x_3 \bar{x}_4 AB$, which is also the generalized consensus of $x_1 \bar{x}_2 A$ and $x_2 x_3 \bar{x}_4 B$.

On the basis of formation trees, let us describe a procedure to derive all irredundant disjunctive forms for a given function f without using minterms. For the sake of simplicity, let us present the conceptually simplest procedure. (Procedure 5.1.2 and variations mentioned in Remark 5.1.1 are more complex but more efficient.) Procedure 5.1.1, which follows, presented in Tison [1965], remarkably resembles the iterated-consensus method, or its Tison version in Section 3.4.

Procedure 5.1.1*—The Tison method: Derivation of All Irredundant Disjunctive Forms for a Given Function

Suppose that all prime implicants of the given function f are obtained (e.g., by one of the procedures in Section 3.4).

1. Derive all the generalized-consensus formation trees from all the prime implicants, as follows:

    (a) Index the prime implicants, and then arrange all the indexed prime implicants in the top level, as illustrated in Fig. 5.1.2.
    Choose the first biform variable, say x.

    (b) With respect to the chosen biform variable x, generate every consensus from the existing indexed products. Write it, along with the composite index, in the new level corresponding to x.
    In this case, whenever a new indexed product is written, it must be compared with any existing products in the same or higher level. If the new product subsumes or is subsumed by any existing one, **interpreting indexes as literals of the products**,† cross out the subsuming product.‡ The crossed-out products need not be included in the further formation of trees. For example, $\bar{x}_1 x_2 x_3 x_5 x_6 ABE$ in Fig. 5.1.2 does not subsume $\bar{x}_1 x_2 x_3 x_5 x_6 EF$ because ABE, the index of the former, does not subsume EF, the index of the latter, even though $\bar{x}_1 x_2 x_3 x_5 x_6$ in the former subsumes $\bar{x}_1 x_2 x_3 x_5 x_6$ in the

---

* If this Tison method is to be differentiated from another Tison method, Procedure 3.4.2, Procedure 5.1.1 may be called the Tison method for the derivation of all irredundant disjunctive forms, and Procedure 3.4.2 the Tison method for the derivation of all prime implicants.
† This extended definition of "subsuming" is used only in this section.
‡ This deletion of subsuming indexed products is done to avoid the generation of unnecessary products in lower levels.

Derivation of Irredundant Disjunctive Forms 199

latter. But $\bar{x}_1 x_2 x_3 x_5 x_6 ABE$ subsumes $\bar{x}_1 x_2 x_3 x_5 AE$ since $\bar{x}_1 x_2 x_3 x_5 x_6$ and ABE in the former subsume $\bar{x}_1 x_2 x_3 x_5$ and AE in the latter, respectively. Thus cross out $\bar{x}_1 x_2 x_3 x_5 x_6 ABE$ in Fig. 5.1.2.

(c) Choose a next biform variable and return to Substep (b). In this case a biform variable need not be repeated, once it is used.

In the above tree formation, some trees may be contained in other trees, but they must be retained. In the lowest level, indexed products that are not prime implicants of f need not be formed, whereas such indexed products must be generated in higher levels. (For example, $\bar{x}_1 x_2 x_3 x_6 DEF$ is not shown in the lowest level in Fig. 5.1.2, since it is not one of the given prime implicants in the top level, whereas $\bar{x}_1 x_2 x_3 x_5 AE$ is shown in the third level, despite its being a nonprime implicant.) Also, in the lowest level, indexed

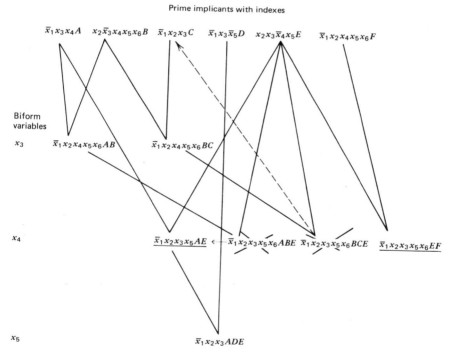

Underlined indexed products are not prime implicants (i.e., not among those in the top level).
Indexed products which subsume others (shown by the dotted lines) are crossed out, interrupting tree formation from them.
Tison function: AB (C ∨ ADE) DE (F ∨ AB ∨ BC).

Fig. 5.1.2 Example for Procedure 5.1.1.

products that subsume others (such as $\bar{x}_1 x_2 x_3 x_4 x_6 ABD$, which subsumes $\bar{x}_1 x_3 x_4 A$) need not be shown.

When you reach the lowest level corresponding to the last biform variable, go to Step 2.

2. From all the indexed products in the trees, form all **inclusion functions** as follows: For each prime implicant P_i, find all indexed products that have this P_i, and then obtain the inclusion function for P_i as the disjunction of the indexes that are contained in these indexed products.

   For example, prime implicant $\bar{x}_1 x_2 x_3$ is indexed with C or ADE in Fig. 5.1.2. Thus the inclusion function for prime implicant $\bar{x}_1 x_2 x_3$ is $(C \vee ADE)$. In Fig. 5.1.2 prime implicant $\bar{x}_1 x_3 x_4$ is indexed with A only. Thus the inclusion function for $\bar{x}_1 x_3 x_4$ is simply A.

3. Obtain the **Tison function** as the conjunction of these inclusion functions. Then derive the complete sum for the Tison function. The terms in the complete sum for the Tison function represent all irredundant disjunctive forms for f.

   For the trees in Fig. 5.1.2, we get the Tison function
   $$AB(C \vee ADE)DE(F \vee AB \vee BC) = ABDE(C \vee ADE)(F \vee AB \vee BC)$$
   $$= ABDE(CF \vee BC \vee ADEF \vee ABDE) = ABDE.$$

   Thus we get only one irredundant disjunctive form corresponding to $ABDE$:
   $$\bar{x}_1 x_3 x_4 \vee x_2 \bar{x}_3 x_4 x_5 x_6 \vee \bar{x}_1 x_3 \bar{x}_5 \vee x_2 x_3 \bar{x}_4 x_5.$$

   If there is more than one irredundant disjunctive form, minimal sums can be found among them. □

Irredundant disjunctive forms or minimal sums for the given function f can be derived by the above procedure without the use of minterms (or, equivalently, maps). When the number of variables of f is large, the procedure is very useful if the formation trees among prime implicants are not complex. For example, when a function of 10 variables is to be processed with four-variable Karnaugh maps, 64 maps must be processed simultaneously; but if the formation trees of f are simple, the function can be processed quickly by Procedure 5.1.1. Of course, if the formation trees are complex, this procedure also is too time-consuming.

The following are examples of cases where minimal sums can be easily obtained from Procedure 5.1.1 (instantly by inspection, for some functions).

Example 5.1.1: If no pair of terms in the complete sum of f has consensus, the complete sum is the unique minimal sum for f. As a special case, this property holds when a function has its complete sum without any biform variable.

Derivation of Irredundant Disjunctive Forms

Since there is no consensus, Procedure 5.1.1 yields no tree, and consequently yields a Tison function that consists of a single term. Examples are $x_1\bar{x}_2 \vee \bar{x}_2 x_3 \vee x_3 x_4$ and $x_1 x_2 x_3 x_5 x_6 \vee \bar{x}_1 \bar{x}_2 x_4 x_7 x_8$, given in their complete sums. The latter function requires 16 maps, if processed by the map method in Section 4.2.

Example 5.1.2: If, in every tree for a function f whose root is a prime implicant, the products in all other levels of the tree are not prime implicants, the disjunction of all prime implicants except those that are the generalized consensuses of others as the roots of the trees is the unique minimal sum of f.

An example is $f = \bar{x}_1 x_2 \vee x_1 x_3 x_4 \vee \bar{x}_2 x_4 \vee x_2 x_3$ in its complete sum, which has the tree shown in Fig. 5.1.3. The Tison function becomes $ABC(D \vee ABC) = ABC$. Thus we obtain the unique minimal sum for f by deleting $x_2 x_3$.

Example 5.1.3: If a function f has only trees such that the root of each tree is not contained in any other tree (i.e., all the trees are one level high), the disjunction of all the prime implicants except these roots (some roots may not be prime implicants) is the unique minimal sum of f.

An example is

$$f = \bar{x}_1 x_3 x_5 \bar{x}_6 \vee x_1 x_2 \bar{x}_7 \vee \bar{x}_1 x_4 x_8 x_9 \vee x_2 x_3 x_5 \bar{x}_6 \bar{x}_7 \vee x_2 x_4 \bar{x}_7 x_8$$

in its complete sum, which has only two trees with roots $x_2 x_3 x_5 \bar{x}_6 \bar{x}_7 AB$ and $x_2 x_4 \bar{x}_7 x_8 x_9 BC$, as shown in Fig. 5.1.4. One of these roots is a prime implicant, $x_2 x_3 x_5 \bar{x}_6 \bar{x}_7 D$. The Tison function becomes

$$ABC(D \vee AB)E = ABCE.$$

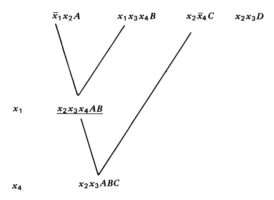

Fig. 5.1.3 Example 5.1.2. (*A nonprime implicant is underlined.*)

202 Advanced Simplification Techniques and Basic Properties of Gates

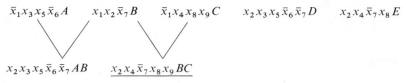

Fig. 5.1.4 Example 5.1.3.

Thus we get the unique minimal sum for f by deleting prime implicant $x_2 x_3 x_5 \bar{x}_6 \bar{x}_7 D$, which is the generalized consensus of others, from the above complete sum for f. If this function is processed by the map method in Section 4.2, 32 maps are needed. (It is cumbersome to draw many maps, and also it is hard to find relationships among loops scattered on many maps.)

Procedure 5.1.1 is convenient when functions have simple trees as in the above example (or slightly more complex trees than in these examples). But when functions have markedly more complex trees, variations based on tables, such as Procedure 5.1.2 (or those in Remark 5.1.1), are more efficient than drawing schematic figures by Procedure 5.1.1, though relationships among indexed products are harder to see intuitively (the variations in Remark 5.1.1 are even more efficient than Procedure 5.1.2, though harder).

Illustration of Procedure 5.1.1 on a Karnaugh map

The difference between Procedure 5.1.1 and the procedures in Chapter 4 can be illustrated on a Karnaugh map. For example, if a function $f = \bar{x}_1 x_3 \vee x_1 x_2 \bar{x}_4 \vee x_3 x_4 \vee x_1 \bar{x}_3 \bar{x}_4 \vee x_2 x_3$ in its complete sum is processed by Procedure 5.1.1, the formation trees in Fig. 5.1.5a are derived. This is illustrated in Fig. 5.1.5b by a map. To derive a Tison function T for this function f based on a Karnaugh map in Fig. 5.1.5b, we have to determine how each prime implicant can be generated as a generalized consensus from other prime implicants. For example, E can be generated as the generalized consensus of A, B, and C. (In other words, loops A and B have the consensus loop shown in the dotted-line loop in Fig. 5.1.5b, and then this consensus loop, which is not a prime-implicant loop, and loop C have the prime-implicant loop E, as their consensus.) Hence we consider the disjunctive form $(E \vee ABC)$ for E. Next, B can be generated as the generalized consensus of D and E, that is, the consensus loop of loops D and E. Hence we consider the disjunctive form $(B \vee DE)$ for B. No other prime implicant, A, C, or D, can be generated as a generalized consensus of other prime implicants. Hence, for A, C, or D, we consider a single-term disjunctive form,

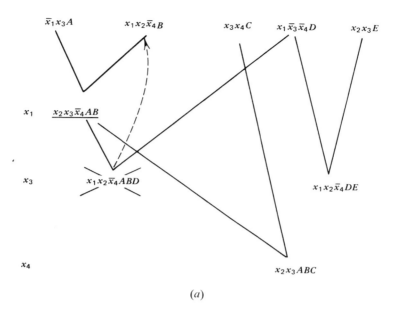

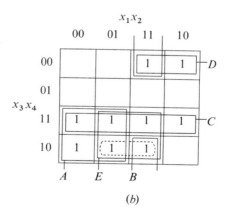

Fig. 5.1.5 Illustration of Procedure 5.1.1 on a Karnaugh map. **(a)** *Formation trees by Procedure 5.1.1 with Tison function* $T = A(B \vee DE)CD(E \vee ABC)$. **(b)** *Illustration of Fig. 5.1.5a on a Karnaugh map.*

203

204 Advanced Simplification Techniques and Basic Properties of Gates

A, C, or D, respectively. The Tison function T is obtained as the conjunction of these disjunctive forms (i.e., inclusion functions),

$$(E \vee ABC)(B \vee DE)ACD.$$

In contrast, we need to find which loops cover each 1-cell, in the methods in Chapter 4. From the map in Fig. 5.1.5b, we obtain the Petrick function $p = A(A \vee C)(A \vee E)(A \vee C \vee E)DC(B \vee D)(B \vee E)(C \vee E)$, where each alterm corresponds to a 1-cell. **In Procedure 5.1.1, however, the 1-cells are not considered, but it is essential to find how each prime implicant can be expressed as a generalized consensus of other prime implicants,** as illustrated on the Karnaugh map in Fig. 5.1.5b.

Why Procedure 5.1.1 Works

Let us now consider why Procedure 5.1.1 works. Arrange the Tison function $AB(C \vee ADE)DE(F \vee AB \vee BC)$ obtained in Step 3 of Procedure 5.1.1 for the example of Fig. 5.1.2 into the following format. Derive alterms from the terms of each inclusion function, as their duals, and write them in one column [e.g., C and ADE in inclusion function $(C \vee ADE)$ are written down as C and $A \vee D \vee E$, respectively, in one column in the following format]:

$$G = A \vee B \vee \begin{Bmatrix} C \\ A \vee D \vee E \end{Bmatrix} \vee D \vee E \vee \begin{Bmatrix} F \\ A \vee B \\ B \vee C \end{Bmatrix}. \quad (5.1.1)$$

This G is called **the generator of irredundant disjunctive forms.*** In each column, a prime implicant and all generalized consensuses that are identical to this prime implicant are listed. For example, the last column consists of the prime implicant F and all the duals $A \vee B$ and $B \vee C$ of consensuses AB and BC, which become F, as seen in Fig. 5.1.2. In the first column, only A is listed since no generalized consensus that becomes A exists. From this generator we can generate all irredundant disjunctive forms for f by the following process, because of Theorem 5.1.1, given below. This is what Procedure 5.1.1 does in Step 3.

Choosing one of the alterms (the alterm at the top consists of a single letter) in each column of (5.1.1), form all possible disjunctive forms. (For the

---

* Although the generator G for this example is based on all the prime implicants of f, generators can be based on fewer prime implicants whose disjunction constitutes any disjunctive form of f, as pointed out by Chang and Mott [1965] and also by Tison [1967]. [Tison never made an explicit statement; but he must be aware of this fact, since his example for an incompletely specified function (p. 453 in his paper) is based on a disjunctive form for f that consists of fewer prime implicants than the complete sum.]

Derivation of Irredundant Disjunctive Forms 205

Tison function illustrated in Procedure 5.1.1, we can form $1 \times 1 \times 2 \times 1 \times 1 \times 3 = 6$ disjunctive forms from G in expression 5.1.1.) Then, in each disjunctive form, delete every term that subsumes another. Among these disjunctive forms, discard those that contain other disjunctive forms. The remainder represents all the irredundant disjunctive forms for f. (For the above example, only one irredundant disjunctive form for f, $A \vee B \vee D \vee E$, is left, shown as $ABDE$ in Step 3 of Procedure 5.1.1.)

Theorem 5.1.1: Suppose that $P_1, P_2, \ldots, P_t$ denote all the prime implicants of a function f. Then:

(i) If a prime implicant P_i has the following irredundant implication relations only:

$$P_i \subseteq P_{ia_1} \vee P_{ia_2} \vee \cdots \vee P_{ia_p} \text{ (irr.)},$$
$$P_i \subseteq P_{ib_1} \vee P_{ib_2} \vee \cdots \vee P_{ib_q} \text{ (irr.)},$$
$$\vdots$$
$$P_i \subseteq P_{ie_1} \vee P_{ie_2} \vee \cdots \vee P_{ie_s} \text{ (irr.)},$$

where each term is some prime implicant of f, prime implicant P_i requires that each irredundant disjunctive form for f contain one of the following:*

$$P_i, \quad (P_{ia_1} \vee P_{ia_2} \vee \cdots \vee P_{ia_p}),$$
$$(P_{ib_1} \vee P_{ib_2} \vee \cdots \vee P_{ib_q}), \quad \ldots, \quad (P_{ie_1} \vee P_{ie_2} \vee \cdots \vee P_{ie_s}).$$

* It is important to note that, although an irredundant disjunctive form contains exactly one of them, when we look at the irredundant disjunctive form with respect to a particular prime implicant P_i, other prime implicants P_k's require that this irredundant disjunctive form contain some prime implicants, and consequently this irredundant disjunctive form may contain two or more of $P_i, (P_{ia_1} \vee \cdots \vee P_{ia_p}), \ldots, (P_{ie_1} \vee \cdots \vee P_{ie_s})$. As an example, consider a function f whose prime implicants are $P_1 = x_1 \bar{x}_3 x_4$, $P_2 = \bar{x}_1 x_2 x_4$, $P_3 = x_2 \bar{x}_3 x_4$, $P_4 = x_1 \bar{x}_2 \bar{x}_3$, $P_5 = \bar{x}_1 x_2 \bar{x}_3$, and $P_6 = \bar{x}_1 x_3 x_4$. Only three of these prime implicants have the following irredundant implication relations:

$$P_1 \subseteq P_3 \vee P_4 \text{ (irr.)},$$
$$P_2 \subseteq P_5 \vee P_6 \text{ (irr.)}, \qquad P_2 \subseteq P_3 \vee P_6 \text{ (irr.)},$$
$$P_3 \subseteq P_1 \vee P_2 \text{ (irr.)}, \qquad P_3 \subseteq P_1 \vee P_5 \text{ (irr.)}.$$

Now let us examine an irredundant disjunctive form $\Psi = P_3 \vee P_4 \vee P_5 \vee P_6$ for f with respect to each prime implicant, according to (i). First, with respect to P_1, which requires that each irredundant disjunctive form contain exactly one of P_1 and $P_3 \vee P_4$, this Ψ contains $P_3 \vee P_4$. Second, with respect to P_2, which requires that each irredundant disjunctive form contain exactly one of P_2, $P_5 \vee P_6$, and $P_3 \vee P_6$, this Ψ contains $P_5 \vee P_6$. But since P_1 requires that Ψ contain $P_3 \vee P_4$, Ψ contains $P_3 \vee P_6$ also. In other words, Ψ contains two of P_2, $P_5 \vee P_6$, and $P_3 \vee P_6$, that is, $P_5 \vee P_6$ and $P_3 \vee P_6$, despite the fact that P_2 requires only one of them.

(ii) If a prime implicant P_i has no implication relation such that its antecedent is P_i, and its consequent consists of some other prime implicants of f, each irredundant disjunctive form contains P_i (i.e., P_i is an essential prime implicant).

In other words, for each prime implicant P_i, each irredundant disjunctive form for f satisfies case (i) or (ii), depending on whether P_i implies disjunctions of some other prime implicants or not.

Proof: (i) (a) Assume that an irredundant disjunctive form for f:

$$\Psi = P_r \vee \cdots \vee P_s$$

does not contain P_i. Since $P_i \subseteq f$,

$$P_i \subseteq P_r \vee \cdots \vee P_s$$

must hold. By removing any prime implicants from the consequent such that an implication relation between P_i and the remainder holds, we obtain

$$P_i \subseteq P_{id_1} \vee P_{id_2} \vee \cdots \vee P_{id_p} \text{ (irr.)},$$

where d is one of $a, b, \ldots, e$ in the theorem statement, since removal of any literal from P_i invalidates this relation (otherwise, P_i is not a prime implicant of f, as can be easily shown). The consequent of this relation is obviously contained in the original irredundant disjunctive form Ψ.

(b) If an irredundant disjunctive form for f contains P_i, the consequent of any irredundant implication relation in the theorem statement is not contained, since it can be deleted without causing the remainder not to express f by Theorem 3.4.1.

Since either (a) or (b) holds, theorem statement (i) is true.

(ii) Assume that there is an irredundant disjunctive form Ψ for f that does not contain P_i:

$$\Psi = P_u \vee \cdots \vee P_v.$$

Since $P_i \subseteq f$,

$$P_i \subseteq P_u \vee \cdots \vee P_v$$

must hold. In other words, P_i has an implication relation whose antecedent is P_i, against the assumption. Q.E.D.

A complete proof of why Procedure 5.1.1 works is lengthy and is omitted here (e.g., we need to prove that all generalized consensuses can be derived by the iteration of forming a consensus of exactly two products).

Features of Procedure 5.1.1.

When the formation trees are complex, Procedure 5.1.1 (also Procedure 5.1.2 and the variations in Remark 5.1.1, which are more efficient, though more complex) is usually much more time-consuming than the procedures in Chapter 4. Even for some functions of four variables, Procedure 5.1.1 (or

Derivation of Irredundant Disjunctive Forms 207

5.1.2) can be very complex and time-consuming, so the procedures such as the Karnaugh map method which were discussed in the preceding chapters are usually preferable. However, if the formation trees are simple, Procedure 5.1.1 can yield minimal sums very quickly, even if the number of variables or the number of prime implicants is very large (so large that the procedures in the preceding chapters cannot be used). Sometimes we can easily obtain minimal sums by inspection, since Procedure 5.1.1 itself becomes very short. Since the Tison function is obtained from the formation trees, its complexity is less dependent on the number of variables, whereas the Petrick function has a rapidly increasing number of alterms for an increase in the number of variables. (On the other hand, if we want to derive a single minimal sum, the Petrick function can be greatly simplified by Procedure 4.6.1, whereas it is harder to simplify the Tison function.) Thus Procedure 5.1.1 (and Procedure 5.1.2 and the variations in Remark 5.1.1) has a unique feature that the procedures in the preceding chapters lack.

Procedure 5.1.1 has been extended to incompletely specified multiple-output functions by Tison [1967].

> ***Remark 5.1.1:*** Reusch's procedure, which is a modification of Tison's procedure combined with the table mentioned in Ghazala [1957], is computationally efficient [Reusch 1975]. When we find a disjunctive form for a given function f which consists of fewer terms than the complete sum (see the footnote on p. 204 in relation to expression 5.1.1), this approach generally requires the generation of fewer inclusion functions. (Inclusion functions can be generated for a smaller number of implicants than all the prime implicants in Procedure 5.1.1.) Also, an inclusion function is generated separately for each individual implicant, thereby processing fewer products. As a result this approach generally requires less processing time, at least for hand processing. (A disadvantage of the approach is possible duplication of the same computation for different implicants, though Procedure 5.1.1 also requires the computation of nonprime implicants that may not be used in the Tison function.)
>
> Other variations that are more efficient are to be published by R. Cutler and S. Muroga. ∎

▲ Tabular Version of Procedure 5.1.1

Procedure 5.1.2 by Tison [1967], which follows, is essentially a formal description of Procedure 5.1.1, using tables. But it contains a simple modification that usually reduces processing time, whether processed by hand or by computer, though the modification makes the procedure slightly more complicated. (In the case of computer processing, the improvement in processing time over Procedure 5.1.1 increases greatly, easily several times or more, as the complexity of formation trees for given functions increases, except for very simple functions, as verified by R. Cutler.)

208   Advanced Simplification Techniques and Basic Properties of Gates

Procedure 5.1.2 — The Tison Method: The Derivation of All Irredundant Disjunctive Forms

Assume that we have all prime implicants of $P, Q, \ldots, T$ for a function f, found by some means, such as the procedures in Section 3.4. Let S denote the set of these prime implicants. We want to find all irredundant disjunctive forms of f.

For example, suppose that all the prime implicants for the function f are as follows:

$$x_1 \bar{x}_3, \; x_1 x_4, \; x_2 \bar{x}_3, \; x_2 x_4, \; x_2 x_5, \; \bar{x}_3 \bar{x}_4, \; \bar{x}_3 \bar{x}_5, \; x_3 x_4, \; x_4 \bar{x}_5.$$

1. Label each prime implicant P with an index γ, denoting it as $P\gamma$ (e.g., $x_1 \bar{x}_3$ is indexed with A).

For our example, all the prime implicants are indexed as follows:

$S = \{x_1 \bar{x}_3 \, A, \; x_1 x_4 \, B, \; x_2 \bar{x}_3 \, C, \; x_2 x_4 \, D, \; x_2 x_5 \, E, \; \bar{x}_3 \bar{x}_4 \, F, \; \bar{x}_3 \bar{x}_5 \, G, \; x_3 x_4 \, H, \; x_4 \bar{x}_5 \, I\}.$

Using Procedure 3.4.2 with the following modification, let us find which pairs of prime implicants generate what products as consensuses.

Choose the first biform variable, say x.

2. Find consensuses with respect to the chosen biform variable x as follows: If $P_1 \gamma_1$ and $P_2 \gamma_2$ generate a consensus with respect to x—in other words, if $P_1 = x p_1$ and $P_2 = \bar{x} p_2$ with $p_1 p_2 \neq 0$—then the consensus $p_1 p_2$ is indexed with a combination of the indices γ_1 and γ_2, followed by the biform variable x, as:

$$p_1 p_2 \quad \gamma_1 \gamma_2 \quad x.$$

Find all such indexed consensuses with respect to x generated from all the prime implicants in S.

In our example, the biform variables are x_3, x_4, and x_5. First choose x_3. In the following table we shall perform the consensus formation with respect to x_3, using only prime implicants that contain x_3 or $\bar{x}_3$. Note that the following table does essentially what Procedure 3.4.2 does, except indexing and addition of a biform variable. The first two columns show the remaining literals in these prime implicants, after deleting x_3 and $\bar{x}_3$, respectively. The last column shows the consensuses.

$R(x_3)$	$R(\bar{x}_3)$	CONSENSUS
$x_4 \; H$	$x_1 \; A$	$x_1 x_4 \; AH \; x_3$
	$x_2 \; C$	$x_2 x_4 \; CH \; x_3$
	$\bar{x}_4 \; F$	$x_4 \bar{x}_5 \; GH \; x_3$
	$\bar{x}_5 \; G$	

Derivation of Irredundant Disjunctive Forms   209

The new set is $S = (x_1 \bar{x}_3 A,\ x_1 x_4 B,\ x_2 \bar{x}_3 C,\ x_2 x_4 D,\ x_2 x_5 E,\ \bar{x}_3 \bar{x}_4 F,\ \bar{x}_3 \bar{x}_5 G,\ x_3 x_4 H,\ x_4 \bar{x}_5 I,\ x_1 x_4 A H x_3,\ x_2 x_4 C H x_3,\ x_4 \bar{x}_5 G H x_3)$.

3. Let us denote each indexed product in S as

$$P \gamma V$$

where P, γ, and V represent a product, an index, and biform variables, respectively (e.g., for the last indexed product in S of the above example, $P = x_4 \bar{x}_5$, $\gamma = GH$, $V = x_3$). Note that V is empty for some indexed products of S (e.g., $x_1 \bar{x}_3 A$), and also that V may contain more than one biform variable after we return to this step.

Choose a new biform variable y among the remaining biform variables. Then compare every two indexed products, $P_1 \gamma_1 V_1$ and $P_2 \gamma_2 V_2$, such that $P_1 = y p_1$ and $P_2 = \bar{y} p_2$. When P_1 and P_2 have the consensus $p_1 p_2$ with respect to y, denote this consensus as

$$p_1 p_2 \quad \gamma_1 \gamma_2 \quad V_1 V_2 y$$

if the following two-part condition holds:

(a) P_1 and V_2 do not contain literals (different or the same) of the same variable, and

(b) P_2 and V_1 do not contain literals (different or the same) of the same variable.

Notice that, if V_2 is empty, (a) is ignored, that is, interpreted as "literals of the same variable do not appear." An empty V_1 is treated similarly. [This condition, (a) and (b), replaces the deletion of subsuming products in Step 1 (b) of Procedure 5.1.1, to avoid the generation of unnecessary consensuses. The condition is more complex to process, but is more effective in speeding up the execution of Procedure 5.1.2.]

For our example, if the next biform variable, x_4, is chosen, we get the following table:

$R(x_4)$	$R(\bar{x}_4)$	CONSENSUS
$x_1\ B$	$\bar{x}_3\ F$	$x_1 \bar{x}_3\ BF\ x_4$
$x_2\ D$		$x_2 \bar{x}_3\ DF\ x_4$
$x_3\ H$		$\bar{x}_3 \bar{x}_5\ FI\ x_4$
$\bar{x}_5\ I$		
$x_1\ AH\ x_3$		
$x_2\ CH\ x_3$		
$\bar{x}_5\ GH\ x_3$		

210 *Advanced Simplification Techniques and Basic Properties of Gates*

In this table the following consensuses are not generated:

$x_1 \bar{x}_3 \, AFH \, x_3 x_4$ from $x_1 \, AH \, x_3$ and $\bar{x}_3 \, F,$

$x_2 \bar{x}_3 \, CFH \, x_3 x_4$ from $x_2 \, CH \, x_3$ and $\bar{x}_3 \, F,$

$\bar{x}_3 \bar{x}_5 \, FGH \, x_3 x_4$ from $\bar{x}_5 \, GH \, x_3$ and $\bar{x}_3 \, F,$

since these cases violate condition (a) and (b) [i.e., in these cases, different literals, x_3 and $\bar{x}_3$, appear in $V_1 = x_3$ and $P_2 = \bar{x}_3$, respectively, violating part (b) of the condition].

Thus the new set is $S = (x_1 \bar{x}_3 A, \; x_1 x_4 B, \; x_2 \bar{x}_3 C, \; x_2 x_4 D, \; x_2 x_5 E, \; \bar{x}_3 \bar{x}_4 F, \; \bar{x}_3 \bar{x}_5 G, \; x_3 x_4 H, \; x_4 \bar{x}_5 I, \; x_1 x_4 A H x_3, \; x_2 x_4 C H x_3, \; x_4 \bar{x}_5 G H x_3, \; x_1 \bar{x}_3 B F x_4, \; x_2 \bar{x}_3 D F x_4, \; \bar{x}_3 \bar{x}_5 F I x_4).$

4. Repeat Step 3 until all biform variables are used. (Once each biform variable is used, we do not need it again.) The expressions in S denote all generalized-consensus relations.

   For our example, after the last biform variable x_5 is chosen, we get the following table:

$R(x_5)$	$R(\bar{x}_5)$	CONSENSUS
$x_2 \, E$	$\bar{x}_3 \, G$	$x_2 \bar{x}_3 \, EG \, x_5$
	$x_4 \, I$	$x_2 x_4 \, EI \, x_5$
	$x_4 \, GH \, x_3$	$x_2 x_4 \, EGH \, x_3 x_5$
	$\bar{x}_3 \, FI \, x_4$	$x_2 \bar{x}_3 \, EFI \, x_4 x_5$

   Thus we have found 19 generalized-consensus relations in $S = (x_1 \bar{x}_3 A, \; x_1 x_4 B, \; x_2 \bar{x}_3 C, \; x_2 x_4 D, \; x_2 x_5 E, \; \bar{x}_3 \bar{x}_4 F, \; \bar{x}_3 \bar{x}_5 G, \; x_3 x_4 H, \; x_4 \bar{x}_5 I, \; x_1 x_4 A H x_3, \; x_2 x_4 C H x_3, \; x_4 \bar{x}_5 G H x_3, \; x_1 \bar{x}_3 B F x_4, \; x_2 \bar{x}_3 D F x_4, \; \bar{x}_3 \bar{x}_5 F I x_4, \; x_2 \bar{x}_3 E G x_5, \; x_2 x_4 E I x_5, \; x_2 x_4 E G H x_3 x_5, \; x_2 \bar{x}_3 E F I x_4 x_5).$

5. Some indexed products in S have identical products, though their indices are different. For each prime implicant,* find all indexed products which contain that prime implicant. Then make an inclusion function, that is, the disjunction of indices in these indexed products.

   For our example, we get the following disjunctive forms for each prime implicant (e.g., $x_1 \bar{x}_3$ appears twice, with indices A and BF, in S):

---

* During iterations of Steps 2 through 4, nonprime implicants may be added to S. They must be kept in S until Step 5, unless they are deleted by condition (a) and (b) in Step 3. Otherwise, an incorrect result may be obtained.

Derivation of Irredundant Disjunctive Forms

PRIME IMPLICANT*	INCLUSION FUNCTION
$x_1 \bar{x}_3$	$A \vee BF$
$x_1 x_4$	$B \vee AH$
$x_2 \bar{x}_3$	$C \vee DF \vee EG \vee EFI$
$x_2 x_4$	$D \vee CH \vee EI \vee EGH$
$x_2 x_5$	E
$\bar{x}_3 \bar{x}_4$	F
$\bar{x}_3 \bar{x}_5$	$G \vee FI$
$x_3 x_4$	H
$x_4 \bar{x}_5$	$I \vee GH$

(i) Then form a Tison function, that is, the conjunction of all these inclusion functions.

For our example, we get the Tison function $T = (A \vee BF)(B \vee AH) \times (C \vee DF \vee EG \vee EFI)(D \vee CH \vee EI \vee EGH)EF(G \vee FI)H(I \vee GH)$.

(ii) Derive the complete sum for T.

(iii) Corresponding to each term of the complete sum, form the disjunction of the prime implicants that have their indices in this term. All disjunctive forms thus obtained are irredundant disjunctive forms for the given function f.

Skipping lengthy intermediate calculations, we find that the above T for our example is reduced to the complete sum, $AEFGH \vee AEFHI \vee BEFGH \vee BEFHI$. Corresponding to each term, we have the following irredundant disjunctive forms for the given function f:

$$AEFGH \rightarrow x_1 \bar{x}_3 \vee x_2 x_5 \vee \bar{x}_3 \bar{x}_4 \vee \bar{x}_3 \bar{x}_5 \vee x_3 x_4,$$
$$AEFHI \rightarrow x_1 \bar{x}_3 \vee x_2 x_5 \vee \bar{x}_3 \bar{x}_4 \vee x_3 x_4 \vee x_4 \bar{x}_5,$$
$$BEFGH \rightarrow x_1 x_4 \vee x_2 x_5 \vee \bar{x}_3 \bar{x}_4 \vee \bar{x}_3 \bar{x}_5 \vee x_3 x_4,$$
$$BEFHI \rightarrow x_1 x_4 \vee x_2 x_5 \vee \bar{x}_3 \bar{x}_4 \vee x_3 x_4 \vee x_4 \bar{x}_5. \quad \square$$

---

* In this table the inclusion functions for all the prime implicants for function f are listed. As noted in the footnote on p. 204 in relation to expression 5.1.1, if we can find a disjunctive form for f that consists of fewer prime implicants, we can form the inclusion functions for these prime implicants only. The Tison function formed by the conjunction of these inclusion functions still yields all irredundant disjunctive forms for f, as can be seen from the property discussed in Chang and Mott [1965].

212 Advanced Simplification Techniques and Basic Properties of Gates

Exercises

5.1.1 (R) By Procedure 5.1.1 find all the irredundant disjunctive forms for one of the following functions given in complete sum:

A. $f = \bar{x}_2\bar{x}_3 x_4 x_6 \vee \bar{x}_1 x_3 x_5 x_6 \vee \bar{x}_1\bar{x}_2 x_4 x_5 x_6 \vee x_2 x_3 \vee x_3 \bar{x}_4 x_5.$

B. $f = x_2 \bar{x}_3 x_4 \vee x_1 \bar{x}_3 \bar{x}_4 x_5 x_6 \vee x_1 x_3 x_4 x_5 \vee x_1 x_2 x_4 \vee x_2 x_4 \bar{x}_5 \vee x_1 x_2 x_5.$

C. $f = \bar{x}_2 x_3 x_5 \vee x_1 \bar{x}_3 x_4 x_5 x_6 \vee x_1 \bar{x}_2 x_3 \vee \bar{x}_2 x_3 \bar{x}_4 \vee x_1 x_3 x_4 \bar{x}_5 \vee x_1 \bar{x}_2 x_4 x_5 x_6.$

5.1.2 (R) Find minimal sums of the following functions (given in the complete sum for each), by drawing generalized-consensus formation trees according to Procedure 5.1.1. Explain why the results are minimal sums.

(i) $\quad x_1 \bar{x}_2 \vee \bar{x}_2 \bar{x}_3 \vee x_1 \bar{x}_3.$

(ii) $\quad x_1 x_2 \bar{x}_3 \vee x_2 x_3 \bar{x}_4 x_5 \vee x_1 \bar{x}_4 \vee x_1 x_5.$

(iii) $\quad x_1 \bar{x}_2 \vee x_2 x_3 \vee x_1 x_3.$

(iv) $\quad x_1 x_2 x_3 \vee \bar{x}_3 x_4 \vee x_1 x_4 \vee x_3 x_5 \vee x_4 x_5.$

(v) $\quad x_1 x_2 \vee \bar{x}_1 x_3 \vee x_1 x_4 \vee x_2 x_3 \vee x_3 x_4.$

(vi) $\quad x_1 x_2 \vee \bar{x}_1 x_3 \vee x_2 x_3 \vee x_3 x_4 x_5 \vee \bar{x}_4 x_6 x_7 \vee x_4 \bar{x}_7 \vee x_5 x_6.$

5.1.3 (R) Find an example of a four-variable function for each of the simple rules in Examples 5.1.1 through 5.1.3, and show it on a four-variable Karnaugh map.

5.1.4 (M−) Derive a Tison function for each of the functions in Figs. E5.1.4a and E5.1.4b directly from the Karnaugh map, as we did with Fig. 5.1.5b.

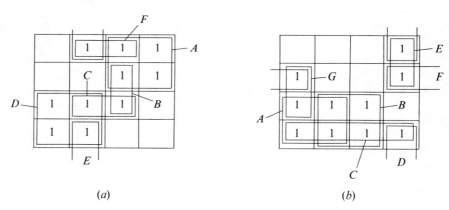

Fig. E5.1.4

5.1.5 (M−) Prove that, if a prime implicant P_i is not essential, the inclusion function for P_i in Procedure 5.1.1 contains at least one term consisting of two or more indices.

5.1.6 (E) Prove that in Procedure 5.1.1 any indexed product whose index consists of exactly two letters (e.g., $\bar{x}_1 x_2 x_4 x_5 x_6 AB$ in Fig. 5.1.2 for Procedure 5.1.1) cannot subsume any other indexed product ("subsume" in the meaning described in Step 1).

Derivation of Irredundant Disjunctive Forms 213

▲ **5.1.7 (RT)** Find, by Procedure 5.1.2, all irredundant disjunctive forms and minimal sums for one of the following functions. The complete sum is given for each function.

A. $f = x_1\bar{x}_2 x_5 x_6 \vee x_2 x_3 x_4 \vee x_1 \bar{x}_4 x_5 \vee \bar{x}_1 x_2 x_4 x_7 \vee x_1 x_3 x_5 \vee x_2 x_4 \bar{x}_5 \vee x_3 x_4 x_5$.

B. $f = x_1 x_3 \bar{x}_5 x_7 \vee x_1 x_4 \bar{x}_5 \bar{x}_6 x_7 \vee \bar{x}_3 x_4 \bar{x}_6 \vee \bar{x}_1 \bar{x}_2 x_4 x_5 \bar{x}_6 \vee \bar{x}_1 x_2 \bar{x}_3 x_5 \vee \bar{x}_1 \bar{x}_3 x_4 x_5$
$\vee \bar{x}_1 \bar{x}_3 x_5 x_6 \vee \bar{x}_1 x_2 \bar{x}_3 \bar{x}_6$.

C. $f = x_1 x_2 x_3 \bar{x}_5 x_7 \vee x_1 x_2 x_4 \bar{x}_5 \bar{x}_6 x_7 \vee \bar{x}_3 x_4 \bar{x}_5 \bar{x}_6 \vee \bar{x}_1 x_2 \bar{x}_3 \bar{x}_6 \vee \bar{x}_1 \bar{x}_3 x_4 \bar{x}_6$
$\vee \bar{x}_1 \bar{x}_2 x_4 \bar{x}_6 \vee \bar{x}_1 x_4 x_5 \bar{x}_6$.

(If the Karnaugh map method is used, eight maps will be needed in each case.)

5.1.8 (R) Prove the irredundant implication relation

$$u\bar{v} \subseteq \bar{x}\bar{y} \vee xz\bar{v} \vee yu \vee \bar{y}\bar{z}u \text{ (irr.)},$$

using some identities in Section 3.1. (We need to prove that, if $u\bar{v} = 1$, the consequent is identically equal to 1, and that, if any literal in the antecedent or any term in the consequent is deleted, the above implication relation does not hold.)

5.1.9 (R) All prime implicants of $f = xy \vee \bar{x}z$ are xy, $\bar{x}z$, and yz. Here $yz \subseteq xy \vee \bar{x}z$ (irr.) holds. According to Theorem 5.1.1, every irredundant disjunctive form for f must contain yz or $xy \vee \bar{x}z$. But if $xy \vee \bar{x}z$ in the above expression for f is replaced by yz alone, yz does not equal f. Therefore a student concluded that Theorem 5.1.1 was wrong.

Explain why his conclusion is wrong.

5.1.10 (M) Suppose that the disjunction of products of literals $A_1, A_2, \ldots, A_p$ is identically equal to 1:

$$A_1 \vee A_2 \vee \cdots \vee A_p = 1.$$

Also suppose that, if any product is deleted, the remainder is not equal to 1. Prove that no monoform variable appears among $A_1, A_2, \ldots, A_p$.

5.1.11 (D) The **generalized consensus** can be defined as follows: The product X of the literals of the monoform variables contained in t products, $A_1, A_2, \ldots, A_t$, is called the **generalized consensus** of $A_1, A_2, \ldots, A_t$ if and only if the disjunction of

$$\frac{A_1}{X}, \frac{A_2}{X}, \ldots, \frac{A_t}{X},$$

that is,

$$\bigvee_{i=1}^{t} \frac{A_i}{X},$$

is identically equal to 1, but none of

$$\frac{A_1}{X}, \frac{A_2}{X}, \ldots, \frac{A_t}{X}$$

can be deleted from the disjunction without invalidating the equality to 1. Here A_i/X is A_i, in which only the monoform variables appearing in X are set to 1. [Tison 1965]

For example, suppose that products $\bar{x}\bar{y}$, $xz\bar{v}$, yu, $\bar{y}\bar{z}u$ are given, where u and v are monoform variables, and the others are biform variables. Thus the product of the literals of the monoform variables in these products is $X = u\bar{v}$. Then

$$\frac{A_1}{X} = \frac{\bar{x}\bar{y}}{u\bar{v}}, \quad \frac{A_2}{X} = \frac{xz\bar{v}}{u\bar{v}}, \quad \frac{A_3}{X} = \frac{yu}{u\bar{v}}, \quad \frac{A_4}{X} = \frac{\bar{y}\bar{z}u}{u\bar{v}}$$

become $\bar{x}\bar{y}$, xz, y, $\bar{y}\bar{z}$, respectively. Their disjunction is identically equal to 1; and if any term is removed from this disjunction, the remainder is not identically equal to 1, as can easily be proved. According to the above definition, $u\bar{v}$ is the generalized consensus of the given products $\bar{x}\bar{y}$, $xz\bar{v}$, yu, and $\bar{y}\bar{z}u$.

Prove that the above definition of the generalized consensus is equivalent to Definition 5.1.2 with $A = X$. (The above example is stated in Exercise 5.1.8 in the format of Definition 5.1.2.)

5.1.12 (M) A designer obtained a two-level network of AND and OR gates (AND gates feeding an OR gate), based on a minimal sum. But the network has the complemented input $\bar{x}_i$ for certain i. The designer wondered whether she could obtain a network that does not have the complemented literal $\bar{x}_i$ of this particular x_i as a network input (but may have the noncomplemented literal x_i), if she derived another minimal sum. Before she started her new effort, she made the following conjecture, stretching her imagination. Prove or disprove it.

If there exists an irredundant disjunctive form of a given completely specified function f that contains the complemented literal $\bar{x}_i$ for a certain variable x_i (note that the noncomplemented literal x_i may or may not be present in this form), every irredundant disjunctive form of that function contains the complemented literal $\bar{x}_i$.

5.1.13 (M) Prove that the expression which results from the following procedure is a minimal product for a given function f. Derive a minimal sum for f^d, the dual of f. Then interchange the disjunctions and conjunctions in the minimal sum.

5.1.14 (E) Obtain a minimal product of $y(x \vee z)(\bar{x} \vee \bar{z}) \vee w(y \vee z)$ by the method of Exercise 5.1.13, and also by the Karnaugh map method based on encircling 0-cells in the map. Then show that the two results are identical.

5.2 Design of a Two-Level Multiple-Output Network with AND and OR Gates

So far, we have discussed the synthesis of a two-level network with a single output. In many cases in practice, however, we need a two-level network with multiple outputs, so let us discuss the synthesis of such a network in this section.

An obvious approach is to design a network for each output function separately. But this approach usually will not yield a less compact network than will synthesis of the functions collectively, because, for example, in Fig. 5.2.1 the AND gate h, which can be shared by two output gates for f_i and f_j, must be repeated in separate networks for f_i and f_j by this approach.

Before discussing a design procedure, let us study the properties of a two-level network that has only AND gates in the first level and only OR gates for given output functions $f_1, f_2, \ldots, f_m$ in the second level, as shown in Fig. 5.2.1. We want to minimize the number of gates as the first objective, and the number of connections as the second objective. The number of OR

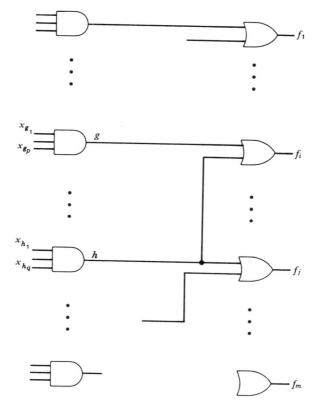

Fig. 5.2.1 A two-level network with multiple outputs.

gates required for this network is at most m. Actually, when some functions are expressed as single products of literals, the number of OR gates can be less than m, since these functions can be realized directly at the outputs of the AND gates without the use of OR gates. Also, when a function f_i has a prime implicant consisting of a single literal, that literal can be directly connected to the OR gate for f_i without the use of an AND gate. (See Remark 5.2.2 for these two possibilities.) However, for simplicity, **assume that every variable input can be connected only to AND gates in the first level and every function f_i, $1 \leq i \leq m$, must be realized at OR gates in the second level**. This is actually required with some electronic implementations of a network (e.g., when every variable input needs to have the same delay to the network outputs, or when PLAs are used to implement a two-level network).

Multiple-Output Prime Implicants

Suppose that we have a two-level multiple-output network that has AND gates in the first level, and m OR gates in the second (i.e., output) level. The network has a minimum number of gates as the primary objective, and a minimum number of connections as the secondary objective. Let us explore the basic properties of such a minimal network.

Property 5.2.1: First consider an AND gate g, which is connected to only one output gate, say for f_i, in Fig. 5.2.1. If gate g has inputs $x_{g_1}, \ldots, x_{g_p}$, its output represents the product $x_{g_1} \cdots x_{g_p}$. Then, if the product assumes the value 1, f_i becomes 1. Thus the product $x_{g_1} \cdots x_{g_p}$ is an implicant of f_i. Since the network is minimal, gate g has no unnecessary inputs, and the removal of any input from gate g will no longer make the OR gate for f_i express f_i. Thus $x_{g_1} \cdots x_{g_p}$ is a prime implicant of f.

Property 5.2.2: Next consider an AND gate h, which is connected to two output gates for f_i and f_j in Fig. 5.2.1. This time, the situation is more complicated. If the product $x_{h_1} \cdots x_{h_q}$ realized at the output of gate h assumes the value 1, both f_i and f_j become 1, and consequently the product $f_i \cdot f_j$ of the two functions also becomes 1. Thus $x_{h_1} \cdots x_{h_q}$ is an implicant of product $f_i \cdot f_j$. Since the network is minimal, removal of any input from this AND gate must let at least one of these OR gates not represent f_i or f_j. Thus $x_{h_1} \cdots x_{h_q}$ is a prime implicant of $f_i \cdot f_j$. (But notice that the product is not necessarily a prime implicant of each single function f_i or f_j, as we will see later.)

Generalizing this, we obtain the following conclusion. Suppose that an AND gate is connected to r output gates for functions $f_1, \ldots, f_r$. Since the network is minimal, this gate has no unnecessary inputs. Then the product realized at the output of this AND gate is a prime implicant of the product $f_1 \cdots f_r$ (but is not necessarily a prime implicant of any product of $r - 1$, or fewer, of these $f_1, \ldots, f_r$).

As in the synthesis of a single-output network, we need to find all prime implicants, and then develop disjunctive forms by choosing an appropriate set of prime implicants. (Notice that each of these disjunctive forms is not necessarily a minimal sum for a single function f_i. See Exercise 5.2.2.) But, unlike the synthesis of a single-output network, in this case all prime implicants not only for each of the given functions $f_1, \ldots, f_m$, but also for all possible products of them, $f_1 f_2, f_1 f_3, \ldots, f_{m-1} f_m; f_1 f_2 f_3, \ldots; \ldots; f_1 f_2 \cdots f_m$, must be considered.

Design of a Two-Level Multiple-Output Network   217

Definition 5.2.1: Suppose that m functions $f_1, \ldots, f_m$ are given. All prime implicants for each of $f_1, \ldots, f_m$, and also all prime implicants for every possible product of these functions, that is, $f_1 f_2, f_1 f_3, \ldots, f_{m-1} f_m; f_1 f_2 f_3, \ldots; \ldots; f_1 f_2 \cdots f_m$, are called the **multiple-output prime implicants** of $f_1, \ldots, f_m$.

When the number of variables is small, we can find all multiple-output prime implicants on Karnaugh maps, as illustrated in Fig. 5.2.2, for the case of three functions of four variables. In addition to maps for given functions f_1, f_2, and f_3, we draw maps for all possible products of these functions, that is, $f_1 f_2, f_2 f_3, f_1 f_3$, and $f_1 f_2 f_3$. Then prime-implicant loops are formed on each of these maps. These loops represent all the multiple-output prime implicants of given functions f_1, f_2, and f_3.

Paramount Prime Implicants

Among multiple-output prime implicants, the same ones correspond to different products of functions. For example, $x_1 \bar{x}_2 x_4$ appears twice in Fig. 5.2.2, as a multiple-output prime implicant for the function f_2 (see the map for f_2), and also as a multiple-output prime implicant for the product $f_2 f_3$. In the two-level network, $x_1 \bar{x}_2 x_4$ for the product $f_2 f_3$ realizes the AND gate with output connections to the OR gates for f_2 and f_3, whereas $x_1 \bar{x}_2 x_4$ for f_2 alone realizes the AND gate with output connection to the OR gate for f_2 only. Thus, if we use $x_1 \bar{x}_2 x_4$ for the product $f_2 f_3$ more often—in other words, if we use AND gates with more output connections—the network will be realized with no more gates. In this sense, $x_1 \bar{x}_2 x_4$ for the product $f_2 f_3$ is more desirable than $x_1 \bar{x}_2 x_4$ for the function f_2. Prime implicant $x_1 \bar{x}_2 x_4$ for the product $f_2 f_3$ is called a paramount prime implicant, as formally defined in the following, and is shown with label J in a bold line in the map for $f_2 f_3$, whereas $x_1 \bar{x}_2 x_4$ for f_2 alone is not labeled, and is also not in a bold line in the map for f_2.

Definition 5.2.2: Suppose that a product of some literals is a prime implicant for the product of k functions $f_{p_1}, f_{p_2}, \ldots, f_{p_k}$ (possibly also a prime implicant for products of $k - 1$ or fewer of these functions), but is not a prime implicant for any other product of more functions, including all these functions $f_{p_1}, f_{p_2}, \ldots, f_{p_k}$. (For the above example, $x_1 \bar{x}_2 x_4$ is a prime implicant for the product of two functions, f_2 and f_3, and also a prime implicant for a single function, f_2. But $x_1 \bar{x}_2 x_4$ is not a prime implicant for the product of more functions, including f_2 and f_3, i.e., for the product $f_1 f_2 f_3$.) Then the prime implicant for the product of $f_{p_1}, \ldots, f_{p_k}$ is called the **paramount prime implicant** ($x_1 \bar{x}_2 x_4$ for $f_2 f_3$ is a paramount prime implicant, but $x_1 \bar{x}_2 x_4$ for f_2 is not). In particular,

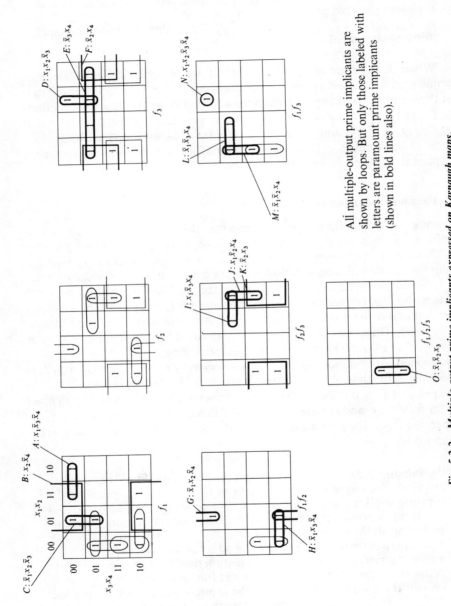

Fig. 5.2.2 *Multiple-output prime implicants expressed on Karnaugh maps.*

Design of a Two-Level Multiple-Output Network

if a product of some literals is a prime implicant for a single function only, it is a paramount prime implicant ($x_1 \bar{x}_3 \bar{x}_4$ for f_1 is such an example in Fig. 5.2.2).

In Fig. 5.2.2 only loops labeled with letters and also shown in bold lines represent paramount prime implicants.

In addition to the above method based on Karnaugh maps, all multiple-output prime implicants can be found by applying the tabular or algebraic methods discussed in the preceding chapters to each of the given functions and their products. (For other methods, see Remark 5.2.1)

> **Remark 5.2.1:** Bartee's method simultaneously generates all paramount prime implicants, using tables, as described in Exercise 5.2.5. The method starts with true input vectors. Kuntzman's method starts with any disjunctive forms and can use Procedure 3.4.2 to generate multiple-output prime implicants, as described on p. 126 of Kuntzman [1965]. ∎

If a two-level network is first designed only with the AND gates that correspond to the paramount prime implicants, we can minimize the number of gates. Then, if we eliminate unnecessary connections from the designed network, we can minimize the number of connections as the secondary objective, as formally stated in the following theorem. This design procedure is simpler than simultaneous consideration of the numbers of gates and connections.

> **Theorem 5.2.1:** Suppose that for given functions $f_1, f_2, \ldots, f_m$ two-level networks with a minimal number of gates, that is, t AND gates in the first level and m OR gates in the second (output) level, are constructed in the following manner, based on certain t paramount prime implicants. For each paramount prime implicant P for the product of functions $f_{j_1}, f_{j_2}, \ldots, f_{j_v}$, we form an AND gate that has the literals of P as its inputs and whose output is connected to the OR gates for functions $f_{j_1}, f_{j_2}, \ldots, f_{j_v}$ (e.g., if paramount prime implicant $L = \bar{x}_1 \bar{x}_3 x_4$ in Fig. 5.2.2 is chosen, we form an AND gate that has inputs $\bar{x}_1$, $\bar{x}_3$, and x_4 and whose output is connected to the OR gates for f_1 and f_3).
>
> Then, by deleting some output connections from AND gates in some of the above networks without changing the network outputs, we can derive all two-level networks each of which has the minimal number $t + m$ of gates (i.e., t AND gates in the first level and m OR gates in the second level) as the primary objective, and a minimal number of connections as the secondary objective.

> **Proof:** Suppose that we have a two-level network with a minimal number of gates as the primary objective, and a minimal number of connections as the secondary objective, as derived in the second paragraph of this theorem statement. Assume

220 *Advanced Simplification Techniques and Basic Properties of Gates*

that some AND gates in this network do not correspond to paramount prime implicants (e.g., even if the network contains an AND gate whose output represents $\bar{x}_1\bar{x}_3 x_4$, the output of this AND gate is connected, not to both OR gates for f_1 and f_3, but to only one of them, though paramount prime implicant $L = \bar{x}_1\bar{x}_3 x_4$ requires connections to the OR gates for f_1 and f_3). Then add extra connections to the outputs of these AND gates so that the AND gates correspond to paramount prime implicants (e.g., the AND gate whose output represents $\bar{x}_1\bar{x}_3 x_4$ corresponds to paramount prime implicant $L = \bar{x}_1\bar{x}_3 x_4$ by being connected to both OR gates for f_1 and f_3). By this conversion the output functions at all the OR gates in the network do not change. Also, the number of AND gates does not change, and consequently the total number of gates (including OR gates) is minimal. Thus we have obtained a network (or networks) as stated in the first paragraph in this theorem statement.

Reversal of the above process is always possible, and the number t of AND gates does not change. Q.E.D.

Design of a Minimal Two-Level Network of AND and OR Gates

In the following procedure we shall first derive a two-level network by finding a minimal number of paramount prime implicants on a prime-implicant table similar to the table used in Procedure 4.5.2. The table reduction can be done in a manner somewhat similar to Procedure 4.5.2. When we reach a cyclic table, a Petrick function can be simplified more quickly by algebraic manipulation similar to that in Procedure 4.6.1. Then, in this network derived by using paramount prime implicants, unnecessary connections from some AND gates to OR gates are deleted (as stated in the second paragraph of Theorem 5.2.1).

Procedure 5.2.1: *Design of a Multiple-Output Two-Level Network that Has a Minimum Number of Gates as the Primary Objective, and a Minimum Number of Connections as the Secondary Objective*

(The number of connections is the total number of inputs to all AND and OR gates.)

We want to design a two-level network that has AND gates in the first level and OR gates in the second level. We shall assume that variable inputs can be connected to AND gates only (not to OR gates), and that the given output functions $f_1, f_2, \ldots, f_m$ are to be realized only at the outputs of OR gates.

Suppose that the three functions of four variables f_1, f_2, f_3 shown in the Karnaugh maps in Fig. 5.2.2 are given.

1. Find the paramount prime implicants among all the multiple-output prime implicants of given functions $f_1, f_2, \ldots, f_m$.

On the Karnaugh maps in Fig. 5.2.2, only the loops labeled with letters are paramount prime implicants.

2. Draw a prime-implicant table* that shows the relationship among the paramount implicants (only those labeled with letters in Step 1) and the minterms of f_1, f_2, f_3, as shown in Table 5.2.1a, in a manner similar to Procedure 4.5.2. The columns are grouped by given functions f_1, f_2, and f_3, separated by lines. The rows are also grouped by the given functions and their products. Notice that some rectangular areas are completely empty.

We want to select the minimum number of rows that satisfies the following conditions:

(a) Every column has a × in at least one of the selected rows.
(b) The number of connections in the network to be designed on the basis of the selected set of rows is minimized.

Since the number of OR gates is a constant m, minimization of the number of rows means minimization of the number of AND gates and, consequently, of the total number of gates. But meeting condition (b) is not easy in Table 5.2.1a, since even if a paramount prime implicant is for the product of f_1 and f_3, for example, the AND gate for this prime implicant is not necessarily connected to the OR gates for f_1 and f_3 in the final network to be derived, as we can see from Theorem 5.2.1. (This will be explained later in detail in Step 5.) So condition (b) will not be considered until Step 5.

From this observation, we have the following steps.

3. Reduce the prime-implicant table without eliminating dominated† rows. In other words, apply the following two operations only.

(i) Select essential rows, at the same time deleting all columns that contain ×'s in these rows. Label these rows with asterisks.

> *For our example* in Table 5.2.1a, rows A, B, G, I, and K are essential rows, since columns 8 and 14 for f_1, columns 4, 10, and 13 for f_2, and column 10 for f_3 (these columns are encircled) have single ×'s in these rows. Therefore select these rows, at the same time deleting columns (check-marked) that contain ×'s in these rows.

(ii) Delete any columns that dominate† other columns. This is done because the dominated column must be covered by rows that have ×'s in this

---
* For the sake of illustration, Steps 2 and 3 are described with prime-implicant tables. But, as we did in Procedure 4.6.1, these steps can be processed completely with switching expressions, without the use of prime-implicant tables, by forming a Petrick function at Step 2. This, in general, greatly reduces the space and time required for hand processing.
† For the definition of "domination" between rows or columns, see Definition 4.5.1 or 4.5.2.

Table 5.2.1a Multiple-Output Prime-Implicant Table for Functions in Fig. 5.2.2

FUNCTIONS	PRIME IMPLICANTS	\multicolumn{9}{c}{MINTERMS FOR f_1}	\multicolumn{8}{c}{f_2}	\multicolumn{9}{c}{f_3}

FUNCTIONS	PRIME IMPLICANTS	1	2	3	4	5	6	⑧	12	⑭	2	3	④	6	9	⑩	11	⑬	1	2	3	5	9	⑩	11	12	13
f_1	A^*				×	×	×	×																			
	B^*				×	×			×	×																	
	C																										
f_3	D																		×	×	×						
	E																		×	×			×	×	×	×	
	F																				×	×	×	×			
$f_1 f_2$	G^*		×				×							×	×												
	H			×							×																
$f_2 f_3$	I^*														×	×	×						×	×	×		×
	J											×									×		×	×	×		
	K^*											×		×				×									
$f_1 f_3$	L	×				×													×	×		×					
	M	×	×																×		×						
	N								×																	×	
$f_1 f_2 f_3$	O	×	×								×	×							×	×							
		✓		✓	✓	✓	✓	✓	✓	✓	✓	✓	✓	✓	✓	✓	✓	✓	✓	✓		✓	✓	✓	✓		✓

column, and the dominating column is then automatically covered by these rows. Notice that this has nothing to do with the number of gates or connections in the network to be obtained.

In Table 5.2.1a, column 1 for f_3 can be deleted, since it dominates column 1 for f_1 (or since it dominates column 5 for f_3). Thus Table 5.2.1b has been obtained.

Notice that here unlike Procedure 4.5.2, dominated rows cannot be deleted because the networks that result may not meet condition (b) of Step 2, that is, may not have a minimal number of connections, as will be illustrated later.

4. **Form a Petrick function*** by forming an alterm of letters representing rows with respect to each column, and then forming the conjunction of all the alterms obtained (in the same way as we did in Procedure 4.5.2).

Next we want to **derive all terms consisting of a minimum number of letters in the complete sum of the Petrick function**, since all other terms consisting of more letters will yield networks with more gates and consequently must be discarded.

These terms consisting of a minimum number of letters can be derived quickly as follows, instead of explicitly deriving the complete sum, as we did in Procedure 4.5.2, a method that is usually very time-consuming. The following calculation method is a modification of Procedure 4.6.1, by taking into consideration that no single letter in a multiletter alterm is deleted here. (In other words, Step 3 of Procedure 4.6.1 is not applied, so we need some technique to speed the calculation, as follows.)

(a) Check whether any letter appears more than once in each product of alterms. [Initially the entire expression for the Petrick function consists of a single product. But after the first application of this operation (a), the entire expression consists of more than one product.] If some letters appear more than once in a product, multiply out some appropriate multiletter alterms in that product [e.g., alterms such that operation (c) is applicable as often as possible, or alterms that share common literals] into a **single** disjunctive form.

For our example in Table 5.2.1b, we obtain the following Petrick function:

$$p = \underline{(L \vee M)}(H \vee O)(M \vee O)\underline{(C \vee L)}(E \vee L)(D \vee N) \quad (5.2.1)$$

(Multiply the two underlined alterms into a single disjunctive form, $L \vee MC$.)

$$= (L \vee MC)(H \vee O)(M \vee O)(E \vee L)(D \vee N). \quad (5.2.2)$$

Then expand each product of alterms into products by distributing out the product with respect to the disjunctive form just obtained.

---

* A Petrick function can be formed as early as at Step 2, instead of Step 4, in order to reduce the processing time.

224 Advanced Simplification Techniques and Basic Properties of Gates

Table 5.2.1b Reduced Table

FUNCTIONS	PRIME IMPLICANTS	MINTERMS FOR					
		f_1				f_3	
		1	2	3	5	5	12
f_1	C				×		
f_3	D						×
	E					×	
	F						
$f_1 f_2$	H		×				
$f_2 f_3$	J						
$f_1 f_3$	L	×		×	×		
	M	×		×			
	N						×
$f_1 f_2 f_3$	O	×	×				

Expression 5.2.2 is expanded into the following two products of alterms, by distributing out with respect to $(L \vee MC)$:

$$p = L(H \vee O)(M \vee O)(E \vee L)(D \vee N)$$
$$\vee\ MC(H \vee O)(M \vee O)(E \vee L)(D \vee N). \qquad (5.2.3)$$

(b) If no letter appears more than once in a product of alterms, count the number of letters to be contained in a term if that product is expanded into its complete sum (notice that the complete sum will be a partially symmetric function consisting of all terms with equal numbers of letters). Then, among such products of alterms in the entire expression of the Petrick function, discard the products whose terms have more letters than other products* (e.g., the second and third products of alterms in expression 5.2.6 in the following are discarded).

* Even when some letters appear more than once in products, we can sometimes discard some products, speeding up the computation further. For example, the second product in (5.2.4) will have terms consisting of five or more letters each, if it is expanded into its complete sum. But the first product will have terms consisting of three or more letters. Hence the second product in (5.2.4) can be discarded before reaching (5.2.6.).

Also, when a single disjunctive form containing many single-letter terms such as $(R \vee S \vee TUV)$ is obtained in operation (a), partially distributing out a product containing such a disjunctive form, say $(R \vee S \vee TUV)(W \vee X)(Y \vee Z)$ into $(R \vee S)(W \vee X)(Y \vee Z) \vee TUV(W \vee X)(Y \vee Z)$, may facilitate the detection of products that can be discarded.

(c) In each new product of alterms, delete multiletter alterms that subsume any single-letter alterms present in that product.

In the first product of (5.2.3), multiletter alterm $(E \vee L)$ can be deleted, since it subsumes the single-letter alterm L in the first product, as illustrated in the following. In the second product, $(M \vee O)$ can be deleted, since it subsumes M:

$$p = L(\underline{H \vee O})(\underline{M \vee O})(\cancel{E \vee L})(D \vee N)$$
$$\vee\ MC(H \vee O)(\cancel{M \vee O})(E \vee L)(D \vee N) \qquad (5.2.4)$$

(d) Repeat (a), (b), and (c) in any order until none can be applied. Then expand the remaining expression into its complete sum.

For the above p, repeat (a), (b), and (c). In the first product of (5.2.4), multiply two underlined alterms into a single disjunctive form. In the second product, every letter appears at most once, so we need not multiply out multiletter alterms according to operation (b).

$$p = L(O \vee HM)(D \vee N) \vee MC(H \vee O)(E \vee L)(D \vee N) \quad (5.2.5)$$

(Distribute the first product with respect to the new single disjunctive form $(O \vee HM)$ into two products.)

$$\begin{aligned}
&= LO(D \vee N) && \text{3 letters,} \\
&\vee\ LHM(D \vee N) && \text{4 letters,} \\
&\vee\ MC(H \vee O)(E \vee L)(D \vee N) && \text{5 letters.} \qquad (5.2.6)
\end{aligned}$$

[Each letter appears only once in each product of alterms. According to operation (b), we count how many letters each term consists of, if each product of alterms is expanded into its complete sum. Each term for the first, second, and third products consists of three, four, and five letters, respectively. From now on, therefore, the second and third products need not be considered. Then we multiply out the first product. Thus p is reduced* as follows: $p \Rightarrow LOD \vee LON$.]

Multiply each term obtained for the complete sum by the letters representing the rows selected in Step 3, if any rows were selected. Then the terms which result represent all sets of the minimum number of paramount prime implicants that cover the given functions.

Thus LOD and LON are all the terms with the minimal number of letters if the Petrick function is expanded into its complete sum.† Since we have already selected rows A, B, G, I, and K in Table 5.2.1a, we have

$$ABGIKLOD \quad \text{and} \quad ABGIKLON \qquad (5.2.7)$$

as all possible sets of the minimal number of paramount prime implicants that cover the given functions f_1, f_2, and f_3.

---
* The symbol "$\Rightarrow$" is used to show that p is reduced to a new expression.
† Instead of doing computations according to operations (a), (b), and (c), if we continue to multiply out alterms without splitting each product into new products [e.g., instead of splitting the product in (5.2.2) into two in (5.2.3), we continue to multiply alterms in a single product, $(L \vee MC)(H \vee O)(M \vee O)(E \vee L)(D \vee N) = (L \vee MC)(O \vee HM)(ED \vee EN \vee LD \vee LN) = \ldots$], the computation will usually become much more time-consuming and may never be ended because of errors.

226 *Advanced Simplification Techniques and Basic Properties of Gates*

5. In the network implemented, on the basis of each of the terms obtained in Step 4, delete any unnecessary connections. Among all the networks that result from all the terms obtained in Step 4, those with the minimum number of connections are the desired networks. (If we need to examine too many networks, we might choose the best among networks examined within a reasonable time limit, since the number of gates is already minimized, and, in practice, networks with the absolutely minimal number of connections may not really be necessary. In this case the technique for partial reduction mentioned in Remark 5.2.5 can be used.)

For our example, let us find networks with a minimal number of connections for the two terms *ABGIKLOD* and *ABGIKLON* obtained in (5.2.7) in Step 4. First, corresponding to *ABGIKLOD*, the network in Fig. 5.2.3 has been obtained. Although the number of gates in this network has been minimized, the number of connections is not necessarily the minimum, and is actually the maximum (for this particular set of AND gates), since paramount prime implicants, which are prime implicants of products of a greatest number of functions, are used. Therefore

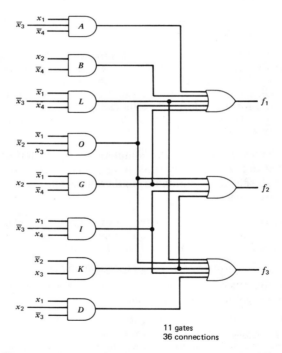

Fig. 5.2.3 The network corresponding to **ABGIKLOD**. (*Since the outputs of the AND gates represent paramount prime implications, this network does not necessarily have a minimum number of connections.*)

Design of a Two-Level Multiple-Output Network 227

let us eliminate redundant connections, following the reversal stated in the second paragraph in Theorem 5.2.1. This can easily be done by using Karnaugh maps* as shown in Fig. 5.2.4 and finding redundant paramount prime implicants for each of the given functions. For f_1, paramount prime implicant G is redundant, since all 1-cells of f_1 can be covered without G. Therefore the connection from the AND gate for G to the OR gate for f_1 can be eliminated, as shown in Fig. 5.2.5. Similarly, O is a redundant paramount prime implicant for f_2 and f_3. By eliminating the corresponding connections from Fig. 5.2.3, we have obtained the network in Fig. 5.2.5, which has the minimal number of connections, starting from the network in Fig. 5.2.3 for $ABGIKLOD$.

For the other term, $ABGIKLON$, obtained in (5.2.7), we have derived the network with the minimal number of connections in Fig. 5.2.6, by deleting some connections, based on the Karnaugh maps in Fig. 5.2.7.

Comparing the two networks in Figs. 5.2.5 and 5.2.6, we conclude that the network in Fig. 5.2.5 has the minimal number of connections, 33. ☐

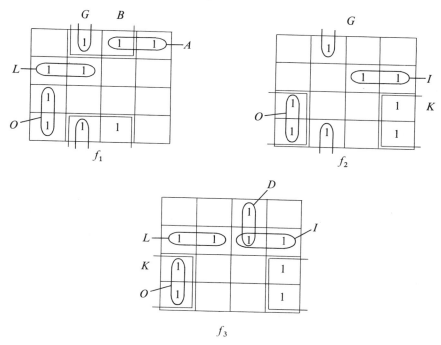

Fig. 5.2.4 Karnaugh maps for **ABGIKLOD**.

---

* Step 5 can also be processed algebraically without the use of Karnaugh maps, as we did in Procedure 4.6.1 (Exercise 5.2.6).

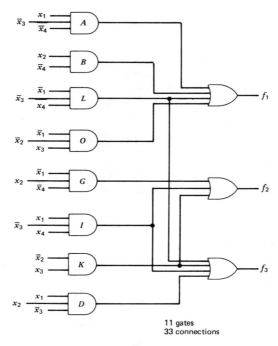

Fig. 5.2.5 The network with a minimal number of connections derived from **ABGIKLOD**.

Each paramount prime implicant in the terms obtained in Step 4 is not redundant for at least one of the given functions (Exercise 5.2.3). For each term, therefore, Step 5 can be processed quickly, though if too many terms are obtained in Step 4, the total time for processing all of them may become excessive.

Procedure 5.2.1* is straightforward because the processing of a prime-implicant table is done separately from the minimization of the connection count. Also, when minimization of the connection count is not mandatory but fewer connections are still desirable, Procedure 5.2.1 has the convenience of totally or partially skipping Step 5 (Remark 5.2.5). Of course, if a network is too big to be processed by this approach, we must give up minimization of the connection count, and use Procedure 5.2.2, to be discussed next.

---

* The description of Procedure 5.2.1 is very long. For a quick review, read only the boldface sentences, which are summaries of the steps.

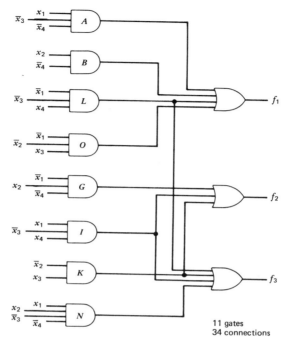

Fig. 5.2.6 The network with a minimal number of connections derived from ABGIKLON.

Why Dominated Rows Are Not Deleted in Procedure 5.2.1

Let us illustrate why we may not be able to minimize the number of connections, even with the number of gates minimized, if dominated rows are deleted, in Procedure 5.2.1 (though sometimes we may be able to minimize; Remark 5.2.4). Suppose that two functions, f_1 and f_2, are given, as shown in Fig. 5.2.8. Using the paramount prime implicants obtained on Karnaugh maps in Fig. 5.2.8, we derive the prime-implicant table in Table 5.2.2a. By selecting rows C and D for essential prime implicants of the product $f_1 f_2$, the reduced table in Table 5.2.2b is obtained. Here, unlike Procedure 5.2.1, let us delete row A, since A is dominated by B, and the prime implicants for A and B both consist of three literals. Thus we have obtained the set of rows CDB, and correspondingly the network of five gates and 15 connections in Fig. 5.2.9a. Actually, the network in Fig. 5.2.9b, which corresponds to CDA, discarded during the procedure, has one less connection. Thus, if we delete dominated rows from a prime-implicant table, we may not be able to obtain

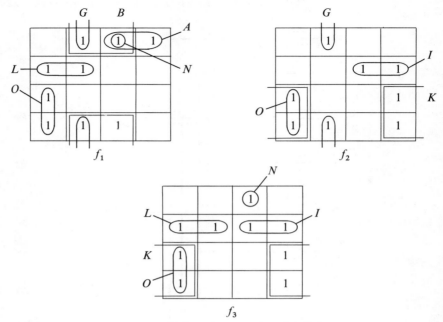

Fig. 5.2.7 Karnaugh maps for **ABGIKLON**.

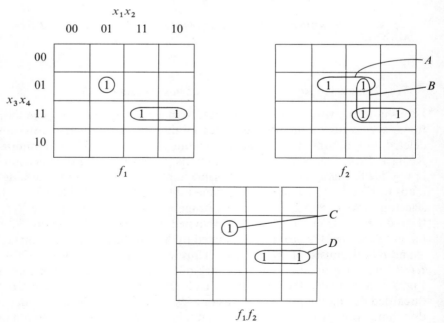

Fig. 5.2.8 Multiple-output prime implicants of given functions f_1 and f_2.

Design of a Two-Level Multiple-Output Network

Table 5.2.2   Prime-Implicant Tables

a.

FUNCTIONS	NUMBER OF LITERALS	PRIME IMPLICANTS	f_1			f_2		
			⑤	⑪	⑮	5	⑪	13	15
f_2	3	A				×	×		
	3	B					×	×	×
$f_1 f_2$	4	C	×			×		×	×
	3	D		×	×		×		×
			✓	✓	✓	✓	✓		✓

b.

FUNCTIONS	NUMBER OF LITERALS	PRIME IMPLICANTS	f_2
			13
f_2	3	A	×
	3	B	×

a network with the minimal number of connections, although the number of gates is minimized. Of course, if we use Procedure 5.2.1, we obtain the Petrick function $CD(A \vee B)$ from Table 5.2.2b; and, after comparing the two networks in Fig. 5.2.9, we can choose the better one.

Remark 5.2.2: In a network to be derived by Procedure 5.2.1, all variable inputs are to be connected to AND gates only (not to OR gates), and output functions to be realized at the outputs of OR gates only (not AND gates). But if variable inputs can be directly connected to OR gates (i.e., some output functions have single-literal prime implicants), or output functions can be realized at AND gates (i.e., some output functions consist of single prime implicants), Procedure 5.2.1 needs to be modified.

(a) If there are single-literal multiple-output prime implicants, these literals can be directly connected to the OR gates for the corresponding functions. But if these prime implicants are included in the prime-implicant table in Step 2 of Procedure 5.2.1, Procedure 5.2.1 may sometimes not yield an minimal network, as the following example illustrates. When three functions are given, as shown in Fig. 5.2.10, Procedure 5.2.1 yields the network in Fig. 5.2.11a (Exercise 5.2.4). However, the network in Fig. 5.2.11b has one less

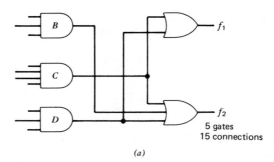

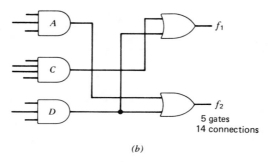

Fig. 5.2.9 *Networks for the functions in Fig. 5.2.8.*

connection. Therefore first derive minimal networks by straightforwardly applying Procedure 5.2.1. If these networks contain AND gates corresponding to single-literal multiple-output prime implicants, delete these AND gates and connect their inputs directly to the corresponding OR gates. But if these networks do not contain AND gates corresponding to single-literal multiple-output prime implicants, apply Procedure 5.2.1, starting with the prime-implicant tables, in which some or all single-literal multiple-output prime implicants and all the columns covered by them are deleted. Then connect the single-literal multiple-output prime implicants to the corresponding OR gates in the networks obtained. Compare these networks with those previously obtained by straightforward applications of Procedure 5.2.1, and choose the best networks among them.

(b) The case of the realization possibility of output functions at AND gates is more cumbersome to treat. First, derive minimal networks by straightforwardly applying Procedure 5.2.1. If these networks contain OR gates that have single inputs (i.e., the outputs of these OR gates realize the corresponding output functions with single prime implicants), delete these OR gates and use their inputs as output functions. In this case, at Step 4 of Procedure 5.2.1, it is necessary to find not only all terms consisting of a minimal number of letters, but also terms consisting of at most k extra

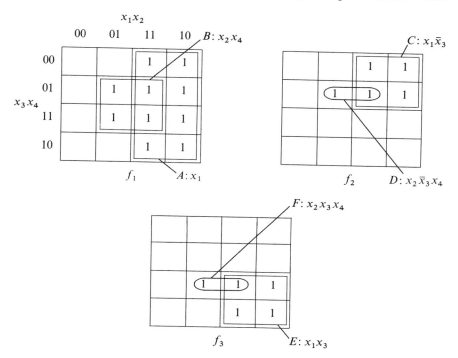

Fig. 5.2.10 Example functions for Remark 5.2.2a.

letters, if k out of m given functions can be directly realized at AND gates without OR gates. (An upper bound on the value of k can be found by examining whether each of the m given functions can be expressed by a single prime implicant.) Next, if some functions that can be expressed as single prime implicants are not realized as single prime implicants at OR gates in the minimal networks derived by the above straightforward application of Procedure 5.2.1, apply Procedure 5.2.1, but start with the prime-implicant tables in which some or all of the single prime implicants (by which some of the given functions can be expressed) and all the columns covered by them are deleted. Then add the AND gates whose outputs realize the single prime implicants to the networks obtained. Compare these networks with those previously obtained and choose the best ones.

Remark 5.2.3: Procedure 5.2.1 has variations (Remarks 5.2.4 and 5.2.5). Use of paramount prime implicants in Procedure 5.2.1 instead of all prime implicants makes the initial tables (or, equivalently, the Petrick functions) simpler, and for this reason, we can have Procedure 5.2.2, which is useful when there is no need to minimize the connection count. The following straightforward procedure, however, may sometimes be useful. All multiple-output prime implicants, including

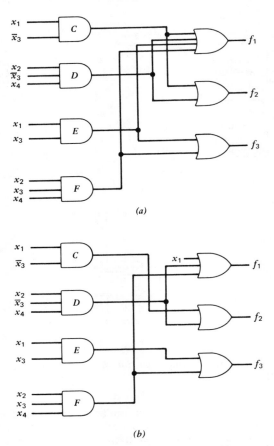

Fig. 5.2.11   Networks for Fig. 5.2.10.

paramount prime-implicants, are used to derive a Petrick function (or, equivalently, a prime-implicant table), and then essential rows, dominating columns, and dominated rows are eliminated by using Procedure 4.6.1. (When we try to eliminate a dominated row, the total number of inputs and fan-out connections of a gate for a dominated row must be compared with that for a dominating row. Only if the former is not smaller, can the dominated row be eliminated.) Since the elimination of connections, as in Step 5 of Procedure 5.2.1, need not be considered separately, this procedure may work quickly for some functions if many dominated rows can be eliminated, though initially we need to deal with many more prime implicants than in the case of Procedure 5.2.1. ∎

Remark 5.2.4: We can have variations of Step 3 of Procedure 5.2.1. For example, we can delete dominated rows without losing a minimal network under certain

Design of a Two-Level Multiple-Output Network 235

conditions. But consideration of such conditions would make Procedure 5.2.1 very complicated, and tables would usually not be greatly simplified. ∎

Remark 5.2.5: When there are too many networks to be examined in Step 5 of Procedure 5.2.1, we may want to examine only some of them in order to save time. In this case we can reduce the number of connections by the following technique.

When we examine some network, say the network in Fig. 5.2.6, we try to replace each loop by a larger prime-implicant loop on the corresponding Karnaugh maps in Fig. 5.2.7, after deleting redundant paramount prime implicants, such as G and N for f_1, and O for f_2 and f_3. In this case, loop N on the map for f_3 can be replaced by larger prime-implicant loop D, shown in Fig. 5.2.4. (In other words, we can get Fig. 5.2.4 from Fig. 5.2.7, and consequently the network in Fig. 5.2.5.) By this technique we can sometimes reduce the number of inputs to AND gates.

Notice that, if all the networks are examined in Step 5, this technique is not necessary, since the networks that can be derived from others by this technique will be contained among all the networks to be examined. ∎

Remark 5.2.6: Design procedures different from Procedure 5.2.1 are discussed in the literature. (See, e.g., McCluskey [1965], Bartee, Lebow and Reed [1962], Booth [1971], or Miller [1965].) ∎

Simplified Design Procedure without Minimization of the Number of Connections

When compactness or reduction of the parasitic capacitance is vitally important for some key networks in digital systems, the number of connections can be reduced by Procedure 5.2.1. But for some electronic implementations (e.g., PLAs that are implemented by some electronic means), we need not minimize the number of connections. In this case we want to derive networks that have a minimum number of gates, without considering the number of connections. Then Procedure 5.2.1 can be greatly simplified as follows. Also, when networks are too large to be processed by Procedure 5.2.1 or other variations, we have to give up minimization of the connection count, and then Procedure 5.2.2 can be used if we still want to minimize the gate count.

Procedure 5.2.2: Design of a Multiple-Output Two-Level Network That Has a Minimum Number of Gates without Minimizing the Number of Connections

By algebraically manipulating switching expressions (as in Procedure 4.6.1), we want to design a multiple-output network that has a minimum number of gates, without considering the number of connections. Variable inputs are to be connected to AND gates only, and output functions are to be realized at the outputs of OR gates only.

236 Advanced Simplification Techniques and Basic Properties of Gates

1. **Derive all paramount prime implicants** for the given functions $f_1, f_2, \ldots, f_m$ by any method. (For example, the tabular method in Exercise 5.2.5 directly yields all paramount prime implicants.)

2. **Form the Petrick function** for $f_1, f_2, \ldots, f_m$, **using only the paramount prime implicants** obtained in Step 1. (How to form a Petrick function is described in Procedure 4.6.1, Step 1.)

   For example, suppose that the Petrick function corresponding to Table 5.2.1a is obtained as follows:
   $$p = (L \vee M)(H \vee O)(M \vee O)(B \vee C \vee G)(C \vee L)(B \vee G \vee H)A$$
   $$(A \vee B \vee N)B(H \vee K \vee O)(K \vee O)G(G \vee H)(I \vee J)K(J \vee K)I$$
   $$(E \vee F \vee L \vee M)(K \vee O)(F \vee K \vee M \vee O)(E \vee L)(E \vee F \vee I \vee J)$$
   $$K(F \vee J \vee K)(D \vee N)(D \vee E \vee I).$$

3. **Simplify the Petrick function** as follows. (This is similar to Procedure 4.6.1, but because there is no need to consider the number of connections, greater simplification of the Petrick function is usually possible, and consequently the processing may be faster.)

   (i) In each product of alterms [initially, the entire expression of the Petrick function consists of a single product. But after the first application of Substep (iii), the entire expression consists of more than one product], delete multiletter alterms that subsume other single-letter or multi-letter alterms. (Of course, delete duplicate alterms as a special case.)

   (ii) In each product of alterms, consider a pair of letters, say L_i and L_j. (The pair need not be in alterms for the same function f_k, where $1 \leq k \leq m$.) If one letter, L_j, appears in every alterm where the other letter, L_i, appears, delete L_i. (The numbers of literals in the prime implicants that L_i and L_j represent need not be compared, because the number of connections need not be minimized.) Repeat this for every such pair. Notice that the possibility of finding such a pair greatly increases now, compared with Step 3 of Procedure 4.6.1, and consequently many more rows can be deleted, since the number of literals need not be considered.

   (iii) In each product of alterms, multiply out appropriate alterms [e.g., alterms such that Substeps (i) and (ii) are applicable as often as possible, or alterms that share common literals] into a single disjunctive form. Then distribute out that product of alterms with respect to this single disjunctive form.

   (iv) Repeat Substeps (i), (ii), and (iii) in any order until no substep can be applied.

   For illustration, let us work on the above Petrick function:
   $$p = (\cancel{L \vee M})(H \vee O)(\cancel{M \vee O})(\cancel{B \vee C \vee G})(\cancel{C \vee L})(\cancel{B \vee G \vee H})A$$
   $$(\cancel{A \vee B \vee N})B(\cancel{H \vee K \vee O})(\cancel{K \vee O})G(\cancel{G \vee H})(\cancel{I \vee J})K(\cancel{J \vee K})I$$
   $$(\cancel{E \vee F \vee L \vee M})(\cancel{K \vee O})(\cancel{F \vee K \vee M \vee O})(E \vee L)(\cancel{E \vee F}$$
   $$\cancel{\vee I \vee J})K(\cancel{F \vee J \vee K})(D \vee N)(\cancel{D \vee E \vee I}).$$

[By (i), $(B \vee C \vee G)$, $(B \vee G \vee H)$, and $(A \vee B \vee N)$ are deleted, since they subsume B. $(G \vee H)$ is deleted, since it subsumes G. $(H \vee K \vee O)$, $(K \vee O)$, $(J \vee K)$, $(K \vee O)$, $(F \vee K \vee M \vee O)$, $(F \vee J \vee K)$, and the duplicated K are deleted, since they subsume K. $(I \vee J)$, $(E \vee F \vee I \vee J)$, and $(D \vee E \vee I)$ are deleted, since they subsume I. $(E \vee F \vee L \vee M)$ is deleted, since it subsumes $(E \vee L)$. By (ii), H, E and N can be deleted. Then, by (i), $(L \vee M)$ and $(C \vee L)$ are deleted, since they subsume L, and $(M \vee O)$ is deleted, since it subsumes O.] Thus p is reduced to the following: $p \Rightarrow OABGKILD$.

4. **For a term with the fewest letters obtained in Step 3** (if more than one such term is obtained), **a unique network is obtained.** (Since the number of connections is not considered, the deletion of redundant connections in Step 5 of Procedure 5.2.1 is not necessary.) This network has the minimal number of gates.

    For our example, the network of 11 gates shown in Fig. 5.2.3 has been uniquely obtained. (Notice that, depending on the order of applying Substeps (i), (ii), and (iii) in Step 3, we may have different networks, though all the networks have the same minimum number of gates.) □

Procedure 5.2.2 saves the time expended in drawing tables or maps, and can also be processed very quickly by hand, since it uses simplification of a switching expression. Furthermore, Procedure 5.2.2 is much simpler than Procedure 5.2.1, since we ignore the number of connections.

Modifications of Procedure 5.2.2

When even Procedure 5.2.2 becomes too time-consuming as the number n of variables and the number m of outputs increase, we can design a network by a less time-consuming approach, such as the following: deriving a network with a minimal number of gates, but without minimizing the number of connections, for each of the given functions $f_1, f_2, \ldots, f_m$, and then sharing as many AND gates as possible for the entire set of functions. Of course, in this case the number of gates in the entire network may not be minimized.

The above procedures can easily be extended to the case where incompletely specified functions [certain combinations of input variable values may not occur, or some outputs may not be important for some combinations of output (or input variable) values] are given. The only modification to be made is the formation of multiple-output prime implicants by taking don't-care conditions into account (Exercise 5.2.16).

For other heuristic procedures, see the end of Section 5.3.

Remark 5.2.7: If L_i is deleted in Step 3(ii) of Procedure 5.2.2 only when the number of literals in the prime implicant corresponding to L_i is no less than that corresponding to L_j, and if in Step 4 we choose such a term that the total number of literals in the prime implicants expressed by the letters constituting this term is the minimum, we can derive a network with a minimal number of inputs to AND gates, but without minimizing the total number of connections including those from AND gates to OR gates. This design procedure is a compromise between Procedures 5.2.1 and 5.2.2. Minimization of the number of inputs to AND gates is useful for reducing the fan-out of the gates (to be outside the network under consideration) that supply the input variables to this network. ∎

Applications of the Preceding Design Procedures

A network designed by any procedure in this section is useful in the following cases. First, it can be converted to a minimal multiple-output network of two levels consisting of NOR gates or NAND gates only, as will be discussed in Section 6.1, when we want the fastest network of NOR or NAND gates under the assumptions that both complemented and noncomplemented variables are available as inputs to the network, that all gates have equal switching times, and that no maximum fan-in or fan-out restriction is imposed on each gate. Second, networks designed by the preceding procedures are important when we implement them with integrated circuits such as PLAs, as will be discussed in Section 9.4.

Networks that Cannot Be Designed by the Preceding Procedures

Notice that the design procedures in this section yield only a network that has all AND gates in the first level and all OR gates in the second level. (If 0-cells on Karnaugh maps are worked on instead of 1-cells in the above procedures, we have a network with all OR gates in the first level and all AND gates in the second level.) Therefore these procedures present the following two problems.

1. Weiner and Dwyer showed an example of a network with multiple outputs that has fewer gates if a mixture of AND and OR gates is permitted in each level [Weiner and Dwyer 1968]. For example, the network in Fig. 5.2.12 has a mixture of AND and OR gates in each level. This network has the minimum number of gates as the primary objective, and the minimum number of connections as the secondary objective. If different gate types are not permitted to be mixed in each level, this network will require more gates or connections.

2. The output of each output gate in a network designed by these procedures is not connected to other gates. If the outputs of some output gates are connected to other gates—in other words, if the network need not be in two levels—we might obtain a network that has fewer gates, though it will be slower (assuming that each gate has the same switching time).

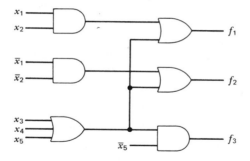

Fig. 5.2.12

These two problems can be solved by the integer programming logical design method, which is complex and sophisticated, as will be mentioned in Section 5.4.

Exercises

When we apply Procedure 5.1.1, a Petrick function can be formed at Step 2, and then processed by algebraic manipulation as we did in Procedure 4.6.1, without using prime-implicant tables (used only for illustration purpose in the text). This, in general, greatly reduces the hand processing time.

5.2.1 (RT) By Procedure 5.2.1, design a multiple-output two-level network that has AND gates in the first level and OR gates in the second level for one of the following sets, A, B, C, or D, of functions f_1, f_2, f_3, as the network outputs, using a minimal number of gates as the primary objective and a minimal number of connections as the secondary objective.

A. $f_1 = \Sigma(0, 1, 2, 4, 9, 10, 11, 12)$,
$f_2 = \Sigma(0, 2, 5, 10, 11, 13, 14, 15)$,
$f_3 = \Sigma(1, 4, 5, 9, 10, 11, 13, 14, 15)$.

B. $f_1 = \Sigma(1, 4, 5, 6, 14)$,
$f_2 = \Sigma(2, 3, 4, 9, 10, 11, 12, 13)$,
$f_3 = \Sigma(1, 2, 3, 4, 5, 9, 10, 11, 13)$.

C. $f_1 = \Sigma(3, 5, 7, 9, 11, 12, 13)$,
$f_2 = \Sigma(3, 7, 9, 10, 11, 12, 14)$,
$f_3 = \Sigma(9, 12, 13, 14, 15)$.

D. $f_1 = \Sigma(7, 8, 10, 12, 13, 15)$,
$f_2 = \Sigma(0, 2, 3, 4, 6, 7, 11, 13, 15)$,
$f_3 = \Sigma(0, 2, 3, 4, 6, 11, 12, 13)$.

(C is more time-consuming than A or B, and D is less.)

240 Advanced Simplification Techniques and Basic Properties of Gates

5.2.2 (M) When a two-level multiple-output network with AND gates in the first level and with OR gates in the second level, using a minimal number of gates as the primary objective and a minimal number of connections as the secondary objective, is designed by Procedure 5.2.1, the corresponding disjunctive form for each output function f_i is not necessarily a minimal sum for f_i, in the sense of Definition 4.1.3.
(i) Show an example. (There are many simple examples even for three-variable functions).
(ii) Prove or disprove that, if the disjunctive form for a particular function f_i is a minimal sum, the OR gate for the f_i does not share AND gates with other OR gates.

5.2.3 (M−) Prove that each paramount prime implicant that is contained in the terms obtained in Step 4 of Procedure 5.2.1 is not redundant (i.e., if it is deleted, some minterms are not covered by any other paramount prime implicants) for at least one of the given functions $f_1, f_2, \ldots, f_m$, when we want to cover each function with the derived paramount prime implicants only. (For example, G for f_1 in Fig. 5.2.4 can be deleted because of the existence of B, but G for f_2 cannot be deleted.)

▲ **5.2.4** (E) Show that application of Procedure 5.2.1 to the functions in Fig. 5.2.10 yields the network shown in Fig. 5.2.11a, but not the one in Fig. 5.2.11b.
Then derive a minimal network by the procedure described in Remark 5.2.2a.

5.2.5 (M − T) Paramount prime implicants for functions $f_1, f_2, \ldots, f_m$ can be found among the multiple-output prime implicants derived by applying Procedure 4.5.1 to these functions and also their products, when these functions are given in minterms. But the following procedure, due to Bartee [1961], is more efficient, since there is no need for forming the products of the functions explicitly, and paramount prime implicants can be directly found. (See McCluskey [1965].)

Table E5.2.5a

DECIMAL REPRESENTATION	INPUT VECTOR x_1	x_2	x_3	x_4	TAG f_1	f_2	f_3	
0	0	0	0	0	0	—	0	✓
4	0	1	0	0	0	—	—	✓
8	1	0	0	0	—	—	0	✓
5	0	1	0	1	0	—	—	✓
9	1	0	0	1	—	0	—	✓
10	1	0	1	0	—	—	0	✓
12	1	1	0	0	—	—	0	✓
7	0	1	1	1	—	—	0	✓
11	1	0	1	1	—	0	—	✓
13	1	1	0	1	—	—	—	✓
14	1	1	1	0	—	—	0	✓
15	1	1	1	1	—	—	—	✓

Design of a Two-Level Multiple-Output Network 241

1. True input vectors of any of the given functions are arranged in a list, as in Table E5.2.5a, which shows an example of three output functions, in such a way that all input vectors in the same group have the same number of 1's, and the group with the least number of 1's is placed at the top of the list, followed by groups of increasing numbers of 1's. Groups are separated by horizontal lines. If an input vector is not for a function f_i, the tag entry for that input vector and the f_i is 0. Otherwise, the tag entry is a dash (—).

2. Compare each input vector in the first group with the input vectors in the second group, in order to find an input vector in the second group that differs from the input vector in the first group in only one bit position. If you find such a pair of input vectors, enter the combined vector in the second list, as shown in Table E5.2.5b, in the following manner: the decimal numbers for the pair of the input vectors combined, then the common bits of these vectors along with the dash replacing the different bits, followed by the tag entries, each of which is 0 when at least one of the original entries is 0 and is a dash otherwise.

Then place a check mark at the end of an input vector in the first list, if the tag entries of this input vector in the first list are identical to those of the input vector generated in the second list in their respective bit positions. For example, when (5, 13) in the second list is generated from vectors 5 and 13 in the first list, the tag entries of vector

Table E5.2.5b

	x_1	x_2	x_3	x_4	f_1	f_2	f_3	
(0, 4)	0	—	0	0	0	—	0	✓
(0, 8)	—	0	0	0	0	—	0	✓
(4, 5)	0	1	0	—	0	—	—	
(4, 12)	—	1	0	0	0	—	0	✓
(8, 9)	1	0	0	—	—	0	0	✓
(8, 10)	1	0	—	0	—	—	0	✓
(8, 12)	1	—	0	0	—	—	0	✓
(5, 7)	0	1	—	1	0	—	0	✓
(5, 13)	—	1	0	1	0	—	—	
(9, 11)	1	0	—	1	—	0	—	✓
(9, 13)	1	—	0	1	—	0	—	✓
(10, 11)	1	0	1	—	—	0	0	✓
(10, 14)	1	—	1	0	—	—	0	✓
(12, 13)	1	1	0	—	—	—	0	✓
(12, 14)	1	1	—	0	—	—	0	✓
(7, 15)	—	1	1	1	—	—	0	
(11, 15)	1	—	1	1	—	0	—	✓
(13, 15)	1	1	—	1	—	—	—	
(14, 15)	1	1	1	—	—	—	0	✓

Table E5.2.5c

	x_1	x_2	x_3	x_4	f_1	f_2	f_3	
(0, 4, 8, 12)	—	—	0	0	0	—	0	
(4, 5, 12, 13)	—	1	0	—	0	—	0	
(8, 9, 10, 11)	1	0	—	—	—	0	0	✓
(8, 9, 12, 13)	1	—	0	—	—	0	0	✓
(8, 10, 12, 14)	1	—	—	0	—	—	0	
(5, 7, 13, 15)	—	1	—	1	0	—	0	
(9, 11, 13, 15)	1	—	—	1	—	0	—	
(10, 11, 14, 15)	1	—	1	—	—	0	0	✓
(12, 13, 14, 15)	1	1	—	—	—	—	0	

5 in the first list are (0 — —), which are identical to those of (5, 13) in the second list. Therefore vector 5 is check-marked. But the tag entries of vector 13 are different, so vector 13 is not check-marked at this time [it will be check-marked when it is combined with vector 15 to generate (13, 15) in the second list]. Next, when vectors 4 and 5 are combined to generate (4, 5), both are check-marked, since they have tag entries identical to those of (4, 5).

If all the tag entries of a new combined vector are 0, we need not enter that vector in the second list, since it is not a prime implicant of any product of functions.

3. Go down to the next group in the first list. In the same manner, compare this group with the next lower group in the first list and enter new combined vectors in the second list, whenever discovered. Repeat this process until the group second from the bottom is compared with the bottom group.

4. In the same manner, compare each group in the second list in Table E5.2.5b with the next lower group in the second list in order to prepare the third list in Table E5.2.5c. But the combination rule is now somewhat different. Only when a pair of vectors that have dashes in the same bit positions differ in only one bit position which is 1 and 0, can they be combined, and the combined vector entered in the third list.

5. Continue to prepare new lists until no new list can be formed. Thus we terminate the procedure at Table E5.2.5d.

The vectors not check-marked throughout the lists represent all paramount prime implicants of the given functions. Decimal representations show minterms that imply

Table E5.2.5d

	x_1	x_2	x_3	x_4	f_1	f_2	f_3
(8, 9, 10, 11, 12, 13, 14, 15)	1	—	—	—	—	0	0

Design of a Two-Level Multiple-Output Network 243

Table E5.2.5e Paramount Prime Implicants Obtained

(8, 9, 10, 11, 12, 13, 14, 15);	x_1	for	f_1
(0, 4, 8, 12);	$\bar{x}_3 \bar{x}_4$	for	f_2
(4, 5, 12, 13);	$x_2 \bar{x}_3$	for	f_2
(8, 10, 12, 14);	$x_1 \bar{x}_4$	for	$f_1 f_2$
(5, 7, 13, 15);	$x_2 x_4$	for	f_2
(9, 11, 13, 15);	$x_1 x_4$	for	$f_1 f_3$
(12, 13, 14, 15);	$x_1 x_2$	for	$f_1 f_2$
(4, 5);	$\bar{x}_1 x_2 \bar{x}_3$	for	$f_2 f_3$
(5, 13);	$x_2 \bar{x}_3 x_4$	for	$f_2 f_3$
(7, 15);	$x_2 x_3 x_4$	for	$f_1 f_2$
(13, 15);	$x_1 x_2 x_4$	for	$f_1 f_2 f_3$

the corresponding paramount prime implicants. Table E5.2.5e shows all paramount prime implicants, along with the minterms (expressed in decimal numbers) that imply them.

(i) By the above procedure, derive all the paramount prime implicants for one of the following sets of functions, along with the corresponding minterms in decimal numbers.

A. $f_1 = \Sigma(1, 2, 3, 9)$,
    $f_2 = \Sigma(2, 3, 5, 7, 9, 10, 11)$,
    $f_3 = \Sigma(1, 2, 3, 7, 9, 10, 11)$.

B. $f_1 = \Sigma(1, 2, 3, 4, 5, 9, 10, 11)$,
    $f_2 = \Sigma(2, 6, 7)$,
    $f_3 = \Sigma(2, 3, 4, 7, 9, 10, 11)$.

(ii) Prove that the above procedure yields all paramount prime implicants for the given functions. (Since only Step 2 is different from Procedure 4.5.1 for deriving prime implicants for a single function, only Step 2 needs to be explained.)

5.2.6 (M) In Step 5 of Procedure 5.2.1, redundant connections in the networks corresponding to the terms obtained in Step 4 are deleted based on the Karnaugh maps in Figs. 5.2.4 and 5.2.7. When the number of variables is five or more, the use of Karnaugh maps (or prime-implicant tables) becomes cumbersome.
Develop an algebraic procedure for this deletion of redundant connections.

5.2.7 (E) By Procedure 5.2.1, design a full adder as a two-level network that has AND gates in the first level and OR gates in the second level, using a minimal number of gates as the primary objective and a minimal number of connections as the secondary objective. The adder has inputs x, y, z and their complements, sum output s, and carry output c. There is no maximum fan-in or fan-out restriction.

5.2.8 (M) The full adder designed in Exercise 5.2.7 has a minimal number of gates as the primary objective and a minimal number of connections as the secondary objective, if a two-level network is to be obtained with AND and OR gates only.

Prove that the full adder designed in Exercise 5.2.7 is still a solution, even if we try to find a full adder in two levels with a minimal number of gates as the primary objective and a minimal number of connections as the secondary objective, under the assumption that a mixture of AND and OR gates is permitted in each level. Assume that x, y, z and their complements are available as the network inputs, and that there is no maximum fan-in or fan-out restriction. (*Hint*: All possible cases must be examined.)

5.2.9 (M−)   By Procedure 5.2.1, design an adder as a two-level network that has AND gates in the first level and OR gates in the second level, using a minimal number of gates as the primary objective and a minimal number of connections as the secondary objective. The adder has inputs x, y, z and their complements, and also sum output s, carry output c, and its complemented output $\bar{c}$. There is no maximum fan-in or fan-out restriction.

5.2.10 (MT)   By Procedure 5.2.1 (if necessary, with modification), design an adder to add four binary variables, x_1, x_2, x_3, and x_4, as a two-level network that has AND gates in the first level and OR gates in the second level, using a minimal number of gates as the primary objective and a minimal number of connections as the secondary objective. The network inputs are x_1, x_2, x_3, x_4 and their complements. The network outputs are f_1, f_2, f_3, where f_1 and f_3 are the most and least significant bits, respectively.

5.2.11 (M − T)   By Procedure 5.2.1 design a multiplier to multiply two binary numbers of two bits each, $X = (x_1, x_2)$ and $Y = (y_1, y_2)$, into the product $F = (f_1, f_2, f_3, f_4)$, as a two-level network that has AND gates in the first level and OR gates in the second level, using a minimal number of gates as the primary objective and a minimal number of connections as the secondary objective. Here x_2, y_2, and f_4 are the least significant bits.

5.2.12 (M − T)   By Procedure 5.2.1 design a two-level network that has AND gates in the first level and OR gates in the second level, using a minimal number of gates as the primary objective and a minimal number of connections as the secondary objective, for the output functions of the network in Fig. 5.2.12.

5.2.13 (M−)   By Procedure 5.2.2 design a two-level network that has AND gates in the first level and OR gates in the second level, using a minimal number of gates (without considering the number of connections), for one of the sets of functions in Exercise 5.2.1.

5.2.14 (M−)   Throughout Procedure 5.2.2 paramount prime implicants rather than implicants were used. Each of the AND gates corresponding to paramount prime implicants has a minimal number of inputs. But Procedure 5.2.2 does not consider the number of connections in a network. Explain why paramount prime implicants rather than implicants are used to find a network with a minimal number of gates (without minimizing the number of connections).

5.2.15 (M−)   Describe a procedure to design a two-level multiple-output network, for a given set of functions, that has OR gates in the first level and AND gates in the second (i.e., output) level, using a minimal number of gates as the primary objective and a minimal number of connections as the secondary objective.

Then apply your procedure to one of the sets of functions in Exercise 5.2.1.

5.2.16 (M)   When a set of functions that are incompletely specified is given, describe a procedure to design a two-level multiple-output network for these functions that has AND gates in the first level and OR gates in the second (i.e., output) level, using a minimal number of gates as the primary objective and a minimal number of connections as the secondary objective.

5.3 Comparison of the Different Methods for Derivation of a Minimal Sum

As means to derive irredundant disjunctive forms or minimal sums for a single function, the Karnaugh map method and the tabular methods are discussed in Sections 4.2 and 4.5, respectively, and the algebraic methods in Sections 4.6 and 5.1. In their application these methods are complementary in usefulness, as follows.

When the number of variables is four or fewer, the Karnaugh map method in Section 4.2 is easy to use for hand processing. Irredundant disjunctive forms and minimal sums can be easily obtained, and don't-care conditions readily dealt with. The Karnaugh map method can probably be used without great difficulty for 5 or 6 variable functions also, when functions do not have many prime implicants. In this sense the map method is extremely powerful and practical, and in many cases logic design would be cumbersome without it.

However, when a function has many prime implicants, even in the case of four variables, it is easy to overlook some of them on a map. For example, the function shown in Fig. 5.3.1 has many prime implicants and many minimal sums. With the map method some of them may be overlooked. (Readers should try to find all minimal sums.) If we want a single minimal sum, application of the map method may be easier. When a function of five or six variables has many prime implicants, however, we can imagine how complex the map method will be, since we have to work on many maps simultaneously. It is usually hard to find a minimal sum, since prime-implicant loops often contain 1-cells scattered in many maps, and it is difficult to compare the sizes of different prime-implicant loops simultaneously.

When the number of variables is five or more, the tabular methods in Section 4.5 and the algebraic methods in Section 4.6 appear to be preferable.

$x_1 x_2$

$x_3 x_4$	00	01	11	10
00	1	1	1	1
01	1	0	1	1
11	1	1	1	1
10	1	1	1	0

Fig. 5.3.1 A function that has many prime implicants.

246 Advanced Simplification Techniques and Basic Properties of Gates

These methods are certainly convenient for processing by computer, a feature that is becoming important for computer-aided design, though further details or modifications must be devised for speed-up because of their strong influence on computational efficiency (a few times, several thousand times, or more speed-up may be possible sometimes even with simple techniques). For hand processing the algebraic methods in Section 4.6 appear to be most efficient.

When the number of variables is much larger, however, none of these methods is useful because all of them depend on the use of minterms, whose number increases very rapidly, as illustrated in Figs. 4.1.1 and 4.1.2. In this case the algebraic methods in Section 5.1, which do not use minterms, appear to be useful, if the generalized-consensus formation trees are not complex.

The foregoing comparison of the different methods is roughly illustrated in Fig. 5.3.2. When a function has many variables and complex generalized-consensus formation trees (even some functions of several variables), the methods in Section 5.1 are also not useful, because of excessively long processing times even with a high-speed computer. This is shown by the shaded area in Fig. 5.3.2. Even before reaching this shaded area, we need to consider the trade-off between minimization and processing time (to be discussed in Section 9.1). If the derivation of a minimal sum is too time-consuming, we have to give up the attempt. Nevertheless, some efforts to try the derivation within a reasonable time limit are usually necessary, and we need to resort

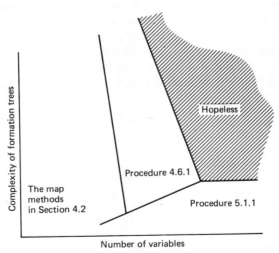

Fig. 5.3.2 Comparison of the different methods for the derivation of a minimal sum.

Design of Networks with AND and OR Gates 247

to heuristic methods that are currently not well explored (e.g., it is not known how close to the minimum the solutions derived by heuristic methods are).

In the case of multiple-output two-level networks, we do not have many choices in design methods, as seen in Section 5.2, because the design problem itself is much more difficult and complex than the problem of single-output two-level network synthesis based on a minimal sum. If the number of connections need not be minimized, Procedure 5.2.2 is efficient for the synthesis of two-level networks with a small number of variables; but if the number of variables is large, Procedure 5.2.2 is not useful and we have to resort to heuristic methods such as those described in Hong, Cain, and Ostapko [1974] and Michalski and Kulpa [1971].

Comparison of two-level networks and multilevel networks (i.e., networks in more than two levels) is discussed in Sections 5.4, 6.6, and 6.7, from the viewpoints of network size, speed, and reliability.

▲ 5.4 Design of Networks with AND and OR Gates under Arbitrary Restrictions

When some conditions in Assumption 4.1.1 or 4.4.1 are violated (e.g., when the restriction of maximum fan-in is imposed, or when a network need not be in two levels), the design methods based on a minimal sum or a minimal product discussed so far cannot be directly used. We must either modify the networks designed by these methods so that the resulting networks satisfy the new assumptions (or restrictions), or use much more sophisticated but complex design procedures, like the integer programming logic design method to be discussed later, which needs to be processed by computers. In the former case no systematic guidelines are known for the modification, so it must be done heuristically. The resulting networks may be reasonably good in many cases, but perhaps not minimal. More precisely speaking, nobody can tell how close to the minimal networks the solutions obtained by heuristic methods are. (Usually, designers conclude only intuitively, based on their experiences, that the solutions obtained must be close to minimal networks, when they have spent a considerable amount of time.) In practice, networks need not be minimal in most cases, though networks with as few gates as possible are desirable for not only economic but also other reasons, as we discussed in Section 4.1. Hence the trade-off between minimization and processing time is important, as will be discussed in Section 9.1.

As an example, let us discuss the problem of designing a network with a minimum number of AND and OR gates in **multilevels (i.e., more than two levels)**. Networks in multilevels usually can be realized with fewer gates than those in two levels, which are designed based on a minimal sum or product.

248 *Advanced Simplification Techniques and Basic Properties of Gates*

When a function of four or fewer variables is realized with a multilevel network, the minimum number of AND and OR gates required is, at most, nine, and the maximum difference in the minimum number of gates between networks in two levels and those in multilevels is, at most, one. Although this difference is numerically small, it occurs for networks with as few as four gates, so the difference of 25 percent is great in terms of percentage. (These figures were derived by the integer programming logic design method, for all functions of four or fewer variables [Culliney, Young, Nakagawa, and Muroga 1979].) Also, it is known that, at least for all functions of four or fewer variables, except the parity function and its complement for which minimal networks are not known, networks that have a minimal number of gates as the primary objective and a minimal number of connections as the secondary objective can be realized in, at most, four levels of AND and OR gates [Culliney, Young, Nakagawa, and Muroga 1979].

When a function of more than four variables is to be realized, the difference in required gate counts can be greater. For example, the function $f = x_1(x_2 \vee x_3 \vee x_4 \vee x_5 \vee x_6 \vee x_7) \vee x_2 x_3 x_4 x_5 x_6 x_7$ can be realized with four AND and OR gates in three levels, as shown in Fig. 5.4.1. (This network is known to have the minimum number of gates for this function, as proved by the integer programming logic design method.) If f is to be realized with a two-level network, eight gates are required, since the minimal sum consists of seven prime implicants, and its minimal product consists of seven prime implicates. Thus the number of gates for this function is reduced to half by realizing a network in multilevels instead of one in two levels.

For some functions of n variables, the difference in gate counts becomes greater. For example, parity function $x_1 \oplus x_2 \oplus \cdots \oplus x_n$ in a two-level network requires the minimum number of gates, $2^{n-1} + 1$, whereas the minimum number of gates in a multilevel network for this function is linearly proportional to n. Also, there are classes of functions of n variables such that the ratio of the minimum number of gates for a two-level network to the minimum number of gates for a multilevel network increases almost linearly as n increases. For example, a function of n variables in a certain class of functions can be realized with a network of only five gates (AND and OR),

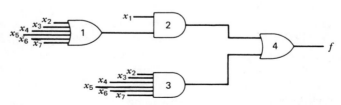

Fig. 5.4.1 A multilevel network with a minimum number of gates.

regardless of the value of n, if implemented in multilevels, whereas $2(n - 2) + 2$ gates are required in a two-level network (Exercise 5.1.C). This means that the latter network requires 3.6 times more gates than the former, for $n = 10$.

In addition to the reduction in gate count, multilevel networks for some functions have the advantage of lower fan-in and fan-out, on an average, since connections tend to be more evenly distributed in the network. Thus multilevel networks may operate more reliably. Two-level networks and multilevel networks will be compared in respect to other aspects in Section 6.7.

No systematic procedure for the design of multilevel networks with a minimum number of AND and OR gates which can be easily processed by hand is known, despite its practical importance. Only heuristic procedures that can be processed by hand but do not guarantee the minimality of the gate count, and also the integer programming logic design method, which can be processed only by computer, though it guarantees the minimality of the gate count, are known thus far.*

An Algebraic Procedure

As an example of the heuristic procedures, let us illustrate the derivation of a multilevel network with relatively few gates (or connections) by rewriting a switching expression for a given function. [Since minimization of the gate (or connection) count is difficult, we try to have as few gates (or connections) as possible.] Many other heuristic procedures are conceivable.

For example, suppose that the function whose minimal sum is

$$f = \bar{x}_1 x_2 x_3 \vee x_2 x_3 \bar{x}_4 \vee x_1 \bar{x}_2 x_4 \vee x_1 \bar{x}_3 x_4 \qquad (5.4.1)$$

is given. A two-level network based on this minimal sum requires five gates. (The minimal product for f requires more gates.) By grouping terms, f may be rewritten as

$$f = x_2 x_3 (\bar{x}_1 \vee \bar{x}_4) \vee x_1 x_4 (\bar{x}_2 \vee \bar{x}_3). \qquad (5.4.2)$$

By adding redundant terms $x_2 x_3 \bar{x}_2$, $x_2 x_3 \bar{x}_3$, $x_1 x_4 \bar{x}_1$, and $x_1 x_4 \bar{x}_4$, which are identically equal to 0, (5.4.2) may be rewritten as

$$f = x_2 x_3 (\bar{x}_1 \vee \bar{x}_2 \vee \bar{x}_3 \vee \bar{x}_4) \vee x_1 x_4 (\bar{x}_1 \vee \bar{x}_2 \vee \bar{x}_3 \vee \bar{x}_4), \qquad (5.4.3)$$

using the common expression $(\bar{x}_1 \vee \bar{x}_2 \vee \bar{x}_3 \vee \bar{x}_4)$.

---

* Ironically, both heuristic procedures and the integer programming logic design method are more complex and time-consuming with networks of AND and OR gates than with those of NOR (or NAND) gates, to be discussed in Chapter 6, though AND and OR gates are conceptually simpler than NOR (or NAND) gates. The reason is that the former networks consist of two different types of gates, requiring examination of which gate type should be used in each gate position in a network, whereas the latter consists of a single type of gate.

We have obtained, corresponding to (5.4.3), the network shown in Fig. 5.4.2, by implementing the expression $(\bar{x}_1 \vee \bar{x}_2 \vee \bar{x}_3 \vee \bar{x}_4)$ common to the two products in (5.4.3) by the OR gate numbered 1. This network consists of four gates, whereas the networks based on (5.4.1) and (5.4.2) require five gates. (The minimalities of gate and connection counts in the network in Fig. 5.4.2 was proved by the integer programming design method in Culliney, Young, Nakagawa, and Muroga [1979].)

The Exhaustive Method

If we are not concerned with the computational efficiency of design, minimal networks under arbitrary restrictions can be designed by the exhaustive method. In other words, we consider all network configurations that have a given number of gates or fewer, and then choose the best one among those that realize the given function. If a network requires only a small number of gates, the exhaustive method does not take much computation time. But if a network requires seven or more gates, the exhaustive method is very time-consuming, as L. Hellerman [1963] has showed. Hellerman worked on NOR or NAND gates only, and the design problem with AND and OR gates would be even more time-consuming. Particularly if a network is to have multiple outputs, the exhaustive method would require much more computation time.

The Integer Programming Logic Design Method

The design of minimal networks under arbitrary restrictions by the integer programming method* (including the branch-and-bound version) is much more efficient than using the exhaustive method (Chapter 14 of Muroga [1971]; also Muroga [1970], Muroga and Ibaraki [1972], Davidson [1969], Lai, Nakagawa, and Muroga [1974]). The number of gates, connections, or levels can be minimized for completely or incompletely specified multiple-output functions under arbitrary restrictions, such as maximum fan-in or

---

* Actually, a few different approaches were developed, using the implicit enumeration and branch-and-bound methods. Since there is no essential difference, theoretically, between the implicit enumeration method and the branch-and-bound method used in these approaches, the approaches here are collectively called "the integer programming logic design method" (though they are sometimes called "the branch-and-bound logic design method"), because, historically, the implicit enumeration method was first developed, followed by the branch-and-bound logic design method (but the papers were not published in chronological order, though the integer programming logic design method had been explored and taught by the author since around 1965), and also the branch-and-bound method is sometimes considered part of integer programming in a broad sense. (The implicit enumeration method may be regarded as a special case of the branch-and-bound method.) These different approaches have different features in their usages (Chapter 14 of Muroga [1971]).

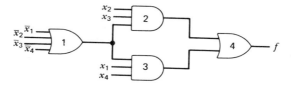

Fig. 5.4.2   Network based on (5.4.3).

fan-out. Comparison of computation time for **the integer programming logic design method** and the exhaustive method is not available for networks with AND and OR gates (statistics for the exhaustive method are lacking), but for networks with NOR gates only, the integer programming logic design method is ten thousand times faster for networks of, say, eight NOR gates than the exhaustive method. On the other hand, the integer programming logic design method is considerably complicated, requiring a large-scale computer, and if a network needs 10 or more gates, the method also usually becomes excessively time-consuming.

The Transduction Method

Since the integer programming logic design method is too time-consuming beyond networks of 10 gates, a heuristic design method, called the **transduction method**, has been developed for designing networks of more gates by analyzing the properties of minimal networks designed by the integer programming logic design method [papers by Muroga, Kambayashi, Lai, Culliney, and Hu, to be published]. The transduction method yields networks that are often minimal if they are small (so the minimality of the networks can be checked by the integer programming logic design method) and can design networks that require a large number of gates, in a short computation time. The computation time is relatively independent of the number of gates, in contrast to the integer programming logic design method.

Remark 5.4.1: So far, the costs of any two gates are assumed to be equal in our discussion. The design methods discussed thus far, however, including those treated in the preceding chapters, are applicable to some extent, with some modifications, to the case of different costs. When a minimal sum is to be chosen among irredundant disjunctive forms, we assign different costs to terms, and then choose a set of terms whose total cost is minimal. ∎

Remark 5.4.2: By generalizing the concept of prime implicant, Lawler developed a systematic procedure to obtain a switching expression that has a minimum number of literals, allowing the expression to contain parentheses. For example, $xy\bar{z} \vee wx \vee wy$ has seven literals; but if the use of parentheses is allowed, this function may be rewritten as $xy\bar{z} \vee w(x \vee y)$, which has six literals. Under certain restrictions on types of gates and other network parameters (e.g., MOS networks; see Exercise 5.6), this type of switching expression gives a network with a minimum number of variable inputs to all gates, no matter how many levels or gates the network has. For example, the above expression $xy\bar{z} \vee w(x \vee y)$ is realized with a network in three levels of AND and OR gates that has six variable inputs [Lawler 1964, De Vries 1970]. The problem, however, appears to be too difficult for us to have efficient, simple procedures. ■

5.5 Gate Types

In this section we discuss important properties of gate types and the significance of the gate concept.

Complete Sets of Gate Types

So far, we have designed networks using only three types of gates: AND gates, OR gates, and NOT gates, though NOR and NAND gates are occasionally used instead. In this section let us study whether a network can be realized with other types of gates. To facilitate our discussion, we define the following.

Definition 5.5.1: Consider a set of different gate types, such that each gate type is the set of all gates that have different numbers of inputs, but perform the same logic operation (e.g., $x_1 \vee x_2, x_1 \vee x_2 \vee x_3, \ldots$ are OR gate type, and $x_1 \cdot x_2, x_1 \cdot x_2 \cdot x_3, \ldots$ are AND gate type). Then the set is called a **complete set of gate types**, or simply a **complete set of gates**, if any switching function can be realized by a loopless network consisting of only gate types in this set, provided that each input of a gate is permitted to connect to the output of another gate, a variable input, constant input* 0, or constant input 1, and two or more inputs of a gate are permitted to connect together to the same output of another gate, a variable input, constant input 0, or constant input 1. Here complemented variables are assumed to be unavailable. Each gate type is called a **primitive gate type** or a **primitive gate**. Each logic operation that each gate performs is called a **primitive logic operation**, and the complete set is called the **complete set of primitive logic operations**.

The set of AND, OR, and NOT gates is certainly a complete set of gate types.

---

* Constant inputs of 0 or 1, as well as variable inputs, are assumed to be available to any gates, since constant inputs of 0 or 1 are usually supplied directly from a power supply. But in some theoretical studies (see Remark 5.5.1), the constants 0 and 1 are possible elements of complete sets, instead of making this assumption.

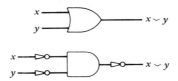

Fig. 5.5.1 Completeness of AND and NOT gates.

The question arises whether complete sets other than this set exist.

Any network consisting of AND gates, OR gates, and NOT gates can be changed into a network consisting of only AND gates and NOT gates, for the following reason.

Every OR gate can be replaced by an AND gate that has NOT gates connected at its output and inputs, because of $x_1 \vee x_2 \vee \cdots \vee x_n = \overline{\bar{x}_1 \cdots \bar{x}_n}$ (De Morgan's theorem). This is illustrated for the case of $n = 2$ in Fig. 5.5.1, where the triangles with small circles denote NOT gates. Thus **the set of AND and NOT gates, even without OR gates, is a complete set**. By duality, every AND gate can be replaced by an OR gate that has NOT gates connected at its inputs and output. Thus **the set of OR and NOT gates is also a complete set**.

Every complete set discussed so far contains two or more gate types. In this sense it is interesting to see that **the NAND gate type alone constitutes a complete set**. This is so because NOT and AND operations can be implemented with NAND gates only, as shown in Fig. 5.5.2. The NAND gate is often called a **Sheffer stroke** for a historical reason in logic, denoted by $x|y$ for $\overline{xy}$. Similarly, **the NOR gate type alone constitutes a complete set**. (Readers should find a reason.) The NOR gate is often called a **Pierce arrow**, denoted by $x \downarrow y$ for $\overline{x \vee y}$.

It has been shown that the majority of switching functions require NOT operation in their expressions, and that only certain functions called positive functions can be expressed without NOT operations [Quine 1953]. Note that NAND and NOR gates contain NOT in their operations. Furthermore, a NOT gate alone cannot express an arbitrary function.

Remark 5.5.1: Definition 5.5.1 is engineering-oriented. In some theoretical studies the following definition is used.

Consider a set of different gates each of which has a prespecified number of inputs. This set is defined as a **complete set of gates or (logic) primitives**, if any switching function can be realized by a loopless network consisting of these gates, provided that each input of a gate is permitted to connect to the output of another gate or a variable input, and two or more inputs of a gate are permitted to connect together to the same output of another gate or a variable input. Each gate is called a **primitive**. Complemented variables and also constant inputs of 0 or 1 are assumed

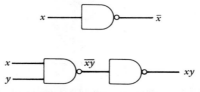

Fig. 5.5.2   Completeness of NAND gate.

to be unavailable. Constants 0 and 1 can be generated from gates in the case of some complete sets. For example, when a complete set contains AND and NOT gates, the constants 0 and 1 can be obtained by $x \cdot \bar{x} = 0$ and $\overline{x \cdot \bar{x}} = 1$. But, as seen later in this section, constant 1 cannot be generated in the case of a complete set consisting of AND and EXCLUSIVE-OR gates. In this case the AND gate, the EXCLUSIVE-OR gate, and the constant 1 constitute a complete set.

For example, the set of an AND gate with two inputs, an OR gate with two inputs, and a NOT gate with a single input is a complete set, as we saw in the preceding chapters that AND and OR are defined as logic operations on two variables and NOT as a logic operation on a single variable. Also, the set of an AND gate with m inputs, an OR gate with m inputs, and a NOT gate with a single input is a complete set, where $m \geq 2$, since an AND (or OR) gate with two inputs can be replaced by an AND (or OR) gate with m inputs that are tied together into two inputs, and since AND and OR gates with two inputs and a NOT gate with a single input constitute a complete set.

Various complete sets were originally discussed by Post [1941], but this book is hard to read and also is out of print. The theoretical paper written by Ibuki, Naemura and Nozaki [1963], without knowledge of Post's publication, is easy to read and interesting. The treatment is different from Post's. There is a related discussion by Mukhopadhyay [1971]. ∎

EXCLUSIVE-OR gate

An EXCLUSIVE-OR gate, which is usually denoted by the symbol shown in Fig. 5.5.3b for three variables, represents the EXCLUSIVE-OR of its input variables, that is, it has the output value 1 only when the number of 1's in its input variables is odd. In other words, the EXCLUSIVE-OR of $x_1, \ldots, x_n$ is the modulo 2 sum of $x_1, \ldots, x_n$, as illustrated for the case of $n = 3$ in Table 5.5.1, and is denoted by

$$x_1 \oplus x_2 \oplus \cdots \oplus x_n.$$

Sometimes it is called the **parity function**. As is easily seen, the EXCLUSIVE-OR operation is associative, that is,

$$(x_1 \oplus x_2) \oplus x_3 = x_1 \oplus (x_2 \oplus x_3),$$

and also distributive with respect to the AND operation, that is,

$$x_1(x_2 \oplus x_3) = x_1 x_2 \oplus x_1 x_3.$$

Table 5.5.1 Truth Table for Parity Function

x_1	x_2	x_3	$x_1 \oplus x_2 \oplus x_3$
0	0	0	0
0	0	1	1
0	1	0	1
0	1	1	0
1	0	0	1
1	0	1	0
1	1	0	0
1	1	1	1

This associative property means that, if a network consists of EXCLUSIVE-OR gates only, it can be replaced by a single EXCLUSIVE-OR gate, as shown in Fig. 5.5.3, where the two gates in Fig. 5.5.3a are reduced to a single gate in 5.5.3b. If an EXCLUSIVE-OR gate has many inputs of the same variable x, as shown in Fig. 5.5.3c, each pair of the same input x can be eliminated, since $x \oplus x = 0$. If we use only EXCLUSIVE-OR gates, we cannot realize all functions of three variables for the following reason. When only $\oplus$ is used, $x_1 \oplus x_2 \oplus x_3$ is the only function of three variables. This function cannot express the function of three variables $\overline{x_1 \vee x_2 \vee x_3}$, for example, because these two functions differ for $x_1 = x_2 = x_3 = 0$. Thus **an EXCLUSIVE-OR gate alone cannot be a complete set.** If the constant 1

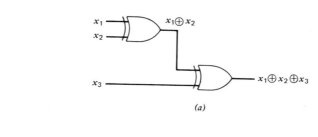

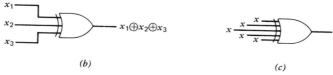

Fig. 5.5.3 *Network of EXCLUSIVE-OR gates.*

256 Advanced Simplification Techniques and Basic Properties of Gates

is available, the EXCLUSIVE-OR gate can realize the NOT operation, since $x \oplus 1 = \bar{x}$. But OR or AND still cannot be expressed by using EXCLUSIVE-OR gates and 1, because a network of EXCLUSIVE-OR gates with constant input 1 available can express only $x_1 \oplus x_2 \oplus \cdots \oplus x_n$ or $x_1 \oplus x_2 \oplus \cdots \oplus x_n \oplus 1 = \overline{x_1 \oplus x_2 \oplus \cdots \oplus x_n}$, each of which differs from $x_1 \vee \cdots \vee x_n$ or $x_1 x_2 \ldots x_n$. But **the set of an EXCLUSIVE-OR gate and an AND gate is obviously a complete set if the constant 1 is available. And so is the set of an EXCLUSIVE-OR gate and an OR gate if the constant 1 is available.** Often $\overline{x_1 \oplus x_2 \oplus \cdots \oplus x_n}$ is denoted by $x_1 \odot x_2 \odot \cdots \odot x_n$ without using the overhead bar, and is called the **EXCLUSIVE-NOR** of n variables. An EXCLUSIVE-NOR gate is usually denoted by the symbol shown in Fig. 5.5.4.

The EXCLUSIVE-OR operation has an interesting property. Any function can be expanded into a unique canonical form (similar to the minterm or maxterm expansion), using EXCLUSIVE-OR, AND, and 1 only, as follows. For the sake of simplicity, let us consider an arbitrary switching function of three variables, $f(x_1, x_2, x_3)$. (Generalization of the following approach to a function of n variables can be easily done in a similar manner.) Here $f(x_1, x_2, x_3)$ can be expressed in the following minterm expansion, using AND, OR, and NOT:

$$f(x_1, x_2, x_3) = f_{000}\bar{x}_1\bar{x}_2\bar{x}_3 \vee f_{001}\bar{x}_1\bar{x}_2 x_3 \vee f_{010}\bar{x}_1 x_2 \bar{x}_3$$
$$\vee f_{011}\bar{x}_1 x_2 x_3 \vee f_{100} x_1 \bar{x}_2 \bar{x}_3 \vee f_{101} x_1 \bar{x}_2 x_3$$
$$\vee f_{110} x_1 x_2 \bar{x}_3 \vee f_{111} x_1 x_2 x_3, \qquad (5.5.1)$$

where f_{000} denotes $f(0, 0, 0)$, and similarly with other f_{ijk}. Let us replace every $\vee$ in (5.5.1) by $\oplus$:

$$f(x_1, x_2, x_3) = f_{000}\bar{x}_1\bar{x}_2\bar{x}_3 \oplus f_{001}\bar{x}_1\bar{x}_2 x_3 \oplus f_{010}\bar{x}_1 x_2 \bar{x}_3$$
$$\oplus f_{011}\bar{x}_1 x_2 x_3 \oplus f_{100} x_1 \bar{x}_2 \bar{x}_3 \oplus f_{101} x_1 \bar{x}_2 x_3$$
$$\oplus f_{110} x_1 x_2 \bar{x}_3 \oplus f_{111} x_1 x_2 x_3. \qquad (5.5.2)$$

Then (5.5.1) and (5.5.2) are equivalent, because only one minterm in each of these expressions becomes 1 for every combination of values of x_1, x_2, x_3, and other minterms become 0. By substituting the identity $\bar{x}_i = 1 \oplus x_i$ into

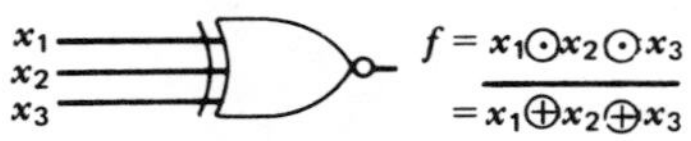

Fig. 5.5.4   EXCLUSIVE-NOR gate.

Gate Types 257

every $\bar{x}_i$ of (5.5.2) for $i = 1, 2, 3$, we obtain the following polynomial form, using the property $x_i \bar{x}_i = 0$:

$$f(x_1, x_2, x_3) = f_{000}(1 \oplus x_1)(1 \oplus x_2)(1 \oplus x_3) \oplus f_{001}(1 \oplus x_1)(1 \oplus x_2)x_3$$
$$\oplus \cdots \oplus f_{111}x_1 x_2 x_3.$$

Skipping the intermediate calculation, we find that this is reduced to the following expression:

$$f(x_1, x_2, x_3) = [f_{000}]$$
$$\oplus [\{f_{000} \oplus f_{100}\}x_1 \oplus \{f_{000} \oplus f_{010}\}x_2 \oplus \{f_{000} \oplus f_{001}\}x_3]$$
$$\oplus [\{f_{000} \oplus f_{100} \oplus f_{010} \oplus f_{110}\}x_1 x_2$$
$$\oplus \{f_{000} \oplus f_{010} \oplus f_{001} \oplus f_{011}\}x_2 x_3$$
$$\oplus \{f_{000} \oplus f_{100} \oplus f_{001} \oplus f_{101}\}x_1 x_3]$$
$$\oplus [\{f_{000} \oplus f_{100} \oplus f_{010} \oplus f_{001} \oplus f_{011} \oplus f_{101}$$
$$\oplus f_{110} \oplus f_{111}\}x_1 x_2 x_3] \qquad (5.5.3)$$
$$= [g_0] \oplus [g_1 x_1 \oplus g_2 x_2 \oplus g_3 x_3] \oplus [g_{12} x_1 x_2$$
$$\oplus g_{23} x_2 x_3 \oplus g_{13} x_1 x_3] \oplus [g_{123} x_1 x_2 x_3], \qquad (5.5.4)$$

where $g_0 = \{f_{000}\}$, $g_{12} = \{\cdots\}$, ..., are constant coefficients that are determined by the value of f for each combination of values of x_1, x_2, x_3.

Since expansion 5.5.4 looks like a polynomial in calculus (no complemented literals are present), (5.5.4) is sometimes called the **polynomial expansion**. This expansion is unique, because expansion (5.5.2) with $\oplus$ and minterms is unique and the conversion to (5.5.4) also gives unique coefficients $g_0, g_1, g_2, \ldots$.

Any function $f(x_1, x_2, \ldots, x_n)$ of n variables can be similarly expanded into the following polynomial:

$$f(x_1, \ldots, x_n) = [g_0] \oplus [g_1 x_1 \oplus g_2 x_2 \oplus g_3 x_3 \oplus \cdots \oplus g_n x_n]$$
$$\oplus [g_{12} x_1 x_2 \oplus g_{13} x_1 x_3 \oplus g_{23} x_2 x_3 \oplus \cdots \oplus g_{n-1,n} x_{n-1} x_n]$$
$$\oplus \cdots \oplus [g_{12 \cdots n} x_1 x_2 \cdots x_n], \qquad (5.5.5)$$

where $g_0, g_1, \ldots$ are constant coefficients.

The reversion from (5.5.4) into (5.5.2) can be done by substituting $1 = x_i \oplus \bar{x}_i$ for all x_i's missing in each term of (5.5.4) [e.g., the term $g_{23} x_2 x_3$ in (5.5.4) should be rewritten as $g_{23}(x_1 \vee \bar{x}_1)x_2 x_3$], and then multiplying out each product.

Parity functions are important for computers and digital communications, being used in logic networks for error correction, arithmetic operations, and code conversions. (For an IC implementation example of EXCLUSIVE-OR, see Fox and Nestork [1971].)

Remark 5.5.2: Polynomial expression 5.5.5 has interesting properties that were independently discovered in relation to error-correcting codes by Muller [1953, 1954] and Mitani [1951]. There are related discussions by Mukhopadhyay and Schmitz [1970] and Kodandapani and Setlur [1977].

The Threshold Function and the Threshold Gate

A threshold gate is defined as a gate with inputs $x_1, \ldots, x_n$, for which there exist a set of real numbers $w_1, \ldots, w_n$, called **weights**, and t, called a **threshold**, such that the output of the gate is

$$f = 1 \quad \text{if} \quad w_1 x_1 + w_2 x_2 + \cdots + w_n x_n \geq t,$$

and

$$f = 0 \quad \text{if} \quad w_1 x_1 + w_2 x_2 + \cdots + w_n x_n < t.$$

Here, notice that $x_1, \ldots, x_n$ assume 0 or 1, whereas $w_1, \ldots, w_n$ and t are real-number constants.

The switching function $f(x_1, \ldots, x_n)$ that the output of a threshold gate represents is called a **threshold function**.

A threshold gate is a generalization of the conventional switching gates mentioned so far, such as AND, OR, NOT, NAND, and NOR gates (except the EXCLUSIVE-OR gate). For example, consider a threshold gate with $w_1 = w_2 = \cdots = w_n = 1$ and $t = n$. As is easily seen, this represents $x_1 x_2 \cdots x_n$, that is, an AND gate with inputs $x_1, x_2, \ldots, x_n$. Similarly, a threshold gate with $w_1 = w_2 = \cdots = w_n = -1$ and $t = 0$ represents $\overline{x_1 \vee x_2 \vee \cdots \vee x_n}$, that is, a NOR gate.

Since AND, OR, NOT, NOR, and NAND gates are special cases of the threshold gate, the **threshold gate itself is a complete set** when the weight and the threshold can be arbitrarily chosen. Also, when a network is synthesized with threshold gates only, the number of required gates is never more, and is often much fewer, than in a network synthesized with the gates discussed so far (except in a mixture of EXCLUSIVE-OR and other gates, which realize some functions with fewer of them, and other functions with more of them). For example, a threshold gate with $w_1 = 2$, $w_2 = w_3 = 1$, and $t = 2$ represents the switching function $x_1 \vee x_2 x_3$, as illustrated in Table 5.5.2, and any single gate mentioned so far, such as the AND or OR gate, does not represent this function. Another example is $x_1 x_2 \vee x_2 x_3 \vee x_1 x_3$, which is realized by a threshold gate with $w_1 = w_2 = w_3 = 1$ and $t = 2$. If this function is to

Table 5.5.2 Function Represented by a Threshold Gate with $w_1 = 2$, $w_2 = w_3 = 1$, and $t = 2$
(This truth table expresses $x_1 \vee x_2 x_3$.)

x_1	x_2	x_3	$\Sigma w_i x_i$		f
0	0	0	0		0
0	0	1	1		0
0	1	0	1		0
0	1	1	2	≥ 2	1
1	0	0	2	≥ 2	1
1	0	1	3	≥ 2	1
1	1	0	3	≥ 2	1
1	1	1	4	≥ 2	1

be realized with AND and OR gates, four gates will be required. Networks for m-out-of-n detectors are greatly simplified by the use of threshold gates [Amarel, Cooke, and Winder 1964; Wooley and Baugh 1974]. For further study of threshold logic, see Muroga [1971] or Winder [1962, 1968a].

In particular, threshold functions that are realized by threshold gates with all equal weights, $w_1 = w_2 = \cdots = w_n$, are called **majority functions**. (With different values of t, there are many majority functions.) The functions $x_1 x_2 \vee x_2 x_3 \vee x_1 x_3$ and $x_1 x_2 x_3 \vee x_1 x_2 x_4 \vee x_1 x_3 x_4 \vee x_2 x_3 x_4$ are examples.

Threshold gates can be implemented without too much engineering difficulty, but the tolerance in regard to the values of components (e.g., resistor values) required to maintain the reliability of operation is somewhat more stringent than that for conventional gates such as AND, OR, or NAND. Consequently, threshold gates are not widely used.* A threshold gate with $w_1 = w_2 = w_3 = 1$ and $t = 2$ was used in the ILLIAC I computer constructed in the early 1950s.

---

* "Micromosaic," available from Fairchild, contains a simple threshold gate based on a p-channel MOS. Motorola also has simple threshold gates based on the CMOS, such as MC 14530AL, MC 14530CL, and MC 14530CP [Garrett 1973].

A threshold gate can be implemented with a magnetic bubble, which can also implement more complex functions than threshold functions [Minnick, Bailey, Sanfort, and Semon 1972]. Research efforts to implement small networks with threshold gates in IC packages have been made. Since networks could be realized with fewer threshold gates, power consumption and cost would be reduced. If such IC packages are compatible with other gate types, they will be useful [Winder 1968b, Hampel and Winder 1971].

Remark 5.5.3: The concept of threshold gates is sometimes useful for theoretical study. A complex network with arbitrary gate types can be conveniently analyzed by means of an equivalent network of threshold gates [Akers 1968, Huffman 1971].

Threshold logic, the theory of threshold gates, has been applied to information retrieval systems. (See, e.g., Brandhorst [1966] or Sommar and Dennis [1969].)

In lattice theory, the output function of a threshold gate that has three inputs, x_1, x_2, x_3, with $w_1 = w_2 = w_3 = 1$ and $t = 2$, is known as "the median" of $x_1, x_2,$ and x_3 [Birkhoff 1948]. ∎

A **multithreshold threshold gate**, an extension of a threshold gate, is a threshold gate that has more than one threshold. A multithreshold threshold gate needs a more complex electronic implementation than a threshold gate, but a single multithreshold threshold gate can express any function, though it is impractical if many inputs and thresholds are required [Ercoli and Mercurio 1962, Muroga 1971]. Integrated circuit implementation of a multi-threshold threshold gate is discussed by Dao, Russell, Preedy, and McCluskey [1977], Dao [1977], and Dao, McCluskey, and Russell [1977].

Gates That Represent Negative Functions

A switching function that has a disjunctive form without complemented literals is called a **positive function** (or frontal function). An example is $x_1 \vee x_2 x_3$, but $x_1 \vee \bar{x}_2 x_3$ is not a positive function because it has no other disjunctive form without $\bar{x}_2$. However, $x_1 \vee \bar{x}_1 x_2$ is a positive function because it can be rewritten as $x_1 \vee x_2$. [As can easily be proved, a function is a positive function if and only if every prime implicant of it contains no complemented literals (Exercise 5.28).] When all variables of a positive function are complemented, we get a **negative function**. An example is $\bar{x}_1 \vee \bar{x}_2 \bar{x}_3$. If a function has a disjunctive form throughout which only one of two literals appears for each variable, it is called a **unate function**. For example, $\bar{x}_1 x_2 \vee \bar{x}_1 x_3$ is a unate function. Positive and negative functions are also unate functions.

An MOS gate, discussed in Section 2.3., realizes a negative function when all inputs of the gate are noncomplemented variables. If a slow switching time of a gate is acceptable, a single MOS gate can contain as many as 40 MOSFETs. Thus a very complex negative function ("complex" here means "many literals" involved in the minimal sum of the function) can be implemented by a single MOS gate, and a MOS gate has great logic capability. But when the number of MOSFETs in an MOS gate increases, its switching time slows, and in the case of a static MOS its size increases rapidly.

Single-Rail and Double-Rail Logic

Sometimes we want to design a network that has both a function f and its complement $\bar{f}$ available as outputs of the network, assuming that variables $x_1, \ldots, x_n$ and their complements $\bar{x}_1, \ldots, \bar{x}_n$ are available as inputs. Such a network is said to be in **double-rail logic**. When complemented and noncomplemented outputs are available but there is no restriction on the availability of noncomplemented or complemented inputs (i.e., noncomplemented inputs, complemented inputs, or both are available), a network is said to be in **double-rail output logic**. When the inputs $x_1, \ldots, x_n$ and their complements are available but there is no restriction on the availability of noncomplemented or complemented outputs, a network is said to be in **double-rail input logic**. A network for which complemented inputs and outputs are not available is said to be in **single-rail logic**.* When complemented inputs are not available but there is no restriction on the availability of noncomplemented or complemented outputs, a network is said to be in **single-rail input logic**.* **Single-rail output logic*** can be defined similarly. All these types of logic are used in practice. Networks in double-rail logic (or double-rail input logic) require no more gates (usually fewer) than those in single-rail logic (or single-rail input logic). The former, however, require more input and output connections (or input connections only) than the latter. Therefore the decision as to which logic to choose depends on this trade-off.

When some types of electronic circuits are used to realize a gate, it is easy electronically to provide double outputs, that is, an output and its complement. An example is a gate that has both NOR and OR as its outputs, as shown in Fig. 5.5.5. (This is a gate in ECL, which is a very important IC logic family for high-speed applications.) This is an example of a complex gate with which IC technology has provided us. If gates with double outputs as shown in Fig. 5.5.5 are used, a network in double-rail logic for given functions can usually be designed with fewer gates than are required with single outputs, and consequently the number of levels is also often reduced, further speeding up the network response.

Implementation of Gates

So far we have discussed different gate types. Certain gate types are preferred to others because of many reasons discussed thus far. The AND and OR operations are directly associated with the mechanical implementation of such devices as relay contacts. Thus they are basic concepts in designing relay

---

* "Single-rail" may literally mean the availability of either complemented or noncomplemented variables, but not both. Usually, however, it means the availability of noncomplemented variables, as defined here.

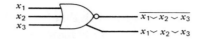

Fig. 5.5.5 Gate with OR and NOR outputs.

contact networks. In the case of electronic circuit implementation, the situation is different. As discussed in Section 2.3, gates whose logic operations do not contain NOT, such as AND gates and OR gates, can be easily implemented with certain types of electronic circuits such as the EFL (emitter follower logic) gate. But these implementations do not always offer desirable electronic performance. Also, these gates are difficult to implement with other types of electronic circuits. Thus NAND and NOR gates are often used in practice and are regarded as simple and reliable, though AND or OR gates can synthesize networks for certain functions with fewer gates than are required with other gate types. If a logic designer wants to use electronic circuits, NOR and NAND gates are more important than AND, OR, or NOT gates, because of their easy implementation with currently favored technology of electronic devices; hence the designer must be able to utilize NOR and NAND gates skillfully. As will be discussed in Chapter 6, however, switching theory based on AND, OR, and NOT is sometimes still an important basis for logic design with NOR or NAND.

However, AND, OR, and NOT gates are commercially available from semiconductor manufacturers, despite the above problems, for people who are more familiar with these gates than with NOR or NAND, since logic design procedures with AND, OR, and NOT are easier.

In our discussion of logic design in this book, networks are designed in most cases simply by connecting simple gates such as AND, OR, NAND, and NOR gates. But generally **"gate" is not a fixed concept**, as explained in the following.

First, although the word "gate" is used, MOS gates and threshold gates are both conceptually different from AND, OR, NOT, and EXCLUSIVE-OR gates above, in the sense that MOS gates and threshold gates each represent any function in a certain class of functions, whereas AND gates, for example, represent the fixed logic operation "AND." In other words, an MOS gate and a threshold gate can each express any operation in a class of logic operations, whereas AND, OR, NOT, NOR, NAND, and EXCLUSIVE-OR gates each express a specific logic operation. An MOS gate can represent any negative function, and a threshold gate any threshold function.

Second, when certain types of electronic circuits are used to implement gates, the OR operation can be realized by simply connecting outputs of

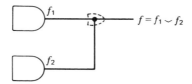

Fig. 5.5.6 Wired-OR.

gates without requiring an extra gate for this OR operation. This is called **wired-OR** and is denoted by a dotted-line small OR gate, as illustrated in Fig. 5.5.6. In other words, wired-OR is a logic operation without gates. Since wired-OR is not always permitted, depending on the electronic circuits used to implement gates, we have to find out whether or not we can use wired-OR in each case. Similarly, the AND operation can be realized by connecting certain points in electronic circuits for gates; this is called **wired-AND**. Implementation of logic operations simply by connections, such as wired-OR and wired-AND, is called **wired logic**.

Third, by connecting an inside point of a gate to an inside point of another gate, it is sometimes possible to realize a much more complex function than the one each gate can realize, or a different function. In particular, this is often done with ECL gates. Also, a **single diode or transistor is sometimes used to perform a fairly complex logic operation that would need many gates if it were implemented with gates only**. Thus, in general, logic design is flexible, and we can sometimes obtain a simpler network by freely connecting diodes and transistors than by making simple connection of gates. In this sense the concept of "gates" is sometimes not essential, and if we want to design good networks in terms of economy or performance without using discrete gates or off-the-shelf IC packages, we need to learn not only the gate-level logic design discussed in this book but also transistor-level logic design, especially in terms of layout on chips. (See, e.g., Motorola [1972b].)

However, since the detailed discussion of such transistor-level logic design is beyond the scope of this book, we will discuss only logic design with simple gates such as AND, OR, NOT, NAND, and NOR gates, and also logic design with complex gates such as MOS gates.

Exercises

▲ **5.1** (M) The network in (i) in each of the following A, B, or C, is designed with a minimum number of gates as the primary objective and a minimum number of connections as the secondary objective (as proved by the integer programming logic design method). By adding new inputs to appropriate gates as stated in (ii), let this network represent a function of n variables, f_n (without changing the network configuration, so that the number of gates remains the same in this network). Then find out the minimum number of gates required when f_n is to be realized in a network of AND and OR gates in exactly two levels. (Both minimal sum and minimal product must be found.)

A.   (i)   Fig. 5.4.1.
    (ii) Inputs $x_8, \ldots, x_n$ are added to each of gates 1 and 3.

B.   (i)   Fig. 4.1.5b.
    (ii) Inputs $\bar{x}_8, \ldots, \bar{x}_n$ are added to gate 1, and inputs $x_8, \ldots, x_n$ are added to gate 3.

C.   (i)   Fig. E5.1.
    (ii) Inputs $x_5, \ldots, x_n$ are added to each of gates 1 and 4.

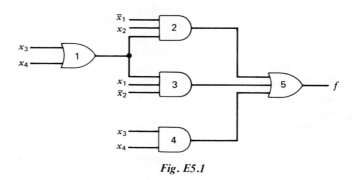

Fig. E5.1

5.2 (M) For one of the functions in Fig. E5.2, design a network with AND and OR gates that has as few gates as possible as the primary objective and as few connections as possible as the secondary objective. The number of levels need not be two. (Since, as mentioned in Section 5.4, there is no simple procedure known except very complex ones, design on a trial-and-error basis.)

A

$x_3 x_4$ \ $x_1 x_2$	00	01	11	10
00				1
01			1	1
11		1		1
10		1	1	

B

$x_3 x_4$ \ $x_1 x_2$	00	01	11	10
00			1	
01			1	1
11		1		1
10		1	1	1

Fig. E5.2

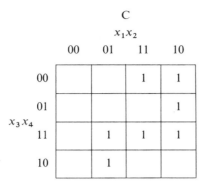

Fig. E5.2. (*Continued*)

5.3 (M) For the function f in Fig. E5.3, design a network with AND and OR gates, reducing the number of gates as much as possible, and assuming that the maximum fan-in of each gate is 3. In each level AND and OR gates may be mixed. Also, the number of levels need not be two. (Since, as mentioned in Section 5.4, there is no simple procedure known except very complex ones, design on a trial-and-error basis.)

$x_1 = 0$

$x_4 x_5$ \ $x_2 x_3$	00	01	11	10
00		1	1	d
01	1			1
11	1	1	1	
10	d	1	1	

$x_1 = 1$

$x_4 x_5$ \ $x_2 x_3$	00	01	11	10
00	1	1	1	d
01	1			1
11	1	1	1	
10		1	1	

Fig. E5.3

5.4 (M−) Design a network using AND and OR gates for the function

$$(f_1, x_2, x_3, x_4) = \Sigma(1, 3, 4, 6, 9, 12, 14) + d(7)$$

under a maximum fan-in of 2. Reduce the number of gates as much as possible, and also, as the secondary objective, reduce the number of connections. The number of levels of the network need not be two. (Since, as mentioned in Section 5.4, there is no simple systematic procedure known except very complex ones, design on a trial-and-error basis.)

5.5 (M) Design a network using AND and OR gates for the following function:

$$f = \Sigma(3, 5, 6, 7, 8, 10, 12, 13)$$

266   Advanced Simplification Techniques and Basic Properties of Gates

under a maximum fan-in of 3. Reduce the number of levels as the primary objective, the number of gates as the secondary objective, and the number of connections as the third objective. (Since, as mentioned in Section 5.4, there is no simple systematic procedure known except very complex ones, design on a trial-and-error basis.)

5.6 (M−) If an MOS gate for a given function f is to be implemented by series or parallel connections of MOSFETs, the MOS gate with the fewest MOSFETs (fewest under the assumption that MOSFETs are connected in series or parallel) can be obtained by deriving a switching expression with the fewest literals for $\bar{f}$, the complement of f, by appropriately using parentheses. For example, Fig. E5.6b, based on the expression

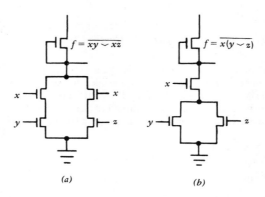

Fig. E5.6

$\bar{f} = x(y \vee z)$, consists of fewer MOSFETs than Fig. E5.6a, based on $\bar{f} = xy \vee xz$ without the use of parentheses.

Derive an MOS gate with as few MOSFETs as possible by series or parallel connections for each of the following functions, assuming that complemented and non-complemented variables are available as inputs. (Since there is no simple procedure known, design on a trial-and-error basis.)

(i) $f(w, x, y, z) = \Sigma(3, 5, 6, 9, 12, 13, 14, 15) + d(0, 1, 7)$.
(ii) $f(w, x, y, z) = \Sigma(3, 6) + d(1, 4, 7, 13)$.

> **Remark:** The number of MOSFETs in a gate can often be reduced by appropriately using bridge connections such as Fig. 2.3.6. But it is usually hard to find appropriate bridge connections.
>
> Functions (i) and (ii) above are A and C, respectively, in Exercise 4.4.15. As seen above, the MOS design problem is considerably different from an AND/OR network design problem, such as Exercise 4.4.15. ∎

5.7 (R)   Prove that the function $\overline{x \vee y}$ is a complete set. (Constants 0 and 1 are available.)

5.8 (R) When constant inputs 0 and 1 are not available, does each of the following operations form a complete set? Explain why. If not, what operations or which of the constant inputs 0 and 1 must be augmented in order to form a complete set?

　　　　　　A. $\bar{x} \vee y$.　　B. $x \cdot \bar{y}$.

Remark: Logic operation $\bar{x} \vee y$ in A is often called "material implication" and denoted by $x \to y$, as mentioned in Section 3.7. ∎

5.9 (M−) Prove that a gate which performs logic operation $\bar{x}yz \vee x\bar{y} \vee \bar{y}z$ is a complete set. (Constants 0 and 1 are available.)

5.10 (R) Derive the polynomial expansion with EXCLUSIVE-OR and AND (and possibly constant 1) only, for one of the following functions without NOT, OR, or parentheses:

　　　　A. $f = x_1 \vee x_2 \bar{x}_3$.　　B. $f = \bar{x}_1 \bar{x}_2 \vee x_1 \bar{x}_3$.

5.11 (M−) Prove or disprove each of the following:
(i) If $x \oplus y = z \oplus y$, then $x = z$.　　(iii) If $x \oplus y = 0$, then $x = y$.
(ii) $x \oplus (y \vee z) = (x \oplus y) \vee (x \oplus z)$.　　(iv) $x \oplus y = \bar{x} \oplus \bar{y}$.

5.12 (R) Derive the complete sum for one of the following functions:

　　　　A. $f = x_1 \oplus x_2 x_3$.　　B. $f = 1 \oplus x_1 x_2 \oplus x_1 x_3$.

5.13 (M−) Prove that any switching function can be expressed, using only EXCLUSIVE-OR, AND, and complemented and noncomplemented variables, without constant 1 and parentheses. Then show, with an example, that such an expression is generally not unique for a given function.

5.14 (M) Prove or disprove that the parity function $x_1 \oplus x_2 \oplus \cdots \oplus x_n$ is self-dual.

5.15 (M) Rewrite the expression

$$gx_{n+1} \vee g^d \bar{x}_{n+1}$$

into an equivalent expression using only EXCLUSIVE-OR, AND, and constant 1 but not OR, NOT, or parentheses, where

$$g = x_1 \oplus x_2 \oplus \cdots \oplus x_n,$$

and g^d is the dual of g.

5.16 (M) Any function f can be expressed using EXCLUSIVE-OR, AND, constant 1, and complemented and noncomplemented variables only, and not using parentheses. Since such an expression is generally not unique, an expression consisting of a minimal number of terms as the primary objective, and a minimal number of literals as the secondary objective, is called a minimal expression for f. For example,

$$f = \bar{x}\bar{y}\bar{z} \vee \bar{x}yz \vee x\bar{y}\bar{z} \vee x\bar{y}z$$

can be expressed as

$$f = \bar{x}\bar{y}\bar{z} \oplus \bar{x}yz \oplus x\bar{y}\bar{z} \oplus x\bar{y}z,$$

but its minimal expression is

$$x \oplus y \oplus \bar{x}\bar{z}.$$

268 *Advanced Simplification Techniques and Basic Properties of Gates*

A minimal expression can be expressed on a Karnaugh map, using rectangular loops, each of which consists of 2^i cells, where i is a nonnegative integer (each loop can contain both 0-cells and 1-cells). Fig. E5.16 shows the minimal expression.

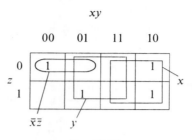

Fig. E5.16

Using a little imagination on this example, develop a procedure to find a simple expression as close to a minimal expression as possible on a Karnaugh map. Then find such an expression for the function $f(w, x, y, z) = \Sigma(3, 4, 6, 9, 11, 12, 13)$. [Kohavi 1970].

Remark: When the number of variables is large, the above map method becomes difficult. Algebraic methods to derive a minimal expression are discussed in Even, Kohavi, and Paz [1967], Marinković and Tošić [1974], and Kodandapani and Setlur [1977]. ■

5.17 (M−) Let $X = (x_2, x_1, x_0)$ and $Y = (y_2, y_1, y_0)$ be two binary numbers of three bits each, where x_0 and y_0 are the least significant bits. Using EXCLUSIVE-NOR gates and AND gates only, design a two-level network whose output is 1 only when the two binary numbers are equal. Use as few gates as possible. Assume that only $x_2, x_1, x_0, y_2, y_1, y_0$ but not their complements are available as the network inputs.

5.18 (M) Let $X = (x_2, x_1, x_0)$ and $Y = (y_2, y_1, y_0)$ be two binary numbers of three bits each, where x_0 and y_0 are the least significant bits. Using EXCLUSIVE-OR gates, AND gates, and OR gates only, design a network whose output is 1 only when X is greater than Y. Use as few gates as possible. Assume that $x_2, x_1, x_0, y_2, y_1, y_0$, and their complements are available as the network inputs.

5.19 (R) Derive, in the complete sum, the function represented by a threshold gate with one of the following sets of weights and threshold:

A. $w_1 = 1, w_2 = -1, w_3 = w_4 = 2$, and $t = 3$.
B. $w_1 = 1, w_2 = 2, w_3 = -2, w_4 = 3$, and $t = 3$.
C. $w_1 = -1, w_2 = 3, w_3 = -3, w_4 = 4$, and $t = 2$.

5.20 (R) Find what function the output of the network of two threshold gates in Fig. E5.20 represents.

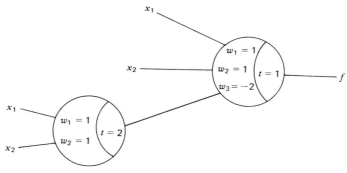

Fig. E5.20

5.21 (M) When a threshold gate that has weights $w_1, \ldots, w_n$ and threshold t represents a function $f(x_1, \ldots, x_n)$, determine what weights and threshold a gate realizing f^d, the dual of f, should have.

5.22 (M) Design a threshold gate network that realizes $x_1 \oplus x_2 \oplus x_3$, using as few gates as possible. Assume that x_1, x_2, x_3 but not their complements are available as the network inputs.

5.23 (M+) Prove that every threshold function is a unate function.

5.24 (M) Prove or disprove that the following gate, called a **residue-threshold gate**, is a complete set. A residue-threshold gate has inputs $x_1, x_2, \ldots, x_n$ and assumes the output value

$$\begin{matrix} 1 & \text{if } (x_1 + x_2 + \cdots + x_n \pmod{m}) \geq t, \\ 0 & \text{if } (x_1 + x_2 + \cdots + x_n \pmod{m}) > t. \end{matrix}$$

Here n, m, and t are integers that we can arbitrarily specify. Each x_i assumes 0 or 1. Also, $(x_1 + x_2 + \cdots + x_n \pmod{m})$ denotes the remainder of $(x_1 + x_2 + \cdots + x_n)$ divided by m. As an example, the value of $(x_1 + x_2 + \cdots + x_n \pmod{m})$ is shown in Table E5.24 for all possible combinations of x_1 and x_2 when $n = 2$ and $m = 2$. Like a threshold gate, this residue-threshold gate can express different functions by choosing different combinations of n, m, and t [Ho and Chen 1972].

Table E5.24

x_1	x_2	$x_1 + x_2$	$x_1 + x_2 \pmod 2$
0	0	0	0
0	1	1	1
1	0	1	1
1	1	2	0

5.25 (E) Prove that a negative function can be expressed as the complement of a positive function.

5.26 (M) Suppose that a loopless network for a function $f(x_1, \ldots, x_n)$ has a minimal number of MOS gates, under no restriction on maximum fan-in, maximum fan-out, or the number of levels. Prove that each MOS gate except the output gate must have at least one input from variables $x_1, \ldots, x_n$ [Liu 1972].

5.27 (M) (i) Prove that $f(\mathbf{x})$ is a positive function if and only if $f(\mathbf{a}) \geq f(\mathbf{b})$ for every pair of input vectors $\mathbf{a}$ and $\mathbf{b}$ such that $\mathbf{a} > \mathbf{b}$. Here $\mathbf{a} > \mathbf{b}$ means that $a_i \geq b_i$ for every $i = 1, 2, \ldots, n$, and $a_i > b_i$ holds for at least one i, where $\mathbf{a} = (a_1, \ldots, a_n)$ and $\mathbf{b} = (b_1, \ldots, b_n)$.

(ii) The condition in (i) can be conveniently expressed on a lattice, as illustrated for $f = x_1 \vee x_2 x_3$ in Fig. E5.27a, where each input vector is shown in parentheses and the corresponding value of f is inside a circle. A function is a positive function if and only if, for every circle for which f is 1, f is 1 for all other circles that can be reached by tracing the lines upward, starting from the circle in question. Using a lattice, determine the don't-care d's in the map in Fig. E5.27b so that the given function f is a positive function.

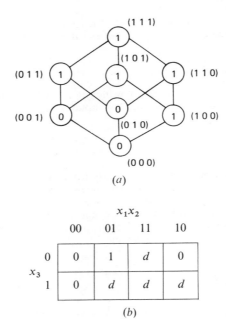

Fig. E5.27

5.28 (M) Prove that there exists a prime implicant of a function f that contains a complemented literal, if and only if f is not a positive function.

5.29 (M−) Determine the values of don't-care d's in the Karnaugh map in Fig. E5.29 so that the function f becomes a positive self-dual function.

x_3x_4 \ x_1x_2	00	01	11	10
00	0	d	d	d
01	0	0	1	d
11	1	d	d	d
10	0	0	1	1

f

Fig. E5.29

CHAPTER 6

Design of NAND (NOR) Networks and Properties of Combinational Networks

The design of networks with NAND or NOR gates is very important in logic design practice, because these gates are used very often, being implemented in many integrated circuits.

In the case of double-rail input logic, two-level networks can be easily designed, based on the minimal sums or products discussed in the preceding chapters. But often, in practice, single-rail input logic is important, since the number of pins on IC packages or the number of connections (consequently the area) on IC chips or pc boards can thereby be reduced, though the number of required gates tends to increase (at most by n gates, where n is the number of variable inputs). Unfortunately, simple procedures to design minimal networks in single-rail input logic are not known (the integer programming logic design method is complex, as mentioned in Section 6.6). Therefore a convenient heuristic procedure called the map-factoring method, which usually yields reasonably good networks, is presented for the design of NAND networks in single-rail input logic. The map-factoring method can easily be extended to the design of multilevel NAND networks in double-rail input logic also.

The three-level NAND networks have the shortest gate delay in single-rail input logic. But no simple procedure to design minimal networks is known, unlike two level AND/OR or NAND minimal networks in double-rail input logic, which can be designed based on minimal sums or products. However, a heuristic design procedure that often yields minimal three-level NAND networks is presented.

Then some important aspects of combinational networks, such as effective gate delay, spurious transient output values, and net gate delay, are examined. Also, we discuss the classification of switching functions by equivalence classes. This is convenient for the preparation of a catalog of networks so that useful networks can be found for functions to be realized, without time-consuming redesign efforts.

6.1 Introduction to Design of NAND (or NOR) Networks

Networks for digital systems are often designed with NAND (or NOR) gates only (more often than with AND and OR gates), since NAND and NOR are intrinsically associated with some important electronic implementations of gates, as explained in Section 2.3. Logic design procedures with NAND or NOR gates are the basis for many important IC logic families such as ECL (one of the two outputs of an ECL gate is NOR, as illustrated in Fig. 5.5.5), VMOS,* and IIL (integrated injection logic; see Exercise 6.10). NOR gates and NAND gates implemented by p-MOS, n-MOS, or CMOS gates are often used in logic design practice.† Also, PLAs are two-level networks of NAND or NOR gates.

Unlike networks of AND and OR gates, which are directly associated with disjunctive or conjunctive forms, networks of NAND or NOR gates are not directly associated with switching expressions that can be easily manipulated. Therefore the design of minimal NAND (or NOR) gate networks by algebraic manipulation is far more difficult than the design of minimal two-level AND/OR gate networks, which can be done on the basis of minimal sums or products.

Often, NAND and NOR gates are not mixed in a network. As seen in Section 5.5, each is a complete gate set. Use of only one gate type is often preferred by users because of convenience and simplicity.

Double-Rail Input Logic and Two-Level NAND (or NOR) Gate Networks

The design of NAND (or NOR) gate networks tends to be easier in double-rail than in single-rail input logic. Also, the former requires no more gates than the latter, though the difference is at most n gates (n is the number of variables).

Two-level NAND (or NOR) gate networks can be easily designed on the basis of the methods discussed so far, under the following assumption, which is similar to Assumption 4.1.1 or 4.4.1.

---

* Unlike p-MOS, n-MOS, or CMOS, only NOR gates can be implemented by VMOS [Rodgers and Meindl 1974].

† Although p-MOS, n-MOS, or CMOS gates can realize negative functions, these gates are often used as NOR or NAND gates only (except, probably, in key standard networks), since implementation of more complex negative functions by the gates requires more careful examination in regard to performance (parasitic capacitance may become too large) or design errors, and in many cases, logic designers in industry need to finish their designs in the shortest time possible, as explained in Section 9.1.

274 Design of NAND (NOR) Networks

Assumption 6.1.1

1. The number of levels is at most two (i.e., unless $\overline{x_1 \vee x_2 \vee \cdots \vee x_n}$ or $\overline{x_1 x_2 \cdots x_n}$ is to be realized, a network consists of two levels).
2. Either NAND gates or NOR gates only (not both) are used.
3. Complemented variable $\bar{x}_i$ and noncomplemented variable x_i for each i are available as network inputs (i.e., double-rail input logic).
4. No maximum fan-in and fan-out restrictions are imposed on any gate in a network to be designed.
5. Among networks realizable in two levels, a network that has a minimal number of gates is chosen; if there is more than one such network, one with a minimal number of connections is selected.

Suppose that by any method in the preceding chapters we have designed a minimal network in two levels of AND and OR gates under Assumption 4.1.1. As an example, consider Fig. 6.1.1a, where the outputs of AND gates are denoted by g_1, g_2, g_3. Then, as shown in Fig. 6.1.1b, insert the cascade of two inverters at each input of the OR gate. Since each output of the AND gates is inverted twice, the output of the network still represents the same function f. Then, as shown in Fig. 6.1.1c, combine the first inverters with the AND gates; these gates then become NAND gates. Combine the second inverters with the inputs of the OR gate. Then the gate becomes a NAND gate, since an OR gate with inverted inputs is equivalent to a NAND gate because of De Morgan's theorem. (Since the OR gate has inputs g_1, g_2, g_3, its output f can be expressed as $g_1 \vee g_2 \vee g_3 = \overline{\overline{g_1 \vee g_2 \vee g_3}} = \overline{\bar{g}_1 \cdot \bar{g}_2 \cdot \bar{g}_3}$. Thus the subnetwork shown inside the dotted line in Fig. 6.1.1b becomes the last NAND gate in Fig. 6.1.1c). Therefore we have obtained a network with NAND gates only. This obviously satisfies Assumption 6.1.1.

Similarly, starting with a minimal network with OR gates in the first level and an AND gate in the second level, we can obtain a minimal network with NOR gates only.

By the above conversion, procedures for designing a minimal multiple-output network with AND and OR gates in Section 5.2 yield minimal multiple-output networks with two levels of NAND gates only, or NOR gates only.

When any condition in Assumption 6.1.1 does not hold, no simple design procedure—tabular, algebraic, or graphical—that guarantees the minimality of the number of NAND or NOR gates in designed networks is known. If other constraints, such as maximum fan-in restriction, are added, the conversion in Fig. 6.1.1 may not yield minimal two-level networks that satisfy these constraints.

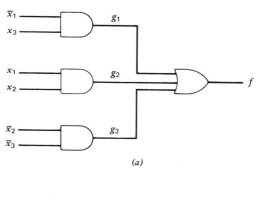
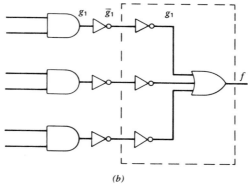
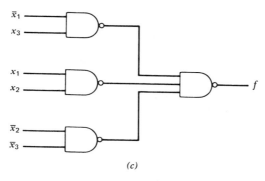

Fig. 6.1.1 (a) *AND and OR gates* (b) *Insertion of inverters* (c) *NAND gates only*

Multilevel NAND (or NOR) gate networks (i.e., those in more than two levels) are often desired for the following reasons. The number of gates in a multilevel network is no more than that in a two-level network. In fact, for some functions the number of gates is greatly reduced in a multilevel network. (For example, there is a class of functions of n variables for which a two-level network implementation requires $2^{n-1} + 1$ gates, whereas the number of gates in a multilevel network implementation is linearly proportional to n.) Also, a two-level network tends to have gates that violate maximum fan-in or fan-out restrictions. A multilevel network often avoids large fan-in in some gates, probably with a tendency toward more uniformly distributed fan-in and fan-out. Also, the power consumption tends to be reduced because of fewer gates, and the network layout on IC chips or pc boards is likely to be compact, yielding economies in realizations and, in some cases, improved speed (due to the reduction of parasitic capacitance).

Multilevel networks, however, have longer gate delays than two-level networks if all gates have equal switching times, and also tend to generate spurious transient output values, as discussed in Section 6.7.

Although there is no known simple procedure to design minimal multilevel networks in double-rail input logic (the integer programming logic design method is complex, as will be mentioned in Section 6.6), a heuristic procedure that usually yields reasonably good networks will be presented in Section 6.5.

Single-Rail Input Logic

If networks are to be in double-rail input logic—in other words, both x_i and $\bar{x}_i$ for every $i = 1, 2, \ldots, n$ are to be available as network inputs—the number of connections on a pc board or an IC chip and the number of pins of IC packages for these network inputs must be doubled, unless $\bar{x}_i$'s are realized with inverters at the end of connections for x_i's. A larger number of connections requires a larger IC chip or pc board, requiring, in turn, more space and increasing cost. Also, the networks that supply these inputs x_i and $\bar{x}_i$ as their outputs must be more complex and usually must deliver greater output power because of double-rail connections for x_i and $\bar{x}_i$. Generally, each IC package can have only a limited number of pins at reasonable cost. If we try to increase the number of pins, the cost rises very rapidly, and an increase beyond a certain number (probably 150 pins) will be impractical.

In conclusion, single-rail input logic is often preferable, though not always because of possible increase in the number of gates (at most n gates).

If a network is to be in single-rail input logic, we generally cannot realize a network of NAND (or NOR) gates in two levels. But any function can be realized in at most three levels, as can be easily seen. The design of three-level NAND (or NOR) gate networks that have the shortest gate delay will be discussed in Section 6.4. If we are allowed to use four or more levels, however, we may be able to reduce the number of gates. As a matter of fact, the number of gates can be greatly reduced for some functions. Simple design procedures to design networks in single-rail input logic with a minimal number of NAND (or NOR) gates are not known (the integer programming logic design method is complex, as mentioned in Section 6.6). But if we are satisfied with reasonably good networks instead of those with a minimal number of gates, some heuristic design procedures, which we shall discuss in Sections 6.2 and 6.3, are available.

6.2 Switching Expressions for NAND (or NOR) Networks

When a network of NAND (or NOR) gates is given, it is somewhat cumbersome to find the switching function that the network represents, unlike the case of networks of AND and OR gates. Therefore let us develop a means to find the function quickly [Maley and Earle 1963].

Suppose that a NAND gate network is given as shown in Fig. 6.2.1a. Find all paths of gates. There are three paths (e.g., path 1 consists of gates 1, 2, 4, and 5). Draw a tree network consisting of all paths of gates discovered, as illustrated in Fig. 6.2.1b. (A **tree network** is a network where the output of each gate is connected to only one gate. Fig. 6.2.1a is not a tree network, since the output of gate 5 is connected to gates 3 and 4.) Then, as shown in Fig. 6.2.1c, replace each NAND gate in the odd-numbered levels (counting from the output gate toward the network inputs; the level numbers are shown with arrows in the bottom of Fig. 6.2.1) by an OR gate with all gate inputs inverted. (This replacement is the reversal of the replacement of the OR gate with inverted inputs in the dotted-line square in Fig. 6.1.1b by the output NAND gate in Fig. 6.1.1c. This is always possible by De Morgan's theorem.) In Fig. 6.2.1c the OR gates with inverted inputs are shown with bubbles at the inputs, where bubbles mean inversions. (These symbols are commonly used in practice, though each of them means an inverter in the dotted-line square in Fig. 6.1.1b.) In Fig. 6.2.1d we have obtained the network of AND and OR gates, by deleting bubbles at the two ends of each connection line (each inversion in a pair cancels the other), and inverting the variable inputs of the OR gates.

Once we have obtained the network of AND and OR gates in Fig. 6.2.1d, we can easily write the switching function that the network represents:

$$f = \bar{u}(\bar{z} \vee x\bar{y}) \vee w \vee v(\bar{z} \vee x\bar{y})(\bar{x} \vee y). \tag{6.2.1}$$

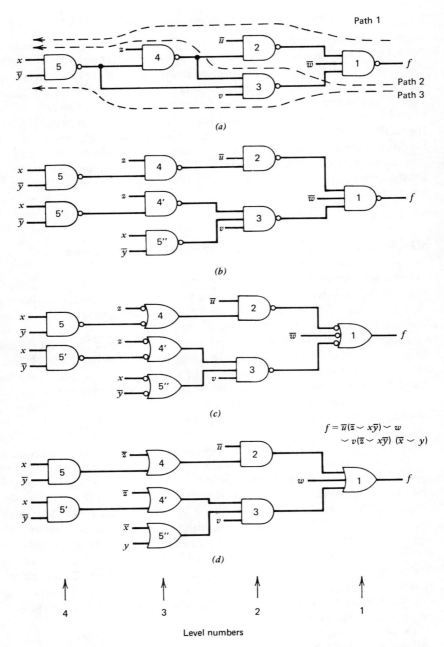

Fig. 6.2.1 *Conversion of a NAND network into an AND/OR network.* (a) *Given NAND network.* (b) *Tree network of NAND gates.* (c) *Replacement of NAND gates in odd-numbered levels by OR gates.* (d) *Merger of paired inversions.*

Switching Expressions for NAND (or NOR) Networks 279

In practice, however, we need not convert given networks as we did in Figs. 6.2.1*b* through 6.2.1*d*. We can immediately write the switching function by the following rule, which can easily be obtained by generalizing the conversion in Fig. 6.2.1 (though the conversion in Fig. 6.2.1 is convenient to recall Rule 6.2.1).

Rule 6.2.1: Derivation of a Switching Expression for a NAND Network

Go through all different paths of gates from the output gate toward the network inputs (e.g., the three paths shown in Fig. 6.2.1*a*). As you go through different levels of gates in each path, write the logic operation on variable inputs, alternating between the AND and OR operations, and inverting variable inputs at every other level—in other words:

1. Consider the OR operation for a NAND gate that is in an odd-numbered level, counting levels from the network output toward the network inputs. Complement every variable that is connected to this NAND gate, if this gate has any variable inputs.
2. Consider the AND operation for a NAND gate that is in an even-numbered level. Variables are not complemented.

Thus we have obtained a switching expression for the given NAND network.

□

Rule 6.2.1 can be extended to networks of NOR gates, or a mixture of NAND and NOR gates [Morris and Miller 1971].

Network Design Based on Rule 6.2.1.

Reversing the above procedure, we can design a network directly from an appropriate switching expression found after trial-and-error efforts, though the result may or may not be a good network.

When a network is of the tree type (i.e., the output of each gate goes to only one gate), it is usually easy to use this approach. But if we want to obtain networks other than tree networks, it may be more difficult to find appropriate switching expressions. For example, $(\bar{z} \vee x\bar{y})$ appears twice in (6.2.1), but it is usually difficult to determine whether such a complex expression yields a better network than a simpler expression. In finding appropriate switching expressions, we can take into account some restrictions such as maximum fan-in, though not very easily. In the network in Fig. 6.2.1*a* corresponding to switching expression 6.2.1, every gate has a maximum fan-in of 3. If (6.2.1) is expanded into its minimal sum without parentheses, the output gate in the

280 Design of NAND (NOR) Networks

corresponding network has a fan-in of 5. Simultaneous consideration of maximum fan-in and fan-out restrictions is difficult.

▲ Conversion of an AND/OR Network into a NAND Network

The conversion of a network of AND and OR gates into the network of NAND gates illustrated in Fig. 6.1.1 can be extended to multilevel networks. Thus, when we have a network of AND and OR gates, we can easily derive a network of NAND gates. This NAND network design procedure, however, presents the following problem.

Suppose that a network of AND and OR gates is given, as shown in Fig. 6.2.2a. Then, by inserting a cascade of two inverters in each input of the OR gates (as we did in Fig. 6.1.1b), we can derive the NAND network shown in Fig. 6.2.2b. In Fig. 6.2.2b, however, we have one extra inverter which must be replaced by a NAND gate having a single input, since gate 6 in Fig. 6.2.2a is in the odd-numbered level in path 1–3–6 but in the even-numbered level in path 1–2–4–6. Of course, if all OR gates are in the odd- (or even-) numbered level in every path in a given network, the NAND (or NOR) network derived will not have extra inverters.

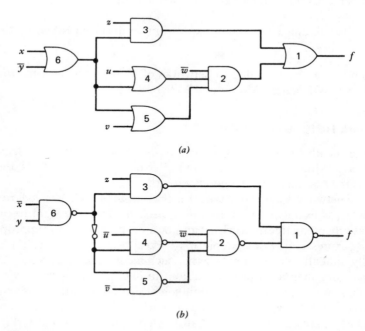

Fig. 6.2.2 *Derivation of a NAND network from an AND/OR network.*

Design of NAND (or NOR) Networks in Single-Rail Input 281

In addition, there is the following problem. Design of an AND/OR network to start with is generally not easy, as discussed in Section 5.4, even if we are content to derive a network with as few gates as possible, instead of with a minimum number of gates. Such an AND/OR network could be obtained by the algebraic procedure in Section 5.4; but if a given function is to be algebraically manipulated, a NAND network could be directly designed by reversing Rule 6.2.1, as described above. Since the AND and OR operations are easier to manipulate than the NAND operation, we might think that the AND/OR network to start with is easier to design than NAND networks. Contrary to this false impression, NAND networks are usually easier, as we will see in Sections 6.3, 6.4, and particularly 6.5, if we want to derive minimal networks. The reason is that only one gate type (i.e., the NAND gate) needs to be considered in each gate position in a network, instead of two gate types, AND and OR. [In the integer programming logic design method also (to be mentioned in Section 6.6), minimal NAND networks can be designed in a much shorter time by computer than minimal AND/OR networks.]

If the NAND networks are to be in single-rail input logic, the complemented variable inputs in Fig. 6.2.2b must be inverted by NAND gates, adding more gates.

6.3 Design of NAND (or NOR) Networks in Single-Rail Input Logic by the Map-Factoring Method

When a network of NAND (or NOR) gates in single-rail input logic is to be designed, there is no simple design procedure, whether tabular, algebraic, or graphical, that guarantees the minimality of the network; the integer programming logic design method in Section 6.6 is complex, requiring computer processing. However, if we are satisfied with reasonably good networks, the **map-factoring method** is useful for hand processing. It is based on a Karnaugh map but, unlike the Karnaugh map method for minimal two-level networks with AND and OR gates discussed so far, it does not guarantee the minimality of the number of gates. By using designer's intuition based on the pictorial nature of a Karnaugh map, at least reasonably good networks can be derived after trial-and-error efforts. As a matter of fact, minimal networks can sometimes be obtained, though the method does not enable us to prove their minimality. A drawback of this method, however, is the difficulty of learning it. A good feeling about how to use it can be gained only after working on many examples. The map-factoring method was first described in Chapter 6 of Maley and Earle [1963].

First let us define **permissible loops** on a Karnaugh map. A permissible loop is a rectangle of 2^i cells that contains the cell corresponding to "the

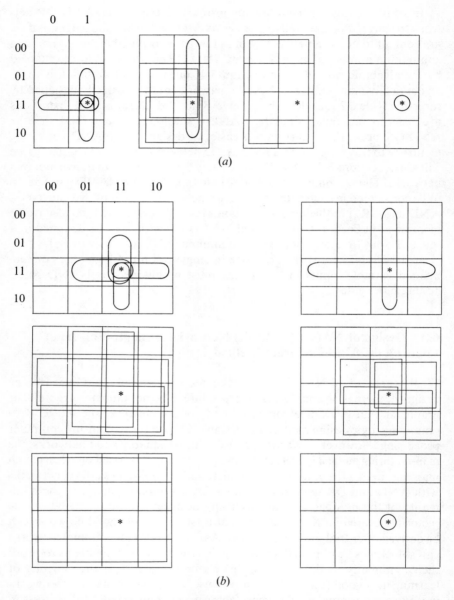

Fig. 6.3.1 Permissible loops. (a) Case of three variables. (b) Case of four variables.

Design of NAND (or NOR) Networks in Single-Rail Input 283

combination of variable values of all 1's" (i.e., the cell marked with the asterisk in Fig. 6.3.1), where i is one of 0, 1, 2, ... In other words, a **permissible loop must contain the particular cell denoted by an asterisk in Fig. 6.3.1, but, unlike the prime-implicant loops discussed in Chapter 4, this permissible loop consists of 1-cells, 0-cells, or a mixture.** All permissible loops are shown in Fig. 6.3.1 for three and four variables.

In the following, let us describe the map-factoring method, modifying the original version of Maley and Earle. (The modification is explained in the footnote for Step 3.)

Procedure 6.3.1—The Map-Factoring Method: Design of a Network in Single-Rail Input Logic with as Few NAND Gates as Possible

We want to design a network with as few NAND gates as possible, under the assumption that x_i, but no $\bar{x}_i$ for each i, is available as a network input. (Readers are advised to read this procedure a few times, though it is not really complicated, once understood.)

1. Make the first permissible loop,* encircling 1-cells, 0-cells, or a mixture of them. Draw a NAND gate corresponding to this loop. To this gate, connect all the literals in the product which this loop represents. Shade this loop.

    For example, when a function $f = x \vee y \vee \bar{z}$ is given, let us choose the first permissible loop, as shown in Fig. 6.3.2a, though there is no reason why we should choose this particular loop. (Another loop could be better, but we cannot guess at this moment which loop is the best choice. Another possibility in choosing the first permissible loop will be shown later in Fig. 6.3.3a.) The loop we have chosen is numbered 1 and is shaded. Then NAND gate 1 is drawn. Since the loop represents x, we connect x to this gate.

2. Make a permissible loop,† encircling 1-cells, 0-cells, or a mixture of them. Draw a NAND gate corresponding to this loop. To this new gate (which the new permissible loop represents), connect literals in the product that this loop represents.

    In the above example, a new permissible loop is chosen as shown in Fig. 6.3.2b, though there is no strong reason why we should choose this particular permissible loop. This loop represents product yz, so y and z are connected to the new gate, which is numbered 2.

---

* There is no guiding principle for finding which loop should be the first permissible loop. Any other permissible loop could be as reasonably good as the first one, and usually it is very hard to determine which permissible loop is the best choice. After deriving a few networks by using the map-factoring method in different ways, you may be able to "feel" which loop is a reasonably good choice at each step.

† Again there is no guiding principle for deciding which permissible loop should be chosen.

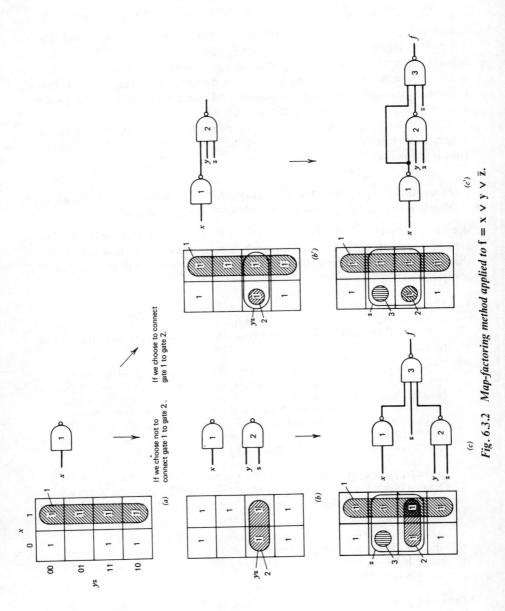

Fig. 6.3.2 Map-factoring method applied to $f = x \lor y \lor \bar{z}$.

284

Design of NAND (or NOR) Networks in Single-Rail Input 285

Up to this point this step is identical to Step 1. Now, to this new NAND gate, we further connect the outputs of some gates already drawn,* if we choose to do so.†

(a) If we choose not to connect any previous gate to the new gate, the new permissible loop is entirely shaded.

> *If we prefer* not to connect gate 1 to gate 2 in the above example, we get the network in Fig. 6.3.2*b*. The shaded loop for gate 1 is completely ignored. The new permissible loop is entirely shaded and is numbered 2, as shown in Fig. 6.3.2*b*.

(b) If we choose to connect some previously drawn gates to the new gate, only some or all of the previously drawn gates‡ whose shaded loops **inhibit** (i.e., intersect or are included in) the new permissible loop are allowed to be connected to the new gate. Encircle and shade the area inside the new permissible loop, excluding the shaded loops of the previously drawn gates connected to the new gate (e.g., we derive shaded loop 2 inside the new permissible loop, which represents yz in Fig. 6.3.2*b'*). The shaded loop thus formed§ is said to be **associated with** the output of this new gate.

> *In the above example*, if we choose to connect gate 1 to gate 2, we obtain the network in Fig. 6.3.2*b'*. The part of the new permissible loop that is not covered by the shaded loop associated with gate 1 is shaded and is numbered 2.

3.‖ Repeat Step 2 until the following condition is satisfied:

> *Termination condition:* When a new permissible loop and the corresponding new gate are introduced, all the 0-cells on the entire map constitute the shaded loop associated with the output of the new

---

* When Step 2 is applied for the first time, we have only the one gate previously drawn at Step 1; but after Step 2 is iterated, we have many gates drawn in the previous iterations.

† There is no guiding principle for deciding whether or not each of these gates should be connected to the new gate.

‡ In this case any previously drawn gates whose shaded loops are outside the new permissible loop are not allowed to be connected to the new gate under any circumstance. (We do not have such previously drawn gates at the first application of Step 2.)

§ Note that this shaded loop is not necessarily a permissible loop. As a matter of fact, the output of the new gate represents the function whose 0-cells are the cells of this shaded loop (Exercise 6.6.9).

‖ This step is different from the original version of the map-factoring method by Maley and Earle, which yields only networks whose output gates have no external variables connected as inputs. It is modified here so that networks with variable inputs at output gates such as those in Figs. 6.3.8 and 6.3.9 can be obtained.

gate (i.e., all the 0-cells on the entire map are contained in the new permissible loop, and the shaded loop associated with the output of the new gate contains only these 0-cells but no 1-cell).

Let us continue Fig. 6.3.2b. If we choose the new permissible loop as shown in Fig. 6.3.2c, and if we choose to connect gates 1 and 2 to the new gate 3 (i.e., the previously shaded loops 1 and 2 are interpreted to inhibit this new permissible loop), the above termination condition has been satisfied; in other words, all the 0-cells on the entire map (i.e., the only 0-cell shaded and numbered 3 in Fig. 6.3.2c in this particular example) constitute the shaded loop associated with the output of the new gate 3.

We have obtained a network of NAND gates for the given function. Depending on what permissible loops we choose and how we make connections among gates, we may get different networks. After several trials we choose the best network.

As an alternative for the network for f obtained in Fig. 6.3.2c, let us continue Fig. 6.3.2b'. If we choose the new permissible loop as shown in Fig. 6.3.2c', and the previously shaded loops for gates 1 and 2 are interpreted to inhibit this new loop, we satisfy the termination condition in Step 3, and we have obtained the network in Fig. 6.3.2c', which is different from the one in Fig. 6.3.2c.

If we choose the first permissible loop 1, as shown in Fig. 6.3.3a, differently from Fig. 6.3.2a, we can proceed with the map-factoring method as shown in Fig. 6.3.3b, and we obtain the third network as shown in Fig. 6.3.3c. Of course, we can continue differently in Figs. 6.3.3b and 6.3.3c. Also, the first permissible loop can be chosen differently from Fig. 6.3.2a or 6.3.3a, but it is too time-consuming to try all possibilities, so we must be satisfied with several trials. We need a few trials to gain a good feeling of how to obtain good selections, and thereafter we may obtain a reasonably good network. For the above example, $f = x \vee y \vee \bar{z}$, the network obtained in Fig. 6.3.3 is, as a matter of fact, the minimal network (the minimality of the number of gates and then, as the secondary objective, the minimality of the number of connections were proved by the integer programming logical design method mentioned in Section 6.6). □

A proof of why the map-factoring method works is left to readers as Exercise 6.6.9. It is not difficult if we see that the output of a new gate actually expresses the function whose 0-cells are the cells inside the shaded loops associated with the new gate.

Notice that, even when a map contains d-cells, that is, cells for don't-care conditions, the map-factoring method, Procedure 6.3.1, can be easily used by appropriately interpreting each d-cell as a 0-cell or a 1-cell only when we examine the termination condition in Step 3.

Examples: Let us design networks for another sample function, $f = \bar{x}\bar{y} \vee \bar{y}z \vee xy \vee y\bar{z}$. Let us choose the first permissible loop as shown in Fig. 6.3.4a. This loop is numbered 1 and is shaded. Then

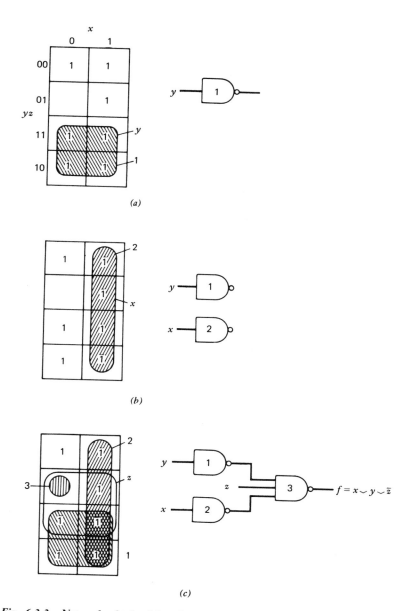

Fig. 6.3.3 Network obtained by choosing the first permissible loop differently from Fig. 6.3.2.

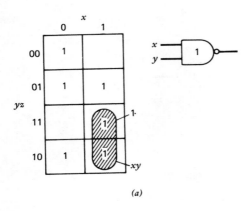

(a)

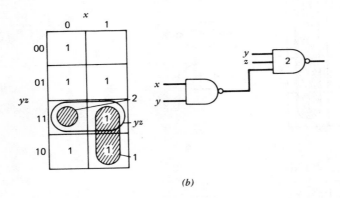

(b)

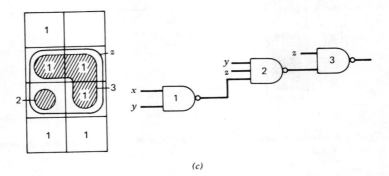

(c)

Fig. 6.3.4 Map-factoring method applied to $f = \overline{x}\overline{y} \vee \overline{y}z \vee xy \vee y\overline{z}$.

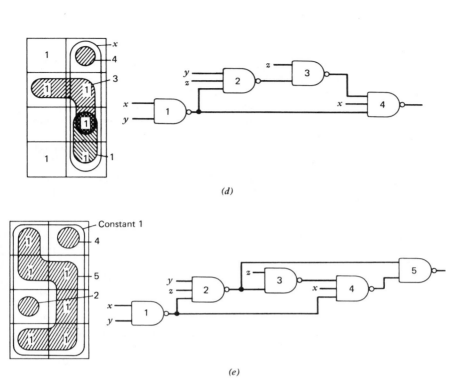

(d)

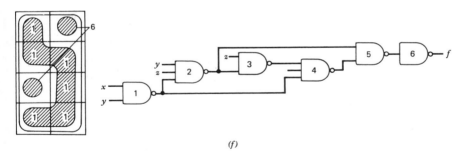

(e)

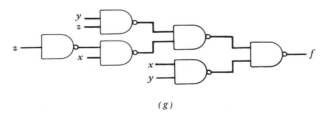

(f)

(g)

Fig. 6.3.4. (*Continued*)

NAND gate 1 is drawn. Since the loop represents the product xy, we connect x and y to this gate.

Let us choose the second permissible loop as shown in Fig. 6.3.4b. Since this loop represents the product yz, y and z are connected to the new gate, numbered 2. The new permissible loop intersects shaded loop 1, so we connect the output of gate 1 to gate 2 (we could have chosen not to make this connection). The remainder inside the new permissible loop is shaded and is numbered 2.

By repeating Step 2 with the above sample function, we draw the next permissible loop representing z, as shown in Fig. 6.3.4c. Gate 3 is correspondingly drawn. Since the shaded loop for gate 2 is contained in this new loop, and the shaded loop for gate 1 (not shown in Fig. 6.3.4c) intersects the new loop, the outputs of gates 1 and 2 could be connected to new gate 3. Let us connect only gate 2, but not gate 1, to gate 3. (We do not know whether this decision is good. **If gate 1 is also connected, if gate 1 is connected but gate 2 is not connected, or if both gates 1 and 2 are not connected, then we will get different networks, correspondingly.**) At the next iteration of Step 2, we draw the new permissible loop representing x, as shown in Fig. 6.3.4d. Correspondingly, gate 4 is drawn. Gates 1 and 3 can be connected to gate 4, since the shaded loop for gate 1 is contained in this new loop, and the shaded loop for gate 3 intersects the new loop. But **we are not allowed to connect gate 2 to gate 4, since the shaded loop for gate 2 neither intersects nor is contained in the new loop**. We do connect gates 1 and 3 to gate 4. At the next iteration of Step 2, the new permissible loop representing the constant 1 is drawn as shown in Fig. 6.3.4e. Gate 5 is correspondingly drawn. The input of the constant 1 (corresponding to the new loop) is not connected to gate 5, since, regardless of whether or not the constant 1 is connected as input, the input-output relation of gate 5 is not changed. Any gate previously drawn could be connected to gate 5, but let us connect gates 2 and 4 only. Still, the termination condition stated in Step 3 is not satisfied (if 0 and 1 were interchanged in Fig. 6.3.4e, the termination condition would be satisfied). At another iteration of Step 2, let us choose the permissible loop representing the constant 1 again, as shown in Fig. 6.3.4f. If we decide to connect only gate 5 to gate 6, the termination condition has finally been satisfied because all the 0-cells on the map are contained in the new loop and are shaded as the loop associated with the new gate 6 (the new shaded loop contains only two 0-cells and no 1-cells). Thus we have obtained a NAND network for the given f in Fig. 6.3.4f.

Fig. 6.3.4g shows a network for the given f that has a minimum number of gates as the primary objective and a minimum number of connections as the secondary objective (proved by the integer programming logic design method). This network has two less connections than Fig. 6.3.4f, though the number of gates remains the same. We should be able to derive this network by the map-factoring method by appropriate selections of permissible loops and inhibitions. (Readers

Design of NAND (or NOR) Networks in Single-Rail Input 291

should try to do so, or draw permissible loops and shaded loops on maps corresponding to this network. Once a network is given, it is easy to draw these loops by tracing the map-factoring method in reverse.)

Another example of logic design by the map-factoring method is shown in Fig. 6.3.5. (Fig. 6.3.5a shows the result after a few iterations of Step 2.) Since Fig. 6.3.5c satisfies the termination condition in Step 3, we have obtained a network.

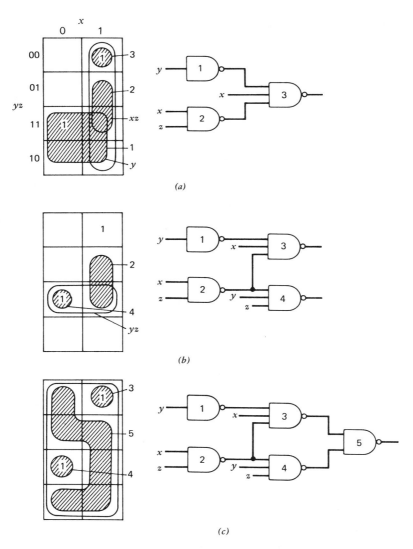

Fig. 6.3.5 Map-factoring method applied to $f = x\bar{y}\bar{z} \lor \bar{x}yz$.

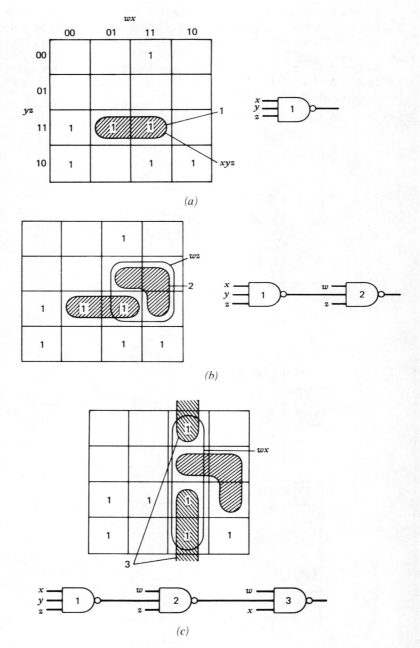

Fig. 6.3.6 Map-factoring method applied to $f = wx\bar{z} \vee \bar{x}y\bar{z} \vee xyz \vee \bar{w}yz$.

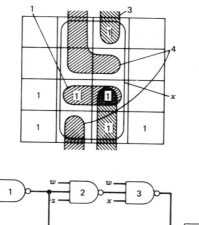

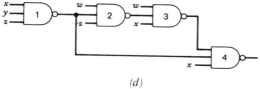

(d)

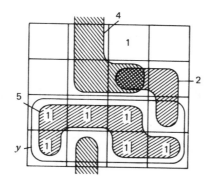

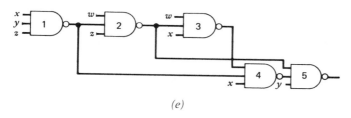

(e)

Fig. 6.3.6. (Continued)

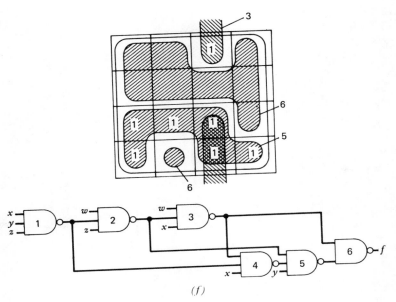

(f)

Fig. 6.3.6. *(Continued)*

An example for four variables is shown in Fig. 6.3.6. If we choose not to connect gates 1 and 3 to 4 in Fig. 6.3.6d, we can proceed as shown in Fig. 6.3.7, reaching a different network.

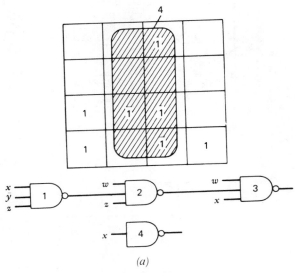

(a)

Fig. 6.3.7. *Variation for Fig. 6.3.6.*

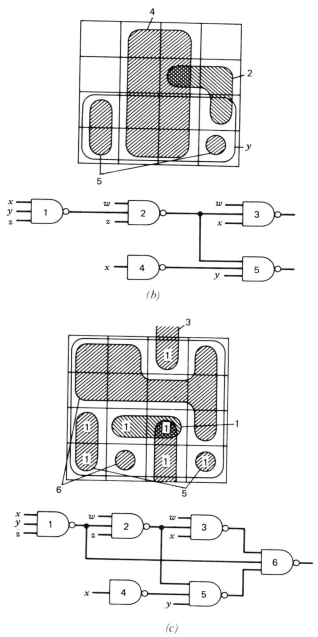

(b)

(c)

Fig. 6.3.7. *(Continued)*

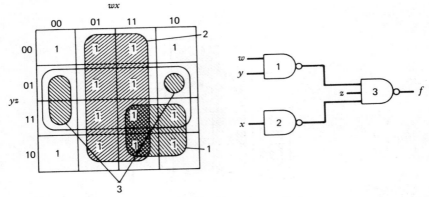

Fig. 6.3.8 *Map-factoring method.*

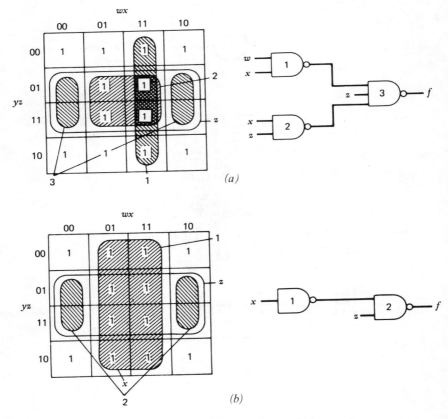

Fig. 6.3.9 *Map-factoring method.*

Design of NAND (or NOR) Networks in Single-Rail Input   297

Other examples for four variables are shown in Figs. 6.3.8 and 6.3.9, without giving intermediate steps. Figs. 6.3.9a and 6.3.9b show different sets of loops for the same function, and obviously Fig. 6.3.9b yields a better result than Fig. 6.3.9a.

Simple Exercises for Self-Study

Before proceeding further, try to derive NAND networks for the following functions by the map-factoring method, reducing the number of gates, and then, as the secondary objective, the number of connections as much as possible (Answers are shown after the exercises for Section 6.6.)

$$f_1 = \bar{x}_1 x_3 \vee \bar{x}_1 x_4 \vee \bar{x}_2 x_3 \vee \bar{x}_2 x_4 \quad \text{and} \quad f_2 = \bar{x}_1 \vee \bar{x}_4 \vee \bar{x}_2 x_3 \vee x_2 \bar{x}_3.$$

Consideration of Restrictions Such as Maximum Fan-In

If the restriction of maximum fan-in or fan-out is imposed, permissible loops and connections must be chosen so as not to violate it. With the map-factoring method, unlike the other design methods discussed so far, it is easy to take such a restriction into consideration. For example, suppose that we have a maximum fan-in and fan-out of 3.

1. As to a maximum fan-in of 3 for a four-variable function, we need to use permissible loops containing two or more cells in a map for the first-level gates, which do not have outputs of other gates as inputs, such as gate 1 in Fig. 6.3.6. For gates in the second or later levels, permissible loops must contain four or more cells, depending on whether the number of connections from the previously drawn gates is one or more. In this manner we must limit the number of connections from the previously drawn gates so that the new gate does not violate the maximum fan-in restriction.

2. As to a maximum fan-out of 3, we need to check whether or not each previously drawn gate already has three output connections, whenever we make a shaded loop associated with a new gate.

Improvement Techniques

Although the above method is relatively simple, at each step there are many choices of permissible loops and connections among gates, and there is no good guideline to finding a best choice. Also, a map does not provide as much pictorial insight as it does in the Karnaugh map method. (Probably for these two reasons, the map-factoring method has not attracted the interest of many people so far.) But if an appropriate sequence of choices is made, we may obtain a good network.

298 Design of NAND (NOR) Networks

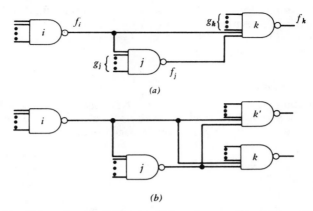

Fig. 6.3.10 *Triangular condition.* (a) *Triangular condition among three gates.* (b) *Generalization of case in Fig. 6.3.10a.*

Very often, we can simplify designed networks by using the following triangular condition.

Theorem 6.3.1: Suppose that a network contains three NAND gates, i, j, k, connected as shown in Fig. 6.3.10a. Gate i has output connections to gates j and k, and gate j has no output connection except the connection to gate k. If the configuration of three gates satisfies this condition, which is called the **triangular condition**, the connection between gates i and j can be removed without changing the output function of gate k.

Proof: Let f_i, f_j, and f_k denote the outputs of gates i, j, and k, respectively. Then $f_k = \overline{f_i f_j g_k}$ holds, where g_k is the conjunction of the functions represented by all inputs of gate k except those from gates i and j.
 Also
$$f_j = \overline{f_i g_j},$$
where g_j is the conjunction of functions represented by all inputs of gate j except the one from gate i. Thus $f_k = \overline{f_i(\overline{f_i g_j})g_k} = \bar{f}_i \vee (f_i g_j) \vee \bar{g}_k = \bar{f}_i \vee g_j \vee \bar{g}_k$.

When the connection between gates i and j is deleted, the output function of gate k becomes
$$\overline{f_i f'_j g_k} = \overline{f_i \bar{g}_j g_k} = \bar{f}_i \vee g_j \vee \bar{g}_k,$$
where f'_j denotes the output function of gate j after the connection between gates i and j is deleted. Since this is identical to f_k, the connection between gates i and j can be deleted without changing the output function of gate k. Q.E.D.

On the basis of Theorem 6.3.1, we can get the following procedure.

Procedure 6.3.2:* Simplification of a NAND Network by the Triangular Condition

If any three gates in a NAND network satisfy the triangular condition as shown in Fig. 6.3.10a, the connection from gate i to gate j can be removed without changing the network output. This can be extended to the more general case in Fig. 6.3.10b, where there is more than one gate k. Also, even if we have a variable x_i, instead of a gate i, which is connected to gates j and k (or k, k', ..., as shown in Fig. 6.3.10b), we can remove the connection from x_i to gate j. (Exercises 3.5.7, 6.6.13, 6.20, and 6.21 are generalizations of Procedure 6.3.2.) □

For example, input z, gate 2, and gate 3 in Fig. 6.3.2c satisfy the triangular condition. Therefore, by deleting z from gate 2, we get the simpler network shown in Fig. 6.3.3c, where gates 1 and 2 are numbered 2 and 1, respectively. Furthermore, in Fig. 6.3.2c' the connection from gate 1 to gate 2 and also the connection of z to gate 2 can be removed, and the network in Fig. 6.3.3c, which is minimal, is obtained. Thus we can sometimes improve a network, even if we obtain one with many gates and connections, by poor use of the map-factoring method.

When the map-factoring method is not processed in an appropriate manner because of the lack of guidelines at each step, we may not be able to derive good networks. (The minimality of the networks obtained cannot be proved by the method itself, so all we can say is that we seem to obtain reasonably good networks by deriving several networks using different sequences of loops, and then choosing the best result.) The triangular condition is not always helpful. One solution to this difficulty is to try to improve a network by simultaneously using the map-factoring method and the design procedures in Section 6.2. This approach may sometimes be useful. (For other approaches, see Remark 6.3.1.)

Of course, if a designer applies the map-factoring method carefully spending a sufficient amount of time, further simplification by the triangular condition or any other simplification rule will not be possible. As a matter of fact, minimal networks are obtained in many cases by the map-factoring method, though the minimality cannot be proved by the method. Usually, it is difficult to use the map-factoring method for five or more variables, but even for four variables there are numerous functions that cannot be easily dealt with by any other methods, but can be handled by the map-factoring method. The

---

* This procedure is also very useful in the integer programming logic design method mentioned in Section 6.6 [Baugh, Ibaraki, Liu, and Muroga 1969; Nakagawa, Lai, and Muroga 1971]. The procedure has also been used by Hellerman [1963].

300 Design of NAND (NOR) Networks

map-factoring method is probably most effective in designing the parity function or similar functions. If a function requires a simple network such as a tree network, however, other methods such as the one based on switching expressions in Section 6.2 may be more effective. (For some experience with the map-factoring method, see Remark 6.3.2.)

The Map-Factoring Method for NOR Networks

A minimal network of NOR gates for a given function f can be designed by the following approach. Use the map-factoring method to derive a minimal network of NAND gates for the dual f^d of f. Then replace NAND gates in the network with NOR gates. The result will be a minimal network of NOR gates for f.

Also, a minimal network of NOR gates can be obtained directly by the map-factoring method with the exchange of 0-cells and 1-cells, and the change of terms to alterms.

Remark 6.3.1: When the network appears to require many gates, the map-factoring method can be applied from the network output toward the network inputs (i.e., starting from the termination condition in Procedure 6.3.1, Step 3), instead of moving from left end toward right end in Figs. 6.3.4 through 6.3.6. Also, for some functions such as parity functions, decomposition of a given function into functions of fewer variables makes the application of the map-factoring method easier.

Also, if on a map we can identify variables that can be directly connected to the output gate, and we change the corresponding cells to don't-care conditions on the map, the use of map-factoring method becomes easier.* This is always possible unless the number of inputs at the output gate violates the maximum fan-in restriction, when the restriction is imposed. (This will be used in Step 1 in Procedure 6.4.1. Also, see Exercise 6.6.18.)

In addition to the possibilities of eliminating some inputs from gates by Exercise 3.5.7, 6.6.13, 6.20, or 6.21, there are many other simplification procedures that can be applied. (See, e.g., Lai, Nakagawa, and Muroga [1974].) ∎

Remark 6.3.2: To find out how good networks designed by the map-factoring method could be, functions of three and four variables were sometimes given as exercises to students. Then the results were compared with minimal networks designed by the integer programming logic design method mentioned in Section 6.6. The students' results were minimal in many cases. Also, the time spent for design is very short once students learn how to use the method.

For example, design of a network with a minimum number of NOR gates under maximum fan-in and fan-out restrictions of 3 for function $\overline{x \oplus y \oplus z}$ is not easy

---

* This has probably been known by many authors. See, for example, [Baugh, Ibaraki, Liu, and Muroga [1969] and Section 8.8 of Torng [1972]].

for hand processing by any method discussed so far, though the function has only three variables. A student who used to be a logic designer in industry first tried to solve this problem with the Karnaugh map method in Chapter 4, modifying it by taking into account the maximum fan-in and fan-out restrictions. A network of eight NOR gates was obtained fairly quickly. He then tried to reduce the number of gates. After spending 3 or 4 hours, he obtained a minimal network of seven NOR gates (as proved by the integer programming logic design method). After learning the map-factoring method, however, he obtained a network of seven NOR gates within about 10 minutes. This was surprisingly fast; the network of seven gates was obtained by the integer programming logic design method within 2 minutes on the IBM 360/75I computer. (The interger programming logic design method obtained this network very quickly, but the major part of the computer time was spent in proving the minimality of the network, as the integer programming logic design method does with many other functions. But notice that the map-factoring method cannot guarantee the minimality of any results.) Other students who did not have logic design experience tended to spend much more time.

The major shortcoming of the map-factoring method is probably the difficulty of learning it. Also, because of the lack of a guiding principle in each step, bad networks can be obtained if an appropriate sequence of selections is not made by stretching the imagination for many possibilities. In this sense it tends to be accidental whether we can obtain very good networks. ∎

▲ 6.4 Design of Three-Level NAND (or NOR) Networks in Single-Rail Input Logic

When a NAND (or NOR) gate network is to be in single-rail input logic, a three-level network has the shortest gate delay, unless a given function f is a positive function, a complement of the product of noncomplemented variables, or a disjunction of them. Any function can be realized in at most three levels, since a network can be designed by the following simple-minded approach. Design a two-level network of AND and OR gates, assuming that both x_i's and $\bar{x}_i$'s are available as inputs to the network. By the procedure discussed at the beginning of Section 6.1, convert it to the corresponding network of NAND gates only (or NOR gates only). Then each complemented input $\bar{x}_i$ can be replaced by a NAND gate (or a NOR gate) having a single input x_i. The network that results, however, may not be minimal.

Despite the importance of three-level NAND (or NOR) gate networks, which are the fastest if all gates have equal switching times, no simple design procedure for hand processing is known, unlike the case of minimal two-level AND/OR gate networks, which can be designed based on minimal sums or minimal products. If computers are available, Gimpel's algorithm, and also the integer programming logic design method mentioned in Section 6.6, can be used. But both methods are considerably complex and are not appropriate for hand processing (Remark 6.4.1).

302   Design of NAND (NOR) Networks

If we do not require minimization of the number of gates, the following procedure, which is based partly on the map-factoring method, is convenient for hand processing. By stretching the imagination in regard to possible loops, we can reduce the number of gates as the primary objective, and the number of connections as the secondary objective, as much as possible. Furthermore, we can often obtain networks with a minimum number of gates as the primary objective and a minimum number of connections as the secondary objective, though the procedure itself cannot prove the minimality.

Procedure 6.4.1: Design of a Three-Level Network in Single-Rail Input Logic with as few NAND Gates as Possible

Let us design a three-level network, using as few NAND gates as possible, under the assumption that only x_i's, but not $\bar{x}_i$'s, for $i = 1, 2, \ldots, n$ are available as the network inputs.

1. **Let us form the output NAND gate (i.e., in the third level of the network).** If prime-implicant loops representing single complemented literals $\bar{x}_i$'s are found on a Karnaugh map (it does not matter whether there are prime-implicant loops that represent noncomplemented single literals x_i's, so ignore them), replace the 1-cells contained in these loops by don't-care condition d's. [Notice that prime-implicant loops consist of only 1-cells (and possible d-cells), as defined in Chapter 4.] Then connect the noncomplemented literals x_i's corresponding to these complemented literals $\bar{x}_i$'s to the output NAND gate in the third level of the network to be obtained (Exercise 6.6.18).

   For example, when the function f shown in the Karnaugh map in Fig. 6.4.1a is given, there is a prime-implicant loop that represents $\bar{x}_1$. Replace the 1-cells in this loop by d's, as shown in Fig. 6.4.1b. Then draw output gate 1 and connect input x_1 to this gate, as shown in Fig. 6.4.1a.

   If no prime-implicant loops for $\bar{x}_i$'s are found, draw the output NAND gate without any inputs.

2. **Let us form NAND gates in the second level that have no connections from the gates in the first level.** If there are permissible loops that consist of only 1-cells and d-cells (but no 0-cells), draw NAND gates corresponding to these permissible loops, and as inputs to these gates connect the literals in the products that these loops represent. Connect these gates to the output gate. Replace the 1-cells contained in these loops by d's. In choosing these permissible loops, it is probably better to choose a minimal number of permissible loops, using as large loops as possible.

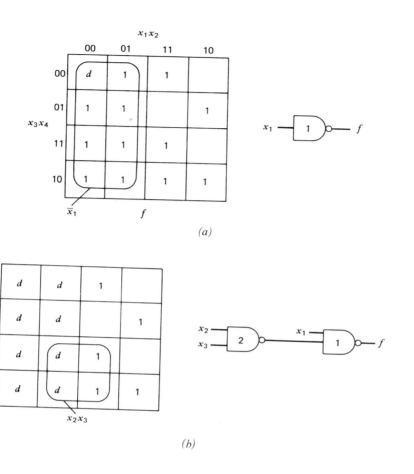

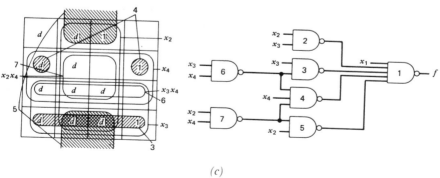

Fig. 6.4.1 *The map-factoring method for a three-level network.*

304   Design of NAND (NOR) Networks

(Notice that these gates have no input connections from the gates in the first level.)

For our example, find the permissible loop representing product $x_2 x_3$, as shown in Fig. 6.4.1b. Then form NAND gate 2 with inputs x_2 and x_3 in the second level of the network. The 1-cells in this loop are changed to d's, as shown in Fig. 6.4.1c.

3. **Let us form NAND gates in the first and second levels in the network, where the gates in the second level have connections from those in the first level.** First, find two sets of permissible loops* such that the remaining 1-cells and possibly some d-cells in the map are expressed by the first set of permissible loops inhibited by the second set. (Notice that some loops in the second set are used to inhibit some, but not others, in the first set.) Encircle and shade these remaining 1-cells and possibly some d-cells, inside the first set of permissible loops, which are not inhibited by the second set.

   (a) Corresponding to the shaded loops, draw NAND gates in the second level, connecting, as inputs, the literals in the products that the first set of permissible loops represent.

   (b) Draw NAND gates in the first level, connecting, as inputs, the literals in the products that the second set of permissible loops represent.

   (c) Then connect the outputs of these first-level gates to the corresponding gates in the second level (i.e., corresponding to the gates whose permissible loops in the first set are inhibited by the second set).

Among all such two sets of permissible loops, it is probably better to choose two sets such that the number of gates in the corresponding network implementation is the fewest and then, as the secondary objective, the number of connections is the fewest. [Notice that Procedure 6.3.1 is applied in Step 3 only throughout Procedure 6.4.1, with the modification that the 0-cells in the termination condition (0-cells and some d-cells if Procedure 6.3.1 is extended to incompletely specified functions) are replaced by 1-cells and d-cells here.]

For our example, the first set of permissible loops representing x_2, x_3, x_4 and the second set of permissible loops representing $x_3 x_4$, $x_2 x_4$ are chosen in Fig. 6.4.1c. Then gate 3 represents the 1-cell and d-cells in the permissible loop for x_3, inhibited by the permissible loop for $x_3 x_4$, which is realized by gate 6. Gate 4 represents the 1-cell and d-cell in the permissible loops for x_4, inhibited

---

* Notice that the permissible loops here may consist of 1-cells, 0-cells, and d-cells, unlike the loops used in Steps 1 and 2, which did not contain 0-cells, and also that these two sets of permissible loops may share some loops in common.

Design of Three-Level NAND (or NOR) Networks   305

by the two permissible loops for $x_3 x_4$ and $x_2 x_4$. The permissible loop for $x_2 x_4$ is realized by gate 7. Similarly, we have gate 5.

Instead of choosing the two sets of permissible loops in Fig. 6.4.1c, we can choose the two sets shown in Fig. 6.4.2, where the first set is the permissible loop representing x_2, x_3, x_4 and the second set is the permissible loop representing x_2, x_3, x_4. (Notice that these two sets are completely identical.) Corresponding to these sets, we need the six gates numbered 3 through 8 in Fig. 6.4.2. Although the number of loops is fewer in Fig. 6.4.2 than in Fig. 6.4.1c, we need more gates, since each loop is used twice in realizing the gates. Therefore the choice of these two sets is not appropriate.

Notice that all the gates that should be in the first level are obtained by Step 3, whereas all the gates that should be in the second level without inputs from the first level are obtained by Step 2. Also notice that there cannot exist any gate in the first level whose output is simultaneously connected to gates in the second level and the output gate in the third level, because if there exists such a gate, its connections to gates in the second level are redundant by the triangular condition of Theorem 6.3.1.

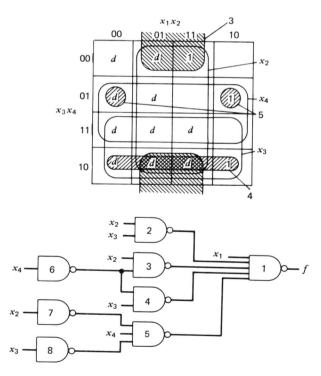

Fig. 6.4.2

4. **Sometimes there are subtle interplays between Steps 2 and 3, and consequently one step may need to be reworked by looking at the outcome of the other step.** For example, some gates obtained in Step 2 may become redundant because of the gates chosen in Step 3, since the 1-cells of the gates in the second level chosen in Step 3 may contain essential 1-cells of some gates chosen in Step 2. ("Essential 1-cells" here means 1-cells not covered by the loops corresponding to other gates in the second level chosen in Step 2.) If so, eliminate such gates chosen in Step 2. Also, the choice of the prime-implicant loops at Step 2 may not be unique; and, depending on the outcome at Step 3, another choice at Step 2 may lead to a better network.

Derive a best network, trying different possibilities.

For example, when the function f shown in Fig. 6.4.3a is given, we form two gates, 2 and 3, by Step 2. Then we add four gates, 4, 5, 6, and 7, by finding two sets of permissible loops in Fig. 6.4.3b by Step 3. Gate 4 has one 1-cell and one d-cell, shaded in the right-hand map in Fig. 6.4.3b, but the d-cell is one of the two 1-cells in the permissible loop for gate 3 in Fig. 6.4.3a (the other 1-cell is covered by the permissible loop representing $x_2 x_3 x_4$ in Fig. 6.4.3a). Therefore gate 3 is redundant and can be deleted, i.e., the network in Fig. 6.4.3b without gate 3 realizes the given function f. □

In this procedure, after the variables that can be directly connected to the output gate are found in Step 1, two types of gates, that is, gates in the second level without inputs from the gates in the first level, and those in the first and second levels that are connected, are processed separately in Steps 2 and 3, respectively. Although there are interplays between these two steps, this may make the use of Procedure 6.4.1 somewhat easier. It is easy to see why Procedure 6.4.1 works (Exercise 6.6.20). If a network to be designed is allowed to be in four or more levels, the number of gates or connections may be further reduced. The networks designed by Procedure 6.4.1 may be used as a basis to explore this possibility further by Procedure 6.3.1.

Remark 6.4.1: Gimpel [1967] developed an algorithm to design a NAND (or NOR) network that has at most three levels and a minimal number of gates, though the number of connections is not necessarily minimized as the secondary objective. (Gimpel's algorithm can be modified so that the number of connections can be minimized, though processing time will considerably increase.) Also, the given function has to be completely specified. If a maximum fan-in or fan-out restriction is imposed, Gimpel's algorithm is not applicable. The algorithm is considerably complicated, requiring computer processing. Improvements of Gimpel's algorithm in terms of computational efficiency are discussed in Hohulin and Muroga [1975] and Wei and Muroga [to be published].

Procedure 6.4.1 can often minimize the number of gates as the primary objective and the number of connections as the secondary objective, without much difficulty.

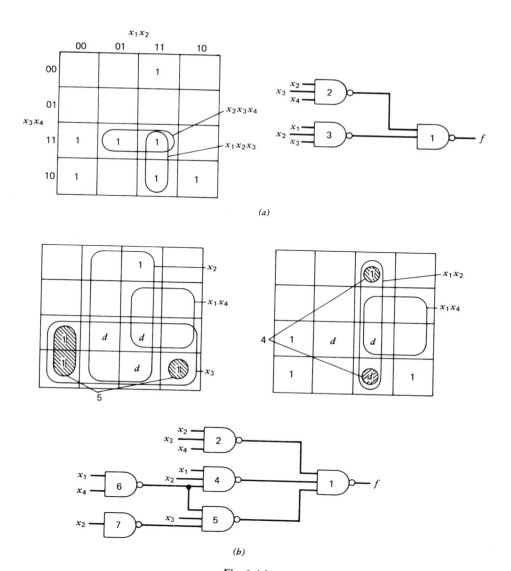

Fig. 6.4.3

308 *Design of NAND (NOR) Networks*

Exercise 6.6.19C is such an example, for which Gimpel's algorithm does not minimize the number of connections as the secondary objective. Of course, Procedure 6.4.1 is useful only for functions of a few variables, whereas Gimpel's algorithm has the advantage of applicability to any number of variables with guaranteed minimality of the number of gates. ∎

▲ 6.5 Design of NAND (or NOR) Networks in Double-Rail Input Logic by the Map-Factoring Method

In Section 6.1 we discussed the design of a two-level NAND (or NOR) network in double-rail input logic that has a minimum number of gates as the primary objective and a minimum number of connections as the secondary objective. If a network need not be in two levels, we may be able to further reduce the number of gates or connections, but there is no known simple systematic procedure (the integer programming logic design to be discussed in Section 6.6 is complex.) By a modification of the map-factoring method in Section 6.3, however, we can design a NAND (or NOR) network that has as few gates as possible (instead of the absolutely minimum number of gates), without considering the number of levels, under the assumption that both x_i's and $\bar{x}_i$'s for each $i = 1, 2, \ldots, n$ are available as network inputs. The number of connections can be reduced as well. In the following, let us discuss this modification.

Procedure 6.5.1—The Map-Factoring Method for a Double-Rail Input Logic Network: Design of a Network in Double-Rail Input Logic with as Few NAND Gates as Possible

Apply Procedure 6.3.1 (the map-factoring method), replacing permissible loops by rectangular loops consisting of 2^i cells that are not necessarily permissible loops, where i is a nonnegative integer (i.e., rectangular loops consisting of 2^i cells that do not necessarily contain the cell with all coordinates of 1). Here the cells in each of these loops are 1-cells, 0-cells, or d-cells.

An example is illustrated in Fig. 6.5.1. Notice that the loop numbered 1 in Fig. 6.5.1a is not a permissible loop, since it does not contain the cell with coordinates $(w, x, y, z) = (1\ 1\ 1\ 1)$. The procedure is completely identical to Procedure 6.3.1, except that the loops used do not necessarily contain this particular cell. If a minimal two-level AND/OR network is first obtained for the function f in Fig. 6.5.1 and is then converted to a two-level NAND network by the conversion in Fig. 6.1.1, there will be more connections than the network in Fig. 6.5.1d, though the number of gates is the same. □

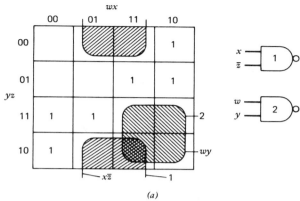
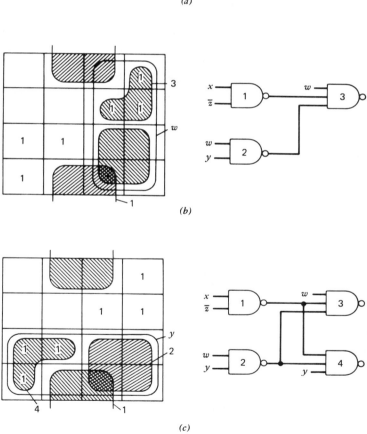

Fig. 6.5.1   Example for Procedure 6.5.1.

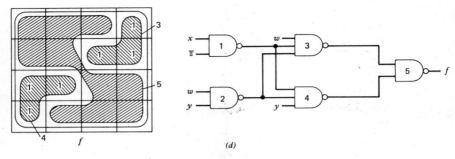

Fig. 6.5.1. (*Continued*)

Notice that Procedure 6.5.1 for double-rail input logic is considerably easier to use than Procedure 6.3.1 for single-rail input logic. Therefore the procedure can be applied to Karnaugh maps of five or more variables. Also notice that Procedure 6.5.1 contains minimal two level NAND networks (discussed in Fig. 6.1.1) as a special case, since if we encircle only 1-cells with a minimum number of rectangular loops of 2^i 1-cells with largest i's, and then choose a permissible loop for the output NAND gate, we can obtain a minimal two-level NAND network.

By choosing loops and their inhibitions in an appropriate manner, we can take into account the maximum fan-in or fan-out restriction, or the number of levels, as well.

▲ 6.6 Other Design Methods of NAND (or NOR) Networks

Let us mention other methods to design networks with NAND (or NOR) gates.

If we want to design a network with a minimum number of NAND (or NOR) gates (or a mixture) under arbitrary restrictions, we cannot do so within the framework of switching algebra, unlike the case of minimal two-level AND/OR gate networks (which can be designed based on minimal sums or minimal products), and the **integer programming logic design method*** mentioned in Section 5.4 is currently the only method available. This method is not appropriate for hand processing, but can design minimal networks for up to about 10 gates within reasonable processing time by computer. The method can design multiple-output networks also, regardless of whether

---

* See Chapter 14 of Muroga [1971], Chapter 5 of Muroga [1970], Muroga and Ibaraki [1972], Davidson [1969], Baugh, Ibaraki, Liu and Muroga [1969], Liu, Hohulin, Shiau, and Muroga [1974], or Lai, Nakagawa, and Muroga [1974].

functions are completely or incompletely specified. Also, networks with a mixture of different gate types (such as NAND, NOR, AND, or OR gates), with gates having double outputs, or with wired-OR can be designed, though the processing time and the complexity of programs increase correspondingly. Also, the number of connections can be minimized as the primary, instead of as the secondary, objective.

Although the primary purpose of the integer programming logic design method is the design of minimal networks with a small number of variables, minimal networks for some important functions of an arbitrary number of variables (i.e., n variables), such as adders and parity functions, have been derived as described in Section 9.2, by analyzing the intrinsic properties of minimal networks designed by the integer programming logic design method for these functions, for small values of n. Also, careful analysis of minimal networks designed by the integer programming logic design method has led to the following efficient heuristic design method.

To design networks with a greater number of gates, a heuristic design procedure called the **transduction method** was developed [Muroga, Kambayashi, Lai, Culliney, and Hu, to be published]. The method does not necessarily guarantee the minimality of designed networks, but it can yield, in a short processing time, networks that appear to be reasonably good. Most of these designed networks prove to be minimal if they are small enough so that the minimality can be checked by the integer programming logic design method. (Also, a four-output network with 25 NOR gates and 42 connections, which was obtained by some heuristic method in literature, was improved to a network with 15 gates and 28 connections by the transduction method.) The method works in the following manner. An initial network for a given function (or functions) is designed by any conventional method, ignoring specified restrictions. Then, starting with the output gate and moving toward the network inputs, we find what functions can replace each gate output. In this case, functions with as many don't-care conditions as possible are usually chosen. Gates whose output functions are expressed only by don't-care conditions and 0's can be deleted. Then the network is **trans**formed into a **reduc**ed network, by taking into account the specified restrictions. By repeating this transformation and reduction, the initial network is usually greatly simplified.

If, however, we do not want to use computers in designing networks with NOR or NAND gates under arbitrary constraints, there is no good systematic method available that guarantees minimality. We have to modify, on a trial-and-error basis, networks designed by any methods or networks found in the literature, and choose the best result. (For networks of NAND (or NOR) gates, see, e.g., Todd [1963], McCluskey [1963], Maley [1970], Baugh, Chandersekaran, Swee, and Muroga [1972], and Liu, Hohulin, Shiau, and Muroga [1974].)

Exercises

6.6.1 (R) (i) Design a two-level network with a minimum number of NAND gates for one of the following functions under Assumption 6.1.1:

A. $f(x_1, x_2, x_3, x_4) = \Sigma(3, 5, 6, 9, 12, 13, 14, 15) + d(0, 1, 7)$.
B. $f(x_1, x_2, x_3, x_4) = \Sigma(3, 5, 6, 8, 9, 10, 11, 13) + d(1, 4, 7)$.

(ii) Solve (i) using NOR gates instead of NAND gates.

6.6.2 (R) Derive the output functions of the NAND networks in Fig. E6.6.2 by Rule 6.2.1.

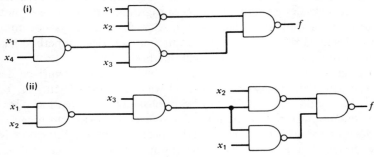

Fig. E6.6.2

6.6.3 (E) (i) Derive, for a network of NOR gates, a rule corresponding to Rule 6.2.1.

(ii) Using the rule you derived, write a switching expression for the network in Fig. 6.2.1a with NAND gates replaced by NOR gates.

6.6.4 (R) Design NAND networks in single-rail input logic by finding appropriate switching expressions for the following functions (i.e., by using Rule 6.2.1 in reverse). Use as few gates as possible.
(i) $\bar{x}_1 x_2 \vee x_3$.   (ii) $x_1 \bar{x}_2 \vee \bar{x}_1 x_2$.

6.6.5 (M) Design NAND networks in single-rail input logic by finding appropriate switching expressions for the following functions (i.e., by using Rule 6.2.1 in reverse). Use as few gates as possible.
(i) $x_1 \bar{x}_2 \bar{x}_3 \vee x_2 x_3$.   (ii) $\bar{x}_1 x_2 \vee \bar{x}_1 x_3 \vee x_2 x_3$.

6.6.6 (RT) Design networks in single-rail input logic with NAND gates only, by the map-factoring method, for one of the sets of functions A, B, and C in Fig. E6.6.6. Try to use as few gates as possible and then, as the secondary objective, as few connections as possible, under each of the following conditions:

1. No maximum fan-in and fan-out restrictions.
2. Maximum fan-in and fan-out are 3. Each of the network inputs x_i also must have at most three fan-out connections.

Whenever a network is obtained, check whether it can be simplified by Procedure 6.3.2.

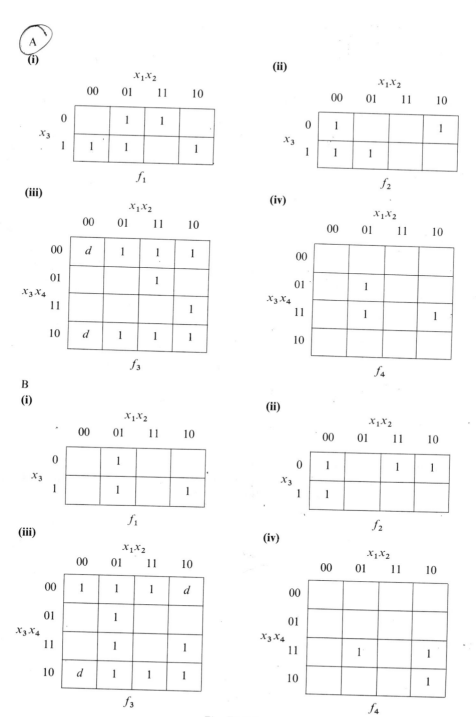

Fig. E6.6.6

C

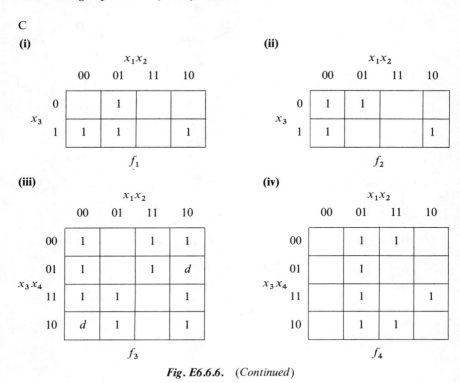

Fig. E6.6.6. (*Continued*)

6.6.7 (RT+)* Design NAND networks in single-rail input logic by the map-factoring method for one of the sets of functions in Fig. E6.6.7. Try to use as few gates as possible and then, as the secondary objective, as few connections as possible.

A

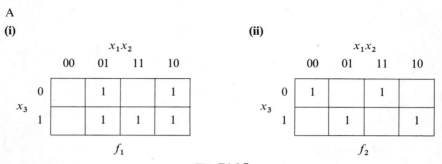

Fig. E6.6.7

* Exercise 6.6.7 is more time-consuming than Exercise 6.6.6.

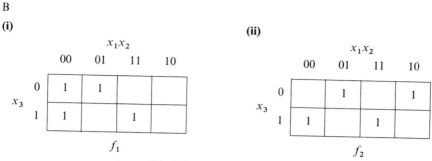

Fig. E6.6.7. *(Continued)*

6.6.8 (RT++)* Design a NAND network in single-rail input logic by the map-factoring method for one of the following functions in Fig. E6.6.8. Try to use as few gates as possible and then, as the secondary objective, as few connections as possible.

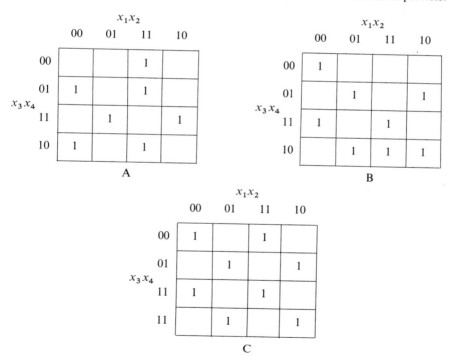

Fig. E6.6.8

* Exercise 6.6.8 is more time-consuming than Exercise 6.6.7.

Design of NAND (NOR) Networks

6.6.9 (M) (i) Explain why loops that contain the cell corresponding to the combination of variable values of all 1's are considered in the map-factoring method. (In other words, why is the concept of permissible loops necessary for the map-factoring method?)

(ii) Prove why the map-factoring method works.

6.6.10 (R) Using Procedure 6.3.2, simplify the network in Fig. E6.6.10.

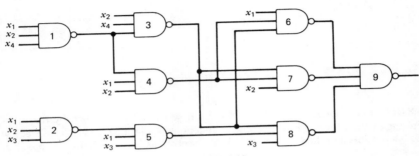

Fig. E6.6.10

6.6.11 (M−) When the three gates, i, j, k, in Theorem 6.3.1 are a mixture of NOR and NAND gates, can we still simplify the configuration? Discuss when we can or cannot, and explain why.

6.6.12 (E) After expressing a given function f as

$$f = \bar{\bar{f}} = \overline{\bar{f}_1 \cdot \bar{f}_2},$$

a student applied the map-factoring method. He thought that f_1 and f_2 can be chosen so that the map-factoring method can be more easily applied to f_1 and f_2 than directly to f.

Discuss the shortcomings of this approach.

6.6.13 (M) Suppose that a network consists of a subnetwork (denoted by the rectangle) and a NOR gate whose output is f, and that g is an input function to the subnetwork, as shown in Fig. E6.6.13a. Here the subnetwork consists of arbitrary gate types (a single gate type or a mixture of different gate types; also, NOR gates may not be contained). Also, $\mathbf{x} = (x_1, \ldots, x_n)$ and $\mathbf{y} = (y_1, \ldots, y_m)$ are other inputs to the subnetwork and the NOR output gate, respectively. Then, if $g \subseteq \bar{f}$, prove that the new network shown in Fig. E.6.6.13b, which is derived by adding g to the NOR gate and replacing g to the subnetwork with constant 0, has the same output function f as the network of Fig. E6.6.13a.

Also prove that the network of Fig. E.6.6.13c, which is derived from E6.6.13a by adding g to the NOR gate in Fig. E.6.6.13a, has the same output function f as the network in Fig. E.6.6.13a.

Remark: After we show that Fig. E.6.6.13c has the same output function f as E.6.6.13a, Fig. E.6.6.13c can be simplified to Fig. E.6.6.13b. This simplification is a **generalization of the triangular condition** stated in Theorem 6.3.1 [Lai, Nakagawa, and Muroga 1974]. ∎

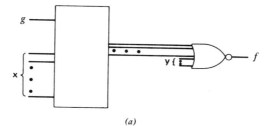

(a)

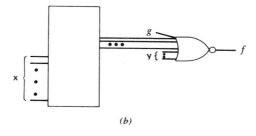

(b)

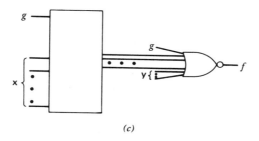

(c)

Fig. E6.6.13

6.6.14 (M) Suppose that we have a network of NOR gates such that the removal of any connection in this network changes the function that the network output realizes. The output of this network is not identically equal to 0. Suppose that there are two NOR gates connected as shown in Fig. E6.6.14, as part of this network. A gate i that is not the output gate is connected to another gate, j, and gates i and j have a common input h, as shown in the figure. Gate i has another input k (or a set of inputs, k). Prove that gate i must have two or more inputs and two or more output connections. (In other words, prove that both the input k and the connection to other gates shown with the arrow in the figure are not void.) [Lai, 1976]

318   Design of NAND (NOR) Networks

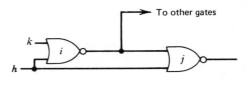

Fig. E6.6.14

6.6.15 (M) Suppose that we want to design a network of NOR gates only for a given function f, assuming that no complemented variables are available as network inputs. Let us design such a network by the following procedure:

(i) Derive the dual f^d of a given function f.

(ii) Derive a network of NAND gates only for f^d, by Procedure 6.3.1 (the map-factoring method).

(iii) In the derived network, replace every NAND gate by a NOR gate.

Prove that the network which results is a NOR gate network whose output expresses f.

6.6.16 (M−) Suppose that we want to design a network of NOR gates only, assuming that no complemented variables are available as network inputs. We can derive such a network by directly working on a Karnaugh map for f (instead of f^d, as we did in Exercise 6.6.15), that is, by considering 0-cells instead of 1-cells, and 1-cells instead of 0-cells, and by considering alterms (instead of terms) for permissible loops.

Describe a formal procedure, modifying Procedure 6.3.1.

6.6.17 (MT) By the Procedure in Exercise 6.6.15 or 6.6.16, design NOR networks in single-rail input logic for one of the sets of functions in Fig. E6.6.6, using as few gates as possible and then, as the secondary objective, as few connections as possible.

6.6.18 (M) We want to design a NAND network under the assumption that no complemented variables are available as network inputs.

Prove that, if and only if a function f has single-literal prime implicants $\bar{x}_i$'s, there exists a NAND network for f that has a minimal number of gates as the primary objective and a minimal number of connections as the secondary objective, and in which the x_i's are connected to the output gate only and not to other gates. (This can be applied in Procedure 6.3.1 and also justifies Step 1 in Procedure 6.4.1.)

▲ **6.6.19 (RT)** By Procedure 6.4.1 design a three-level NAND network in single-rail input logic for one of the functions in Fig. E6.6.19, using as few gates as possible and then, as the secondary objective, as few connections as possible.

▲ **6.6.20 (M)** Explain why Procedure 6.4.1 works.

▲ **6.6.21 (RT)** By Procedure 6.5.1, design a NAND network in double-rail input logic for one of the functions in Fig. E6.6.21, using as few NAND gates as possible as the primary objective, and as few connections as possible as the secondary objective.

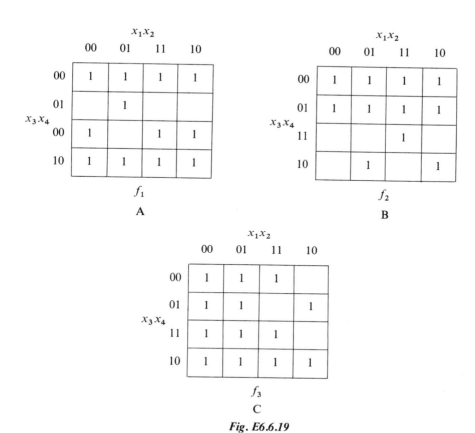

Fig. E6.6.19

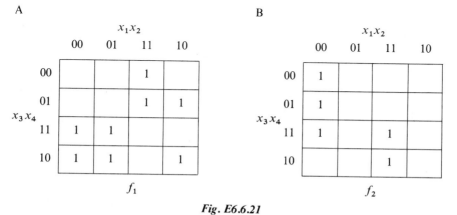

Fig. E6.6.21

320 *Design of NAND (NOR) Networks*

▲ **6.6.22** (M) Design a NOR network in double-rail input logic for one of the functions in Fig. E6.6.21, using as few NOR gates as possible and then, as the secondary objective, as few connections as possible.

Answers for simple exercises for self-study in Section 6.3

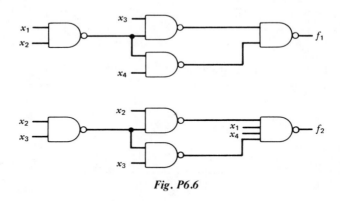

Fig. P6.6

6.7 Transient Response of Combinational Networks

If the inputs of a combinational network change, the outputs of the network will also change. It is important to find out how the outputs change, since the network has to perform reliably. Also, it is very important to determine when the outputs change, since this determines the speed of the network.

Spurious Response of a Network

A network generates spurious output values when the network inputs change, depending on the network configuration implemented.

For example, the network shown in Fig. 6.7.1a has two outputs,* f_1 and f_2. Suppose that x changes from 1 to 0 while y is kept at 1. From the switching expression $f_1 = x \vee y$, f_1 must be kept at 1, since $y = 1$. Actually, f_1 shows the spurious output value 0 for the transient period, as the waveforms in Fig. 6.7.2 illustrate. In complex networks the spurious output values may last longer than τ, the switching time of each gate.

Let us call this type of hazard an **intrinsic hazard**, since it occurs even if every gate has the same switching time.† Intrinsic hazards occur because of

---

* The connection from gate 1 to gate 2 in this network cannot be deleted by the triangular condtion (Theorem 6.3.1), since the output f_2 will be changed.

† Or even if the operations of all gates are synchronized by a clock, as discussed in Chapter 8.

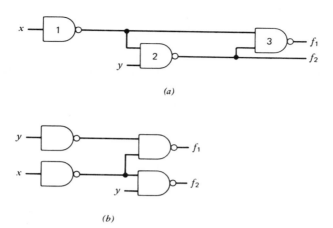

Fig. 6.7.1 Intrinsic hazard and remedy. (a) *Network with intrinsic hazard.* (b) *Remedied network.*

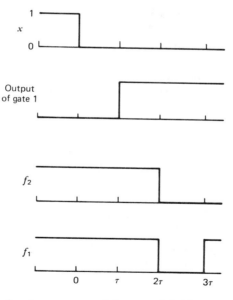

Fig. 6.7.2 Spurious response of the network in Fig. 6.7.1a, while y = 1.

multiple paths of different lengths between a network input and the network output, and different inversions along these paths. [The number of gates for some functions can be greatly reduced by using multilevel, instead of two-level (or three-level network implementation with NAND or NOR gates), network implementation, as mentioned in Sections 4.1, 5.4, and 6.1. But multilevel networks tend to have more chances of multiple paths of different lengths.] As the number of levels in a network increases, the chances of intrinsic hazards also increase.

A remedy for intrinsic hazards is reconfiguration of the network, possibly with extra gates, so that critical paths of different lengths are eliminated. One solution for the above example is shown in Fig. 6.7.1b, where x has a single path to f_1 instead of two paths in Fig. 6.7.1a. Another solution is to insert a gate for a delay between gates 1 and 3 in Fig. 6.7.1a. (A NAND gate cannot be inserted because the NOT operation will be added. An AND or OR gate may be inserted.)

When gates have different switching times, even the simplest network may generate spurious output values. For example, if $x = y = 1$ in the network in Fig. 6.7.3a for $f = xz \vee y\bar{z}$, the network output f must be 1 no matter what value z has. Assume that the switching time of AND gate 1 is longer than that of AND gate 2. Then, when z changes from 0 to 1, the new output value of gate 2 and the old output value of gate 1 are simultaneously present at the inputs of OR gate 3 during a very short transition period. Thus the output f shows impulsively the erroneous value 0 for a short period.

This type of hazard is called an **extrinsic hazard**. Extrinsic hazards occur because of the different switching times of gates. When the difference between switching times of different gates is small, the spurious output values last for a proportionally short period. Thus spurious output values due to extrinsic hazards last for a much shorter period than those due to intrinsic hazards. A remedy for extrinsic hazards is the use of gates with equal switching times by adjusting the electronic implementation,* if this is possible. Another remedy is synchronization of the operations of gates by clocks, as will be discussed in Section 8.1. A third remedy, which is probably simpler, is the addition of an extra AND gate whose output represents the consensus of the output functions of two other AND gates, as illustrated by gate 4 in Fig. 6.7.3b.

If we use a Karnaugh map, we can easily see what kinds of networks tend to have extrinsic hazards and why the third remedy works. In Fig. 6.7.3c, loops 1 and 2 denote the outputs of gates 1 and 2, respectively. When $x = y = 1$, the value 1 changes between loops 1 and 2 in the same column, $(xy) = (1\ 1)$, by a change in the value of a single variable, z. (If the loops in the map

---

* Sharp spurious pulses can be suppressed by small capacitors.

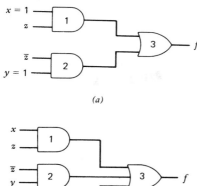

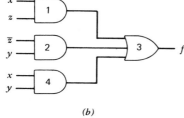

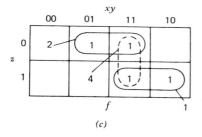

Fig. 6.7.3 *Extrinsic hazard and remedy.* (a) *Network with extrinsic hazard.* (b) *Remedied network.* (c) *Map to illustrate the operation of Fig. 6.7.3b.*

are not adjacent, the value 1 changes between the loops by a simultaneous change of two or more variables, which occurs with much smaller probability.) When the consensus loop shown in the dotted line is added, f is 1 because of the output of gate 4. The consensus loop does not change the function f, so Fig. 6.7.3b still expresses the same function as Fig. 6.7.3a.

Notice that the speed of the network is usually not sacrificed, since Fig. 6.7.3b still has two levels. (Hence this remedy is often used in networks for high speed applications. See, e.g., [Fujioka et al. 1973].)

When a network that has many levels contains many different paths of many gates, the situation is more complex than in the simple network used above as an example. Even if the difference in switching times of gates is small, spurious outputs due to extrinsic hazards may last much longer, and may show complex transient changes (they may change values many times before settling to final values). This is so because spurious signals over different paths

324 Design of NAND (NOR) Networks

tend to arrive at the network outputs at slightly different times. Such complex and longer extrinsic hazards are harder to eliminate. (One remedy for extrinsic hazards, synchronization of the operations of gates by clocks, will be discussed in Section 8.1.)

When a network that is supposed to retain the same output value (0 or 1) shows a spurious output value, this is called a **static hazard** (e.g., f_1 in Fig. 6.7.2), regardless of whether it is intrinsic or extrinsic. When a network is supposed to have different output values before and after the transient period, and it shows oscillating values before settling to a new output value, this is called a **dynamic hazard**. (See, e.g., McCluskey [1965].)

Whether an entire digital system that contains networks with hazards malfunctions depends on what are connected after these networks. If only combinational networks are connected thereafter, correct output values can be obtained after the transient period. If the output displays or printers, which are insensitive to spurious values for a short period, are connected, the hazards of these networks are not important. If sequential networks are connected, however, the entire digital system may malfunction, changing the local hazards to global ones. This hazard problem with sequential networks will be discussed in Chapters 7 and 8.

Effective Gate Delays

When a network contains paths of different lengths from its inputs to its outputs, it is sometimes not easy to immediately find when the network outputs change for an input change. Consider, for example, the network in Fig. 6.7.4 which has six as the (maximum) number of levels, since the longest path, 1–2–3–4–5–7, consists of six gates. When the input u changes, it looks as if the output f will have a new value 6τ time later, assuming that all gates have the same switching time τ. But, as illustrated in Fig. 6.7.5 for the case of $x = y = z = 1$, f changes after only 5τ, when input u changes from 0 to 1. For other combinations of values of x, y, z, or change of u from 1 to 0, however, a spurious output value may appear for a while (as explained later, the output f in Fig. 6.7.6, for example), and then the

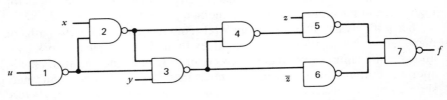

Fig. 6.7.4

network output will settle down to a final output value. The response time for any change of the network output is defined as effective gate delay, in contrast to net gate delay defined later. In other words, **effective gate delay** is defined as a response time when the network output changes for an input change, whether the new output value is spurious or not. Thus the number of levels from the input under consideration to the network output is simply an upper bound on the effective gate delay, when all gates have the same switching time (as discussed in Section 4.1). The size of the effective gate delay is usually not obvious from the network configuration. Of course, there are many cases where effective gate delays are equal to the number of levels.

We can find the effective gate delay as follows. Calculate what multiple of τ is required until the network output changes for the change of a network input (say u in Fig. 6.7.4), by drawing waveforms such as those in Fig. 6.7.5 for a specific combination of values of other network inputs (i.e., x, y, and z).

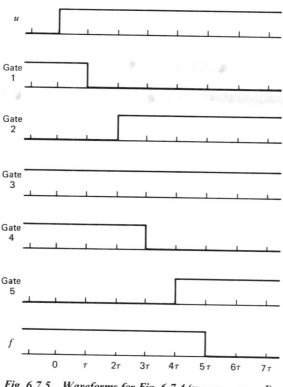

Fig. 6.7.5 Waveforms for Fig. 6.7.4 ($x = y = z = 1$).

Then we can find the effective gate delay as a multiple of τ. This approach, however, is very time-consuming if the given network is complex and has many network inputs at different gates.

Although effective gate delay from an input to an output of a network is defined above for a specific change of an input (i.e., from 0 to 1, or from 1 to 0) and a specific combination of values of other network inputs, let us use the following convention for the sake of simplicity. **If the way of change of an input under consideration (i.e., from 0 to 1, or from 1 to 0) and also a combination of values of other inputs are not specified, the effective gate delay from that input to a particular network output** is defined as the maximum of all effective gate delays for all possible combinations of values of all inputs. Also the **effective gate delay of a network** is defined as the maximum of all effective delays for all pairs of an input and a network output for all possible combinations of values of inputs.

For example, we can find that the effective gate delay from network input u to network output f in Fig. 6.7.4 is 5, despite the longest path which has delay 6.

Net Gate Delay of a Network

When a network has an intrinsic hazard, spurious output values also appear after its effective gate delay. In this case we need to wait until the network output settles to its final output value. This time is called the **net gate delay** of a network.

For example, the network in Fig. 6.7.4 has a spurious output for the change of u from 1 to 0, while keeping $x = y = 1$ and $z = 0$, at the effective gate delay 4τ, as illustrated in Fig. 6.7.6. We need to wait for 5τ time. Thus the net gate delay is 5. (If the output value does not change for an input change, the effective and net gate delays are both undefined.)

The net gate delay of a network is the real response time of the network, after which we have final new output values. The net gate delay is equal to or greater than the effective gate delay, and equal to or smaller than the number of levels. In other words, the number of levels and the effective gate delay are the upper and lower bounds, respectively, on the net gate delay. (Notice that if new output values that are not spurious values appear, then the effective gate delay and net gate delay are identical.) The net gate delay can also be calculated by drawing waveforms for combinations of different values of inputs.

The net gate delay from a certain input to a network output, when the way of change of the input and a combination of values of other inputs are not specified, and also the net gate delay of a network can be defined in a manner similar to those for effective gate delay.

Classification of Switching Functions and Networks 327

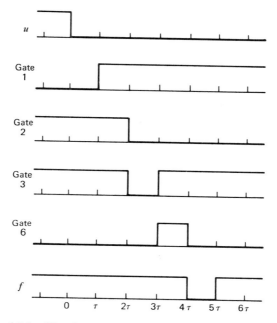

Fig. 6.7.6 *Waveforms for Fig. 6.7.4* (x = y = 1 and z = 0).

The net gate delay of a network determines the speed of the network, if all gates have the same switching time. So it is very important for logic designers to determine what the effective and net gate delays are for what combinations of values of inputs, in order to use the networks in the most effective ways. (An adder example which is more complex than Fig. 6.7.4 will be shown in Section 9.2.).

▲ 6.8 Classification of Switching Functions and Networks

By simple transformations, functions can be grouped into classes. This concept is useful for preparing catalogs of networks for functions.

Classification of Functions

Assuming that two switching functions

$$f_1(x_1, x_2, \ldots, x_n) \quad \text{and} \quad f_2(x_1, x_2, \ldots, x_n)$$

are given. Then f_2 may be made identical to f_1 by permuting or complementing some variables. For example, consider $f_1(x_1, x_2, x_3) = x_1\bar{x}_2 \vee \bar{x}_1 x_2 x_3$ and $f_2(x_1, x_2, x_3) = x_1 x_3 \vee \bar{x}_1 \bar{x}_3 x_2$. If x_2 and x_3 are permuted in f_2 and then x_2 is complemented, the new function becomes identical to f_1. Next, compare the third function $f_3 = x_1 \vee x_2 x_3$, with the above f_1. This does not become identical to f_1, no matter which variables are complemented or permuted.

Thus we have reached the concept of NP-equivalence, based on permutation and complementation (i.e., negation) of variables, as defined in Definition 6.8.1. f_1 and f_2 belong to the same NP-equivalence class, but f_1 and f_3 do not.

When the complementation of a function itself is introduced in addition to the two operations of permuting and complementing variables, some other functions may become identical to f_1. For example,

$$f_4 = x_1 \bar{x}_3 \vee \bar{x}_1 x_3 \vee \bar{x}_1 \bar{x}_2 \vee \bar{x}_2 \bar{x}_3 = \overline{(x_1 x_3 \vee \bar{x}_1 \bar{x}_3 x_2)} = \bar{f}_2$$

results from f_2, and consequently from f_1, if the function is complemented in addition to the two operations of permutation and complementation of variables. If variables are only permuted or complemented, f_1 cannot be converted to f_4. Thus we have reached the concept of NPN-equivalence, as defined in Definition 6.8.1. The functions f_1, f_2, and f_4 belong to the same NPN-equivalence class.

Definition 6.8.1: Suppose that an arbitrary switching function f is given. Any functions that can be obtained from f by any combination of the following three operations (some combinations may not contain some of these three operations) are said to be **NPN-equivalent** to f:

1. Negation of one or more variables of f.
2. Permutation of variables.
3. Negation of f, that is, $\bar{f}$.

The given f and all functions that are NPN-equivalent to f are termed an **NPN-equivalence class** of functions, with f as its **representative function**. The relation* that is defined by this is called an **NPN-equivalence**† (N and P stand for "negation" and "permutation," respectively). Switching functions obtained from f by any combination of operations 1 and 2 (excluding operation 3) are said to be **NP-equivalent** to f, and the relation† is called

---

* These relations are special cases of "equivalence," defined in mathematics. (See, e.g., Liu [1977] or Preparata and Yeh [1973].)
† The NPN-equivalence was introduced by E. Goto and H. Takahashi in relation to the study of threshold logic [Goto and Takahashi 1962].

Classification of Switching Functions and Networks

NP-equivalence. The terms **P-equivalence** (permutation only) and **N-equivalence** (negation of variables only) are similarly defined. (Although PN-equivalence and Nf (i.e., negation of f)-equivalence could be similarly defined, these terms are usually not used because the concepts are usually not very useful.)

For example, when $f(x_1, x_2) = x_1 \vee x_2$ is given, $\bar{x}_1 \vee x_2$, $x_1 \vee \bar{x}_2$, $\bar{x}_1 \vee \bar{x}_2$, $\overline{x_1 \vee x_2}$, $\overline{\bar{x}_1 \vee x_2}$, $\overline{x_1 \vee \bar{x}_2}$, $\overline{\bar{x}_1 \vee \bar{x}_2}$, and f itself constitute an NPN-equivalence class represented by f, since no other switching function is NPN-equivalent to f.

It is known that all functions of n or less variables can be partitioned into NPN-equivalence classes. Similarly, we can classify all functions of n variables by P-equivalence or NP-equivalence. Table 6.8.1 shows the classification of switching functions of up to three variables in the P-, NP-, and NPN-equivalence classes.

Table 6.8.2* shows the numbers of P-, N-, NP-, and NPN-equivalence classes for different values of the number of variables, n. For large n, counting the number of functions in each class, and consequently the number of equivalence classes, is not at all simple.†

Classification of Networks

Networks for switching functions can be classified by the equivalence classes introduced above. This is convenient for the preparation of catalogs of minimal or good networks for different functions.

Suppose that we want to design networks in double-rail input logic with a minimum number of NOR gates as the primary objective and a minimum number of connections as the secondary objective (no maximum fan-in or fan-out restriction), for all functions of exactly two variables. If we derive a minimal network for a function f_1, then a minimal network for any function f_2 that can be obtained by negating some or all variables of f_1 can be easily derived by negating the corresponding variable inputs in the minimal networks for f_1 without changing the network configuration. (As an example, a

---

* Table 6.8.2 is taken from Muroga [1971]. (In Table 2.3.2 in Muroga [1971], entries in the rows for "NP-equivalence classes of unate functions of n or fewer variables" and "NPN-equivalence classes of unate functions of n or fewer variables" should be shifted right by one column, adding 2 and 1 for $n = 0$, respectively, in these rows. Also "200, 263, 951, 911, 058" in the first column on p. 39 should read "200, 253, 951, 911, 058.")

† This requires the knowledge of advanced mathematics. For the mathematical theory for counting these numbers, see Chapters 5 and 6 of Harrison [1965] and Chapter 4 of Harrison [1971]. In the case of equivalence classes with permutation of variables, the size of each class is related to the symmetry of functions. [Lazär, Almer, and Donciu 1972; Lazär and Almer 1972].

Table 6.8.1 Representative Function of Each Equivalence Class and Number of Functions in Each Class

NUMBER OF VARIABLES	ALL FUNCTIONS	P-EQUIVALENCE CLASS REPRESENTATIVE FUNCTION	P-EQUIVALENCE CLASS NUMBER OF FUNCTIONS	NP-EQUIVALENCE CLASS REPRESENTATIVE FUNCTION	NP-EQUIVALENCE CLASS NUMBER OF FUNCTIONS	NPN-EQUIVALENCE CLASS REPRESENTATIVE FUNCTION	NPN-EQUIVALENCE CLASS NUMBER OF FUNCTIONS
0	0	0	1	0	1	1	2
	1	1	1	1	1		
1	x_1	x_1	1	x_1	2	x_1	2
	$\bar{x}_1$	$\bar{x}_1$	1				
2	$x_1 x_2$	$x_1 x_2$	1	$x_1 x_2$	4	$x_1 \vee x_2$	8
	$\bar{x}_1 x_2$	$\bar{x}_1 x_2$	2				
	$x_1 \bar{x}_2$						
	$\bar{x}_1 \bar{x}_2$	$\bar{x}_1 \bar{x}_2$	1				
	$x_1 \vee x_2$	$x_1 \vee x_2$	1	$x_1 \vee x_2$	4		
	$\bar{x}_1 \vee x_2$	$\bar{x}_1 \vee x_2$	2				
	$x_1 \vee \bar{x}_2$						
	$\bar{x}_1 \vee \bar{x}_2$	$\bar{x}_1 \vee \bar{x}_2$	1				
	$x_1 \oplus x_2$	$x_1 \oplus x_2$	1	$x_1 \oplus x_2$	2	$x_1 \oplus x_2$	2
	$\bar{x}_1 \oplus x_2$	$\bar{x}_1 \oplus x_2$	1				

Table 6.8.1 (continued) (The columns for "All Functions" and "P-Equivalence Class" are omitted here because of many entries)

NUMBER OF VARIABLES	NP-EQUIVALENCE CLASS		NPN-EQUIVALENCE	
	REPRESENTATIVE FUNCTION	NUMBER OF FUNCTIONS	REPRESENTATIVE FUNCTION	NUMBER OF FUNCTIONS
3	$x_1 x_2 \vee x_2 x_3 \vee x_1 x_3$	8	$x_1 x_2 \vee x_2 x_3 \vee x_1 x_3$	8
	$x_1 \oplus x_2 \oplus x_3$	2	$x_1 \oplus x_2 \oplus x_3$	2
	$x_1 x_2 x_3$	8		
	$x_1 \vee x_2 \vee x_3$	8	$x_1 \vee x_2 \vee x_3$	16
	$x_1(x_2 \vee x_3)$	24		
	$x_1 \vee x_2 x_3$	24	$x_1(x_2 \vee x_3)$	48
	$x_1 x_2 x_3 \vee \bar{x}_1 \bar{x}_2 \bar{x}_3$	4		
	$(x_1 \vee x_2 \vee x_3)(\bar{x}_1 \vee \bar{x}_2 \vee \bar{x}_3)$	4	$x_1 x_2 x_3 \vee \bar{x}_1 \bar{x}_2 \bar{x}_3$	8
	$\bar{x}_1 x_2 x_3 \vee x_1 \bar{x}_2 \vee x_1 \bar{x}_3$	24	$\bar{x}_1 x_2 x_3 \vee x_1 \bar{x}_2 \vee x_1 \bar{x}_3$	24
	$x_1(x_2 x_3 \vee \bar{x}_2 \bar{x}_3)$	12		
	$x_1 \vee x_2 \bar{x}_3 \vee \bar{x}_2 x_3$	12	$x_1(x_2 x_3 \vee \bar{x}_2 \bar{x}_3)$	24
	$x_1 x_2 \vee x_2 x_3 \vee \bar{x}_1 x_3$	24	$x_1 x_2 \vee x_2 x_3 \vee \bar{x}_1 x_3$	24
	$\bar{x}_1 x_2 x_3 \vee x_1 \bar{x}_2 x_3 \vee x_1 x_2 \bar{x}_3$	8		
	$x_1 x_2 \vee x_2 x_3 \vee x_1 x_3 \vee \bar{x}_1 \bar{x}_2 \bar{x}_3$	8	$\bar{x}_1 x_2 x_3 \vee x_1 \bar{x}_2 x_3 \vee x_1 x_2 \bar{x}_3$	16
	$x_1 \bar{x}_2 \bar{x}_3 \vee x_2 x_3$	24		
	$(x_1 \vee \bar{x}_2 \vee x_3)(x_2 \vee x_3)$	24	$x_1 \bar{x}_2 \bar{x}_3 \vee x_2 x_3$	48

331

Table 6.8.2 Numbers of Equivalence Classes under Different Equivalence Relations

n	0	1	2	3	4	5	6	7	8
Functions of up through n variables	2	4	16	256	65,536	About 4.3×10^9	About 1.8×10^{19}	About 3.4×10^{38}	About 1.16×10^{77}
Functions of exactly n variables	2	2	10	218	64,594	About 4.3×10^9	About 1.8×10^{19}	About 3.4×10^{38}	About 1.16×10^{77}
P-Equivalence classes of functions of up through n variables	2	4	12	80	3,984	37,333,248	—	—	—
P-Equivalence classes of functions of exactly n variables	2	2	8	68	3,904	37,329,264	25,626,412,300,941,056	—	—
NP-Equivalence classes of functions of up through n variables	2	3	6	22	402	1,228,158	400,507,806,843,728	—	—
NP-Equivalence classes of functions of exactly n variables	2	1	3	16	380	1,227,756	400,507,805,615,570	—	—
NPN-Equivalence classes of functions of up through n variables	1	2	4	14	222	616,126	200,253,952,527,184	—	—
NPN-Equivalence classes of functions of exactly n variables	1	1	2	10	208	615,904	200,253,951,911,058	—	—

Classification of Switching Functions and Networks 333

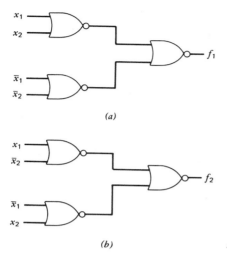

Fig. 6.8.1 Negation of variables.

minimal network for $f_1 = x_1 x_2 \vee \bar{x}_1 \bar{x}_2$ is given in Fig. 6.8.1a. When $f_2 = x_1 \bar{x}_2 \vee \bar{x}_1 x_2$ that can be obtained from f_1 by negating x_2 is given, a minimal network for f_2 can be derived from Fig. 6.8.1a, as shown in Fig. 6.8.1b.) The minimality of the new network is easy to see for the following reason: If the new network does not have a minimum number of gates or connections, f_2 has a network with fewer gates or connections; thus, corresponding to this network, f_1 has a network with fewer gates or connections than the original minimal network, but this contradicts the assumption that the original network is minimal for f_1. If we are required to derive a minimal network under different restrictions, say in single-rail input logic instead of double-rail input logic, then we cannot obtain a network for the above f_2 simply by negating some variable inputs in the minimal network for f_1, and we inevitably change the network configuration by adding more gates or changing the connections among gates.

Even if we consider the permutation of variable inputs in addition to the above negation of variable inputs, the situation does not change. In other words, if we try to design minimal networks in double-rail input logic, then, once we obtain a minimal network for one function, minimal networks for all functions that are NP-equivalent to the original function can be immediately derived, simply by negating or permuting some or all variable inputs in the minimal network for the original function, without changing the network configuration. This means that the design effort is greatly reduced, since we need to derive a minimal network for only one function instead of for many NP-equivalent functions.

Table 6.8.3 Catalog of Minimal NOR Networks in Double-Rail Input Logic for All Two-Variable Functions Classified by NP Equivalence

REPRESENTATIVE OF EACH NP-EQUIVALENCE CLASS	MINIMAL NOR NETWORK	OTHER FUNCTIONS FOR WHICH MINIMAL NETWORKS CAN BE OBTAINED BY NEGATION OR PERMUTATION OF VARIABLES
$f^1 = x_1 x_2$	[NOR gate with inputs $\bar{x}_1, \bar{x}_2$]	$f^1_1 = x_1 \bar{x}_2$ $f^1_2 = \bar{x}_1 x_2$ $f^1_3 = \bar{x}_1 \bar{x}_2$
$f^2 = x_1 \vee x_2$	[Two NOR gates in series with inputs x_1, x_2]	$f^2_1 = x_1 \vee \bar{x}_2$ $f^2_2 = \bar{x}_1 \vee x_2$ $f^2_3 = \bar{x}_1 \vee \bar{x}_2$
$f^3 = x_1 \bar{x}_2 \vee \bar{x}_1 x_2$	[NOR network with inputs $x_1, x_2, \bar{x}_1, \bar{x}_2$]	$f^3_1 = x_1 x_2 \vee \bar{x}_1 \bar{x}_2$

As a simple example, a catalog of minimal NOR networks for all functions of exactly two variables, classified by NP-equivalence, is shown in Table 6.8.3. In the first column a representative function for each NP-equivalence class is shown, and all other functions in the same NP-equivalence class are given in the third column. In the middle column a minimal network for each representative function is shown. For any function in the third column, a minimal network can be obtained from this network by negating or permuting corresponding variable inputs.

If P-equivalence is used, the catalog in Table 6.8.3 needs more entries, as can be seen from Table 6.8.1 (e.g., $x_1 \bar{x}_2$ and $\bar{x}_1 \bar{x}_2$ must be in different rows instead of in the same row as $x_1 x_2$). Similarly, the use of N-equivalence will increase the number of entries. But NPN-equivalence cannot be used, since a minimal network for a function that is NPN-equivalent to a given function f cannot be obtained from a minimal network for f without changing the network configuration (e.g., a minimal network for $x_1 \vee x_2$ which is NPN-equivalent to $\overline{x_1 \vee x_2} = \bar{x}_1 \bar{x}_2$ cannot be obtained from a minimal network for $\bar{x}_1 \bar{x}_2$ without changing the network configuration, as can easily be seen from Table 6.8.3). Thus the use of NP-equivalence minimizes the number of entries in the catalog in Table 6.8.3.

Notice that NPN-equivalence puts a function f and its dual f^d into the same equivalence class, since $f^d(x_1, x_2, \ldots, x_n) = \bar{f}(\bar{x}_1\bar{x}_2, \ldots, \bar{x}_n)$ is NPN-equivalent to $f(x_1, x_2, \ldots, x_n)$. Thus NPN-equivalence can serve to classify networks of different gate types also, as follows. When we have a network with an arbitrary gate type or a mixture of different gate types, (as in Fig. 3.1.2a), we replace each gate with a new gate whose output represents the dual of the function that the original gate represents (e.g., we replace AND gates by OR gates, OR gates by AND gates, NOR gates by NAND gates, and so on). As can easily be proved, the entire network represents the dual of the function represented by the original network. In this case the network configuration does not change (e.g., compare Figs. 3.1.2a and 3.1.2c), so the number of gates and the number of connections do not change and the new network is minimal for f^d, if the original network is minimal for f (Exercise 6.22). Thus the dualization of gate types does not change the network configurations or, consequently, the minimality of the networks.

In conclusion, by the use of equivalence classes, cost and time for design efforts are greatly reduced. For example, when we want to derive minimal NAND gates in single-rail input logic for all functions of three or less variables, we need to design for only 80 representative functions instead of 256 functions, as found in Table 6.8.2, by using P-equivalence. Which equivalence should be used depends on single-rail input logic, double-rail input logic, gate types, or other network restrictions.

Notice that, even if the catalogs consist of good networks that are not necessarily minimal, the reduction of cost and time for design efforts does not change.

Remark 6.8.1: Logic designers usually look up good networks that were designed in the past by somebody else who made time-consuming efforts, in literature like professional publications or semiconductor manufacturers' catalogs, though the networks are often modified to meet individual design situations. Catalogs of minimal networks can be found, for example, in the following: Culliney, Young, Nakagawa, and Muroga [1976], Smith [1965], Baugh, Chandersekaran, Swee, and Muroga [1972], and Muroga and Lai [1976].

Also see Hellerman [1963]. In this paper all NOR (or NAND) gate networks with a minimal number of gates as the primary objective and a minimal number of connections as the secondary objectives are listed, under single-rail input logic and maximum fan-in and fan-out of 3, for representative functions of 80 P-equivalence classes of all functions of three or fewer variables, except for the parity function $x_1 \oplus x_2 \oplus x_3$. As shown in Muroga and Ibaraki [1972], the minimality of these networks does not change, even when a maximum fan-in and fan-out of 3 is not imposed (i.e., each gate can have any number of inputs and any number of output connections). The minimal NOR gate networks for the parity function $x_1 \oplus x_2 \oplus x_3$ under the same restriction (i.e., single-rail input logic and a maximum fan-in and fan-out of 3) are obtained in Muroga and Ibaraki [1972] and also in Fig. 14.3.2.2 of Muroga [1971].

Minimal relay contact networks are collected in Appendix 5 of Harrison [1965] and also in Moriwaki [1972] and Higonnet and Grea [1958]. ∎

6.9 General Comments on Combinational Networks

So far we have discussed the design of combinational networks, using map methods, tabular methods, and algebraic methods. These methods are complementary in their usefulness.

Switching algebra is the basis not only of the algebraic methods but also of the map and tabular methods. (If we understand switching algebra well, we may be able to develop other methods.) In terms of design of combinational networks, however, the usefulness of switching algebra, or of these methods, is demonstrated mainly in the design of minimal two-level networks of AND and OR gates, or NAND (or NOR) gates only, under a strong assumption such as Assumption 4.1.1, 4.4.1, or 6.1.1. These minimal two-level networks are very important in themselves (not only for cost reduction but also for performance improvement, due to network size reduction, as discussed in Section 4.1), and also for possible modifications such that restrictions on networks can be satisfied. We have no simple procedure (the integer programming logic design method is complex) to design minimal networks, if any condition in these assumptions is violated. Restrictions imposed on networks in logic design practice are often very complex, as discussed in Section 4.1, and a realistic approach is for designers to first design minimal (or reasonably good) networks without considering the restrictions and then to modify them on a trial-and-error basis so that the restrictions are satisfied, though the minimality of the original networks is somewhat lost (as there is generally no better approach). When restrictions are simple (such as maximum fan-in), they can be taken care of during the processing of some heuristic procedures (such as the map-factoring method). When networks are too large to be designed by the minimization procedures, networks that satisfy only functional requirements can be derived by switching algebra (e.g., lookahead adders in Section 9.2).

Switching algebra is very useful in other aspects of logic design also, such as detection of the dependence of a function on some variables, in a much shorter time than is possible with other straightforward approaches. Also, concise expressions such as minimal sums are convenient for the description of the complex performance of networks, whether the network consists of AND and OR gates or other gates, since diagrams of complex logic networks are hard to understand. Also, switching algebra is useful in implementing computer-aided design (CAD) programs, since tabular or map methods are inappropriate for programming, and complex conditions need to be incorporated into CAD programs. Switching algebra is powerful for some aspects of simulation, test, or diagnosis (Section 9.6), when computer programs for them are to be implemented. Negative functions (e.g., MOS gate) or other complex logic operations are used in implementing some IC logic families. But for human minds simpler logic operations (i.e., AND, OR, and NOT operations) are conceptually easier to manipulate, so switching algebra is still handy.

In the following two chapters on sequential networks, we shall interrupt the discussion of combinational networks until we resume it again in Chapter 9 from the viewpoint of logic design practice. For complex functions, minimization of the number of gates or connections is excessively time-consuming, and in practice absolute minimization is not really necessary in many cases. (In some cases, however, it is critically important. For example, the number of IC packages may double, as we see in Exercise 6.9.) Therefore a trade-off between processing time and network size must be considered. Also. not every switching function is used with the same frequency in practice, but certain functions such as parity functions, majority functions, and multiplexers are used relatively often in logic design practice. Hence, in designing networks, consideration of requirements that are unique to these functions are often important. These problems of combinational networks, as well as others, will be discussed in Chapter 9.

Exercises

6.1 (R) Find the effective and net gate delays from input x to output f in the network in Fig. E6.1, when we set $v = y = z = 1$. (Notice that gate delays for the changes of x from 0 to 1 and also from 1 to 0 must be compared and the greater must be found, to determine the effective and net gate delays.)

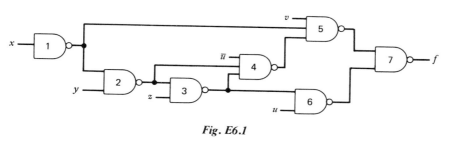

Fig. E6.1

6.2 (R) Find the effective gate delay from input x to output f in the network in Fig. E6.2.

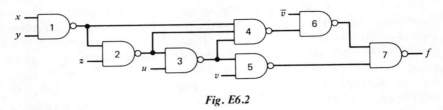

Fig. E6.2

6.3 (R) Find the effective gate delay from variable input x to each of the network outputs, f_1 and f_2, in Fig. E6.3. Explain why.

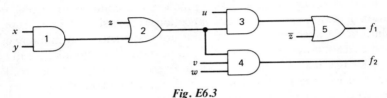

Fig. E6.3

6.4 (M−) The network* in Fig. E6.4 is an adder similar to the full adder shown in Fig. 1.3.3.

(i) Find the effective and net gate delays from input c_{in} to the network outputs.

(ii) Find out whether the network outputs have intrinsic hazards when the value of c_{in} changes. Also, determine how long the hazard, if any, lasts. (Find the time in a multiple of τ, where τ is the switching time of each gate.)

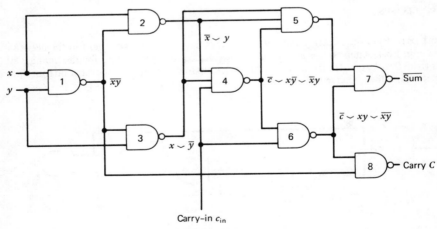

Fig. E6.4   Adder.

* This adder is in Maley [1970]. Notice that the network has the complement of the sum instead of the usual sum $x \oplus y \oplus c_{in}$ in the full adder.

Exercises 339

▲ **6.5 (R)** We want to prepare a catalog of minimal networks. What equivalence relation is useful, in one of the following sets of cases, to minimize the number of entries in the catalog? Also explain why.

A. (i) NOR gates only, under the restriction that no complemented variables are available as network inputs.
   (ii) NOR and AND gates only, under the restriction that no complemented variables are available as network inputs.
   (iii) AND and OR gates only, under the restriction that complemented and non-complemented variables are available as network inputs.

B. (i) NAND gates and OR gates only, under the restriction that no complemented variables are available as network inputs.
   (ii) NOR and OR gates only, under the restriction that complemented and non-complimented variables are both available as network inputs.
   (iii) EXCLUSIVE-OR gates and AND gates only, under the restriction that the constant value 1 is available as an input, but no complemented variables are available as network inputs.

▲ **6.6 (M)** We want to prepare a catalog of minimal networks in which AND, OR, and NOT gates only are used, under the restriction that complemented variables are not available as network inputs. Explain what equivalence relation is useful to minimize the number of entries in the catalog. Also explain why NPN-equivalence cannot be used in this case, although it seems that it can be, since a network for the dual function f^d can be obtained by replacing AND and OR gates by OR and AND gates, respectively, in a network for a function f.

▲ **6.7 (R)** List all functions of exactly three variables that belong to each of the following equivalence classes:

(i) The NP-equivalence class represented by $x_1 x_2 x_3 \lor \bar{x}_1 \bar{x}_2 \bar{x}_3$.
(ii) The NPN-equivalence class represented by $x_1 x_2 x_3 \lor \bar{x}_1 \bar{x}_2 \bar{x}_3$.

▲ **6.8 (R)** List representative functions for all of the P-equivalence classes that correspond to the NP-equivalence class $x_1 \oplus x_2 \oplus x_3$ in Table 6.8.1. Also count the number of functions in each class.

6.9 (M) Assume that we have two kinds of SSI packages, as illustrated in Fig. E6.9a. The first kind contains four NOR gates, each of which has three inputs. The second kind contains four NAND gates, each of which has three inputs. Design a network to replace a block g in one of the following cases, A, B, and C, by using a minimum number of IC packages. (Some inputs to g may not be utilized.) Assume that both complemented and noncomplemented variables are available as the network inputs. You may use any number of packages of only one of these two kinds or both.

A. In Fig. E6.9b, $f(w, x, y, z) = \Sigma(0, 1, 3, 5, 6, 7, 9, 12, 13, 14, 15)$, gate 1 has inputs $\bar{w}, \bar{x}$, and $\bar{y}$ only, and gate 2 has $\bar{w}, x, y$, and z only.
B. In Fig. E6.9b, $f(w, x, y, z) = \Sigma(0, 3, 4, 5, 6, 9, 11, 12, 13, 14, 15)$, gate 1 has inputs $\bar{w}, \bar{x}, \bar{y}$, and $\bar{z}$ only, and gate 2 has $\bar{w}, x, y$, and $\bar{z}$ only.
C. In Fig. E6.9b, $f(w, x, y, z) = \Sigma(3, 4, 5, 6, 7, 9, 10, 11, 12, 13, 15)$, gate 1 has inputs $w, \bar{x}$, and y only, and gate 2 has $w, x, \bar{y}$, and z only.

340 *Design of NAND (NOR) Networks*

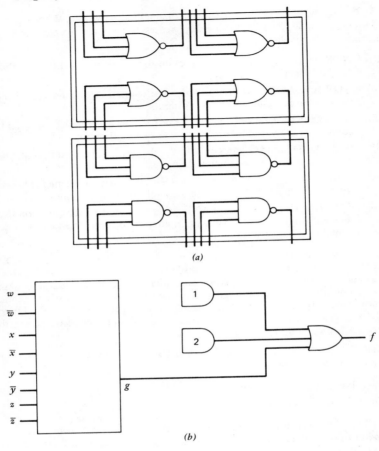

Fig. E6.9

6.10 (M) The following type of gate [the so-called IIL* (integrated injection logic) gate] is available for logic design. The gate has only one input and one or more outputs. When the input, which is 0 or 1, is represented by x, the output of the gate (also, each output if the gate has more than one output) is its complement, that is, $\bar{x}$, as shown in Fig. E6.10a, where the gate is denoted by a circle. When the output of one gate with input x is connected to the output of another gate with input y, the connected point represents $\bar{x}\bar{y}$, as shown in Fig. E6.10b. In other words, function $\bar{x}\bar{y}$ is realized with only the connection, and without any component. (This is **wired-AND**.) When the output

---

* The ILL is based on bipolar transistors. But, unlike other IC logic families based on bipolar transistors, such as ECL or TTL, IIL consumes very little power and also occupies a very small area on IC chips [Berger and Wiedman 1972; Hart and Slob 1972].

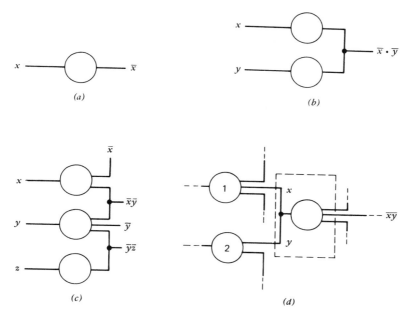

Fig. E6.10

of a gate implements wired-AND with the output of another gate, the outputs of these two gates do not represent their original output functions $\bar{x}$ and $\bar{y}$, but represent $\bar{x}\bar{y}$ instead. If the outputs of three gates are connected, $\bar{x}\bar{y}\bar{z}$ can be realized. Similarly, the outputs of more gates can be connected. Each IIL gate can have more than one output, and each output can be connected to the next gate independently of the other outputs of the same gate, as shown in Fig. E6.10c.

(i) An IIL gate network can be regarded as a NAND gate network by partitioning it as shown in Fig. E6.10d. On the basis of this interpretation of the IIL gate network, describe a formal procedure to design a combinational network for an arbitrary function, using as few IIL gates as possible and, as the secondary objective, as few connections as possible. In other words, discuss under what objectives and restrictions a NAND gate network should be designed, and then how such a NAND gate network can be converted into an IIL gate network.

Assume that complemented variables are not available as network inputs. Also, for simplicity, assume that variable inputs can be wired-ANDed inside an IIL gate network, if necessary.

(ii) If we restrict the maximum number of outputs at each IIL gate to L, how should we modify the design procedure developed in (i)?

(iii) Design a network for $f = \bar{x}_1 x_3 \vee x_1 \bar{x}_2 \bar{x}_3$, using as few IIL gates as possible and, as the secondary objective, as few connections as possible, after designing the corresponding NAND gate network.

342 Design of NAND (NOR) Networks

6.11 (MT) Design a full adder with NAND gates only. Minimize the number of gates as the primary objective and the number of connections as the secondary objective. Assume that only complemented variables $\bar{x}, \bar{y}, \bar{z}$ are available as the adder inputs. (*Hint*: Modify Procedure 6.3.1.)

Then, corresponding to the adder designed above, design a full adder with IIL gates* only, which has only noncomplemented variables x, y, z as network inputs.

6.12 (M−) **(i)** For one of the networks in Fig. E6.12, derive a minimal sum for a function f. (A switching expression for f may first be found by Rule 6.2.1 and then converted into a minimal sum.)

(ii) Design a network for f, using only EXCLUSIVE-OR gates and AND gates, and assuming that the maximum fan-in of each gate is 3 and that both complemented and noncomplemented variables are available as inputs. Reduce the number of gates as the primary objective and the number of connections as the secondary objective, as much as possible.

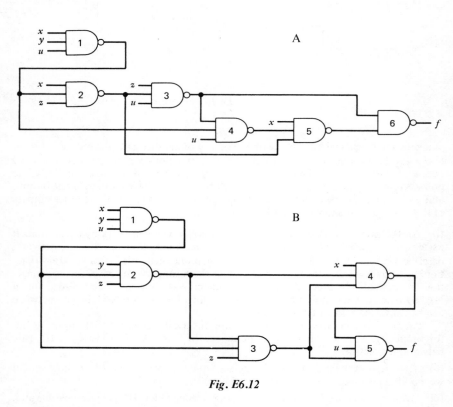

Fig. E6.12

---

* See Exercise 6.10 for the definition of "IIL gates."

Exercises 343

6.13 (E) Suppose that we have a three-level network with a minimum number of NAND gates. The network has noncomplemented variables as network inputs, but no complemented variables. Explain why the following statements hold:

(i) If a literal x_i appears in some prime implicant of the output function f, then x_i must be an input to a gate in the second level.

(ii) If a literal $\bar{x}_i$ appears in some prime implicant of the output function f, then x_i must be an input to a gate in either the first or the third level.

6.14 (M) Using the MOS packages shown in Fig. E2.6a in Exercise 2.6, design a network in single-rail input logic for one of the following functions, using as few packages as possible and then, as the secondary objective, as few outside connections as possible. (Since there is no simple procedure, design on a trial-and-error basis.)

A. $x_1 x_2 \vee \bar{x}_1 \bar{x}_2$.   B. $x_1 \bar{x}_2 \vee \bar{x}_1 x_2$.   C. $\bar{x}_1 \bar{x}_2 \vee \bar{x}_2 x_3$.

6.15 (M) Using the MOS packages shown in Fig. E2.6a in Exercise 2.6, design a network in double-rail input logic for one of the following functions, using as few packages as possible and then, as the secondary objective, as few outside connections as possible. (Since there is no simple procedure, design on a trial-and-error basis.)

A. $f = \Sigma(1, 4, 5, 7, 8, 11) + d(2, 3)$.
B. $f = \Sigma(0, 2, 4, 11, 12, 14, 15) + d(7)$.
C. $f = \Sigma(1, 4, 5, 6, 8, 10) + d(2, 3)$.

6.16 (M−) Design a decoder* for three variables, using NOR gates only and assuming that no complemented variables are available as network inputs. Use as few levels as possible and then, as the secondary objective, as few gates as possible.

6.17 (M) Using NAND gates only, design a network that has data inputs x_1 and x_2, control inputs y_1 and y_2, and outputs z_1 and z_2 and that has the following performance.

For one combination of values of y_1 and y_2, $z_1 = z_2 = x_1$. For a second combination, $z_1 = z_2 = x_2$. For a third combination, $z_1 = 0$ and $z_2 = x_1$. For a fourth combination, $z_1 = x_2$ and $z_2 = 0$. Use as few gates as possible. (*Hint*: The particular combination of values of y_1 and y_2 that is assigned to each of the above cases is important to reduce the number of NAND gates.)

6.18 (M) Suppose that we have a NOR gate network (loop-free) in single-rail input logic that realizes the parity function $x_1 \oplus x_2 \oplus \cdots \oplus x_n$.

Prove that, if some input variable x_k is connected to only one gate, i, throughout this network, gate i cannot have any other input from variable x_j, with $j \neq k$, or from another gate.† (Notice that the network does not necessarily have a minimum number of gates or connections.)

Also prove that this property holds with a NOR gate network that realizes $\overline{x_1 \oplus x_2 \oplus \cdots \oplus x_n}$. [Lai 1976]

---

* See Exercise 2.10 for the definition of "decoder."
† If the output of another gate is connected but constantly 0, this is regarded as no connection to gate i.

344 Design of NAND (NOR) Networks

6.19 (M) Suppose that we have a NOR gate network which realizes the parity function of n variables, $x_1 \oplus \cdots \oplus x_n$, with a minimum number of gates. Prove that, if variable x_k is connected to only one gate, i, gate i has two or more output connections to other gates. (*Hint*: Use the properties stated in Exercise 6.18.)

6.20 (E) Suppose that we have two NOR gates connected as shown in Fig. E6.20a as part of a network, where a set of inputs whose disjunction represents a function $g(\mathbf{x})$ is connected to gates 1 and 2 simultaneously, another set of inputs whose disjunction represents a function $h(\mathbf{x})$ is connected to gate 1, and gate 1 has no output connection other than that to gate 2. Prove that the removal of inputs $g(\mathbf{x})$ to gate 1, as shown in Fig. E6.20b, does not change the output function $f(\mathbf{x})$ of gate 2.

Remark: This is another generalization of the triangular condition in Theorem 6.3.1. ∎

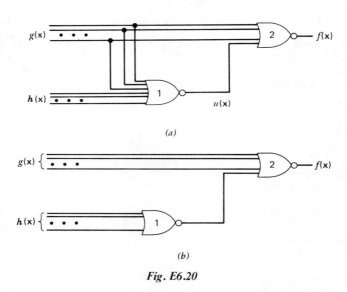

Fig. E6.20

6.21 (M) Suppose that we have a network which consists of a subnetwork and two NAND gates, as shown in Fig. E6.21a, where the subnetwork consists of any gate types. (The subnetwork consists of a single gate type or a mixture of different gate types. NAND gates may not be present.) Gate 1 has one input that represents a function $g(\mathbf{x})$, and other inputs whose conjunction represents a function $h(\mathbf{x})$. Gate 2 has inputs from the subnetwork, and other inputs whose conjunction represents a function $i(\mathbf{x})$.

(i) Prove that, if $\bar{g}(\mathbf{x}) \subseteq \overline{h(\mathbf{x}) \cdot i(\mathbf{x})}$ (i.e., if $g = 0$, then $h \cdot i = 0$), we can eliminate the input $g(\mathbf{x})$ without changing the output f of the entire network in Fig. 6.2.1a.

(ii) As a special case, if the same input x is connected to gates 1 and 2, as shown in Fig. 6.21b, we can disconnect x from gate 1 as shown in Fig. 6.21c without changing the network output, even though no implication relation holds between x and other inputs $u_1, \ldots, u_s, v_1, \ldots, v_t$ to gates 1 and 2. Explain why.

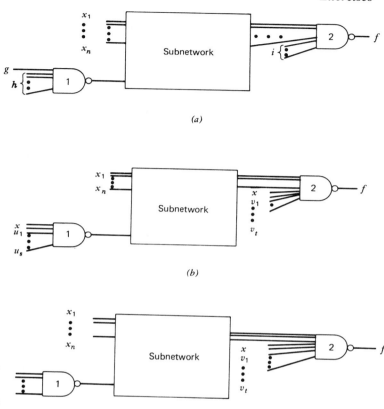

Fig. E6.21

Remark: This is the property dual to Exercise 3.5.7 [Lai, Nakagawa, and Muroga 1974] and is another generalization of the triangular condition in Theorem 6.3.1. ∎

6.22 (M) Suppose that a network with a minimum number of gates for a function f is given, where, possibly, different gate types are mixed. Then prove that the network derived by replacing each gate by its dual type (e.g., a NOR gate by a NAND gate, an AND gate by an OR gate, etc.) has a minimum number of gates among the networks for f^d.

6.23 (M) Suppose that we have a network with NAND gates only, for parity function $x_1 \oplus x_2 \oplus \cdots \oplus x_n$. No gate has a constant input of 1 or 0. Discuss whether the network output still expresses $x_1 \oplus x_2 \cdots \oplus x_n$ when every NAND gate is replaced by a NOR gate.

346 Design of NAND (NOR) Networks

6.24 (MT) Gates are available, each of which has double outputs, NOR and OR (see Fig. 5.5.5), and has the restriction of maximum fan-in and fan-out of 3. Using only these gates, design a network that has the two outputs

$$x_1 \oplus x_2 \oplus x_3 \quad \text{and} \quad \overline{x_1 \oplus x_2 \oplus x_3}.$$

Assume that only x_1, x_2, x_3, but not their complements, are available as inputs to the network. Use as few gates and as few connections as possible.

(i) A network with eight gates is easy to obtain. Show it. If you do not succeed, show your best network.

(ii) Can you reduce the number of gates to less than eight? If you can, show your network.

CHAPTER 7

Asynchronous Sequential Networks

The outputs of sequential networks depend not only on the current values of inputs but also on the past values, complicating the analysis and synthesis of these networks. When the timing relationship among the operations of gates is critical for the reliable operation of the networks, the gate operations are synchronized by clocks, that is, trains of periodic pulses. Such clocked networks, which are called **synchronous networks**, will be discussed in Chapter 8; networks without clocks, that is, **asynchronous networks**, are the subject of this chapter ("synchronous" and "asynchronous" will be defined in detail in Section 8.1).

First, the performance of asynchronous sequential networks is analyzed, after discussing the simplest type of flip-flop, a commonly used logic building block to remember information about the past values of inputs. Important concepts such as internal variables, states, and fundamental mode are introduced. Then we discuss hazards, which is a much more serious problem here than in the case of combinational networks.

In the rest of this chapter we discuss the synthesis of sequential networks, which is generally much more complicated and difficult than that of the loopless combinational networks discussed in the preceding chapters.

7.1 Latches

A combinational network is a network whose outputs assume values depending only on the current values of the inputs, whereas in a sequential network the outputs assume values depending not only on the current values but also on the past sequence of values of the inputs. Thus a sequential network must remember information about the past values of its inputs. (The past values need not entirely be remembered; we need to remember only essential information about them.) Simple networks called **flip-flops** are usually used as memories for this purpose. These flip-flops are synthesized with gates, and consequently flip-flops work with a speed comparable to that of gates used for logic operations. Memory devices such as semiconductor (or magnetic core) memories can also serve as memories for sequential networks, but flip-flops are used* if higher speed is necessary to match the speed of gates, even though semiconductor memories are cheaper.

In this section let us consider the simplest flip-flops, which are called **latches**.†

The network in Fig. 7.1.1a, which is called an **S-R latch**, consists of two NOR gates. Assume that the values at terminals S and R are 0, and the value at terminal Q is 0 (denote them by $S = R = 0$ and $Q = 0$). Since gate 1 has inputs of 0 and 0, the value at terminal $\bar{Q}$ is 1 (denote it by $\bar{Q} = 1$). Since gate 2 in the network has two inputs, 0 and 1, the output is $Q = 0$. Thus signals 0 and 1 are maintained at terminals Q and $\bar{Q}$, respectively, as long as S and R remain 0. Now change the value at S to 1. Then $\bar{Q}$ is changed to 0, and Q becomes 1 after a short time delay. Even if $Q = 1$ and $\bar{Q} = 0$ were original values, the change of the value at S to 1 still yields $Q = 1$ and $\bar{Q} = 0$. In other words, Q is set to 1 by supplying 1 to S, even if we originally had $Q = 0$, $\bar{Q} = 1$ or $Q = 1$, $\bar{Q} = 0$. When 1 is supplied to R with S remaining at 0, Q and $\bar{Q}$ are set to 0 and 1, respectively, after a short time delay, no matter what values they had before. Thus we get the first three combinations of the values of S and R in the table in Fig. 7.1.1b. In other words, as long as $S = R = 0$, the values of Q and $\bar{Q}$ are not changed. If $S = 1$, Q is set to 1. If $R = 1$, Q is reset to 0. Thus S and R are called **set** and **reset terminals**, respectively.

Here notice that, when the latch is to be set or reset to new output values, the value 1 at S or R must be maintained sufficiently long so that new output values appear at Q and $\bar{Q}$ after a short transition period. Assume that $Q = 0$ and $\bar{Q} = 1$ with $S = R = 0$. Let us change S to 1. If S returns to 0 too quickly after going to 1, the new output values $Q = 1$ and $\bar{Q} = 0$ may not be established. Suppose that each NOR gate in Fig. 7.1.1a has a switching time of τ. Suppose that we apply a signal 1 at S which lasts only time τ, as shown in Fig. 7.1.2a ($S = 0$, $\bar{Q} = 1$, and $Q = 0$ before time 0). At time τ, gate 1 responds, changing its output $\bar{Q}$ from 1 to 0. Then at time 2τ, gate 2 responds, changing its output Q from 0 to 1. During time τ to 2τ, gate 1 has inputs 0 only, so at time 2τ it responds, changing its output $\bar{Q}$ from 0 to 1. During

---

* However, since semiconductor memories are becoming cheaper, faster and more compact, these memories, particularly read-only memories, are now widely used to substitute for some gates in combinational networks or for flip-flops in sequential networks if high speed is not required, as will be discussed in Section 9.4.

† According to the catalogs of semiconductor manufacturers, at least, the flip-flops discussed in this section are called "latches," in contrast to raceless flip-flops such as master-slave and edge-triggered flip-flops, to be discussed in Chapter 8. The term "flip-flops" in a narrow sense usually means "raceless flip-flops." But some authors still call latches "flip-flops."

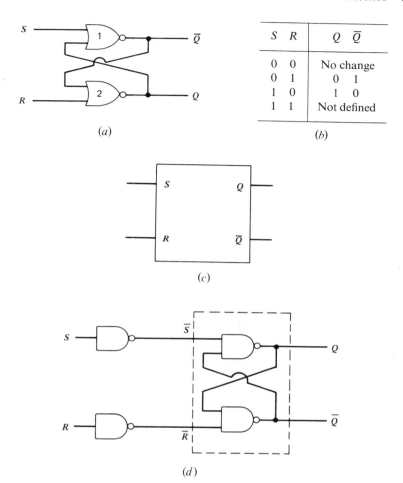

Fig. 7.1.1 **S-R latch. (a) S-R latch with NOR gates. (b) Input-output relationship. (c) Symbol for S-R latch. (d) S-R latch with NAND gates.**

time 2τ to 3τ, both gates have input 1, so at time 3τ both gates respond, changing $\bar{Q}$ and Q from 1 to 0 simultaneously. Thereafter the outputs $\bar{Q}$ and Q oscillate between 0 and 1 at every τ time. However, this argument is based on an idealized performance of gates: that each gate establishes its new output value exactly at time τ later, as illustrated in the solid lines in Fig. 7.1.2a. In reality the signals change more gradually, and each gate also establishes its output gradually, as shown in the dotted line for the first pulse (similarly,

350 *Asynchronous Sequential Networks*

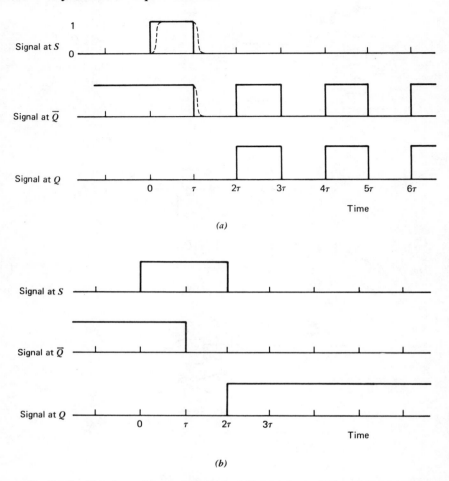

Fig. 7.1.2 Signal transition in S-R latch of Fig. 7.1.1a. (a) When S = 1 lasts for only time τ (unstable operation of latch). (b) When S = 1 lasts for time 2τ (stable operation of latch).

with more distortion, for the waveforms after time 2τ). Thus $\bar{Q}$ and Q may not oscillate indefinitely, but instead eventually settle down to certain output values. We do not know the state in which the network will settle down. If the transition is not short, the latch performs differently from what we expected. As can easily be seen, whenever $S = 1$ lasts less than 2τ, the situation is similar. The network may not settle down quickly, and the output values it will eventually have are not predictable [Barna and Porat 1973].

If, however, the signal at S lasts 2τ, as shown in Fig. 7.1.2b, or longer, $\bar{Q}$ and Q settle to new stable values quickly because gate 1 receives at least one input of 1 without interruption until time 2τ. But notice that even in this case both $\bar{Q}$ and Q have the same output 0 during the period from time τ to 2τ. The situation is similar for other cases for Q, $\bar{Q}$, S, and R. In conclusion, **in order to let the latch work properly, the value 1 at S or R must be maintained until new values of Q and $\bar{Q}$ are established.** In this case Q and $\bar{Q}$ assume the **same value for a short transition period of about τ before they assume different values steadily**, though this identical value may differ depending on the implementation of a flip-flop (e.g., 0 in the case of the flip-flop of NOR gates in Fig. 7.1.1a, and 1 in the case of flip-flop of NAND gates in Fig. 7.1.1d). Simultaneous identical values of Q and $\bar{Q}$ may be regarded as an intrinsic hazard (defined in Section 6.7) and, in some cases, can be a cause for the malfunction of sequential networks, as we will see later.

Notice that latch outputs Q and $\bar{Q}$ always assume complementary values (i.e., if one is 0, the other is 1, and vice versa), except during the short transition period. This is the reason why $\bar{Q}$, the notation denoting the complement of Q, is used here.

When $S = R = 1$ occurs, the outputs Q and $\bar{Q}$ are both 0. If S and R simultaneously return to 0, these two outputs cannot maintain 0. Actually, a simultaneous change of S and R to 0 or 1 is physically impossible* unless we make the network more sophisticated (e.g., if every gate is synchronized by a clock, as explained in Chapter 8, we may be able to avoid this problem). If S returns to 0 from 1 before R does, we have $Q = 0$ and $\bar{Q} = 1$. If R returns to 0 from 1 before S does, we have $Q = 1$ and $\bar{Q} = 0$. Thus it is not possible to predict what values we will have at the outputs after having $S = R = 1$. The output of this network is not defined for $S = R = 1$, as this combination is not used.

For simplicity, let us **assume, throughout this book, that only one of the inputs to any network changes at a time**, unless otherwise noted (such as the cases in Chapter 8). This is a reasonable and important assumption.

Notation for *S-R* Latch

The *S-R* latch is usually denoted as in Fig. 7.1.1c (notice that the positions of Q and $\bar{Q}$ in Fig. 7.1.1a are interchanged in Fig. 7.1.2c). The *S-R* latch with

---

* Actually, signals change gradually, as illustrated in dotted lines in Fig. 7.1.2a, instead of the instantaneous changes shown in solid lines. Thus many signals may simultaneously change gradually. This can often cause the network to malfunction. But for the sake of simplicity, all signals are assumed to have instantaneous changes.

NAND gates is shown in Fig. 7.1.1d. This is sometimes convenient for modification, as we will see later. The dotted-line rectangle in Fig. 7.1.1d is called an $\bar{S}\text{-}\bar{R}$ latch and is often used by itself.

The **S-R latch** is also called an **R-S latch**, **R-S flip-flop**, **S-R flip-flop**,* or **set-reset flip-flop**.

Usually, S-R latches are used in designing asynchronous sequential networks, though sequential networks can be designed without them (i.e., with longer loops of gates than the loop in the latch, or with memory devices). As many different IC logic families are developed, flip-flops other than S-R latches are used.† Also, in the synchronous sequential networks discussed in Chapter 8, many different types of flip-flops are used. (These flip-flops have clock terminals. Since the clock terminals need not necessarily be connected to clocks, these flip-flops can be used not only in synchronous but also in asynchronous networks. For the sake of simplicity, the use of flip-flops with clock terminals in asynchronous networks will not be discussed until Section 9.3.)

7.2 Sequential Networks in Fundamental Mode

In a sequential network we are interested in finding out what sequence of outputs will be produced for each sequence of inputs, instead of for each individual combination of the current input values, since the network outputs depend on both the current input values and some functions of past input values stored in the network. In this section we discuss how to do this. It can be done by examining step by step what values the outputs have for each individual combination of input values, depending on what signal values are stored in the network. Essentially, this means interpreting the signals stored inside the network as new input variables, and then interpreting the entire network as a combinational network of the new input variables, plus the original inputs. Much of this approach was developed by Huffman [1954].

Let us consider the sequential network in Fig. 7.2.1. **Assume that the inputs are never changed unless the network is in a stable condition, that is, unless none of the internal signals is changing.** Also **assume that, whenever the inputs change, only one input changes at a time**, as stated in Section 7.1. Let y_1, $\bar{y}_1$, y_2, and $\bar{y}_2$ denote the outputs of the two S-R latches.

---

* Some authors define S-R flip-flops (set dominant) and R-S flip-flops (reset dominant) differently. When $S = R = 1$, Q of an S-R flip-flop is set to 1 and Q of an R-S flip-flop is set to 0 [Engineering Staff of AMI 1972]. See Exercise 7.3.13.

† For example, a loop consisting of a pair of inverters and a special gate called a transmission gate is used in an asynchronous CMOS network (see, e.g., RCA [1972]).

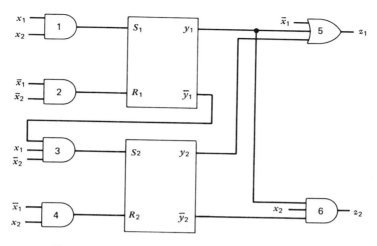

Fig. 7.2.1 *A sequential network in fundamental mode.*

Let us assume that $y_1 = y_2 = 0$ (accordingly, $\bar{y}_1 = \bar{y}_2 = 1$) and $x_1 = x_2 = 0$. Then, as can easily be found, we have $z_1 = 1$ and $z_2 = 0$ for this combination of values. Since x_1 and x_2 are 0, the inputs of the latches have values $R_1 = 1$ (accordingly, $y_1 = 0$) and $S_1 = S_2 = R_2 = 0$. Then y_1 and y_2 remain 0. As long as x_1, x_2, y_1, and y_2 remain 0, none of the signal values in this network changes, and consequently this combination of values of x_1, x_2, y_1, and y_2 is called a **stable state**.

Next assume that x_1 is changed to 1, keeping $x_2 = y_1 = y_2 = 0$ (see Table 7.2.1a). For this new combination of input values, we get $z_1 = 0$ and $z_2 = 0$ after a time delay of τ, where τ is the switching time of each gate. The two latches have new inputs $S_2 = 1$ and $S_1 = R_1 = R_2 = 0$, after a time delay of τ. Then they have new output values $y_1 = 0$, $\bar{y}_1 = 1$, $y_2 = 1$, and $\bar{y}_2 = 0$, after a delay due to the response time of the latches. The lower latch had $y_2 = 0$ previously, and output z_1 changes from 0 to 1. After this change the network does not change any further. Summarizing the above, we can say that, when the network has the combination $x_1 = 1$, $x_2 = y_1 = y_2 = 0$, it does not remain in this combination, but changes into the new combination $x_1 = 1$, $x_2 = y_1 = 0$, $y_2 = 1$, as shown by an arrow in Table 7.2.1a. Also, outputs z_1 and z_2 change into $z_1 = 1$ and $z_2 = 0$, after assuming the values $z_1 = z_2 = 0$ temporarily. After the network changes into the combination $x_1 = 1$, $x_2 = y_1 = 0$, $y_2 = 1$, it remains there. The combination $x_1 = 1$, $x_2 = y_1 = y_2 = 0$ is called an **unstable state**. The transition from an unstable to a stable state, such as the above transition to the stable

Table 7.2.1 Analysis of the Sequential Network in Fig. 7.2.1

a.							b.								
x_1	x_2	y_1	y_2	z_1	z_2		x_1	x_2	y_1	y_2	Y_1	Y_2	z_1	z_2	
0	0	0	0	1	0		0	0	0	0	0	0	1	0
0	0	0	1	1	0		0	0	0	1	0	1	1	0
0	0	1	0	1	0		0	0	1	0	0	0	1	0
0	0	1	1	1	0		0	0	1	1	0	1	1	0
0	1	0	0	1	0		0	1	0	0	0	0	1	0
0	1	0	1	1	0		0	1	0	1	0	0	1	0
0	1	1	0	1	1		0	1	1	0	1	0	1	1
0	1	1	1	1	0		0	1	1	1	1	0	1	0
1	0	0	0	0	0		1	0	0	0	0	1	0	0
1	0	0	1	1	0		1	0	0	1	0	1	1	0
1	0	1	0	1	0		1	0	1	0	1	0	1	0
1	0	1	1	1	0		1	0	1	1	1	1	1	0
1	1	0	0	0	0		1	1	0	0	1	0	0	0
1	1	0	1	1	0		1	1	0	1	1	1	1	0
1	1	1	0	1	1		1	1	1	0	1	0	1	1
1	1	1	1	1	0		1	1	1	1	1	1	1	0

state $x_1 = 1$, $x_2 = y_1 = 0$, $y_2 = 1$, is the key to the analysis of a sequential network.

In Table 7.2.1a, all possible combinations of values of x_1, x_2, y_1, and y_2 are listed. All stable states are encircled, and the transitions from unstable to stable states are shown with arrows.

Transition Table

Another method of describing these transitions is to list the next values of y_1 and y_2 for each combination of the values of x_1, x_2, y_1, and y_2. Table 7.2.1a is redrawn as Table 7.2.1.b, listing the next values of y_1 and y_2, denoted by Y_1 and Y_2. The variables y_1 and y_2 (also Y_1 and Y_2) are usually called **internal variables**, denoting their present values by small letters and their next values by capital letters. Thus, more specifically, variables y_1 and y_2 are called **present-state (internal) variables**, and Y_1 and Y_2 are called **next-state (internal) variables**. As can easily be seen, when the values of Y_1

Table 7.2.2 Tables Derived from Table 7.2.1

a. Transition Table

$y_1 y_2$ \ $x_1 x_2$	00	01	11	10
00	⓪⓪	⓪⓪	10	01
01	⓪①	00	11	⓪①
11	01	10	⑪	⑪
10	00	⑩	⑩	⑩

$Y_1 Y_2$

b. Output Table

$y_1 y_2$ \ $x_1 x_2$	00	01	11	10
00	10	10	00	00
01	10	10	10	10
11	10	10	10	10
10	10	11	11	10

$z_1 z_2$

c. Transition-Output Table

$y_1 y_2$ \ $x_1 x_2$	00	01	11	10
00	⓪⓪, 10	⓪⓪, 10	10, 00	01, 00
01	⓪①, 10	00, 10	11, 10	⓪①, 10
11	01, 10	10, 10	⑪, 10	⑪, 10
10	00, 10	⑩, 11	⑩, 11	⑩, 10

$Y_1 Y_2, z_1 z_2$

and Y_2 are identical to those of y_1 and y_2, respectively, a state containing these values of Y_1, Y_2, y_1, and y_2 is stable, since there is no transition of y_1 and y_2 into a new, different state. Next-state variables in stable states are encircled in Table 7.2.1b. Let us rewrite Table 7.2.1b further, into a more compact form as shown in Table 7.2.2a, where coordinates are arranged in the same manner as in a Karnaugh map. (Since there are only four rows, they can be arranged in a Karnaugh map so that adjacent coordinates differ in only one bit position. But if the number of coordinates is more than four, the coordinates cannot be arranged in this manner and may be arranged in an arbitrary order, as we will see in the example in Section 7.7.) The values of Y_1 and Y_2 are entered in cells. The table is called a **transition table**. Each column in this table corresponds to a combination of values of network inputs x_1 and x_2, that is, an **input state**. Each row corresponds to a combination of values of internal variables y_1 and y_2, that is, a present **internal state**. Thus each cell corresponds to a **total state**, that is, a combination of values of $x_1, x_2, y_1,$ and y_2. The entry in each cell shows the coordinate values of the next internal state for a total state.

Fundamental Mode

A change in input variables of the network causes a change from one column of the transition table to another, without a row change. If, in the new column, the values of Y_1 and Y_2 disagree with those of y_1 and y_2 in the same row, the $Y_1 Y_2$ values are an unstable state, and the network state moves to a new row, where the $y_1 y_2$ values are the same as these $Y_1 Y_2$ values, that is, a transition to a stable state takes place. For example, when inputs (x_1, x_2) change from (0 1) to (1 1), a transition occurs from the stable state (x_1, x_2, y_1, y_2) = (0 1 0 0) to the stable state (1 1 1 0) through the unstable state (1 1 0 0). This transition is shown by the arrow in Table 7.2.2a. The stable states in Table 7.2.2a are encircled. Notice that an entry $Y_1 Y_2$ is in a stable state if the $Y_1 Y_2$ values are identical to the $y_1 y_2$ values in the same row.

A sequential network is said to be operating in **fundamental mode** when the transition between states occurs in this manner—in other words, when the transition occurs horizontally in the transition table corresponding to each network input change and then, unless the new state in the same row is stable, the transition continues vertically (not diagonally*), settling in a new stable state. (Thus the fundamental mode could be called the horizontal-vertical transition mode.) The transitions of the network take place under the assumption that only one of the network inputs changes at a time, only when the network is not in transition, that is, only when the network is settled in a stable state.

The output values corresponding to the total states are shown in the **output table** in Table 7.2.2b. Often transition and output tables are combined into a single **transition-output table**, as shown in Table 7.2.2c. By using such a table, the output sequence corresponding to any input sequence can be easily obtained. In other words, the output sequence can be obtained by choosing columns corresponding to the input sequence, and then moving to stable states whenever states in these columns are unstable.

---

* In this book, "pulse mode," which is often discussed in literature for asynchronous sequential networks, is not discussed, since its realization is impractical, as explained in the footnote on p. 468 in Section 8.3.

Key Point in the Above Analysis

Since the example network in Fig. 7.2.1 is very simple, it is easy to find out what will be a next stable state for each transition. Using the switching expressions for the S and R terminals of each latch,

$$S_1 = x_1 x_2, \qquad R_1 = \bar{x}_1 \bar{x}_2, \qquad S_2 = x_1 \bar{x}_2 \bar{y}_1, \qquad \text{and} \qquad R_2 = \bar{x}_1 x_2,$$

we can easily find out which latch (or latches) receives signal 1 at which terminal, and then determine into which state the network will move. In some more complex networks, however, we cannot find a next state for each transition by using switching expressions. **When a network contains multiple paths of different lengths, a combinational network that has inputs x_i's, y_i's, and their complements may have output values (those at the S_i and R_i terminals) different from those given by switching expressions during the transition periods. Then latches may be set to new values different from those that we expect from the switching expressions. Although this phenomenon was discussed as "intrinsic hazards" in Section 6.7, the network may be designed based on it, not treating the phenomena as hazards.** Fig. 6.7.2 shows a spurious output that lasts only time τ, but more complicated networks may generate spurious outputs that last longer, where τ is the switching time of each gate. For example, the combinational network consisting of the leftmost three gates in Fig. 7.2.2a as part of a sequential network generates a spurious output lasting 2τ at output f, for the change of x_1 from 0 to 1, with x_2 and x_3 being kept at 1, as illustrated in Fig. 7.2.2b. This is certainly long enough to set the latch, as discussed in Section 7.1. But output f has the switching expression $f = x_1 x_2 \bar{x}_3$, and for $x_2 = x_3 = 1$, f is identically 0, not setting the latch.

This means that, when we prepare a transition-output table for a given sequential network, **we may obtain an incorrect result if we use switching expressions derived from the network to find next-state Y_i's.** Therefore we need to find what value each gate assumes at every time interval of length τ, to determine next-state Y_i's. (Of course, if every path from a network input, or an internal variable, to input terminals of latches has the same length, we do not have this problem, and we can rely on switching expressions.) **Once we obtain a transition table, switching expressions for Y_i's can be obtained from this table by treating the transition table as a Karnaugh map** (an example will be illustrated with Table 7.3.4), and can be conveniently used as an alternative to the transition table.

The values of the network outputs (e.g., z in Fig. 7.2.1) during the transitions can be found in the same manner, by finding what value each gate assumes at every τ time, but the values of the network outputs for transition periods are not important unless we are worried about the malfunction of

358 *Asynchronous Sequential Networks*

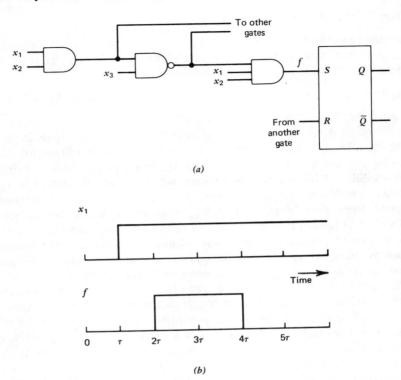

Fig. 7.2.2 *Hazardous network.* (a) $f = x_1 x_2 \bar{x}_3$. (b) *Waveform at f when* $x_2 = x_3 = 1$.

other networks that are connected to the outputs of this network. Once the network settles down in the next stable state, the corresponding output values can be easily found from the new values of x_i's and Y_i's, using switching expressions for x_i's, Y_i's, and their complements. (Of course, the output values can be found by determining what value each gate assumes for every interval of time τ.) This is so because the network outputs may be regarded as the outputs of a combinational network (part of the entire network) which has inputs x_i's and Y_i's (e.g., z_1 and z_2 in Fig. 7.2.1 are the outputs of the combinational network consisting of gates 5 and 6 with inputs $\bar{x}_1$, x_2, y_1, y_2, and $\bar{y}_2$), and we can obtain correct output values of the combinational network if we wait for a while. Thus the correct output values after a transition period can be obtained from switching expressions, and it is not necessary to find out what value each gate assumes at every τ time.

7.3 Malfunctions and Performance Descriptions of Sequential Networks

Unless internal variables are set to proper values for each input change, a sequential network does not work properly. Therefore the timing relationship among the operations of gates is very important. If the timing relationship is improper, the sequential network may malfunction. Even if the network does not malfunction, we may have transition tables with more complicated transitions than those in Section 7.2.

In this section we first discuss network malfunctions, and then transition-output tables that have complicated transitions. Then we discuss how to concisely describe the performances of networks.

Racing Problem of Sequential Networks

A difference in time delays of signal propagation along different paths may cause a malfunction of a sequential network that is called a **race**. Since this is caused by a difference in the switching times of gates, it may be regarded as a type of extrinsic hazard.

Let us consider the sequential network shown in Fig. 7.3.1. The transition table that this network is to perform is shown in Table 7.3.1a, and the output table in Table 7.3.1.b.

Suppose that the network is in the stable state $(x_1, x_2, y_1, y_2) = (1\ 1\ 1\ 0)$. If the inputs change from $(x_1, x_2) = (1\ 1)$ to $(1\ 0)$, the network changes into

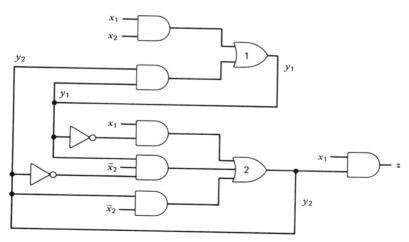

Fig. 7.3.1   A sequential network.

Table 7.3.1 Tables for the Network in Fig. 7.3.1
(The network does not enter the state denoted by a dash.)

a. **Transition Table**

y_1y_2	x_1x_2 00	01	11	10
00	⓪⓪	⓪⓪	11**	01
01	⓪①	00	11	⓪①
11	⑪	10	10	⑪
10	—	00	⑩	01*

Y_1Y_2

b. **Output Table**

y_1y_2	x_1x_2 00	01	11	10
00	0	0	0	0
01	0	0	1	1
11	0	0	1	1
10	—	0	0	0

z

the unstable state $(x_1, x_2, y_1, y_2) = (1\ 0\ 1\ 0)$, marked with an asterisk in Table 7.3.1*a*. Since $y_1 = 1$, $y_2 = 0$, $Y_1 = 0$, $Y_2 = 1$, this means that OR gate 1 in Fig. 7.3.1*a* changes its output value from 1 to 0, but OR gate 2 changes its output value from 0 to 1. In other words, the two OR gates must have simultaneous changes. However, it is extremely difficult for the two gates to finish their changes at exactly the same time, since no two paths of gates that lead to OR gates 1 and 2 will have identical time delays. Actually, one of the two OR gates finishes its change before the other does. In other words, we have the following:

(1) Either y_2 changes first, making the transition of (y_1, y_2) from (1 0) to (1 1) and leading the network into stable state $(x_1, x_2, y_1, y_2) = (1\ 0\ 1\ 1)$; or

(2) y_1 changes first, making the transition of (y_1, y_2) from (1 0) to (0 0), and then y_2 changes, making the further transition of (y_1, y_2) from (0 0) to (0 1).

As illustrated in Fig. 7.3.2, the network will go to either state $(y_1, y_2) = (1\ 1)$, in case (1), or state $(y_1, y_2) = (0\ 0)$, in case (2), instead of going directly to $(y_1, y_2) = (0\ 1)$. If $(y_1, y_2) = (1\ 1)$ is reached, state (1 1) is stable, and the network stops here.* If (0 0) is reached, this is an unstable state, and another

---

* Actually, we may have a more complex multiple transition, since y_1 may change to 0 immediately after y_2 becomes 1 (and then to 1, and so on, depending on the lengths of paths in a network).

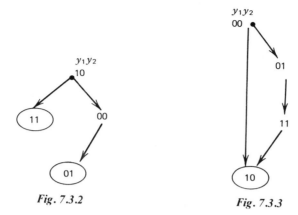

Fig. 7.3.2   Fig. 7.3.3

transition to the next stable state, (0 1), occurs. State (0 1) is the desired state, but (1 1) is not. Thus, depending on which path of gates works faster, the network may malfunction. This situation is called a **race**. The network may or may not malfunction, depending on which case actually occurs.

Next, suppose that the network is in the stable state $(x_1, x_2, y_1, y_2) = (0\ 1\ 0\ 0)$ and that inputs $(x_1, x_2) = (0\ 1)$ change to $(1\ 1)$. The cell with two asterisks in Table 7.3.1a has $(Y_1 Y_2) = (1\ 1)$. Thus y_1 and y_2 must have simultaneous changes. Depending on which OR gate changes its output faster, there are two possible transitions, as shown in Fig. 7.3.3, but in this case both end up in the same stable state, $(y_1, y_2) = (1\ 0)$. This is another race, but the network does not have a malfunction depending on the order of change of internal variables. Hence this is called a **noncritical race**, whereas the previous race is termed a **critical race** because the performance is unpredictable (it is hard to predict which path of gates will transmit the signal change faster), possibly causing the network to malfunction.

Remedies for the Racing Problem

Whenever a network has critical races, we must eliminate them for reliable operation. One approach is to make the paths of gates have definitely different time delays. In the example of Fig. 7.3.1, the critical race can be eliminated by making the path of gates leading to OR gate 2 have a longer delay than the other path of gates leading to OR gate 1. This approach, however, may not be most desirable for the following reasons. First, the speed may be sacrificed by adding gates for delay. Second, there are cases where this approach is impossible if a network contains more than one critical race. In other words,

Table 7.3.2

	$x_1 x_2$			
$y_1 y_2$	00	01	11	10
00	⓪⓪	⓪⓪	11	01
01	⓪①	00	11	⓪①
11	①①	10	10	①①
10	—	00	①⓪	00
	$Y_1 Y_2$			

Table 7.3.3

	$x_1 x_2$			
$y_1 y_2$	00	01	11	10
00	⓪⓪			
01	⓪①			
11	①①			
10		01	①⓪	
	$Y_1 Y_2$			

by eliminating a critical race in one column by using a path of different time delay, a critical race in another column may be aggravated.

A better approach is to choose the entries in a transition table so that no critical races occur. Then, on the basis of this new table, we synthesize a network with the desired performance, according to the synthesis method to be described in the following sections. A change of entries in unstable states without changing stable states, such that only one internal variable changes its value at a time, is one method for eliminating critical races. The critical race discussed above can be eliminated from Table 7.3.1a by replacing the entry marked with an asterisk by (0 0), as shown in Table 7.3.2. If every entry that causes the network to malfunction can be changed in this manner, a reliable network with the performance desired in the original is produced, because every next stable state to which the network should go is not changed, and the output table is also not changed. (We need not worry about noncritical races, since they cause no malfunctions.) However, sometimes there are entries for some unstable states that cannot be changed in this manner. For example, consider Table 7.3.3 (some states are not shown, for the sake of simplicity). The entry (0 1) for $(x_1, x_2, y_1, y_2) = (0\ 0\ 1\ 0)$ is such an entry* and causes a critical race. Since this entry requires simultaneous transitions of two internal variables, y_1 and y_2, we need to change it to (0 0) or (1 1). Both lead the network to stable states different from the desired stable state (0 1).

* State $(x_1, x_2, y_1, y_2) = (0\ 0\ 1\ 0)$ in Table 7.3.1a appears to have the same property, but, actually, the network never enters this state, because inputs x_1 and x_2 do not change simultaneously.

Malfunctions and Performance Descriptions of Sequential Networks 363

When a change of entries in unstable states does not work, we need to redesign the network completely, as discussed in the following sections of this chapter. This redesign may include the reassignment of binary numbers to states, or the addition of intermediate states through which the network goes from one stable state to another.

A race can occur even if the feedback loops in a sequential network contain latches as we will see later. Even in this case, the remedies discussed above are useful.

Remark 7.3.1: Sequential networks of relays also have racing problems and remedies similar to those described above, though some remedies for electronic gate networks may be too complicated for relay contact networks.

Analysis of the hazards of relay contact networks can be treated mathematically by assuming that each variable assumes other values in addition to 0 and 1 [Yoeli and Rinon 1964, and Chapter 13 in Perrin, Denouettes, and Daclin 1972]. For further discussion, see, for example, Chapter 7 of McCluskey [1965], Dlugatch [1971] (this article discusses the design of a network that works reliably even if there are deviations in time delay among gates), Hlavička [1970], and Brzozowski and Yoeli [1976]. ∎

Other Malfunctions

Extrinsic or intrinsic hazards (including simultaneous identical values of Q and $\bar{Q}$ of a latch, discussed in Section 7.1) may contribute to network malfunctions in ways different from the racing problem, though all are caused by the difference in time delays from the network inputs or internal variables to the latch inputs over different paths. The elimination of network malfunctions due to all these factors will be discussed in Section 7.8.

Concise Description of the Performance of Sequential Networks

Transition tables (or transition-output tables) are a convenient means to concisely express the performance of sequential networks, which is often hard to understand from their logic network configurations.

Switching expressions are another convenient means. Once we have transition-output tables, the transition relationship among states can be expressed in switching expressions from the transition-output tables (though these tables themselves cannot necessarily be obtained by switching expressions, as we have explained at the end of Section 7.2). For example, we can treat the transition table in Table 7.2.2a as a Karnaugh map for Y_1 and Y_2 in Table 7.3.4. Thus the next states can be expressed as

$$Y_1 = x_1 x_2 \vee x_1 y_1 \vee x_2 y_1,$$
$$Y_2 = x_1 y_2 \vee \bar{x}_2 y_2 \vee x_1 \bar{x}_2 \bar{y}_1.$$

Table 7.3.4 Karnaugh Maps for Table 7.2.2a

y_1y_2 \ x_1x_2	00	01	11	10
00	0	0	1	0
01	0	0	1	0
11	0	1	1	1
10	0	1	1	1

$$Y_1 = x_1 x_2 \vee x_1 y_1 \vee x_2 y_1$$

y_1y_2 \ x_1x_2	00	01	11	10
00	0	0	0	1
01	1	0	1	1
11	1	0	1	1
10	0	0	0	0

$$Y_2 = x_1 y_2 \vee \bar{x}_2 y_2 \vee x_1 \bar{x}_2 \bar{y}_1$$

Similarly, we can derive switching expressions for outputs z_1 and z_2. Switching expressions sometimes enable us to perceive algebraic relationships among internal variables or output values.

When we look at a given sequential network from the outside, we usually cannot observe the values of the internal variables, that is, the inside of the network (e.g., if the network is implemented in an IC package and no internal variables appear at its pins). Also, binary numbers are not very convenient to use. Therefore binary numbers for the internal states of the network may be replaced in a transition table by arbitrary letters or decimal numbers. The table that results is called a **state table**. For example, Table 7.3.5 is the

Table 7.3.5 State Table

s	00	01	11	10
A	(A)	(A)	C	B
B	(B)	A	C	(B)
C	(C)	D	D	(C)
D	—	A	(D)	A

S

Table 7.3.6 State-Output Table (Obtained from Table 7.2.2*c*)

s \ x_1x_2	00	01	11	10
A	(A), 10	(A), 10	D, 00	B, 00
B	(B), 10	A, 10	C, 10	(B), 10
C	B, 10	D, 10	(C), 10	(C), 10
D	A, 10	(D), 11	(D), 11	(D), 10

S, z_1z_2

state table for the transition table in Table 7.3.2. When a state table is combined with the output values, as shown in Table 7.3.6 (this is obtained from Table 7.2.2*c*), the result is called a **state-output table**.

When state-output tables such as Table 7.3.6 are redrawn in a diagram such as Fig. 7.3.4, the result is called a **state diagram** or a **transition diagram**. Because of their pictorial nature, state diagrams are sometimes preferred to

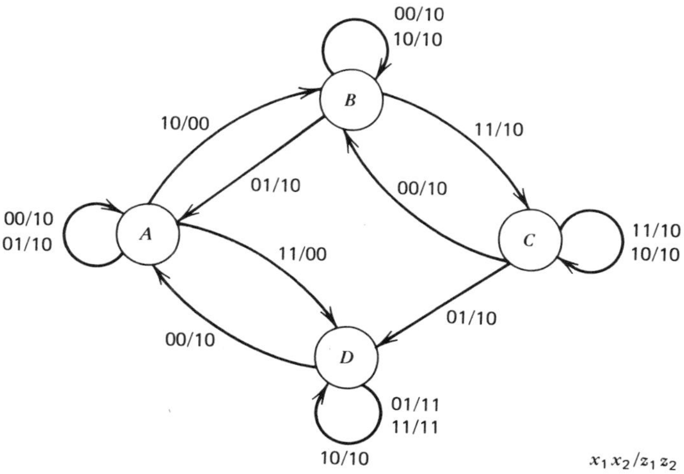

Fig. 7.3.4 State diagram.

Asynchronous Sequential Networks

state-output tables. The transitional relationship among states can be easily understood, and isolated states or transient states (i.e., states from which the network goes out but never returns) can be easily identified. Each transition from one internal state to another is represented by a line connecting circles that represent the internal states. The lines are labeled with the corresponding values of the inputs, a slash, and the corresponding values of the outputs.

Flow Table

In some state tables, networks go from one stable state to another through more than one unstable state, instead of exactly one unstable state. For example, in the state table in Table 7.3.5, the network goes from stable state $(x_1, x_2, s) = (1\ 1\ D)$ to another stable state, $(1\ 0\ B)$, by first going to unstable state $(1\ 0\ D)$, then to intermediate unstable state $(1\ 0\ A)$, and finally to stable state $(1\ 0\ B)$, when inputs (x_1, x_2) change from $(1\ 1)$ to $(1\ 0)$. Here state $(1\ 0\ A)$ is called an **intermediate unstable state**. Such a multiple transition from one stable state to another in a state table cannot be observed well from the outside of the network, and is not important as far as the external performance of the network is concerned. Even if each intermediate unstable state occurring during multiple transitions is replaced by the corresponding ultimate stable state, it does not make any difference if the network performance is observed from outside. Such a table is called a **flow table**. The flow table corresponding to the state table in Table 7.3.5 is shown in Table 7.3.7 [e.g., the intermediate unstable state A in $(x_1, x_2, s) = (1\ 0\ D)$ in Table 7.3.5 is replaced by B in Table 7.3.7].

Table 7.3.7 Flow Table Obtained from Table 7.3.5

s	$x_1 x_2$			
	00	01	11	10
A	Ⓐ	Ⓐ	D	B
B	Ⓑ	A	D	Ⓑ
C	Ⓒ	A	D	Ⓒ
D	—	A	Ⓓ	B
		S		

Malfunctions and Performance Descriptions of Sequential Networks 367

When a state table contains no multiple transition, the flow table is not different from the original state table.

When a flow table is combined with the corresponding output table, it is called a **flow-output table**.

State-output tables, flow-output tables, and state diagrams (we can draw state diagrams for flow-output tables also) are useful for concise description of the performance of sequential networks, since it is hard to understand network performance from complex diagrams of logic networks alone (particularly those with NAND, NOR, or MOS gates). It is often convenient in practice to accompany logic diagrams of networks with transition-output tables (or the corresponding switching expressions), state-output tables, flow-output tables, or state diagrams. (They are also important later for the synthesis of sequential networks.)

Remark 7.3.2: A sequential network whose output value depends on both the current state and the current values of the inputs is called a **Mealy machine**, whereas in another machine, called a **Moore machine**, all output values depend on the current state only. This terminology is used often in the abstract mathematical study of networks (see, e.g., Kohavi [1970]). ∎

Exercises

7.3.1 (E) Analyze the transient performance of the S-R latch in Fig. 7.1.1d by showing waveforms of the inputs and outputs in the same manner as in Fig. 7.1.2 (i.e., discuss how long new values of the inputs should be maintained for reliable operation, and what values Q and $\bar{Q}$ assume simultaneously).

7.3.2 (E) By looking at Fig. 7.1.1b, a student derived the switching expression

$$Y = S\bar{R}$$

to express the next value of the output Q of the S-R latch, when the current value of Q is y. He thought that his expression applies even when no input of the latch has the value 1 (so, the value of Q does not change).

Discuss whether he is correct. If you conclude that he is incorrect, derive a correct switching expression that is valid even when no input of the latch has the value 1.

7.3.3 (R) Suppose that the network expressed in Table 7.2.2a is in the stable state $(x_1, x_2, y_1, y_2) = (0\ 1\ 0\ 0)$. Show a sequence of inputs x_1 and x_2 that leads the network to stable state $(1\ 1\ 1\ 1)$. (Show this sequence by a path on this transition table.) Notice that inputs x_1 and x_2 cannot change simultaneously.

7.3.4 (R) We want to analyze the network of the S-R latches and gates in Fig. E7.3.4.

(i) Draw the transition-output table in fundamental mode.

(ii) Derive switching expressions that show the relationship among the present-state internal variables, the next-state internal variables, the input variables, and the corresponding output values.

(iii) Draw the state table and state diagram.

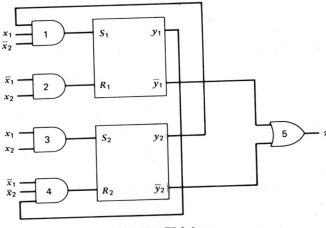

Fig. E7.3.4

7.3.5 (M—) In the flow table in Table E7.3.5, the network cannot enter the unstable state marked with an asterisk from stable state B unless inputs x_1 and x_2 make a simultaneous change.

Nevertheless we may not be able to delete this entry for reliable operation, depending on situation. Explain why.

Table E7.3.5

	\\ $x_1 x_2$			
s	00	01	11	10
A	Ⓐ	B	D	Ⓐ
B	A	Ⓑ	C	A*
C	D	Ⓒ	Ⓒ	A
D	Ⓓ	C	Ⓓ	A

S

7.3.6 (R) We want to analyze the network of the S-R latches and gates in Fig. E7.3.6.
(i) Draw the transition table in fundamental mode. When the network is initially placed in state $(x_1, x_2, y_1, y_2) = (0\ 0\ 0\ 0)$, examine whether there are some total states (x_1, x_2, y_1, y_2) that the network cannot enter, for any sequence of values of inputs x_1 and x_2. If there are such states, identify them.

Malfunctions and Performance Descriptions of Sequential Networks   369

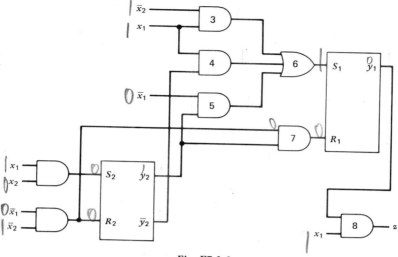

Fig. E7.3.6

(ii) Draw the output table.

Examine whether output z assumes spurious values during transition periods (i.e., when the network moves from one stable state to another, and the values of output are identical for these two stable states, z may assume the different value for the unstable state between them). If so, identify these values.

(iii) Derive switching expressions that show the relationships among the present-state internal variables, the next-state internal variables, the input variables, and the corresponding output values.

(iv) Draw the state table.

7.3.7 (M) For the network in Fig. E7.3.7:

(i) Derive the transition-output table.

(ii) Derive the flow-output table for this network.

(iii) Are there any static or dynamic hazards with this network due to the change in the value of y?

(iv) If there are, eliminate the hazards by modifying the network. Show two different networks (one may be obtained by the remedy discussed in Section 6.7).

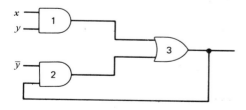

Fig. E7.3.7

7.3.8 (R) (i) Find all the races in the transition table in Table E7.3.8, and indicate whether they are critical or noncritical.
(ii) Then find remedies to eliminate the critical races.

Table E7.3.8

		\multicolumn{4}{c}{$x_1 x_2$}
		00	01	11	10
$y_1 y_2$	00	(00)	11	(00)	11
	01	00	11	11	(01)
	11	00	10	00	(11)
	10	(10)	(10)	00	(10)

$Y_1 Y_2$

7.3.9 (MT) For the network in Fig. E7.3.9:
(i) Derive the transition-output table.
(ii) Are there any races in the table? Are they critical or noncritical? If there is a critical race, try to find a remedy by adding gates or changing connections. (If you cannot find a simple remedy, you may need to redesign the entire network, as discussed in the rest of this chapter. If you feel that this will be necessary, do not spend too much time to find a remedy.) Every gate is assumed to have an equal switching time.
(iii) Derive the flow table.

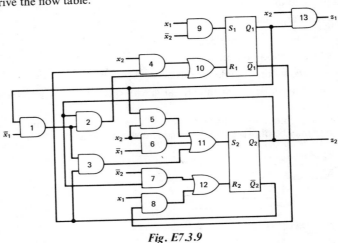

Fig. E7.3.9

7.3.10 (MT) Let us analyze the network in Fig. E7.3.10 with the two S-R latches of Fig. 7.1.1a.

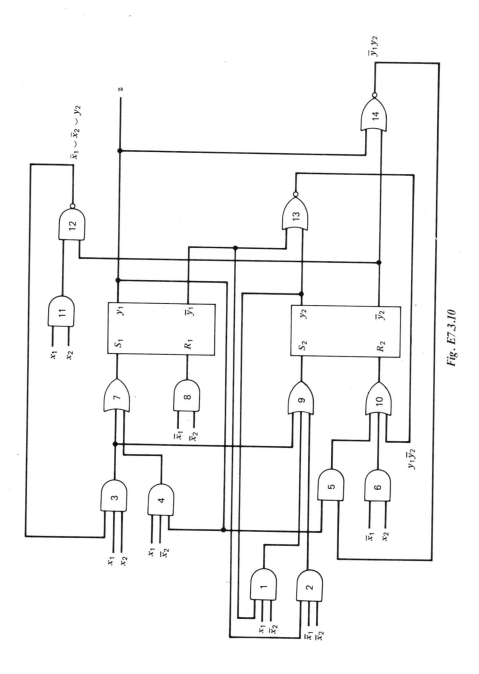

Fig. E7.3.10

372   Asynchronous Sequential Networks

(i) Derive the transition table, ignoring time delays in all the gates (i.e., derive switching expressions that express the outputs of gates 7, 8, 9, and 10, and derive the transition table by calculating the next states, based on these expressions).

(ii) Next, suppose that every gate (including those in the latches) has switching time τ. Then derive the transition table by deriving the output values of all the gates at every time interval of length τ. Especially observe that state $(x_1, x_2, y_1, y_2) = (1\ 1\ 0\ 0)$ is unstable, contradicting its being stable as found in (i); in other words, when inputs (x_1, x_2) of the network placed in stable state $(x_1, x_2, y_1, y_2) = (0\ 1\ 0\ 0)$ change into $(1\ 1)$, the network moves to a state different from $(1\ 1\ 0\ 0)$. (Recall that the S-R latch has two outputs, y_t and $\bar{y}_t$, which are simultaneously 0 for a short interval, as shown in Fig. 7.1.2b.)

7.3.11 (M) Switches or relay contacts that have mechanical springs usually do not close or open the contacts instantly. When a switch x connected to a network is closed at time t_1 (x becomes 1), the signal reaches the value 1 after the signal current fluctuates for a short while, as shown in Fig. E7.3.11a. When the switch is opened at time t_2 (x becomes 0), the signal reaches the value 0 after the fluctuation. This phenomenon, which is due to the metal spring's bouncing, is called **bouncing** or **chattering** (see, e.g., Harper [1977]). Often, this bouncing problem can be eliminated by inserting a simple logic network, as shown in Fig. E7.3.11b.

Design a network to eliminate bouncing, using NAND gates that have a switching time τ and assuming that bouncing does not last more than 3τ at each change in the value of x.

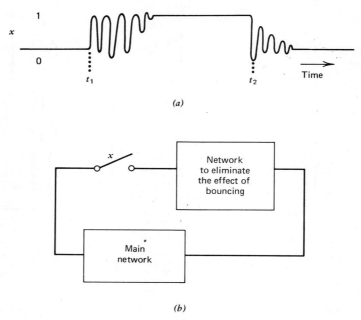

Fig. E7.3.11

Introduction to the Synthesis of Sequential Networks 373

7.3.12 (M—) Fig. E7.3.12 shows part of a network where NAND gates 5 and 6 constitute the $\bar{S}$-$\bar{R}$ latch shown with the dotted-line rectangle in Fig. 7.1.1d.

The designer intended that, when $y = 1$, the output Q of the latch does not change, no matter how x changes, since the output of gate 3 represents $x \vee y$.

Examine whether this is true. If the network has any possibility of malfunction, discuss it. Then discuss whether we can correct the design so that the network works as intended. Assume that each gate has the same switching time τ.

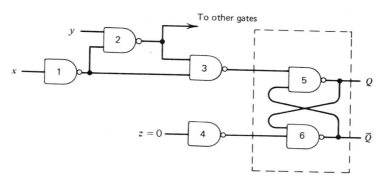

Fig. E7.3.12

7.3.13 (E) The network in Fig. E7.3.13, which consists of an S-R latch and a NOR gate, is called a **set-dominant S-R latch**.

(i) Find out what values Q and $\bar{Q}$ assume for each combination of values of x_1 and x_2, including the combination $x_1 = x_2 = 1$.

(ii) Derive a reset-dominant R-S latch, by modifying this set-dominant S-R latch.

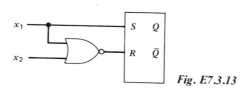

Fig. E7.3.13

7.4 Introduction to the Synthesis of Sequential Networks

Thus far, we have discussed the analysis of sequential networks. In the rest of this chapter, we will discuss the synthesis of sequential networks. Synthesis is essentially a reversal of the analysis discussed so far, but it is a considerably more complicated process (especially since networks without malfunctions due to the hazards discussed in Section 7.3 must be designed). Therefore, in this section a simple synthesis example is discussed with an introduction of the basic concepts, and then, in the following sections, more complex realistic examples are examined with a formal procedure for each step of the synthesis.

Word Statements for Performance Requirements

The performance requirements of a sequential network to be synthesized are usually given in a word statement. The performance requirements can be given in other forms such as truth tables, as mentioned in Section 1.4, but word statements are usually preferred, since in most cases the performance requirements are very complicated when expressed in other forms. As an example, let us consider a simple synthesis problem with the following word statement.

> **Word Statement Example 7.4.1:** We want to design a network that has two inputs, x_1 and x_2, and a single output, z, as shown in Fig. 7.4.1, and has the following performance:
>
> 1. Whenever x_1 and x_2 both assume value 0, output z assumes 0. After this, until $x_1 = x_2 = 1$ occurs, z assumes 0 when $x_1 = x_2 = 0$, and 1 when either $x_1 = 0, x_2 = 1$ or $x_1 = 1, x_2 = 0$.
> 2. Whenever x_1 and x_2 both assume the value 1, output z assumes 1. After this, until $x_1 = x_2 = 0$ occurs, z assumes 1 when $x_1 = x_2 = 1$, and 0 when either $x_1 = 0, x_2 = 1$ or $x_1 = 1, x_2 = 0$.
>
> *As illustrated in Fig. 7.4.2*, suppose that x_1 and x_2 both assume the value 0 at time t_1. Then z is 0 according to condition 1. At t_2, x_1 changes to 1. (Note that inputs are assumed not to change at the same time in Section 7.1, so even if one of x_1 and x_2 changes, the other does not change, at least for a while.) Then $z = 1$. At t_3, both x_1 and x_2 are 0 again, so $z = 0$ again, and so on. Condition 1 applies from time t_1 through t_5, and at t_5 both x_1 and x_2 assume value 1. Then, from t_5 through t_{11}, condition 2 applies, and z assumes value 1 only when both x_1 and x_2 assume value 1. Then, from t_{11}, condition 1 applies.

Let us design a network for this problem given in the word statement. If we want to reverse the analysis done in Section 7.2, we should first convert the above word statement into a transition-output table. But we do not know yet how many internal variables are required, what binary numbers these

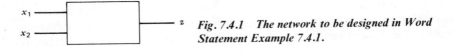
Fig. 7.4.1 The network to be designed in Word Statement Example 7.4.1.

Introduction to the Synthesis of Sequential Networks 375

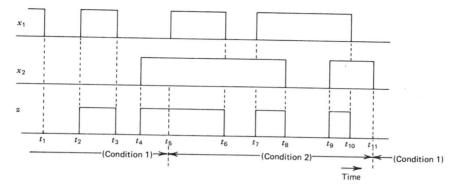

Fig. 7.4.2 Input changes that can occur for the network in Fig. 7.4.1

internal variables should assume, and so on. In other words, reversal of the analysis is not immediately possible. First, we need to have a general model for sequential networks as follows, and then we need to determine, one by one, the parameters of such a general model of a sequential network according to the given word statement.

General Model of Sequential Networks

A sequential network in fundamental mode may be generally expressed in a schematic form, as shown in Fig. 7.4.3. A large block represents a loopless network of gates only (without any flip-flops). All loops and flip-flops (i.e., both loops that do not contain flip-flops and loops that contain flip-flops) are drawn outside the large block.*

Notice that Fig. 7.4.3, where flip-flops are shown on some loops, is a special case of the general model where no flip-flops are on any loops. This is so because the flip-flops can be decomposed into gates and loops, and these gates can be placed inside the loopless network in Fig. 7.4.3, leaving the loops outside. But we will use the model in Fig. 7.4.3, where flip-flops are shown on some (or all) loops, because we want to deal with flip-flops differently from gates, as we will see in the synthesis procedure later. We shall regard flip-flops as basic building blocks of networks, like gates, because

---

* Sometimes for abstract mathematical studies, the delays in the gates inside the loopless network are pushed to the outside of the network. In other words, the delays in these gates are summarized on the loops, and the switching time in each gate in the loopless network is assumed to be zero. For this idealized model the loops shown in Fig. 7.4.3 contain a time delay. But for our design purposes this model will not be used, because the network performance due to this idealized time delay is sometimes too different from reality.

376 *Asynchronous Sequential Networks*

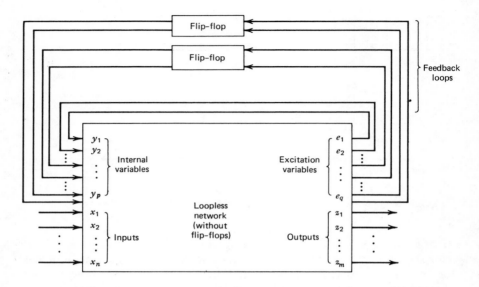

Fig. 7.4.3   A general model of a sequential network in fundamental mode.

keeping memory elements concentrated in flip-flops rather than in many gates connected in loops is advantageous in terms of the reliable operation and diagnosis of sequential networks, as we will see.

This loopless network has input variables $x_1, \ldots, x_n$, and internal variables $y_1, \ldots, y_p$, as its inputs. It also has output variables $z_1, \ldots, z_m$, and **excitation variables** $e_1, \ldots, e_q$, as its outputs. Some of the excitation variables $e_1, \ldots, e_q$ are inputs to the flip-flops; they serve to excite the flip-flops. The remainder of the excitation variables are starting points of the loops without flip-flops, which end up at some of the internal variables. For loops without flip-flops, $e_i = y_i$ holds for each i. As an example, the network in Fig. 7.2.1 is redrawn in the form of Fig. 7.4.3 in Fig. 7.4.4. In this particular example every feedback loop has a latch.

Synthesis as a Reversal of Network Analysis

Since the network in the general model in Fig. 7.4.3 has no loop within it, it is a combinational network, discussed in the preceding chapters. The outputs $e_1, \ldots, e_q, z_1, \ldots, z_m$ of this network express switching functions that have $y_1, \ldots, y_p, x_1, \ldots, x_n$ as their variables. Thus, if these switching functions are given, the loopless network can be designed by the methods

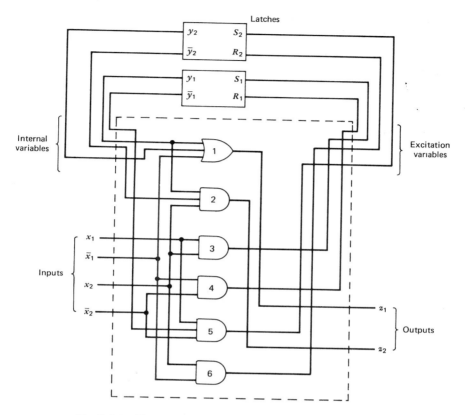

Fig. 7.4.4 *The network of Fig. 7.2.1 in the form of Fig. 7.4.3.*

discussed in the preceding chapters. This means that we have designed the sequential network, since the rest of the general model in Fig. 7.4.3 is simply loops with or without flip-flops. But we cannot derive these switching functions $e_1, \ldots, e_q, z_1, \ldots, z_m$ directly from the above word statement of the given design problem, so let us find, in the following, what gap to fill in.

Among these switching functions $e_1, \ldots, e_q, z_1, \ldots, z_m$, for the moment consider only $e_1, \ldots, e_q$ ($z_1, \ldots, z_m$ will be considered later). A table that shows the relationship between $e_1, \ldots, e_q$ and $y_1, \ldots, y_p, x_1, \ldots, x_n$ is called an **excitation table**. An excitation table is prepared in the same format as is a transition table, except that its entries are values of inputs to the flip-flops, instead of $Y_1, Y_2, \ldots$. For example, if $S_1 = x_1 x_2$, $R_1 = \bar{x}_1 \bar{x}_2$, $S_2 = x_1 \bar{x}_2 \bar{y}_1$, and $R_2 = \bar{x}_1 x_2$ are given as $e_1, \ldots, e_q$ as switching functions

Table 7.4.1 Excitation Table for part of the Network in Fig. 7.2.1

	$x_1 x_2$			
$y_1 y_2$	00	01	11	10
00	01, 00	00, 01	10, 00	00, 10
01	01, 00	00, 01	10, 00	00, 10
11	01, 00	00, 01	10, 00	00, 00
10	01, 00	00, 01	10, 00	00, 00

$S_1, R_1, S_2 R_2$

of $y_1, \ldots, y_p, x_1, \ldots, x_n$, for Fig. 7.4.4 (i.e., Fig. 7.2.1), then we have the excitation table shown in Table 7.4.1. If these switching expressions or this excitation table is given, we can design part of the network (i.e., the subnetwork with outputs S_1, R_1, S_2, R_2 and inputs $y_1, \bar{y}_1, y_2, \bar{y}_2, x_1, \bar{x}_1, x_2, \bar{x}_2$, which consists of gates 3 through 6) in Fig. 7.4.4.

An excitation table with output values is called an **excitation-output table**.

For the performance description of a sequential network, the changes of states or internal variables are more directly important than excitation variables $e_1, \ldots, e_q$ because, if we know what the current state is, we can predict the next state for any input change, whereas excitation variables are of a transient nature (i.e., their values are usually not kept long, and they may or may not change the state of a network). Therefore a transition table is more directly important than an excitation table, in terms of describing the network performance requirements. But unless binary numbers are known for states, we cannot draw a transition table. Hence we need to start with a state table. It is not wise, however, to concern ourselves with intermediate unstable states in a state table, if any, in the beginning of the synthesis, so we had better start with a flow table.

In conclusion, in designing a sequential network, we take the reverse of the above sequence—in other words, a given word statement is converted into a flow-output table, then a state-output table, a transition-output table, and finally an excitation-output table. Then, once we have an excitation-output table, we can design a combinational network as the loopless network in the general model in Fig. 7.4.3, by the methods discussed in the preceding

Introduction to the Synthesis of Sequential Networks

chapters. Since the other part of the general model is prefixed, this completes the synthesis of a sequential network with the given performance requirements.

Synthesis Example

Now let us design a sequential network for Word Statement Example 7.4.1. The synthesis will be done in the following sequence: the word statement—a flow-output table—a state-output table—a transition-output table—an excitation-output table—logic design of a combinational network by the methods in the preceding chapters. Here each step requires consideration of certain aspects.

First, let us translate the word statement into a flow-output table. Since the value of z is different for conditions 1 and 2 when exactly one of x_1 and x_2 assumes the value 1, the network must have at least two internal states. First, let us try to construct a flow-output table using exactly two states. (If two states turn out to be insufficient for expressing the required network performance, we will try three or more states later.) Let A and B be the two states corresponding to conditions 1 and 2, respectively.

From the required performance given above, we can derive the flow-output table shown in Table 7.4.2. If the network has inputs $(x_1, x_2) = (0\ 0)$, it works under condition 1. Thus the network will be in row A and in column $(x_1, x_2) = (0\ 0), (0\ 1)$, or $(1\ 0)$, until the inputs (x_1, x_2) become $(1\ 1)$. Thereafter, the network works under conditions 2 and remains in row B until the inputs (x_1, x_2) become $(0\ 0)$. In each cell in Table 7.4.2, the corresponding value of z is entered. This flow-output table expresses the required performance, so two internal states are sufficient.

We can check whether this table satisfies the performance requirements given in Word Statement Example 7.4.1, by drawing the state diagram in Fig. 7.4.5. If the table contains errors, we may be able to easily detect them

Table 7.4.2 Flow-Output Table

Condition	State s	Inputs x_1 and x_2			
		00	01	11	10
1	A	A, 0	A, 1	B, 1	A, 1
2	B	A, 0	B, 0	B, 1	B, 0

Next state S, output z

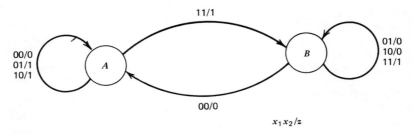

Fig. 7.4.5 *State diagram.*

by the state diagram. Generally, the simultaneous preparation of state diagrams tends to facilitate the preparation of flow-output tables.

Now we need to convert the flow-output table into a state-output table. Since the table consists of only two states, there is no intermediate unstable state like A in $(x_1, x_2, s) = (1\ 0\ D)$ in Table 7.3.5. The state-output table is identical to the flow-output table. (More complex examples in which state-output tables are different from flow-output tables will be shown in Section 7.7.)

Next, we need to convert the state-output table into a transition-output table. To do this, we must assign binary numbers to states as internal variables. This is called the **state assignment**. Since we have only two rows in Table 7.4.2, binary numbers with a single bit are sufficient to differentiate these rows, so let us assign $y = 0$ to A and $y = 1$ to B. (We also can assign $y = 1$ to A and $y = 0$ to B, yielding a different transition-output table.) We then have the transition-output table shown in Table 7.4.3.

Now we need to convert the transition-output table into an excitation-output table. At this stage, however, we have to decide whether we should use S-R latches on the feedback lines in the general model in Fig. 7.4.3. Since

Table 7.4.3 Transition-Output Table

		Inputs x_1 and x_2			
		00	01	11	10
y	0	0, 0	0, 1	1, 1	0, 1
	1	0, 0	1, 0	1, 1	1, 0

Y, z

Table 7.4.4 Input-Output Relationship of an S-R Latch

S	R	y	Y
0	0	0	0
0	0	1	1
1	0	0	1
1	0	1	1
0	1	0	0
0	1	1	0

Table 7.4.5 Output-Input Relationship of an S-R Latch

y	Y	S	R
0	0	0	d
0	1	1	0
1	0	0	1
1	1	d	0

y: Current output value.
Y: Next output value.

the use of latches on the feedback lines is probably more common than their omission, let us use a latch to store the internal variable y. (We can synthesize a sequential network without latches, however, as we will discuss later in this section.)

An excitation table shows the relationship between the current value y of the internal variable and the inputs S and R of the latch, whereas the transition table shows the relationship between the current value y and the next value Y. Hence we need to derive the output-input relation of the S-R latch, as shown in Table 7.4.5, by reversing the input-output relationship in Table 7.4.4. In Table 7.4.5 d's mean don't-care conditions. For example, $y = Y = 0$ in Table 7.4.5 results from $S = R = 0$ and also from $S = 0$, $R = 1$ in Table 7.4.4. Therefore the first row in Table 7.4.5, $y = Y = 0$, $S = 0$, $R = d$, is obtained, since $y = Y = 0$ results from $S = 0$ only, and R can be 0 or 1, that is, don't-care, d. By using Table 7.4.5, the transition-output table in Table 7.4.3 is converted into the excitation-output table in Table 7.4.6. For example, corresponding to $y = Y = 0$ in the first row and the first column in Table 7.4.3, $S = 0$, $R = d$ is obtained in the first row and the first column of Table 7.4.6, because the first row in Table 7.4.5 corresponds to this case.

As the excitation-output table shows the relationship between the inputs and outputs of the combinational network without loops inside the rectangle in the general model in Fig. 7.4.3, we can now design the combinational network by the methods discussed in the preceding chapters. Then Table 7.4.6

382 Asynchronous Sequential Networks

Table 7.4.6 Excitation-Output Table

	x_1x_2			
	00	01	11	10
y 0	0d, 0	0d, 1	10, 1	0d, 1
y 1	01, 0	d0, 0	d0, 1	d0, 0

SR, z

is divided into the three Karnaugh maps shown in Table 7.4.7, to find a minimal sum for each of S, R, and z in Table 7.4.7. [The map method is an appropriate approach, since Karnaugh maps can be easily obtained from the excitation-output table. Strictly speaking, a minimal multiple-output network should be designed (e.g., a two-level network in Section 5.2). But for the sake of simplicity, let us design a minimal two-level network for each output function, separately.] Thus we have obtained the network shown in Fig. 7.4.6, in the form of the general model of Fig. 7.4.3.

Table 7.4.7 Karnaugh Maps Derived from Table 7.4.6

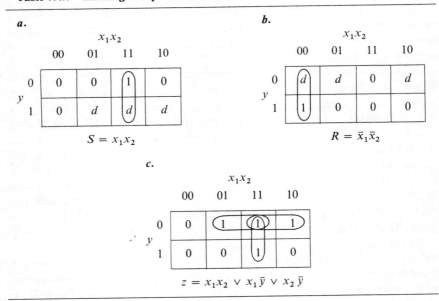

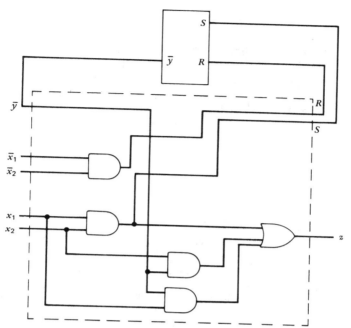

Fig. 7.4.6 *Network designed with a latch for Word Statement Example 7.4.1.*

Now let us design a sequential network for Word Statement Example 7.4.1, without flip-flops on any loops in the general model in Fig. 7.4.3. As explained already, we need not prepare an excitation table. Directly from the transition-output table in Table 7.4.3, Karnaugh maps for Y and z are obtained in Table 7.4.8. Finding a minimal sum for each of Y and z, we have obtained the network in Fig. 7.4.7. As mentioned already, networks designed

Table 7.4.8 Karnaugh Maps Derived from Table 7.4.3

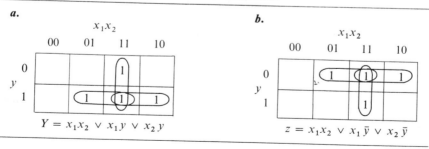

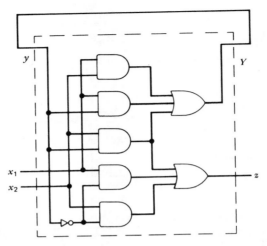

Fig. 7.4.7 Network designed without a latch for Word Statement Example 7.4.1.

without flip-flops are likely to malfunction, since new values of internal variables tend to feed back to the inputs of the networks that are still in transition, and also the networks tend to contain extrinsic hazards (including races). We need to carefully examine whether the designed networks work without malfunctioning, often with electronic prototypes, though simple networks such as the one in Fig. 7.4.7 can be easily found not to malfunction. Remedies for malfunctioning will be discussed later in this chapter.

Design Steps for Synthesis of Sequential Networks

As we explained above, the word statement of the required performance of a network to be designed is first converted into a flow-output table, as illustrated in Fig. 7.4.8. Then this table is converted into a state-output table, and next into a transition-output table, by choosing an appropriate state assignment. Then the transition-output table is converted into an excitation-output table if the loops contain flip-flops. If the loops contain no flip-flops, the excitation-output table need not be prepared, since it is identical to the transition-output table. Finally, a network is designed, using the logic design procedures discussed in the preceding chapters.

The network designed in the above example is simple (in particular, it has only one latch). When sequential networks are complex (particularly when there are many latches), some problems arise, such as malfunction of operation and minimization of the number of latches. In the rest of this chapter,

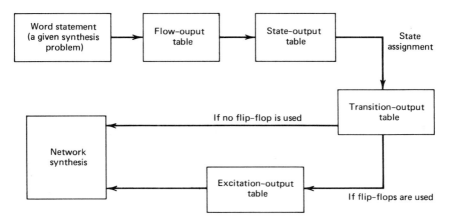

Fig. 7.4.8   Steps for designing a sequential network.

we will discuss these problems in detail, and then design procedures that avoid these difficulties.

7.5   Translation of Word Statements into Flow-Output Tables

Usually the performance of a sequential network to be synthesized is given in a word statement. (As discussed in Section 1.4, performances of simple sequential networks can be expressed in other forms, but word statements are probably most convenient for most sequential networks because of the complexity of their performances.) For the synthesis of sequential networks, the word statements must be converted into more formal expressions that are convenient for synthesis, that is, flow-output tables.

In this section we discuss the translation into flow-output tables of more complicated word statements than the simple example in Section 7.4, as a starting point of the systematic design procedure of sequential networks. (In many cases in practice, networks are synthesized directly from a word statement without going through a flow-output table. The design procedure by heuristic approaches to be discussed in Chapter 9, and Exercise 7.19, are examples.)

The translation of a given word statement into a flow-output table is difficult to describe in a formal and systematic manner, since a word statement itself is not a rigorous formal way to describe a network performance

386 Asynchronous Sequential Networks

and there are many word statement expressions to describe the same performance. The best approach to explain the translation is the presentation of an example.

Example 7.5.1—A sequential network with two inputs x_1 and x_2, and one output z in fundamental mode: We want to design a sequential network in fundamental mode that has two inputs, x_1 and x_2, and one output z and has the following performance:

1. Whenever signal value 1 is repeated at the same terminal where this signal value appeared last time, the output is $z = 1$ only during the presence of input signal value 1. Otherwise $z = 0$ (i.e., when signal value 1 alternates between two different inputs or when 1 is not present at any input).

2. $x_1 = x_2 = 1$ cannot occur at any time.

Generally flow-output tables for fundamental-mode networks are cumbersome to prepare. To simplify the task, flow-output tables can be prepared in a special format, called the **primitive form**. In a fundamental-mode flow-output table in the primitive form, there is only one stable state in each row.

Table 7.5.1 A Flow-Output Table in Primitive Form for Example 7.5.1

a.

Last 1 at		$x_1 x_2$ 00	01	11	10
x_1	A	Ⓐ, 0			
x_2	B	Ⓑ, 0			

S, z

b.

Last 1 at		$x_1 x_2$ 00	01	11	10
x_1	A	Ⓐ, 0	C		
x_2	B	Ⓑ, 0	D		
	C		Ⓒ, 0		
	D		Ⓓ, 1		

S, z

Table 7.5.1 (*Continued*)

c.

Last 1 at		$x_1 x_2$ 00	01	11	10
x_1	A	Ⓐ, 0	C	—	E
x_2	B	Ⓑ, 0	D	—	F
	C		Ⓒ, 0	—	—
	D		Ⓓ, 1	—	—
	E		—	—	Ⓔ, 1
	F		—	—	Ⓕ, 0

S, z

d.

Last 1 at		$x_1 x_2$ 00	01	11	10
x_1	A	Ⓐ, 0	C	—	E
x_2	B	Ⓑ, 0	D	—	F
	C	B	Ⓒ, 0	—	—
	D	B	Ⓓ, 1	—	—
	E	A	—	—	Ⓔ, 1
	F	A	—	—	Ⓕ, 0

S, z

The flow-output table in the primitive form may contain unnecessary states, but these states can be removed by the procedure discussed in the next section, though this elimination may be time-consuming for large tables. Flow-output tables in other forms can also be prepared, but those in the primitive form for the above performance requirement are easier to prepare, since only one stable state needs to be considered in each row. Of course, if flow-output tables are simple enough, we do not need to use the primitive form.

388 Asynchronous Sequential Networks

Let us prepare the flow-output table in the primitive form for the above performance requirement.

1. As shown in Table 7.5.1a, column $(x_1, x_2) = (0\ 0)$ must have at least two stable states, that is, one state corresponding to the last signal 1 at x_1 and the other corresponding to the last signal 1 at x_2, because the value of z at the occurrence of the next signal 1 is determined by considering which input terminal had the last signal 1. Let us denote these states by A and B. Encircle them as stable states. From the performance requirement, $z = 0$ for these states, since no signal 1 is present at any input.

2. Suppose that the network is in row A and column $(x_1, x_2) = (0\ 0)$. When the next signal 1 appears at x_2, the next state in column $(x_1, x_2) = (0\ 1)$ in row A cannot be A or B, respectively, because the primitive form requires only one stable state in each row. (If we have A in row A, this will be the second stable state in row A. If we have B in row A, we need stable state B in row B in this column, and row B will have two stable states.) Therefore let us introduce a new unstable state C in row A, as shown in Table 7.5.1b. Accordingly, in this column there must be a stable state C in a new row labeled as C. Let us encircle this C. For this entry, z must be 0, since the signal 1 alternates between x_1 and x_2.

Next, for the same reason, the next state in column $(x_1, x_2) = (0\ 1)$ in row B cannot be A or B but can be either C or a new state D. If C is chosen, z will be 0. This is a contradiction, however, since $z = 1$ must hold when signal 1 appears again at x_2. Therefore the new unstable state D must be chosen. Then there must be a new row labeled as D which contains a stable state D encircled. Accordingly, z must be 1, since signal 1 is repeated at x_2. Thus Table 7.5.1b has been obtained.

3. The next states, E and F, in column $(x_1, x_2) = (1\ 0)$ are entered in Table 7.5.1c for the same reason. A dash is entered in column $(x_1, x_2) = (1\ 0)$ in row C because the transition of the network from column $(x_1, x_2) = (0\ 1)$ in the same row (this is only way to enter this state) requires a double change of inputs from $(x_1, x_2) = (0\ 1)$ to $(1\ 0)$ and consequently cannot take place. All the other dashed entries in columns $(x_1, x_2) = (1\ 0)$ and $(0\ 1)$ in Table 7.5.1c must be dashes for the same reason. All the entries in column $(x_1, x_2) = (1\ 1)$ are dashes, since this column is assumed not to occur.

4. In column $(x_1, x_2) = (0\ 0)$, the entries in rows C and D must be B, because they can be reached only from the two stable states C and D in column $(x_1, x_2) = (0\ 1)$, and the network must settle in stable states with the last signal 1 at x_2 after signal 1 at x_2 disappears. Similarly, the next states in rows E and F in column $(x_1, x_2) = (0\ 0)$ must be A. Thus we have obtained Table 7.5.1d.

5. Let us assign the values of z for unstable states.

One possibility is to assign to each unstable state the value z of the stable state to which the network goes, as shown in Table 7.5.2. For example, z in

Translation of Word Statements into Flow-Output Tables

Table 7.5.2 A Flow-Output Table in Primitive Form for Example 7.5.1

s	00	01	11	10
A	Ⓐ, 0	C, 0	—	E, 1
B	Ⓑ, 0	D, 1	—	F, 0
C	B, 0	Ⓒ, 0	—	—
D	B, 0	Ⓓ, 1	—	—
E	A, 0	—	—	Ⓔ, 1
F	A, 0	—	—	Ⓕ, 0

$x_1 x_2$ (column header); S, z (bottom)

$(x_1, x_2, s) = (0\ 1\ B)$ is assigned 1, since $z = 1$ in the stable state $(x_1, x_2, s) = (0\ 1\ D)$.

There are other possibilities for entering the values of z for unstable states. Let us describe an assignment that will usually yield the best result later. Consider the unstable state $(x_1, x_2, s) = (1\ 0\ B)$, for example. The network can enter this state only from the stable state $(x_1, x_2, s) = (0\ 0\ B)$ in the same row. And the only state which the network can reach directly from the unstable state $(x_1, x_2, s) = (1\ 0\ B)$ is the stable state $(x_1, x_2, s) = (1\ 0\ F)$. In other words, the unstable state $(x_1, x_2, s) = (1\ 0\ B)$ is sandwiched between two stable states, $(0\ 0\ B)$ and $(1\ 0\ F)$. These two stable states both have the same output value, $z = 0$. Therefore z at the unstable state $(1\ 0\ B)$ must be 0, because, if $z = 1$, the network has $z = 1$ only during a short transition period whose duration is determined by electronic circuit parameters, and accordingly the network shows a spurious output. In other unstable states also, if the values of z at the preceding and succeeding stable states are identical, z at an unstable state must also be identical to this value. On the other hand, if two values of z at the stable states differ, z at the unstable state can have any value because $z = 0$ or 1 is simply a matter of the appearance of the next output value a little earlier or later. Thus we enter d as the value of z. The z for $(1\ 0\ A)$ is such an example. Whether $z = 0$ or 1, the network does not show any spurious output. Thus we have obtained Table 7.5.3.

As we will see later, keeping d in such unstable states is advantageous, since it adds flexibility in designing networks (recall that don't-care conditions d add design flexibility in using Karnaugh maps).

Table 7.5.3 A Flow-Output Table in Primitive Form for Example 7.5.1 (with don't care for some z).

s	x_1x_2 00	01	11	10
A	Ⓐ, 0	C, 0	—	E, d
B	Ⓑ, 0	D, d	—	F, 0
C	B, 0	Ⓒ, 0	—	—
D	B, d	Ⓓ, 1	—	—
E	A, d	—	—	Ⓔ, 1
F	A, 0	—	—	Ⓕ, 0
				S, z

Physical Interpretation of States

When a flow-output table is being obtained, we need not be concerned about minimization of the number of rows, at least at this stage of the design procedure. Even if the table contains unnecessary rows, they can be eliminated by the formal procedure described in Section 7.6. Therefore it is most important to derive a flow-output table accurately from a given word statement, without worrying about minimization of the number of rows in the table. However, if a flow-output table with as few rows as possible can be easily obtained, it is very desirable, since the efforts involved in using the procedure in Section 7.6, which are very great for large tables, can be avoided. For this purpose the following considerations are important.

An appropriate physical interpretation of states is a key point in preparing flow-output tables, though it is easy to find in many cases. In Example 7.5.1 states A and B are physically interpreted to mean "last signal 1 at x_1" and "last 1 at x_2," respectively. But we could start with other physical interpretations of states, since a network generally has numerous parameters. Appropriate physical interpretation of states is very important to keep the size of a flow-output table small, since if we start the preparation of a flow-output table with inappropriate physical interpretation of states [e.g., if "previously we had the sequence (1 0 1 0) at x_1" is used to mean state A], the flow-output table may become much larger than it needs to be.

When the performance requirement of a sequential network to be designed is given, we are not always able to write a flow-output table corresponding to the requirement. Some requirements may be contradictory; even if not contradictory, they may be impossible to implement physically. For example, network performance requirements that are not contradictory cannot be implemented if an infinite number of internal states are required, or if the output values of the networks are determined by future values of inputs.

Remark 7.5.1: Flow-output tables are essentially a representation of a timing chart of state-output changes in the form of a table. Of course, we can design a loopless network based on the initial part of a timing chart and modify it depending on how the timing chart changes. In contrast to this heuristic approach, the design approach based on the flow-output tables in this chapter may be regarded as the simultaneous consideration of all aspects of a timing chart. The decision as to which approach is better depends on the complexity of the networks and the time required for design. ■

Exercises

7.5.1 (R) In Fig. 7.2.1, $\bar{y}_1$ is fed back to the input of gate 3. At first thought, this looks as if it requires an extra feedback loop, in addition to the two loops with the flip-flops outside the loopless network in the general model in Fig. 7.4.3. But actually there is no need for an extra loop, as shown in Fig. 7.4.4.

For each of the following cases, state whether extra loops are needed in addition to the two loops with the flip-flops already shown in Fig. 7.4.4:

(i) When we add the connection from the output of gate 2 to the input of gate 5, and the connection from the output of gate 5 to the input of gate 4.

(ii) When we insert an extra gate in the connection from $\bar{y}_1$ to the input of gate 3.

7.5.2 (R) We want to design a network with two inputs, x_1 and x_2, and one output z. When $z = 0$, z changes to 1 at $(x_1, x_2) = (01)$ only, and when $z = 1$, z changes to 0 at $(x_1, x_2) = (10)$ only. But z does not change its value at any other time.

(i) Show a flow-output table with as few states as possible in fundamental mode. Draw a state diagram.

(ii) Design the network with an S-R latch (or S-R latches), and, if necessary, any types of gates.

7.5.3 (R) We want to design a network whose output z changes at each occurrence of signal 1 at its input x.

(i) Show a flow-output table with as few states as possible, in fundamental mode. Draw a state diagram.

392   Asynchronous Sequential Networks

(ii) At the network output z, we need to show only two output states, $z = 0$ and 1. If your flow-output table requires more internal states (at first thought, this looks like a contradiction), explain why.

(iii) Design the network with S-R latches and, if necessary, any types of gates. (Examine whether the network works as required.)

7.5.4 (M) An **up-down counter** is a counter that has signal input x and control input t. Its count increases by 1 at each occurrence of $x = 1$, if $t = 0$, and decreases by 1 at each occurrence of x, if $t = 1$. Assume that t can change only during the absence of $x = 1$. We want to design such a counter whose count is displayed by two outputs, z_1 and z_2, where z_1 is the least significant bit.
Prepare a flow-output table in fundamental mode, using as few states as possible. Show a state diagram.

7.5.5 (M) We want to design a network in fundamental mode with three inputs, x_1, x_2, and x_3, whose output value is 1 only when signal 1 appears at these inputs in sequence in the order of x_1, x_2, and x_3. Assume that signal 1 can appear at only one of these inputs, and that once signal 1 appears at one of these inputs, signal 1 does not appear for a while until it appears again at some input.
Prepare a flow-output table with as few states as possible. Show a state diagram.

7.5.6 (M) Suppose that a candy vending machine has a slot to accept only a nickel or a dime. When one dime or two nickels are put into the slot, a candy is released. If the total sum of deposited coins exceeds 10 cents (i.e., a dime is put in after a nickel), the deposited dime must be returned. We want to design a network with two inputs, x_1 and x_2, and two outputs, z_1 and z_2, to control the vending machine in this manner. Input x_1 or x_2 is 1 only when a nickel or a dime, respectively, passes the slot, and output z_1 and z_2 becomes 1 only when a candy or a deposited dime, respectively, is to be released.
Prepare a flow-output table in fundamental mode. (This need not be in primitive form.) Show a state diagram.

7.5.7 (M) We want to design a network, for parity checking, which has signal input x, reset input t, and output z. Whenever $t = 1$, z is reset to 0. After t is set to 0 after this resetting, z changes its value at each occurrence of $x = 1$. If we have $t = x = 1$ and then t becomes 0, still keeping $x = 1$, then z becomes 1, and after this, z changes its value whenever $x = 1$ reappears while $t = 0$.

(i) Derive a flow-output table in primitive form, using as few states as possible.

(ii) Derive a flow-output table, ignoring the primitive form (i.e., each row can have two or more stable states), and using as few states as possible.

(iii) Design a network based on the table obtained in (ii), using S-R latches. Use as few extra gates of any types as possible. (Binary numbers can be assigned to the states in an arbitrary manner.)

7.5.8 (M) Find out whether you can implement a flow-output table with fewer states than Table 7.5.2 has, if the requirement for the table to be in primitive form is ignored. If you can, show such a table.

7.5.9 (M + T) Suppose that you are a logic designer working with an elevator-installation company. You are assigned to design a network to control an elevator for a three-story building. The second floor has a pair of buttons U_2 and D_2 for "up" and "down," respectively. The first and third floors have buttons U_1 and D_3 for "up" and "down," respectively. An elevator motor is controlled by three S-R latches for "move upward," "move downward," and "stop" (e.g., if the "move upward" latch is on, the elevator moves upward). Hence at any time exactly one of the three latches is on. The elevator has an input to sense the floor position at each floor.

Inside the elevator we need three buttons, E_1, E_2, E_3, for the three floors. (Ignore buttons for "door open" or "door close" for simplicity.)

We want to design a digital network to control the elevator by utilizing appropriately and economically the electric signals from buttons and floor sensors.

Assume that, once the elevator starts to move, it cannot move in the other direction unless it stops at the next floor.

When the elevator moves up (or down), it must stop at floors it passes, if there are up (or down) requests at these floors. But if the requests are down (or up), that is, in the opposite direction, it does not stop.

Any conditions that are not explicitly stated above but, you judge, are necessary can be arbitrarily set up according to your commonsense judgment.

Write a flow-output table in fundamental mode, using as few states as possible. In this case, if we use $U_1 \vee U_2$, instead of U_1 and U_2 separately, to represent table columns (similarly, $D_2 \vee D_3$), then the table may be smaller. Consider this possibility. (In other words, unlike the discussions in the text, network inputs need not be directly used to represent table columns. There may be other possibilities to reduce the table size.)

Remark: When networks to be designed have complex performance requirements, flow charting may be helpful, as will be discussed in Chapter 9. ∎

7.6 Minimization of the Number of States

In Section 7.5 we discussed how to prepare a flow-output table from a given performance requirement of a sequential network to be synthesized. For any given performance requirement, it is possible to obtain many different flow-output tables (i.e., flow-output tables are not unique). Although these tables are different, all of them represent the same given network performance; this is true for example, of two tables in Tables 7.6.1a and 7.6.1b. In Table 7.6.1a the network that has input x moves between the states, A and B, whenever signal values 1 and 0 alternate at input x of the network. In Table 7.6.1b the network moves among four states, C, D, E, F, going to a next state whenever signal values 1 and 0 alternate at input x. If the network in Table 7.6.1a is initially placed in state A and the network in Table 7.6.1b is initially placed in state C, the two networks have always the identical outputs 1 and 0 alternately, whenever 1 and 0 alternate at network input x. The two flow-output tables or state diagrams in Tables 7.6.1a and 7.6.1b look different, but the two networks represented by them show identical performances. These two networks are said to be **indistinguishable**. Also, these two flow-output tables or state diagrams are said to be **indistinguishable**.

Table 7.6.1 Indistinguishable Flow-Output Tables and State Diagrams

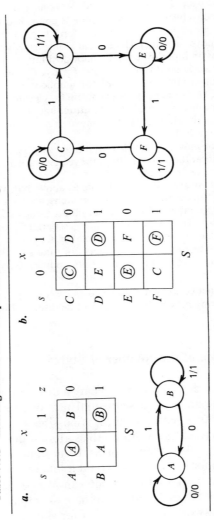

(A formal definition is given later.) Whenever the network in Table 7.6.1*b* is in state *C* or *E*, the network in Table 7.6.1*a* is always in state *A*. And whenever the former is in state *D* or *F*, the latter is always in state *B*. In this sense, state *A* is **indistinguishable** from *C* or *E*, and then *C* and *E* are indistinguishable. Also, state *B* is indistinguishable from *D* or *F*. States *D* and *F* are also indistinguishable. (A formal definition is given later.) Here notice that, if these two networks are initially placed in other states, that is, they are placed for example, in initial states *A* and *D* (or *F*), respectively, they are **distinguishable**, in other words, they show different network performance. **When we discuss the indistinguishability of sequential networks, initial states must be specified.**

Significance of the State Minimization

In this section we want to derive a flow-output table that has a minimum number of states and is still indistinguishable from any flow-output table obtainable for a given performance requirement of a sequential network. If we synthesize a network based on this flow-output table with a minimum number *R* of states, we will have a network that has a minimum number of internal variables, $\lceil \log_2 R \rceil$, where $\lceil x \rceil$ denotes the smallest integer greater than or equal to x. A network that can be synthesized from this flow-output table with a minimal number of states is not unique; as a matter of fact, many networks can be synthesized. But all of these networks have the same minimum number of internal variables, $\lceil \log_2 R \rceil$. Since internal variables correspond to flip-flops or loops in the case of electronic gate networks and to internal relays in the case of relay contact networks, the networks have the minimum number of flip-flops, loops, or internal relays. This is a desirable feature, since flip-flops are complex,* and also since diagnosis of sequential networks is important in IC implementation of networks but is very time-consuming (if a network contains fewer flip-flops, the diagnosis is easier).

If we want to minimize the total number of gates or relay contacts in a network, however, no matter how many states the network has, no efficient procedure is known to date. **If we synthesize a network based on a flow-output table with more states than the minimum in the flow-output table explained above, the total number of gates in the network synthesized can sometimes be smaller than that in the network based on the flow-output table with the minimum number of states. If we minimize the number of states, however, this**

---

* The flip-flops that will be discussed in Section 8.2 are much more complicated than *S-R* latches. In particular, in the case of PLAs in Section 9.4, minimization of the number of flip-flops is desirable, since flip-flops occupy a specific share of the chip area and consequently only a small number of flip-flops are usually provided in a PLA. The use of gates, however, is less restricted, because a large number of gates arranged in matrix configuration is provided.

tends to minimize the number of gates in networks, though there are exceptions. As a rule of thumb, therefore, we may try to minimize the number of states in order to indirectly minimize the network size in most cases.

Completely or Incompletely Specified Flow-Output Table

Before discussing the procedure, let us define terminology.

If a next state S and also an output value z for a stable state are specified for every total state (x_1, x_2, s) considered in a flow-output table, such a flow-output table (e.g., Table 7.6.1) is said to be **completely specified**. On the other hand, if next states or output values for some stable states (not for unstable states) are not specified for some total states [e.g., the next state for total state $(x_1, x_2, s) = (1\ 0\ C)$ in Table 7.5.3 is not specified], such a flow-output table is said to be **incompletely specified**. In this case notice that it is not important whether output values for unstable states are specified or not, since they last for only a transient period. Simplification of completely specified flow-output tables is discussed in the rest of this section except the last two subsections. Minimization of the number of states of incompletely specified flow-output tables, to be discussed in the last two subsections, is usually excessively time-consuming, in spite of the fact that we encounter the latter case in practice. **But if we do not attempt to minimize the number of states, the number of states for an incompletely specified table can often be reduced without spending an excessively long time, and then the networks can be designed by the procedure in Section 7.7 or 7.8.**

Indistinguishability Classes of States in Completely Specified Networks

If we can minimize the number of states, some states must be merged. Let us see what states can be merged.

> **Definition 7.6.1:** Suppose that two completely specified sequential networks (or flow-output tables), P and Q, have the same sets of inputs and outputs. Networks P and Q are said to be **indistinguishable** if any input sequence* applied to the inputs $x_1, x_2, \ldots, x_n$ of both networks (or the networks corresponding to these flow-output tables) generates an identical output sequence† from both networks, assuming that they are placed in their respective appropriate initial states. This relation is denoted as $P \sim Q$. Networks that are not indistinguishable are said to **be distinguishable**.

---

* Here an input sequence at x_1, an input sequence at $x_2, \ldots$, and an input sequence at x_n are collectively called an input sequence to the inputs $x_1, x_2, \ldots, x_n$.

† Output values during transitions (i.e., those for unstable states) are not considered.

Minimization of the Number of States 397

Definition 7.6.2: A state p_i of a completely specified sequential network P is said to be **indistinguishable** from a state q_i of a completely specified sequential network Q if any input sequence applied to both networks generates an identical output sequence from both networks when P and Q are initially placed in states p_i and q_i, respectively. This relation is denoted as $p_i \sim q_i$. Here P and Q may be two different networks ($P \neq Q$), or the same network ($P = Q$). States that are not indistinguishable are said to be **distinguishable**.

For example, when the flow-output table P in Table 7.6.1a and the flow-output table Q in Table 7.6.1b are compared, $A \sim C$, $A \sim E$, $B \sim D$, and $B \sim F$. When only states in the same network in Table 7.6.1b are compared (this means $P = Q$), we have also $C \sim E$ and $D \sim F$.

Obviously, from these definitions, **two sequential networks, P and Q, are indistinguishable if and only if their respective initial states are indistinguishable.** Also, **P and Q are indistinguishable if and only if every state p_i of P and the corresponding state q_i of Q in which the two networks are placed at any moment by an incoming input sequence are indistinguishable.**

The "indistinguishable" relationship between states introduced above is an equivalence relation in algebra (see, e.g., Liu [1977] or Preparata and Yeh [1973]), since it satisfies the three laws of equivalence relations. In other words, let us consider states p_i, q_j, and r_k of the same network or different networks. Then the following hold:

$p_i \sim p_i$ (Reflexive law).
If $p_i \sim q_j$, then $q_j \sim p_i$ (Symmetric law).
If $p_i \sim q_j$ and $q_j \sim r_k$, then $p_i \sim r_k$ (Transitive law).

The first two laws are easy to see. The transitive law can be proved in the following manner. Suppose that we have three networks, P, Q, and R (the argument is the same even if two or three of them are the same network), and they are placed in initial states p_i, q_j, and r_k, respectively. When an arbitrary input sequence is applied to all three networks, the output sequences from P and Q must be identical because $p_i \sim q_j$. The output sequences from Q and R must also be identical because $q_j \sim r_k$. Thus the output sequences from P and R must be identical. This means that $p_i \sim r_k$, and the transitive law is satisfied.

Since the indistinguishable relation of states is an equivalence relation, as shown above, it is known that all internal states of a sequential network can be grouped into disjoint equivalence classes if the given sequential network, that is, the corresponding flow-output table, is completely specified. Let us call them **indistinguishability (equivalence) classes**. All states in each indistinguishability class are indistinguishable from one another, and every state in one class is distinguishable from any state in any other class. When we consider several networks that are indistinguishable from one another, all internal states in these networks are also grouped into disjoint indistinguishability classes, and the same relationship as above holds between states.

Let us observe the following important properties.

Lemma 7.6.1: Suppose that two flow-output tables, P and Q, are indistinguishable. Then any two states in Q are indistinguishable if and only if they are indistinguishable from the same state in P.

Proof: Assume that two states, q_i and q_j, of Q are both indistinguishable from state p_k of P—in other words, $q_i \sim p_k$ and $q_j \sim p_k$. Indistinguishable states belong to the same equivalence class, as explained already. Therefore the three laws of the equivalence relations are applicable. Thus, by the symmetric law of the definition of "equivalence relation," $p_k \sim q_j$. Then, from $q_i \sim p_k$ and $p_k \sim q_j$ by the transitive law, we get $q_i \sim q_j$.

Conversely, assume that $q_i \sim q_j$. Since P and Q are indistinguishable, there must be some state of P, p_k, such that $q_j \sim p_k$. Then, by the transitive law, $q_i \sim p_k$.
Q.E.D.

When a flow-output table Q that may contain many unnecessary states is given, let us try to derive a flow-output table P that is indistinguishable from Q and has a minimum number of states. The following theorem immediately results from Lemma 7.6.1, and is the basis of the state minimization procedure in the following.

Theorem 7.6.1: Given a flow-output table Q, there exists a flow-output table P with a minimum number of states, which has as many states as there are indistinguishability classes in Q, and where the states in every pair are distinguishable, whereas any corresponding states in Q belong to different indistinguishability classes and are distinguishable.

This is illustrated in Fig. 7.6.1.

State Minimization Procedure for Completely Specified Flow-Output Tables

Let us describe a procedure to derive a flow-output table of a minimum number of states that is indistinguishable from a given flow-output table, by eliminating unnecessary states in the given table [Paul and Unger 1959].

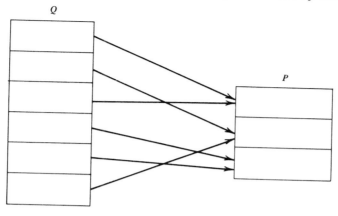

Fig. 7.6.1 Derivation of a flow-output table **P** with a minimum number of states from a given flow-output table **Q** (each indistinguishability class of states in a given flow-output table **Q** is in one-to-one correspondence to each state of **P**).

Procedure 7.6.1: State Minimization for a Completely Specified Flow-Output Table

When a flow-output table such as Table 7.6.2 is given, draw a table to compare states pairwise, as shown in Table 7.6.3a. Let us call this the **state-comparison table**. The coordinates of the table show states.

1. In Table 7.6.2 **find every pair of states such that their output values conflict in some columns** and **enter × in the corresponding cell in the state-comparison table**. In this case, if in some column one or both entries in two rows under comparison represent unstable states, we need to compare* the output values after settling in the stable states in the different rows (still in the same column), since the output values for unstable states last only during the short transition period.

   Every pair of states thus found is distinguishable. This is so because a pair of states is indistinguishable only when any input sequence started from this pair of states generates identical output sequences, as defined in Definition 7.6.2, and consequently the different output values for any combination of input values (i.e., for any part of an input sequence) make the pair of states distinguishable.

---

* Even if we do not compare here, this will be taken care of in Step 4. But comparison of these outputs after settling in Step 1 tends to simplify processing.

Table 7.6.2 Flow-Output Table

s	00	01	11	10
1	①, 1	6, 1	①, 1	2, d
2	3, d	②, 0	②, 1	②, 0
3	③, 1	5, 1	③, 1	2, d
4	④, 1	6, 1	④, 1	2, d
5	1, 1	⑤, 1	4, 1	⑤, 0
6	1, 1	⑥, 1	3, 1	⑥, 0

$x_1 x_2$

S, z

Table 7.6.3 State-Comparison Tables for Minimization of the Number of States

***a.* State-Comparison Table at Step 1**

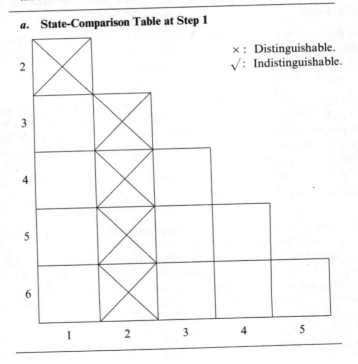

×: Distinguishable.
√: Indistinguishable.

b. State-Comparison Table at Step 2

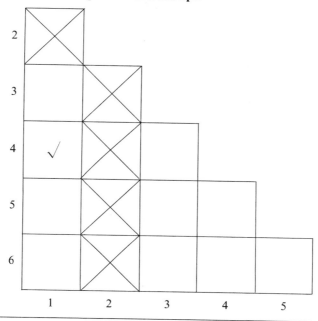

c. State-Comparison Table at Step 3

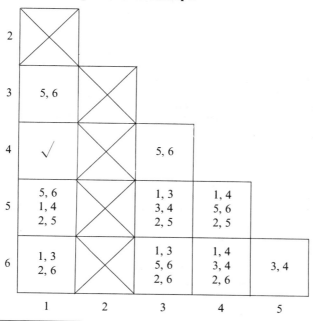

Table 7.6.3 (Continued)

d. State-Comparison Table at Step 4

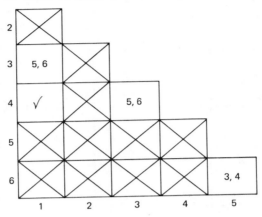

For example, compare two cases: where the network is placed in state 1, and where the network is placed in state 2. In the column $(x_1, x_2) = (1\ 1)$, the output values are identical. In column (0 0) the output values after settling in the stable states are still identical. [In column (0 0) we have stable state 1 with its output $z = 1$. But in the row for state 2, we have an unstable state with output d. Since the output value for this unstable state appears only for a short transient period, it does not matter whether we have 0 or 1 for d. After settling in stable state 3, we have the output value $z = 1$, which should be compared with $z = 1$ for state 1.] But in column (0 1), the row for state 2 has $z = 0$, whereas we have $z = 1$ after settling in state 6 in row 6 from row 1. Thus the two states 1 and 2 are distinguishable. We enter × in the cell of Table 7.6.3a which has coordinates 1 and 2. In a similar manner we can find that pairs of states (2, 3), (2, 4), (2, 5), and (2, 6) are distinguishable, so we enter ×'s in the corresponding cells in Table 7.6.3a.

2. In Table 7.6.2 **find every pair of states such that, in every column, these two states go to the same state, or stay in themselves, and the output values do not conflict. Then enter ✓ in the corresponding cell in the state-comparison table**. In this case also, if in some column one or both entries in two rows under comparison represent unstable states, the output values after settling in the stable states must be compared.

Every pair of states thus found is indistinguishable, since any sequence started with this pair of states generates identical output sequences, as can easily be seen, if the pair of states go to the same state, or stay in themselves.

Minimization of the Number of States   403

It is usually more difficult to determine whether or not a pair of states is indistinguishable than to determine whether or not the pair is distinguishable. The reason is that a pair of states is distinguishable if their output values are different for any input value, whereas a pair of states is indistinguishable only if their output values are not different for any input sequence, as defined in Definition 7.6.2.

Compare two cases: where the network is placed in state 1, and where the network is placed in state 4. When (x_1, x_2) is (0 0) or (1 1), the networks stay in their own current states 1 and 4, respectively, with identical values of z. When (x_1, x_2) changes from (0 0) to (0 1) or (1 0), or from (1 1) to (0 1) or (1 0), the network goes to the same state 6 or 2, respectively, and henceforth the values of z must be identical for any input sequence. Thus we must conclude that states 1 and 4 are indistinguishable, and we enter a ✓ in the corresponding cell in Table 7.6.3*b*. No other indistinguishable pairs can be found easily in Table 7.6.2.

3. The remaining pairs of states must be examined more carefully, since it is not obvious whether they are distinguishable or indistinguishable. In each remaining cell of the state-comparison table, **enter the conditions under which the corresponding pair of states is indistinguishable.**

   For example, compare two states, 5 and 6. As long as the network stays in state 5 or 6 in column (0 1) or (1 0), the values of z are identical. In column (0 0) the network goes to state 1 in both cases, and the values of z are identical. In column (1 1) the network goes to different states 3 and 4 in the two cases. Therefore states 5 and 6 are indistinguishable if and only if states 3 and 4 are indistinguishable. (If states 3 and 4 had different values of z, states 5 and 6 would be obviously distinguishable.) Enter the pair 3 and 4 in the cell of states 5 and 6 in Table 7.6.3*c*. As another example, compare states 1 and 5. In each column in Table 7.6.2, whenever the network is in a stable state in either row 1 or 5, it is in an unstable state in the other row. But states 1 and 5 are indistinguishable if and only if each pair of states (5, 6), (1, 4), and (2, 5) is indistinguishable. Therefore we enter these pairs in the corresponding cell of the state-comparison table. Similar entries are placed in some other cells.

4. **Check whether or not any of the pairs found for each cell in Step 3 was already concluded to be distinguishable (i.e., ×) in the earlier steps. If any pair was found to be ×, replace the entry in the cell by ×. Since replacement of the entry in one cell by × may in turn force the entries in other cells to be replaced by ×'s, we may need to repeat this more than once for each cell until no entries can be replaced by ×'s.**

   For example, cell (1, 5) has pairs (5, 6), (1, 4), and (2, 5). Since states 2 and 5 were found to be distinguishable in Step 1, replace the entry in cell (1, 5) by ×. After checking every cell with conditions entered, Table 7.6.3*d* is obtained.

5. **In the cells that could not be changed in Step 4, change the entries entered into √'s.** Let us consider the following two cases, in order to learn why we can do this.

   (a) If all the pairs of states in a cell among the remaining cells were previously concluded to be √, that is, indistinguishable, we can obviously replace the pairs in that cell by √.

   Among the remaining cells in Table 7.6.3d, we cannot find any such cell.

   It is important to notice that the cell under consideration is distinguishable if *any* of the pairs of states entered in that cell in Step 3 is distinguishable [e.g., even if the pair of states 1 and 4 was indistinguishable, cell (1, 5) is distinguishable because the pair of states 2 and 5 was distinguishable], whereas the cell is indistinguishable only if **all** the pairs in that cell are indistinguishable.

   (b) We can enter √ also in the remaining cells, in each of which all the pairs of states were previously not concluded to be indistinguishable. This is so because the cells that contain these conditions correspond to indistinguishable pairs of states. This can be seen as follows.

   Compare two states, 5 and 6. As already observed, as long as the networks placed in states 5 and 6 go to the same state, 1, or stay in states 5 and 6, respectively, the values of z are identical. If states 5 and 6 are to be indistinguishable, they are indistinguishable because the values of z are identical only after reaching states 3 and 4. If the networks placed in states 5 and 6 go to states 4 and 3, respectively, the networks can further go to the same state, 2, together, go to states 6 and 5, respectively, or stay in states 4 and 3, respectively. Since the values of z are identical for states 5 and 6 and also for states 3 and 4, the networks initially placed in states 5 and 6 have the same sequence of the values of z for any input sequence. Thus states 5 and 6 are indistinguishable. This is a collective comparison of more than one pair of states. By the same reasoning, the entries in other cells are replaced by √'s.

6. Make indistinguishability classes based on √'s. Then **merge each set of indistinguishable states in Table 7.6.2 into the corresponding indistinguishability class, as shown in Table 7.6.4, which is the desired flow-output table with a minimum number of states.**

   For the above example, states 1, 3, and 4 are indistinguishable from one another and are distinguishable from the other states. Thus they constitute one indistinguishability class, B. Similarly, A consists of state 2, and C consists of indistinguishable states 5 and 6. □

Table 7.6.4 Flow-Output Table with a Minimum Number of States and the Corresponding State Diagram

a. Flow-Output Table

s	00	01	11	10
A; (2)	B, d	Ⓐ, 0	Ⓐ, 1	Ⓐ, 0
B; (1, 3, 4)	Ⓑ, 1	C, 1	Ⓑ, 1	A, d
C; (5, 6)	B, 1	Ⓒ, 1	B, 1	Ⓒ, 0

S, z

b. State Diagram

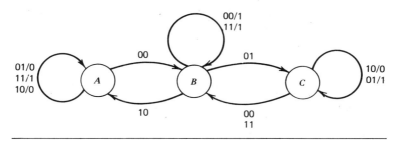

In the case of a completely specified flow-output table, Procedure 7.6.1 is reasonably simple unless the table has too many rows or columns. But state minimization is practical only for small tables, though reduction of the number of states as much as possible, instead of minimization, can be tried for large tables.

State Minimization for Incompletely Specified Flow-Output Tables

In an incompletely specified flow-output table, output values for some stable states, or next states to go from some stable states, are not specified. If we specify the values of these output values to be 1 or 0, or next states, we obtain a completely specified flow-output table. Since each don't-care condition can be replaced by 0 or 1, and unspecified next states can be specified in different ways, many different, completely specified flow-output tables can be generated.

Then Procedure 7.6.1 may generate many different reduced tables from them. We can choose from them the flow-output table with the least number of states.

However, we still cannot minimize the number of states, since state minimization of incompletely specified flow-output tables is much more complex than that of completely specified flow-output tables because of **state-splitting**, as follows.

Suppose that an incompletely specified flow-output table is given as shown in Table 7.6.5a, where the output value for stable state 2 in column $(x_1, x_2) = (0\ 0)$ is not specified. If we reduce the number of states in this table, only state 2 is a candidate to be merged with state 1 or 3, because of the different output values for stable states. States 1 and 2 can be equivalent only if states 2 and 3 are equivalent, by looking at column (0 1). Also, states 2 and 3 can be equivalent only if states 1 and 2 are equivalent, by looking at

Table 7.6.5 State-Splitting

a.

s	$x_1 x_2$			
	00	01	11	10
1	①, 0	3, 0	4, 1	①, 1
2	②, —	②, 0	4, 1	②, 1
3	③, 1	③, 0	4, d	1, 1
4	2, 0	3, d	④, 1	④, 0

S, z

b.

s	$x_1 x_2$			
	00	01	11	10
1	①, 0	3, 0	4, 1	①, 1
2a	②ⓐ, 0	2b, 0	4, 1	②ⓐ, 1
2b	②ⓑ, 1	②ⓑ, 0	4, 1	2a, 1
3	③, 1	③, 0	4, d	1, 1
4	2, 0	3, d	④, 1	④, 0

S, z

Minimization of the Number of States 407

Table 7.6.5 (*Continued*)

c.

s	$x_1 x_2$			
	00	01	11	10
(1, 2a) = A	Ⓐ, 0	B, 0	4, 1	Ⓐ, 1
(2b, 3) = B	Ⓑ, 1	Ⓑ, 0	4, 1	A, 1
4	A, 0	B, d	④, 1	④, 0

S, z

d.

s	$x_1 x_2$			
	00	01	11	10
(1, 2a) = A	Ⓐ, 0	B, 0	4, 1	Ⓐ, 1
(2b, 3) = B	Ⓑ, 1	Ⓑ, 0	4, 1	A, 1
4	B, 0	B, d	④, 1	④, 0

S, z

column (10). But the unspecified value "—" for state 2 must be 0 if states 1 and 2 are to be equivalent, and 1 if states 2 and 3 are to be equivalent. Since this is contradictory, it does not seem possible to reduce the number of states in Table 7.6.5a. But this conclusion is wrong.

By splitting state 2 into new states 2a and 2b, the new flow-output table in Table 7.6.5b is obtained, and this table generates the same output sequence as the original in Table 7.6.5a for any input sequence, whenever output values in these sequences are specified. From this completely specified flow-output table in Table 7.6.5b, we can obtain the completely specified flow-output table in Table 7.6.5c or 7.6.5d, depending on whether unstable state 2 in column (0 0) and row 4 in Table 7.6.5b is 2a or 2b, respectively. These two new tables, Tables 7.6.5c and 7.6.5d, have one less state than does Table 7.6.5a and, as a matter of fact, have the minimum number of states, 3, since in Table 7.6.5a state 2 is the only candidate to be merged with others.

This example demonstrates that, in order to minimize the number of states, assigning 0 or 1 to unspecified values is not sufficient and we need consider state-splitting, and also that, from a given table, we may obtain more than one table, with a minimum number of states, where states in one

408 *Asynchronous Sequential Networks*

table correspond to the states in the given table in a manner different from that for the states of another table [unlike the case of a completely specified flow-output table (Theorem 7.6.1)].

Because of state-splitting, the state minimization procedure that is outlined in the rest of this section is very complex and time-consuming even for computer processing, since the number of comparisons increases very rapidly as the number R of states and the number n of inputs in the original flow-output table increase, despite some speed-up techniques. State minimization of completely specified flow-output tables is time-consuming, but state

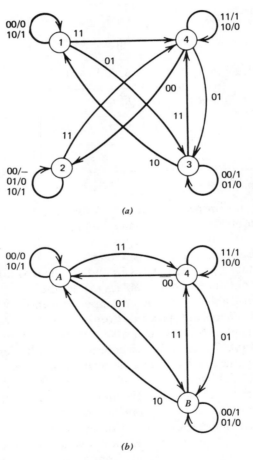

Fig. 7.6.2   State reduction ($x_1 x_2/z$).

Minimization of the Number of States 409

minimization of incompletely specified flow-output tables is much worse. Thus, in the case of incompletely specified flow-output tables, the state minimization procedure is rarely used for hand or computer processing, since processing time is reasonably short only for small tables.

When a given flow-output table is small, we can try to reduce the number of states by visual inspection of the table and the state diagram. (If the number of unspecified values, denoted by dashes, is small, we may be able to minimize.) For example, Table 7.6.5a can be reduced to Table 7.6.5c or 7.6.5d by visually comparing states in Table 7.6.5a and the corresponding state diagram, Fig. 7.6.2a. (Fig. 7.6.2b corresponds to Table 7.6.5c.) Also, the state-comparison table is still helpful. If we apply Procedure 7.6.1 to as many pairs of states that do not contain dashes (as outputs for stable states or next states) as possible, we can obtain information about which states can or cannot be candidates for merging. If we reduce the number of states in this manner on a trial-and-error basis, we had better check whether the new flow-output table still satisfies the given network performance requirement (usually expressed in a word statement).

▲ State Minimization Procedure for Incompletely Specified Flow-Output Tables

To find candidates for states (or split states) to be merged, we need to define the following.

Definition 7.6.3: Two states, p and q, of a flow-output table are said to be **compatible** if and only if the network, represented by the table, that is initially placed in p generates the same output sequence for any input sequence as the network that is initially placed in q, whenever output values in these sequences are specified.

For example, states 1 and 2 in Table 7.6.5a are compatible, and also states 2 and 3 are compatible. In this case the network that is initially placed in state 2 sometimes generates unspecified output values due to the dash in column (0, 0), and, corresponding to these values, the network initially placed in 1 or 3 can generate the output value 1 or 0. (We do not care about the output values for unstable states, since they are short, transient values.)

Definition 7.6.4: A set of states is called a **compatible** if every pair of states in the set is compatible. A compatible is **maximal** if it is not a subset of any other compatible.

Notice that every subset of a compatible is also a compatible.

410   Asynchronous Sequential Networks

The compatible relations for incompletely specified flow-output tables in Definition 7.6.3 correspond to the equivalence relations of indistinguishable classes for completely specified tables. But even if one state is compatible with a second state and also with a third state, the second and third states may not be compatible, unlike the case of indistinguishability classes for completely specified tables, which are equivalence classes as known in mathematics. (In the case of completely specified tables, if one state is equivalent to a second state and also to a third state, the second and third states are equivalent. Thus all three states belong to the same indistinguishability class.) For example, state 2 in Table 7.6.5a is compatible with states 1 and 3, but states 1 and 3 are not compatible. Thus the compatible relations cannot be equivalence relations, since the transitivity law of equivalence relations stated earlier in Section 7.6 (after Definition 7.6.2) is not satisfied. This difference between compatible relations and equivalence relations makes state minimization for incompletely specified tables very complicated.

Procedure 7.6.2: State Minimization for an Incompletely Specified Flow-Output Table

Let us minimize the number of states in a given incompletely specified flow-output table.

1. Find compatible pairs of states, by comparing every pair of states on the state-comparison table. The state-comparison table used in Procedure 7.6.1 is applied here in a somewhat different manner. Suppose that the incompletely specified flow-output table in Table 7.6.6a is given. If a pair of states is not compatible because of conflicting output values (e.g., states 1 and 5), a cross ($\times$) is entered in the corresponding cell in the state-comparison table in Table 7.6.6b. If a pair of states is compatible (e.g., states 1 and 3), a check mark ($\checkmark$) is entered in the cell in Table 7.6.6b. If a pair of states can be compatible under the condition that other pairs of states are compatible, we enter the other pairs in the corresponding cell in Table 7.6.6b. [For example, states 1 and 2 can be compatible if other pairs, 1 and 6, and 2 and 3, are compatible. Hence (1, 6) and (2, 3) are entered in the corresponding cell.] After comparing every pair of states as described above, we need to check whether the pairs of states entered in cells in Table 7.6.6b are already concluded to be compatible or not compatible. In each cell, if any of the pairs was previously concluded not to be compatible in another cell, we enter a cross in the cell [e.g., the cell for the pair of states 4 and 6 is entered with a cross (dotted-line), since state pair 5 and 6 was not compatible].

   Now we have obtained compatible pairs, corresponding to cells that do not have crosses.

Table 7.6.6 Step 1 of Procedure 7.6.2

a. Given Flow-Output Table

s \ x_1x_2	00	01	11	10
1	①, 0	3,	①, 1	—
2	6,	②, 1	—	②, —
3	③, —	③, 1	1,	—
4	—	④, 1	④, 1	5,
5	1,	2,	⑤, 0	⑤, 0
6	⑥, 0	2,	4,	⑥, 1

S, z

b. State-Comparison Table

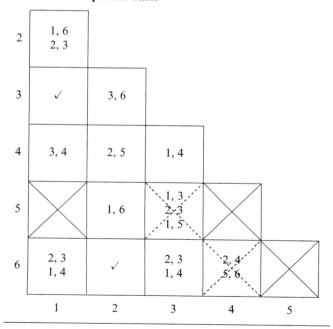

2. Find maximum compatibles, using the state-comparison table obtained in Step 1.

   (a) Starting with the rightmost column in the state-comparison table, move left until you reach a column that contains compatible pairs of states (i.e., a column that contains cells without crosses). List all compatible pairs for that column (not the pairs entered in cells).

   In our example, the first such column is for state 3, and we list pairs (3, 4) and (3, 6).

   (b) Move left to the next column, which contains compatible pairs of states. If the state that this column represents is compatible with all the states in some previously listed compatible, add this state to that compatible. If the state is not compatible with all the states in a previously listed compatible, but is compatible with all the states in a subset of such a compatible, form a new compatible that consists of the subset and the state under consideration. Next, list all compatible pairs that are not included in any previously listed compatible.

   Continuing the above example, the next column that contains compatible states is for state 2. Since state 2 is compatible with all the states in compatible (3, 4), we list (2, 3, 4). Similarly, we list (2, 3, 6). Next we list (2, 5).

   (c) Repeat Substep (b) until all the columns are considered. Thus we have obtained all maximal compatibles.

   Continuing the above example, we find that (1, 2, 3, 4), (1, 2, 3, 6), and (2, 5) are all maximal compatibles.

3. Find a closed set of a minimum number of compatibles that contains all the states of the given flow-output table. Here a set of compatibles is said to be **closed** if every compatible that the network reaches from any compatible in the set for any input sequence is a subset of a compatible in the set. When we find such a set, we regard each compatible as a new state, and then we have obtained a flow-output table with a minimum number of states for the given incompletely specified flow-output table. When we find such a set of a minimal number of compatibles, every state in the given flow-output table must be contained in some compatible, since the given table must be expressed by the new table.

   For example, suppose that we have obtained maximal compatibles (1, 2, 3, 4), (5, 6, 7), (2, 5, 7), and (3, 6) by Step 2. If we choose two maximal compatibles, (1, 2, 3, 4) and (5, 6, 7), all states in the given flow-output table are contained in these two maximal compatibles, and we apparently have a new flow-output table with a minimum number of states (two). But if compatibles (2, 5, 7) and (3, 6) are reached by the network for some input sequences, they also must be chosen, increasing the number of new states. In other words, (1, 2, 3, 4) and (5, 6, 7) are not closed, so we cannot use them alone. Another complication is that, if we choose nonmaximal compatibles, a flow-output table with a minimum number of states may be obtained. In other words, we need to consider subsets of maximal compatibles as possible candidates. (In this sense the maximal compatibles obtained in Step 2 are useful only for finding a lower bound on the minimum number of compatibles and are not directly useful for Step 3.)

Thus Step 3 is very complicated if there are many compatibles. No efficient procedure is known, though some procedures are available. □

State minimization for incompletely specified flow-output tables was developed mostly by Paull and Unger [1959]. For different approaches and detailed techniques, see Kohavi [1970], Grasselli and Luccio [1965, 1966, 1968], Unger [1962, 1965, 1969], McCluskey [1962, 1965], Kella [1970] (Kella uses the branch-and-bound method to cut processing time; he avoids the generation of all maximal compatibles, but adds one state at a time to reduce a flow-output table whose size gradually increases, without making the reductions to the full extent), Meisel [1967], Bouchet [1968], Mealy [1955], Ginsburg [1962], and Luccio [1966].

7.7 State Assignment

We need to prepare a transition-output table from a flow-output table by assigning binary numbers to states, according to the design procedure illustrated in Fig. 7.4.8. As an example, let us consider the flow-output table in fundamental mode in Table 7.7.1. As this flow-output table is incompletely specified, we could apply the procedure to reduce the number of states discussed at the end of Section 7.6. But since it is very cumbersome and time-consuming, let us skip this procedure* and proceed to the design of a network without reducing the number of states.

We need to replace letters representing states by binary numbers. Let us assign binary numbers $0\ 0\ 0$, $0\ 0\ 1$, ..., $1\ 0\ 1$ to states A, B, C, D, E, F in this order, as shown in Table 7.7.2. Dashes are replaced by don't-care conditions, d's. How binary numbers are assigned to states usually influences the network to be designed, in terms of hazards and also the number of gates,

---

* Even if we apply the procedure, the number of states in the particular example in Table 7.7.1 cannot be reduced.

Table 7.7.1 Flow-Output Table in Fundamental Mode

	x_1x_2			
s	00	01	11	10
A	(*A*), 0	*C*, 0	—	*B*, *d*
B	(*B*), 1	*C*, *d*	—	(*B*), 1
C	*D*, 0	(*C*), 0	—	—
D	(*D*), 0	*E*, *d*	—	*F*, 0
E	(*E*), 1	(*E*), 1	—	*F*, *d*
F	*A*, 0	—	—	(*F*), 0
		S, z		

Table 7.7.2 Transition Table for Table 7.7.1

	x_1x_2			
$y_1 y_2 y_3$	00	01	11	10
A 000	(000)	011	*ddd*	001
B 001	(001)	011	*ddd*	(001)
C 011	010	(011)	*ddd*	*ddd*
D 010	(010)	100	*ddd*	101
E 100	(100)	(100)	*ddd*	101
F 101	000	*ddd*	*ddd*	(101)
		Y_1, Y_2, Y_3		

Table 7.7.3 Number of Distinct State Assignments to R States
($\lceil \log_2 R \rceil$ is the number of internal variables.)

NUMBER OF STATES R	$\lceil \log_2 R \rceil$	NUMBER OF DISTINCT ASSIGNMENTS
1	0	1
2	1	1
3	2	3
4	2	3
5	3	140
6	3	420
7	3	840
8	3	840
9	4	10,810,800

connections, and levels. It is desirable to assign binary numbers so that the resulting network is free of races due to multiple change of internal variables and also meets a desired design objective such as a minimum number of gates (also, secondarily, a minimum number of connections or levels). This is called the **state-assignment problem**. Since there are numerous ways* to assign binary numbers for even a small number of rows, as shown in Table 7.7.3, it is essentially impossible to try all possible state assignments. We can only try different state assignments as time permits and find the best solution among them.

The transition table in Table 7.7.2 contains races due to simultaneous changes of more than one internal variable. Suppose that the network is in state $(x_1, x_2, y_1, y_2, y_3) = (1\ 0\ 1\ 0\ 1)$, for example, and (x_1, x_2) changes from $(1\ 0)$ to $(0\ 0)$. Then $(y_1 y_2 y_3)$ must change from $(1\ 0\ 1)$ to $(0\ 0\ 0)$, requiring a double change in y_1 and y_3. Since simultaneous changes are not possible physically, $(1\ 0\ 1)$ will first change to stable state $(1\ 0\ 0)$ or $(0\ 0\ 1)$. Both states, $(1\ 0\ 0)$ and $(0\ 0\ 1)$, are stable and different from the intended next stable state, $(0\ 0\ 0)$. Thus this state assignment makes the network

---

* McCluskey [1965], McCluskey and Unger [1959], Weiner and Smith [1967], Torng [1968], Rhyne and Noe [1977].

malfunction because of the critical race. Hence we must find another state assignment such that hazardous multiple changes of internal variables do not occur.

Notice that noncritical races here can cause malfunction of the network if the network settles down in stable states that are not intended states, unlike the case of network analysis, in which noncritical races are for networks that we know perform correctly.* When we try to find a state assignment, we need not worry about noncritical races in which the network settles down in intended stable states. But we must try not to have critical races, or noncritical races in which the network settles down in non-intended stable states. Let us call these two possibilities, collectively, **hazardous races** or **hazardous multiple changes of internal variables.** (Of course, even if a transition table contains hazardous races, the networks actually designed may not contain races, since different paths of gates in the networks may be implemented with appropriate delay times, as we will see later. Therefore another design approach is to design networks without trying to find a state assignment that is free of hazardous races, and later to adjust the lengths of different paths only when hazardous races occur.)

In the following, we try to eliminate all multiple changes of internal variables for the sake of simplicity, though we actually need to avoid only hazardous multiple changes of internal variables.

Feasibility of State Assignment

The transition relationship among states can be conveniently shown by the state diagram. In the following, arrows, loops, and input-output labelings are not necessary. Such a diagram is called a **state-adjacency diagram**.

The state adjacency diagram for the flow-output table in Table 7.7.1 is shown in Fig. 7.7.1. It can easily be seen from this diagram that we cannot have a state assignment free of multiple changes of internal variables. Assign (0 0 0) to B first, since if B is the first state to be considered, any binary number may be assigned to B. Then A and C must be among (1 0 0), (0 1 0) and (0 0 1). Then A and C differ in two bit positions, though A and C should be adjacent. This is a contradiction.

Generalizing this, we can say that, **if any state-adjacency diagram contains a triangle, we cannot have a state assignment that is free of multiple changes of internal variables**.

---

* For example, in the transition table in Table 7.3.1a, by the noncritical race from stable state $(x_1, x_2, y_1, y_2) = (0\ 1\ 0\ 0)$, the network ends up at $(Y_1, Y_2) = (1\ 0)$, which is different from (1 1) for unstable state (1 1 0 0). When we know that the designed network works correctly, this is not a hazard. But when we are designing a network, contradiction of an intended next state from a stable state that the network will actually reach by non-critical race will be a hazard.

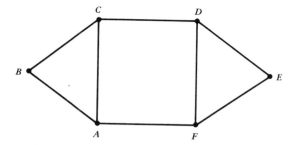

Fig. 7.7.1 State-adjacency diagram for Table 7.7.1.

Derivation of State Assignments

If the given flow table cannot have a state assignment that is free of multiple changes of internal variables, we have to change the transition relationship among the stable states. We had better work on state tables, as follows, since these tables, unlike flow tables, may have intermediate unstable states that we can manipulate. Whenever we have candidate state tables, we can try to find appropriate state assignments by using Karnaugh maps.

First, if we derive a state table from the given flow table, by adding intermediate unstable states, we can sometimes have a state assignment that is free of multiple changes of internal variables.

The state table shown in Table 7.7.4 is constructed from the flow table in Table 7.7.1 by creating intermediate unstable states B and E, labeled with an asterisk, through which the network still can reach the original destination, stable states. For example, when the network placed in stable state A receives $x_2 = 1$, the network goes to stable state C after passing through intermediate unstable state B. (Table 7.7.4 will have the same stable states as Table 7.7.1. Thus the network performance stays unchanged except during transition periods.) Corresponding to Table 7.7.4, the state-adjacency diagram is formed in Fig. 7.7.2a. Notice that A and C in Fig. 7.7.2 are not adjacent now, since the network does not go directly to C from A. Similarly, F and D are not adjacent. Since there is no triangle in this figure, it may be a good candidate. Therefore let us try to find a state assignment that is free of multiple changes of internal variables.

A state assignment that has no multiple changes of internal variables can be conveniently found on a Karnaugh map, since each cell in the map is surrounded by its neighbouring cells, whose coordinates are all binary numbers differing from the binary number of that cell by exactly one bit position. For the state-adjacency diagram in Fig. 7.7.2a, we can find, on

Table 7.7.4 State Table Obtained from Table 7.7.1

	x_1x_2			
S	00	01	11	10
A	Ⓐ	B*	—	B
B	Ⓑ	C	—	Ⓑ
C	D	Ⓒ	—	—
D	Ⓓ	E	—	E*
E	Ⓔ	Ⓔ	—	F
F	A	—	—	Ⓕ

S

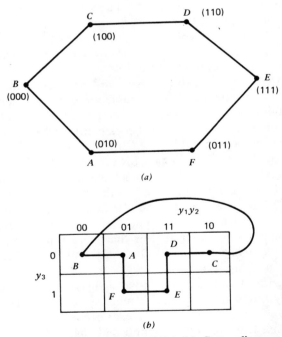

Fig. 7.7.2 State assignment for Table 7.7.4. (a) State-adjacency diagram. (b) Karnaugh map.

418

the map in Fig. 7.7.2b, a state assignment that has no multiple changes of internal variables.

Thus, by working on a state table instead of on the given flow table, our network now has a state assignment that is free of multiple changes of internal variables.

Another approach to derive a state assignment free of multiple changes of internal variables is the addition of new states to a given table. We learned that, if a state-adjacency diagram contains a triangle (e.g., *ABC* in Fig. 7.7.1), it cannot have a state assignment such that binary numbers assigned to any adjacent pair of states differ in one bit position. Therefore let us eliminate triangles by adding new states. Fig. 7.7.3a is such a state-adjacency diagram, where new states *G* and *H* are added. A state assignment that is free of multiple changes of internal variables is found on the map in Fig. 7.7.3b. From Table 7.7.1 by Fig. 7.7.3, we obtain the state table in Table 7.7.5 and the transition table in Table 7.7.6 (entries in the rows for *G* and *H* can be easily found).

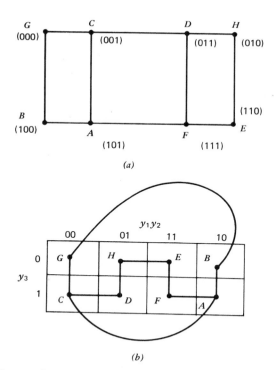

Fig. 7.7.3 **State assignment. (a) State-adjacency diagram with additional states. (b) Karnaugh map.**

Table 7.7.5 State Table Obtained from Table 7.7.1 by Adding States G and H

s	00	01	11	10
A	Ⓐ	C	—	B
B	Ⓑ	G	—	Ⓑ
C	D	Ⓒ	—	—
D	Ⓓ	H	—	F
E	Ⓔ	Ⓔ	—	F
F	A	—	—	Ⓕ
G	—	C	—	—
H	—	E	—	—

$x_1 x_2$ (column header); S (below table)

Table 7.7.6 Transition Table for Fig. 7.7.3.

s	$y_1 y_2 y_3$	00	01	11	10
A	101	101	001	ddd	100
B	100	100	000	ddd	100
C	001	011	001	ddd	ddd
D	011	011	010	ddd	111
E	110	110	110	ddd	111
F	111	101	ddd	ddd	111
G	000	ddd	001	ddd	ddd
H	010	ddd	110	ddd	ddd

$x_1 x_2$ (column header); $Y_1 Y_2 Y_3$ (below table)

There are other state-assignment approaches, such as assigning two or more different binary numbers to one stable state (i.e., an operation opposite to the merger of states). For other approaches or detailed techniques, see Hill and Peterson [1974], Liu [1963], Tracy [1966], Hazeltine [1965], or Saucier [1967]. It is known that any state table that consists of R states and consequently $\lceil \log_2 R \rceil$ internal variables can be converted into a table free of multiple changes of internal variables, using at most $2\lceil \log_2 R \rceil - 1$ internal variables, and also that for any value of $\lceil \log_2 R \rceil$, the number of internal variables in the original table, there are some state tables that require exactly $2\lceil \log_2 R \rceil - 1$ internal variables [Huffman 1955].

7.8 Design of Sequential Networks in Fundamental Mode

Now we are ready to design a loopless network for a desired sequential network. We can apply the design methods discussed in the preceding chapters, since from the transition-output table obtained in Section 7.7, we can obtain the excitation-output table as the input-output relationship of the loopless network in the general model in Fig. 7.4.3.

Let us discuss a design procedure based on the example used in Section 7.7. For our network synthesis we can use either one of the two transition tables, Table 7.7.6 and Table 7.8.1, derived from Table 7.7.4 based on the state assignment shown in Fig. 7.7.2.

Table 7.8.1 Transition-Output Table for Table 7.7.4 by the State Assignment in Fig. 7.7.2

s	$y_1 y_2 y_3$	$x_1 x_2$ 00	01	11	10
A	010	010, 0	000, 0	ddd, d	000, d
B	000	000, 1	100, d	ddd, d	000, 1
C	100	110, 0	100, 0	ddd, d	ddd, d
D	110	110, 0	111, d	ddd, d	111, 0
E	111	111, 1	111, 1	ddd, d	011, d
F	011	010, 0	ddd, d	ddd, d	011, 0

$Y_1 Y_2 Y_3, z$

Table 7.8.2 Output-Input Relationship of an S-R Latch

y	Y	S	R
0	0	0	d
0	1	1	0
1	0	0	1
1	1	d	0

Suppose that we want to use flip-flops to retain the values of internal variables, since the use of flip-flops is a very common practice because it tends to yield a more reliable network (without flip-flops, the next values of internal variables tend to feed back to the network inputs even before the network settles down), and also it is convenient to have memory elements separately. Both Tables 7.7.6 and 7.8.1 require three internal variables, that is, three flip-flops, so it does not matter which of these tables we use, in terms of the number of flip-flops. Let us use Table 7.8.1, since it may possibly re-require fewer gates because of its fewer rows. (On the other hand, Table 7.7.6 may require fewer gates. It is hard to predict until we actually design networks for both tables: Exercise 7.10.)

Since flip-flops are to be used in the network, we need to prepare an excitation table which is different from the transition table. Suppose that we use S-R latches, shown in Fig. 7.1.1a (other types of flip-flops can be similarly treated). Using the output-input relationship of the S-R latch shown in Table 7.8.2 (i.e., Table 7.4.5), we convert Table 7.8.1 into Table 7.8.3, which shows the relationship among y_1, y_2, y_3, x_1, x_2, and the inputs of S-R latches, S_1, R_1, S_2, R_2, S_3, R_3.

From Table 7.8.3, Karnaugh maps* for S_1, R_1, S_2, R_2, S_3, R_3 are obtained as shown in Table 7.8.4. (If there are many variables, methods other than the Karnaugh map method may be preferred, by expressing Table 7.8.4 in switching expressions. Usually don't-care conditions must be considered.) On the maps we derive minimal sums for the latch inputs. In Table 7.8.5 we derive output z. Based on the minimal sums for z, S_1, R_1, S_2, R_2, S_3, R_3, the network shown in Fig. 7.8.1 has been synthesized.

When we want to synthesize a network with gates only and without latches, we form Karnaugh maps for Y_1, Y_2, and Y_3 directly from the tran-

---

* Strictly speaking, minimal multiple-output networks should be used. (Hence, if we want two-level networks, the procedures in Section 5.2 should be used.) But for the sake of simplicity, networks based on minimal sums for individual functions independent of one another are obtained here, on the basis of Karnaugh maps.

Design of Sequential Networks in Fundamental Mode

Table 7.8.3 Excitation Table Derived from Table 7.8.1, Using Table 7.8.2

		$x_1 x_2$			
		00	01	11	10
	010	0d, d0, 0d	0d, 01, 0d	dd, dd, dd	0d, 01, 0d
	000	0d, 0d, 0d	10, 0d, 0d	dd, dd, dd	0d, 0d, 0d
	100	d0, 10, 0d	d0, 0d, 0d	dd, dd, dd	dd, dd, dd
$y_1 y_2 y_3$	110	d0, d0, 0d	d0, d0, 10	dd, dd, dd	d0, d0, 10
	111	d0, d0, d0	d0, d0, d0	dd, dd, dd	01, d0, d0
	011	0d, d0, 01	dd, dd, dd	dd, dd, dd	0d, d0, d0

$$S_1 R_1, S_2 R_2, S_3 R_3$$

sition-output table in Table 7.8.1. (Excitation tables like Table 7.8.3 are not necessary.) Then a network is synthesized without latches.

Postanalysis

Generally, after synthesizing a network, we need to check whether the designed network malfunctions. This is very cumbersome and time-consuming if the network is complicated, and frequently can be done only by computer program simulation (as discussed in Section 9.6). Often bread-boarding (i.e., making a prototype electronic realization) is necessary, because the intricate waveforms that electronic circuits implementing gates generate during transient periods cannot be well simulated by computer programs.

By using a state assignment that is free of multiple changes of internal variables (Section 7.7), the designed network is free of races due to the timing relationship of internal variables y_1, y_2, and y_3. But there are other hazards that may lead to the malfunction of the entire designed network, as follows.

In other words, during the design procedure we assumed that y_i and $\bar{y}_i$ always assume complementary values* (i.e., if one of them assumes the value 0, the other assumes 1, and vice versa), and also that the loopless network

---

* In the state tables, on the basis of which the networks are designed, only y_i's have been considered as internal variables; $\bar{y}_i$'s have not been explicitly treated as internal variables.

Table 7.8.4 Karnaugh Maps for S_1, R_1, S_2, R_2, S_3, R_3 Derived from Table 7.8.3

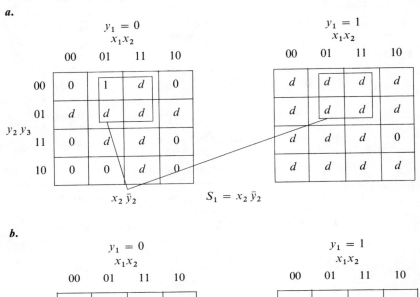

a.

$y_1 = 0$, $x_1 x_2$

$y_2 y_3$	00	01	11	10
00	0	1	d	0
01	d	d	d	d
11	0	d	d	0
10	0	0	d	0

$x_2 \bar{y}_2$

$y_1 = 1$, $x_1 x_2$

	00	01	11	10
00	d	d	d	d
01	d	d	d	d
11	d	d	d	0
10	d	d	d	d

$S_1 = x_2 \bar{y}_2$

b.

$y_1 = 0$, $x_1 x_2$

$y_2 y_3$	00	01	11	10
00	d	0	d	d
01	d	d	d	d
11	d	d	d	d
10	d	d	d	d

$y_1 = 1$, $x_1 x_2$

	00	01	11	10
00	0	0	d	d
01	d	d	d	d
11	0	0	d	1
10	0	0	d	0

$R_1 = x_1 y_3$ $x_1 y_3$

c.

$y_1 = 0$, $x_1 x_2$

$y_2 y_3$	00	01	11	10
00	0	0	d	0
01	d	d	d	d
11	d	d	d	d
10	d	0	d	0

$y_1 = 1$, $x_1 x_2$

	00	01	11	10
00	1	0	d	d
01	d	d	d	d
11	d	d	d	d
10	d	d	d	d

$S_2 = \bar{x}_2 y_1$ $\bar{x}_2 y_1$

d.

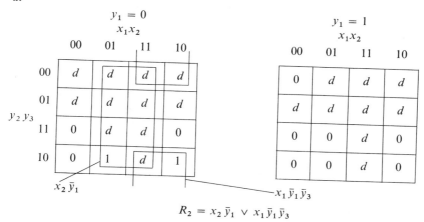

$R_2 = x_2 \bar{y}_1 \vee x_1 \bar{y}_1 \bar{y}_3$

e.

$y_1 = 0$
$x_1 x_2$

y_2 y_3 \	00	01	11	10
00	0	0	d	0
01	d	d	d	d
11	0	d	d	d
10	0	0	d	0

$y_1 = 1$
$x_1 x_2$

y_2 y_3 \	00	01	11	10
00	0	0	d	d
01	d	d	d	d
11	d	d	d	d
10	0	1	d	1

$S_3 = x_2 y_1 y_2 \vee x_1 y_1$ $x_2 y_1 y_2$ $x_1 y_1$

f.

$y_1 = 0$
$x_1 x_2$

y_2 y_3 \	00	01	11	10
00	d	d	d	d
01	d	d	d	d
11	1	d	d	0
10	d	d	d	d

$\bar{x}_1 \bar{y}_1$ $R_3 = \bar{x}_1 \bar{y}_1$

$y_1 = 1$
$x_1 x_2$

y_2 y_3 \	00	01	11	10
00	d	d	d	d
01	d	d	d	d
11	0	0	d	0
10	d	0	d	0

Table 7.8.5 Karnaugh Maps for z Derived from Table 7.8.1

$y_1 = 0$
$x_1 x_2$

$y_2 y_3$	00	01	11	10
00	1	d	d	1
01	d	d	d	d
11	0	d	d	0
10	0	0	d	d

$\bar{y}_1 \bar{y}_2$

$y_1 = 1$
$x_1 x_2$

	00	01	11	10
00	0	0	d	d
01	d	d	d	d
11	1	1	d	d
10	0	d	d	0

$y_1 y_3$

$$z = \bar{y}_1 \bar{y}_2 \vee y_1 y_3$$

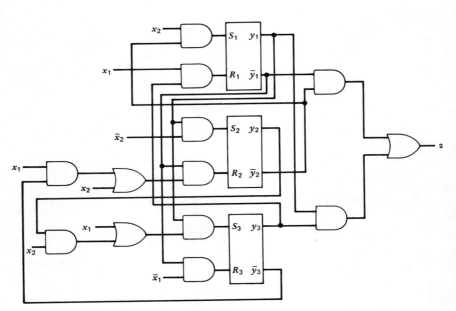

Fig. 7.8.1 *Synthesized network in fundamental mode based on Table 7.8.1.*

contained in the designed sequential network does not change its output values when its outputs are supposed to keep the same values. Actually, however, even when all the gates and latches work perfectly and all the gates (including those inside the latches) have the same switching time τ, y_i and $\bar{y}_i$ assume the same output values for time τ as illustrated in Fig. 7.1.2b, and also the combinational network contained as the loopless network may generate spurious output values because of multiple paths of different lengths when the outputs are supposed to keep the same values, as discussed in Section 6.7. When these spurious signals are fed back to the inputs of the latches (their durations may be stretched to a multiple of τ because of multiple paths of different lengths), the latches may be set to wrong output values. If the duration of these spurious signals is the critical time length τ, the latches may start to oscillate, as illustrated in Fig. 7.1.2a, though the oscillation will eventually die down. The extrinsic hazard of the loopless network due to the difference in the switching times of gates is probably less hazardous, since it usually is of much shorter duration, as discussed in Section 6.7 (but its duration may be also stretched out after going through many gates).

We need to eliminate these causes if the designed network malfunctions. We can eliminate them by adjusting the length of different paths in the network, by using the remedies in Section 6.7, or by reconfiguring the entire network, possibly with extra gates.

Notice that, even if a transition table contains multiple changes of internal variables, a network actually implemented on the basis of such a table may not have any hazard because of the different propagation times of internal variables y_i to the inputs of the latches over different paths.

When a network is synthesized without latches, the transient performance is usually much more complicated and its analysis is much more cumbersome.

By examining output waveforms of all gates (using the latch of Fig. 7.1.1a as those here) for every change of the inputs,* we can find fortunately that the network obtained in Fig. 7.8.1 works† without malfunctioning, as intended in Table 7.8.1.

7.9 General Comments on Asynchronous Sequential Networks in Fundamental Mode

In this chapter we have discussed the analysis and synthesis of asynchronous sequential networks in fundamental mode. Networks in fundamental mode are fundamental in designing sequential networks in this chapter and the

---

* It is assumed that exactly one of x_1 and x_2 changes at a time.
† The network works correctly under the assumption that x_i and $\bar{x}_i$, for $i = 1$ and 2, change simultaneously, or under the assumption that x_i and $\bar{x}_i$ are supplied from the latch in Fig. 7.1.1a.

following ones, in the sense that they are easy to handle, since the network inputs change only after the networks settle down and consequently we need not treat networks during transitions.

The design of asynchronous sequential networks in fundamental mode, however, is complex, and even after the design we need cumbersome postanalysis because of possible malfunction of the networks.

To solve this problem, the use of raceless flip-flops, which are much more complex than latches, along with a clock to synchronize gate operations (gates are synchronized by pulses supplied from a clock), is becoming more common, as we will discuss in Chapter 8. Raceless flip-flops and clocks completely eliminate the malfunction of sequential networks discussed above, which otherwise is often cumbersome to avoid. Also, design procedures are greatly simplified, making redesigning absolutely unnecessary, and designed networks are often simpler than those in fundamental mode. On the other hand, the speed of networks must be sacrificed. In a sense the simplicity of the design procedures and the reliability of network operation are gained at the expense of the network speed and the network simplicity that are due to simple flip-flops and no use of clocks.

Advantages of Asynchronous Sequential Networks

Asynchronous sequential networks in fundamental mode are still useful when high speed is required, or when networks are so simple that the malfunctions discussed above cannot be a problem (such simple networks can be used as part of large networks). Another advantage is that there is no need for complex provision of clocks. (Clock connections require significant areas on pc boards or IC chips.)

Synthesis of Large Networks

We need to design increasingly complex networks every year. The design of such complex networks by the approach in this or the next chapter is usually too cumbersome and time-consuming. Therefore by the use of flow charts, which is a common practice in computer programming and is becoming popular in the logic design, a large, complex network is broken down, as discussed in Chapter 9, and only the small subnetworks that result may be designed by the procedures in this or the next chapter.

Application of Design Procedures Discussed in This Chapter

The design approaches, particularly with flip-flops, in Section 7.8 are useful when we design synchronous networks in Chapter 8, and also when we design

Publications on Sequential Networks

There are numerous papers and many books on sequential networks. Bibliographies are convenient for finding appropriate papers; see Miller and Pugsley [1970] and pp. 236–261 of Moore [1964]. Some books serve this purpose to some extent, such as Langdon [1974], Kohavi [1970], and Miller [1965]. There are papers on some aspects of sequential networks, for example, Armstrong, Friedman, and Menon [1968], Bredeson and Hulina [1971a], Ettinger and Jacob [1968], Senders and Lucchesi [1971], Schultz [1969]. Networks that are independent of gate speed are discussed in Muller [1955, 1963] and Muller and Bartky [1957, 1959].

Exercises

7.1 (R) Minimize the number of states in one of the flow-output tables in Table E7.1 by Procedure 7.6.1 (show a state-comparison table). Then show the resultant flow-output table, along with a state diagram.

Table E7.1

A		$x_1 x_2$		
s	00	01	11	10
1	5, d	①, 1	①, 0	4, d
2	②, 0	3, 0	②, 1	8, 0
3	6, d	③, 1	③, 0	4, d
4	7, d	④, 1	④, 0	④, 1
5	⑤, 0	⑤, 0	1, 0	4, d
6	⑥, 0	⑥, 0	1, 0	4, d
7	⑦, 0	1, d	2, d	⑦, 0
8	2, 0	3, 1	⑧, 1	⑧, 0

S, z

Table E7.1 (*Continued*)

B

s \ x_1x_2	00	01	11	10
1	6, 1	5, d	①, 0	①, 1
2	②, 0	②, 1	6, 1	3, d
3	6, 1	2, d	③, 0	③, 1
4	6, 1	2, d	④, 0	④, 1
5	⑤, 0	⑤, 1	6, 1	1, d
6	⑥, 1	5, 1	⑥, 1	⑥, 0

S, z

Table E7.2

s \ x_1x_2	00	01	11	10
1	2, 1	①, 1	①, 1	7, d
2	②, 1	②, 0	1, d	7, d
3	③, 1	③, 0	1, d	7, d
4	④, 1	5, 1	④, 0	8, 1
5	3, 1	⑤, 1	⑤, 1	7, d
6	⑥, 1	1, 1	4, d	⑥, 1
7	6, 1	⑦, 1	⑦, 1	⑦, 0
8	4, 1	5, d	⑧, 0	⑧, 1

S, z

7.2 (M−) **(i)** When state A appears in a certain column of a state-output table in fundamental mode but state B does not, can you conclude that states A and B are distinguishable?

Explain the reason, regardless of whether your answer is yes or no.

For example, in Table E7.2 state 8 does not appear in column (0 0) but states 2, 3, 4, and 6 appear in this column. Can you say that state 8 is distinguishable from state 4?

(ii) Minimize the number of states in Table E7.2 by Procedure 7.6.1, showing the resultant flow-output table.

7.3 (M) We want to design a network in fundamental mode that has inputs x_1 and x_2 and a single output z. The value of z changes only in the following case: z changes from the current value to its complement when inputs (x_1, x_2) change from (0 0) to (0 1).

(i) Prepare a flow-output table in primitive form. (It is easier to use a primitive form.) Then prepare a flow-output table (not necessarily in primitive form) with a minimum number of states, discussing why the table has this minimum number of states. (Since the table in primitive form will be incompletely specified, derive the flow-output table by visual inspection of the table in primitive form and the corresponding state diagram on a trial-and-error basis.) Then show a state diagram for the new flow-output table.

(ii) Make a state assignment that is free of multiple changes of internal variables.

7.4(M—) From the incompletely specified flow-output table in Table E7.4, derive a flow-output table with as few states as possible, on a trial-and-error basis (possibly by visual inspection of a state diagram but without the use of Procedure 7.6.2).

Table E7.4

	x_1x_2			
	00	01	11	10
A	Ⓐ, 11	B, —	—, —	D, —
B	A, —	Ⓑ, 00	C, —	—, —
C	—, —	B, —	Ⓒ, 10	D, —
D	A, —	—, —	E, —	Ⓓ, 00
E	—, —	B, —	Ⓔ, 01	D, —

$S, z_1 z_2$

▲ 7.5 (M) Minimize the number of states in the incompletely specified flow-output table in Table E7.5 by Procedure 7.6.2 (show a state-comparison table for its Step 1 and maximal compatibles for its Step 2). Then show the resultant flow-output table along with a state diagram.

Table E7.5

s	\multicolumn{4}{c}{$x_1 x_2$}			
	00	01	11	10
A	—	Ⓐ, 1	F, —	Ⓐ, —
B	Ⓑ, 1	C, —	Ⓑ, 0	—
C	B, —	Ⓒ, 1	Ⓒ, —	—
D	Ⓓ, 1	Ⓓ, 1	—	E, —
E	Ⓔ, 0	A, —	B, —	Ⓔ, 0
F	D, —	A, —	Ⓕ, 0	Ⓕ, 1

S, z

▲ 7.6 (M) Minimize the number of states in Table 7.6.5a by Procedure 7.6.2, showing a result at each step.

▲ 7.7 (M) The maximal compatibles described in Step 2 of Procedure 7.6.2 can also be obtained by the following procedure:
Form an alterm of every pair of states that is not compatible. Then form the conjunction of all these alterms. Derive its complete sum. Then, from each term in this expression, form the product of states missing in this term. All these products constitute all maximal compatibles.
For example, suppose that we have nine states. Then the conjunction of alterms, $(3 \lor 4)(3 \lor 6)(3 \lor 8)(4 \lor 9)(6 \lor 7)(6 \lor 9)(7 \lor 8)$, is formed, where states 3 and 4, for example, are not compatible, yielding the first alterm $(3 \lor 4)$. Its complete sum is $3467 \lor 3689 \lor 379 \lor 468$. Then, from these terms in this order, we obtain maximal compatibles (1, 2, 5, 8, 9), (1, 2, 4, 5, 7), (1, 2, 4, 5, 6, 8), and (1, 2, 3, 5, 7, 9).
Prove why this procedure works. [Marcus 1964].

7.8 (M) For one of the two sets of flow tables in Table E7.8, find a transition table with a state assignment that does not contain multiple changes of internal variables. Avoid the addition of new states as much as possible. Even if you need to add some, use as few new states as possible.

Table E7.8

A (i)

s	\multicolumn{4}{c}{$x_1 x_2$}
	00	01	11	10
A	Ⓐ	C	D	Ⓐ
B	D	Ⓑ	Ⓑ	A
C	A	Ⓒ	B	A
D	Ⓓ	B	Ⓓ	A

S

A (ii)

s	\multicolumn{4}{c}{$x_1 x_2$}
	00	01	11	10
A	Ⓐ	C	Ⓐ	C
B	A	Ⓑ	Ⓑ	D
C	D	Ⓒ	B	Ⓒ
D	Ⓓ	B	A	Ⓓ

S

B (i)

s	\multicolumn{4}{c}{$x_1 x_2$}
	00	01	11	10
A	Ⓐ	B	Ⓐ	D
B	D	Ⓑ	Ⓑ	D
C	D	Ⓒ	A	—
D	Ⓓ	C	A	Ⓓ

S

B (ii)

s	\multicolumn{4}{c}{$x_1 x_2$}
	00	01	11	10
A	Ⓐ	Ⓐ	B	E
B	C	—	Ⓑ	—
C	Ⓒ	A	E	D
D	E	—	—	Ⓓ
E	Ⓔ	A	Ⓔ	Ⓔ

S

7.9 (M) An **up-down counter*** is a counter that has two inputs, x and y, and whose count increases by 1 at each occurrence of $x = 1$ and decreases by 1 at each occurrence of $y = 1$. Assume that $x = 1$ and $y = 1$ cannot occur simultaneously. When the count

---

* There is another type of up-down counter. (See Exercise 7.5.4.)

is (0 0 · · · 0) (i.e., all 0's), it will be (1 1 · · · 1) (i.e., all 1's) at the next occurrence of $y = 1$. We want to design such a counter whose count is displayed by two outputs, z_1 and z_2, where z_1 is the least significant bit.

(i) Show a flow-output table for such a counter in fundamental mode.

(ii) Discuss whether four internal variables are sufficient when we want a state assignment free of multiple changes of internal variables. (If four internal variables are not sufficient, you do not need to show a state assignment with five or more internal variables.)

7.10 (MT) Design a network in double-rail input logic, based on the transition table in Table 7.7.6. The values of the network output z are given in Table 7.7.1. Use S-R latches and any types of gates. Then compare the result with the network in Fig. 7.8.1, designed on the basis of Table 7.8.1, and determine which is simpler.

7.11 (M) Design a network in single-rail logic, based on the flow-output table in Table 7.5.3. Design with NAND gates only, without flip-flops.

7.12 (M) **(i)** Reduce the number of states as much as possible in the state-output table in Table E7.12, given in primitive form.

(ii) Find a state assignment free of multiple changes of internal variables for the resultant table.

Table E7.12

	x_1x_2				
s	00	01	11	10	z
A	—	H	(A)	D	0
B	F	—	E	(B)	1
C	F	(C)	A	D	1
D	G	—	A	(D)	0
E	—	C	(E)	B	1
F	(F)	C	—	B	1
G	(G)	H	E	B	0
H	G	(H)	A	—	0

S

Exercises 435

(iii) Then design a network in double-rail input logic, using any types of gates but no flip-flops.

7.13 (M) We want to install an automatic traffic control system at the intersection of roads 1 and 2. Counters 1 and 2 show the number of cars waiting at the intersection on roads 1 and 2, respectively, the cars being detected by loop wires embedded in the concrete coverings of the roads. A comparator of the two counters has two outputs, x_1 and x_2. If counter 1 shows a count greater by 3 or more than counter 2, we have $x_1 = 1$; otherwise $x_1 = 0$. If counter 2 shows a count greater by 5 or more than counter 2, we have $x_2 = 1$; otherwise $x_2 = 0$. Traffic light z_1 becomes green when x_1 changes from 0 to 1, and traffic light z_2 becomes green when x_2 changes from 0 to 1. If traffic light z_1 becomes green by $x_1 = 1$, z_1 remains green until we have $x_2 = 1$, and vice versa. We want to design a network for the lights that has inputs x_1 and x_2 and outputs z_1 and z_2 ("green" and "red" are to be denoted by 1 and 0, respectively).

(i) Draw a flow-output table in fundamental mode in primitive form.

(ii) Reduce the number of states as much as possible on a trial-and-error basis, and show the resultant table.

(iii) Find a state assignment that is free of multiple changes of internal variables.

(iv) Design a relay contact network, using as few contacts as possible. (This is a simple relay contact network, so try to design it on a trial-and-error basis.)

7.14 (MT) A toy automobile is remotely controlled. The control box has two buttons. When the two buttons are not depressed, the automobile does not move. If either is depressed and remains depressed, the automobile moves straight. If the red button is depressed, and then the black button is depressed (i.e., both buttons are depressed after the red button was depressed first), the automobile turns right; and if either is released, it moves straight. If the two buttons are depressed in the reverse order, the automobile turns left; and if either is released, it moves straight. When both buttons are released, it stops. (Notice the following. If two buttons are depressed, the automobile turns right or left, depending on the order in which the two buttons are depressed. If the red button is released and, after a short while, is depressed again while the black button is held down, the automobile turns left. Similarly, if the black button is released and, after a short while, is depressed again while the red button is held down, the automobile turns right.)

The control box has two outputs, z_R and z_L, such that:

$z_R = z_L = 0$ when the automobile is to stop,
$z_R = 1$ and $z_L = 0$ when the automobile is to turn right,
$z_R = 0$ and $z_L = 1$ when the automobile is to turn left, and
$z_R = z_L = 1$ when the automobile is to move straight.

Write a flow-output table in the primitive form in fundamental mode for the control box. Draw a state diagram.

Design a network in double-rail input logic for the control box, using S-R latches and gates of any types. Use as few latches as possible.

436 Asynchronous Sequential Networks

7.15 (MT) A sequential network is controlled by two keys, k_1 and k_2, and has a single output z.

Output z may change only when a key is depressed; it does not change when a key is released. By depressing k_1, z changes to or remains at value 0 (i.e., if $z = 0$, z remains at 0; if $z = 1$, z changes to and remains at 0). By depressing k_2, z changes to or remains at value 1. By depressing k_1 and then k_2 without releasing k_1 (i.e., both k_1 and k_2 are depressed sometime after k_1 is depressed), z changes to and remains at value 1, after assuming value 0. By depressing k_2 and then k_1 without releasing k_2, z changes to and remains at value 0, after assuming value 1.

(i) Draw a flow-output table in fundamental mode in primitive form.

(ii) Reduce the number of states as much as possible. (Notice that this is an incompletely specified case. Try to reduce on a trial-and-error basis.)

(iii) Find a state assignment free of multiple changes of internal variables.

(iv) Design a network in double-rail input logic, by using S-R latches and any types of gates.

7.16 (MT) We want to design a counter that counts up for each occurrence of signal 1 at input x, in the count sequence 0, 1, 4, 7, and then returns to count 0 at the next occurrence of signal 1.

(i) Show a flow-output table in fundamental mode, with a minimum number of states.

(ii) Show a transition-output table that is free of multiple changes of internal variables.

(iii) Design such a counter, using the minimum number of S-R latches and, as the secondary objective, as few gates of any types as possible.

7.17 (MT) A **divide-by-N counter** is a counter that returns to the original count when it receives the Nth 1 at input x of the counter. (For example, if $N = 2^p$, where p is a positive integer, the counter is a binary counter. If $N = 10$, it is a decimal counter.)

We want to design a divide-by-3 counter that counts up consecutively from count 0 to 2.

(i) Show a flow-output table in fundamental mode, with a minimum number of states.

(ii) Show a transition-output table that is free of multiple changes of internal variables.

(iii) Design such a counter with the minimum number of S-R latches, using as few gates of any types as possible.

7.18 (MT) For the counter stated in Exercise 7.5.4; do the following:

(i) Derive a transition-output table that is free of multiple changes of internal variables.

(ii) Design the counter with a minimum number of S-R latches, using as few gates of any types as possible.

7.19 (M) A logic designer in a firm is trying to design sequential networks without the use of flow-output or state-output tables. His approach is based on the use of truth tables. He first converts a given word statement into a truth table which has the next output value of the network in the output column of the table, where the current network output value (if necessary, along with additional internal variables) is contained in the input columns of the table. Then he constructs a network based on this table, using gates or building blocks such as flip-flops.

His approach can be best explained with an example. Suppose that we want to design a network in single-rail input logic with the following performance, using NAND gates only. Reduce the number of gates as much as possible.

1. If input x_1 equals 1 and input x_2 equals 0, the output f is set to 1.
2. If inputs x_1 and x_3 both equal 0, the output f is set to 0.
3. For all other values of the input variables x_1, x_2, x_3, the value of the output f does not change (i.e., the output f keeps the value which it assumed at the previous occurrence of 1 or 2).

Literally translating the above problem requirement, we obtain the truth table shown in Table E7.19, where f and F denote the current and next values of the network output, respectively. In other words, rows 0, 1, 4, and 5 are given by case 2, and rows 8, 9, 10, and 11 are given by case 1. Other rows are obtained by setting $f = F$ to 0 or 1. For example, rows 2 and 3 are obtained by setting $f = F = 0$ and $f = F = 1$, respectively, since if $f = 0$, $F = 0$ and if $f = 1$, $F = 1$ because the network must retain what it had so far, by case 3. On the basis of this truth table, we can get the following switching

Table E7.19

	x_1	x_2	x_3	f	F
0	0	0	0	0	0
1	0	0	0	1	0
2	0	0	1	0	0
3	0	0	1	1	1
4	0	1	0	0	0
5	0	1	0	1	0
6	0	1	1	0	0
7	0	1	1	1	1
8	1	0	0	0	1
9	1	0	0	1	1
10	1	0	1	0	1
11	1	0	1	1	1
12	1	1	0	0	0
13	1	1	0	1	1
14	1	1	1	0	0
15	1	1	1	1	1

438   Asynchronous Sequential Networks

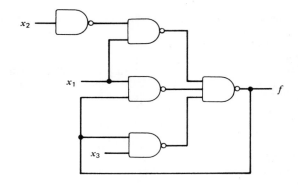

Fig. E7.19

expression, which is transformed into the last expression suitable for implementation with NAND gates, as shown in Fig. E7.19:

$$F = \bar{x}_1\bar{x}_2 x_3 f \vee \bar{x}_1 x_2 x_3 f \vee x_1\bar{x}_2\bar{x}_3 \bar{f} \vee x_1\bar{x}_2\bar{x}_3 f \vee x_1\bar{x}_2 x_3 \bar{f} \vee x_1\bar{x}_2 x_3 f$$
$$\vee\, x_1 x_2 \bar{x}_3 f \vee x_1 x_2 x_3 f = x_1\bar{x}_2 \vee x_1 f \vee x_3 f = \overline{(x_1\bar{x}_2)(\overline{x_1 f})(\overline{x_3 f})}.$$

(i) In the table the value of F in row 1 differs from that of f. This means that this row is an unstable state. This state changes into the stable state of row 0, as indicated by an arrow in the table. If there were no such stable state, the network would oscillate. There are three other unstable states that change into stable states, as shown with arrows.

When there is an oscillation, how should we modify the above design approach in order to derive a network that has no oscillation?

(ii) Although only one internal variable f is sufficient for this problem, we might, in general, need more internal variables for other problems. If so, what should we do to derive a workable network?

(iii) The designer's heuristic approach might be convenient in certain cases because of simplicity, but it appears to lack generality. Discuss the conceivable shortcomings of his design approach, particularly in comparison with the approaches discussed in this chapter based on flow-output or state-output tables.

7.20 (MT) We want to design a network to control an elevator for a two-story building. The first and second floors have buttons B_1 and B_2 for "up" and "down," respectively. The elevator motor is controlled by three S-R latches for "move upward," "move downward," and "stop." (For example, if the "move upward" latch is on, the elevator moves upward.) Hence at any time exactly one of these three latches is on. The network has an input to sense the floor position at each floor.

Inside the elevator is an "on" button, C. When C is pushed, the elevator closes its door and starts to move up or down.

Assume that, once the elevator starts to move, it cannot move in the other direction unless it stops at a floor. (Any extra conditions that you consider necessary can be arbitrarily set up according to your commonsense judgment.)

(i) Draw a flow-output table in fundamental mode, with as few states as possible.

(ii) Design the network, using as few S-R latches and, if necessary, gates of any types as possible.

CHAPTER **8**

Synthesis of Synchronous Sequential Networks

In Chapter 7 we discussed the basic concepts of sequential networks, and then asynchronous sequential networks, that is, sequential networks without clocks.

In this chapter synchronous sequential networks, that is, sequential networks with clocks, is discussed. Although the speed of networks is sacrificed by the use of clocks, some hazards of asynchronous sequential networks can be eliminated, since the operations of some or all gates in a network are synchronized by clocks. It is important to notice that sequential networks with clocks can be in fundamental mode. Sequential networks with clocks in fundamental mode can be synthesized in exactly the same manner as the case in Chapter 7, treating the clock as an additional input.

Sequential networks can be synthesized using more sophisticated flip-flops than latches, along with clocks. With these flip-flops, which are called raceless flip-flops, network malfunctioning due to hazards (including races) can be completely eliminated. Design of sequential networks with raceless flip-flops and clocks is much simpler than that of networks in fundamental mode, since we can use simpler flow-output and state-output tables, which are said to be in skew mode, and multiple changes of internal variables need not be considered in state assignment.

8.1 Clocked Networks

A **clock**, which is a periodic series of pulses, as shown in Fig. 8.1.1a, is usually used in a computer, or digital system, to control the operations of gates and flip-flops. A clock is fed to the input c of the AND gate, for example, as shown in Fig. 8.1.1b. (This is one way to control the operation of a gate by a clock. There are many other ways, some of them based on elaborate electronic means.) Then, even if signal inputs x_1 and x_2 have noise before or after the presence of each clock pulse, the output f of the AND gate shows a clean

Clocked Networks 441

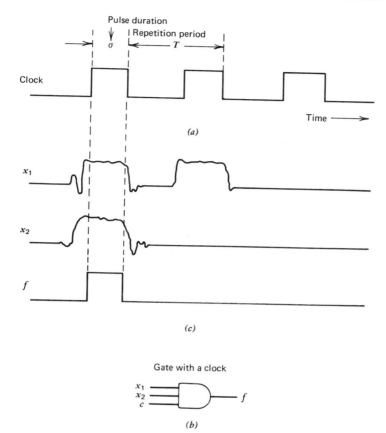

Fig. 8.1.1 Clock.

waveform at the exact time specified by the clock, as shown in Fig. 8.1.1c. In other words, clock pulses are used to sample or probe signal waveforms at specified times. **The operation of gates in a network is thus synchronized**, even though these gates have input signals that rise and fall at different times. Also, **the possibility of generating spurious output values of the gate, such as the hazards discussed in Section 6.7, can be eliminated**, as we will explain later in this section (Fig. 8.1.4). In addition to these two advantages, **networks sometimes can be simplified by proper use of clock pulses**, as we will see later, because gates that are necessary to avoid network malfunction sometimes can be eliminated.

442 Synthesis of Synchronous Sequential Networks

Clocked Latches

When networks are to be clocked, latches whose operations are controlled by clocks are useful.

The S-R latch controlled by a clock* is shown in Fig. 8.1.2, where Fig. 8.1.2a is an example of implementation by NAND gates (a clock is connected to the latch of Fig. 7.1.1d), and Fig. 8.1.2b shows its symbolic representation. In addition to this clocked S-R latch, there are many variations of flip-flops with different additional functions. These flip-flops are often constructed based on S-R latches. Here let us discuss another variation of flip-flop.

The outputs of the S-R latch are not defined for input combination $R = S = 1$. Since this is inconvenient for some applications, flip-flops whose outputs are defined for every combination of inputs are desired. The **J-K latch** shown in Fig. 8.1.3a is such a flip-flop. Since the two AND gates are gated by the outputs of the S-R latch inside this J-K latch, only one of the two

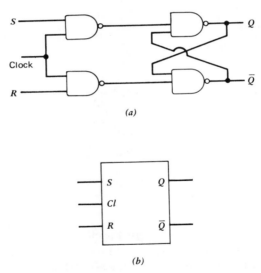

Fig. 8.1.2 Clocked S-R latch.

---

* In reality, the signal value at S or R and the clock pulse change gradually, as shown in the dotted line in Fig. 7.1.2a. Thus, if signal value 1 at S or R disappears when a clock pulse appears (i.e., the signal at S or R and the clock pulse change simultaneously), the clocked S-R latch may malfunction. (See Chaney and Molnar [1973] and other papers cited in Remark 8.2.1.) Designers usually avoid this timing relation.

AND gates can have output value 1, and consequently the S-R latch never has $R = S = 1$ for any combination of values of inputs J and K (except possibly during the transition period of the S-R latch). When $J = K = 1$, only one of the AND gates has output signal 1 (e.g., if $\bar{Q} = 1$, the AND gate with J has output signal 1), and the output Q assumes the complement of the previous value of Q (accordingly, $\bar{Q}$ also assumes the complement of its previous value). The S-R latch is clocked in order to probe the input signals at S and R, avoiding the transition period of Q and $\bar{Q}$. Similarly, the new output values can be found for other combinations of values of J and K. The input-output relationship of the J-K latch, which is identical to that of an S-R latch, except for $J = K = 1$, is shown in Fig. 8.1.3b. The clocked J-K latch is denoted as Fig. 8.1.3c.

However, as we can easily detect, the clocked J-K latch in Fig. 8.1.3a oscillates if the clock pulse and $J = K = 1$ last too long (Exercise 8.2.2). Since this is usually the case, it is not easy to use this clocked J-K latch unless we have a short clock pulse or a signal source that generates a short signal 1 (or unless an auxiliary network that shortens signal 1 is attached to the output of the signal source). Instead, we usually use raceless flip-flops, to be discussed later, which have more complicated structures but do not present this problem.

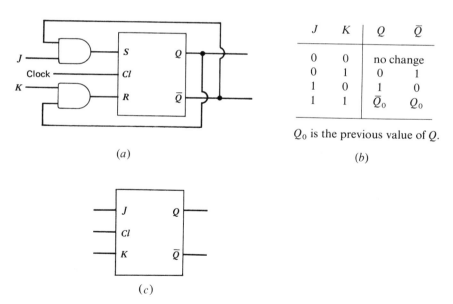

J	K	Q	$\bar{Q}$
0	0	no change	
0	1	0	1
1	0	1	0
1	1	$\bar{Q}_0$	Q_0

Q_0 is the previous value of Q.

(a)　　(b)

(c)

Fig. 8.1.3 *Clocked J-K latch.* (a) *Clocked J-K latch.* (b) *Input-output relationship.* (c) *Symbol for clocked J-K latch.*

444 Synthesis of Synchronous Sequential Networks

Elimination of Spurious Signals by Clock

Some hazards can be eliminated by controlling the output gate of a network by a clock. For example, if gate 3 in Fig. 6.7.3a generates spurious output values because of the extrinsic hazard (caused by the difference in switching times of gates 1 and 2), we can eliminate them by connecting a clock, as shown in Fig. 8.1.4a. This remedy is different from those discussed in Section 6.7. Only during the presence of clock pulses can the output of the NOR gate be 1; during the absence of clock pulses it is always 0. If the transition at f occurs only during the absence of clock pulses, the S-R latch is not set erroneously. Another remedy is the use of a clocked latch, as shown in Fig. 8.1.4b, which responds to inputs only during the presence of clock pulses. In other words, any spurious signals that may occur from combinational or sequential networks because of intrinsic or extrinsic hazards (e.g., the short, identical output values at a latch explained with Fig. 7.1.2b) can be prevented, by a clock, from setting flip-flops to wrong output values.

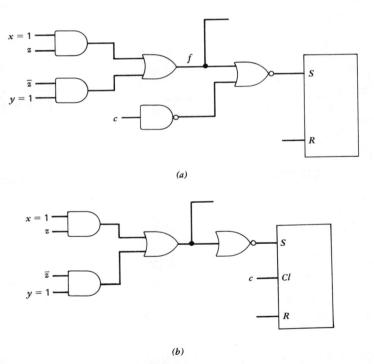

Fig. 8.1.4 Elimination of hazards by clock.

Clocked Networks

Every computer that is currently used, with almost no exception, has a clock (whether or not some or all gates in the computer are clocked), and the use of a clock is a simple remedy for hazards (as a matter of fact, a major reason why nearly every digital system has a clock is to avoid hazards). But when a very high speed network is to be realized, the pulse repetition period T in Fig. 8.1.1a must be shortened. Since it is hard to shorten the pulse duration beyond a certain limit (if the pulse duration is very short, gates must work very rapidly, but this is not always possible), this means that the intervals between adjacent pulses must be shortened. Then the transient behavior of a network, which is to be finished during the interval between adjacent clock pulses, may show up during the presence of the next clock pulse. Also, the distortion of waveforms of signals or clock pulses complicates the problem. In other words, the avoidance of hazards by the use of clock pulses is difficult if we want to have a very high speed network. In this case we must use the remedies discussed in Section 6.7 without the use of clocks.

Synchronous and Asynchronous Networks

Clock pulses must be distributed throughout a computer or a digital system, usually supplied from a central clock generator. (In the following figures, the central clock generator is not shown for the sake of simplicity.) Clock pulses must be distributed with equal time delays to different places in the computer, because if the time delays are different, the computer may malfunction.

Not all the gates in a network in a computer or digital system need to be provided with clock pulses. Only gates at important positions (as a matter of fact, usually some flip-flops) in the network are provided with clock pulses, as illustrated in Fig. 8.1.5. Gates or subnetworks that are not synchronized by clock pulses are inserted between these clocked gates (or clocked flip-flops). Almost all computers with bipolar transistors (some computers with MOS also) work in this manner. Sometimes, however, all the gates in a network are provided with clock pulses. An example is a **dynamic MOS network**, as shown in Fig. 8.1.6a, where all the gates are synchronized by two different sequences of clock pulses shown in Fig. 8.1.6b, that is, a 2-phase clock, through clock terminals c_1 and c_2. (Load and driver MOSFETs in dynamic MOS networks are connected to clock pulse sources but not to any direct-current power source.) The number of different clock waveforms is called the **number of phases**. (See, e.g., Carr and Mize [1970].)

A network in which all the gates are synchronized with a clock, such as the above dynamic MOS network, has the following advantages, though its disadvantage is the need for more connections and sometimes also for more gates. First, since every gate is synchronized, there is much less chance of malfunction (the racing problems discussed in Section 7.3 do not occur).

446 *Synthesis of Synchronous Sequential Networks*

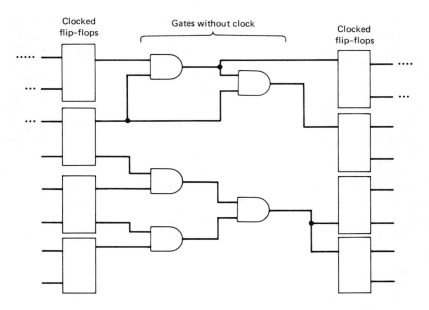

Fig. 8.1.5   Distribution of clock.

Second, electronic performance is sometimes improved, though the extent of the improvement is different depending on the devices and electronic circuits (or IC logic families) used to implement the gates. For example, MOS networks whose gates are all clocked are often used because of low power consumption. The dynamic MOS network in Fig. 8.1.6 is such an example. This differs from the **static MOS networks** in the preceding chapters, where not all gates are clocked.

In a network where some or all gates are clocked, the operation of the entire network is synchronized by clock pulses, and such networks are called **synchronous networks**. (Notice that, even if not all gates in a network are clocked, the operation of the entire network can be completely synchronized.) In contrast to synchronous networks, networks that have no clock are called **asynchronous networks**. Synchronous and asynchronous computers can be similarly defined.

However, the definitions of "synchronous" and "asynchronous" networks are often ambiguous. Sometimes only networks in which all the gates are clocked are called synchronous, and those in which only part of the gates or no gates are clocked are termed asynchronous (especially in the case of static MOS networks). Also, a computer that consists of networks with nonsynchronized clocks may be called an asynchronous computer.

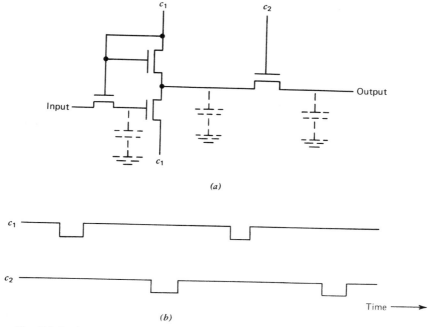

Fig. 8.1.6 (a) *Dynamic MOS network with 2-phase clock. (This is a one-bit position of a shift register. Parasitic capacitances, shown in the dotted line, are essential for the operation of this network.)* (b) *Clock pulse waveforms to be supplied at terminals c_1 and c_2 in Fig. 8.1.6a.*

Although asynchronous computers (defined above as those that have no clock) can usually be designed with higher speed than synchronous computers, where clock cycles set their speeds, asynchronous computers are rarely manufactured, since the timing relationship in the operations of gates is very complicated and therefore design is cumbersome. But notice that asynchronous networks are often used in otherwise synchronous computers.

Clock Pulse Durations and Intervals

The ratio of the **pulse duration** (σ in Fig. 8.1.1a) to the **pulse repetition period** (T in Fig. 8.1.1a) is called the **duty cycle**.* In some cases clock pulses in a clock waveform may not be equally spaced, or clock durations in a clock waveform may not be equal.

---
* Many large computers have a single-phase clock with a duty cycle of 50 percent. Many minicomputers have multiphase clocks with lower duty cycles.

448 Synthesis of Synchronous Sequential Networks

In particular, let us consider networks where some but not all gates are clocked, since such networks are often used. In this case the duration of a clock pulse is usually much longer than the switching time of a gate. (This is so because clock pulses must be distributed throughout a digital system, and consequently their waveforms will be too distorted unless the durations are wide enough.) Furthermore, intervals between adjacent clock pulses must be longer than the propagation time of a signal from a clocked gate to the next clocked gate after traveling through unclocked gates, because, in order to avoid supplying spurious output values to the inputs of the next clocked gates, the outputs of the unclocked gates must settle down before the next clock pulse is applied to initiate the action of the next clocked gate.

Analysis of Clocked Networks in Fundamental Mode

An example of a clocked sequential network that contains an *S-R* latch is shown in Fig. 8.1.7. Clock pulses, denoted by c, serve mainly to probe output values of some gates at a fixed interval. Thus let us assume that **level signals x_1 and x_2 cannot change during the presence of the clock pulses**. In other words, not only is the duration of the input signals longer than that of the clock pulses, but also the input signals do not change during the clock pulses. Assume that the network operates in fundamental mode. In other

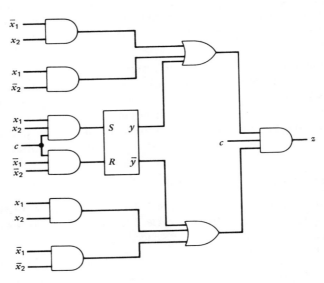

Fig. 8.1.7 A clocked sequential network in fundamental mode.

words, at each leading edge of a clock pulse the network can make a transition to a new stable state, and it stays there as long as $c = 1$. Then, at the trailing edge of the clock pulse, the network can make another transition to a new stable state. (This is usually the case in practice, since the clock pulse duration is usually much longer than the switching time of the gates, as already explained.) The output $z = 1$ can appear only during the presence of a clock pulse at the output gate, since the output gate is synchronized by the clock.

The analysis of the network is shown in Table 8.1.1. The columns for $c = 0$ are separated from those for $c = 1$ by the double lines, where $c = 0$ and $c = 1$ represent the absence and the presence of a clock pulse, respectively. As we can see easily by examining the network in Fig. 8.1.7, internal variable y does not change during $c = 0$, so all the entries in the four columns for $c = 0$ in Table 8.1.1a must be stable states. Thus all the entries in each row in the columns for $c = 0$ are identical to the value of y in that row. Consequently, all the columns for $c = 0$ in Table 8.1.1a can be condensed into a single column, as shown in Table 8.1.1b. Also, in the output table in Table 8.1.1c, all the columns for $c = 0$ are again condensed into a single column, since the output $z = 1$ can appear only during the presence of clock pulses because the output gate is clocked and all the entries in all the columns for $c = 0$ are $z = 0$.

Notice, however, that **generally this condensation of all columns for $c = 0$ into a single column may not be possible**, because in some other networks the change of the network state or outputs may take place because of a change of input variables x_i's, even during $c = 0$, if for example, gates without clock inputs are connected to the inputs of the latches, or output gates have no clock inputs.

Let us illustrate how Table 8.1.1b works. Suppose that the network is in the state $(c, y) = (0\ 0)$ and we have $x_1 = x_2 = 1$. When a clock pulse $c = 1$ arrives, the network moves to the state $(x_1, x_2, y) = (1\ 1\ 0)$ in the same row, as shown by the dotted-line arrow in Table 8.1.1b. Since this state is unstable, the network goes to the stable state $(x_1, x_2, y) = (1\ 1\ 1)$ in the same column, as shown by the solid-line arrow, and stays in this new stable state as long as $c = 1$. When the clock pulse disappears, the network moves to the stable state $(c, y) = (0, 1)$ in the same row.

The performance of the clocked network in Fig. 8.1.7 can now be described by the state-output table (or flow-output table) and state diagram shown in Fig. 8.1.8. In the state diagram in Fig. 8.1.8b, the clock input c is omitted for the sake of simplicity, since transitions obviously take place only during the presence of clock pulses. The output z during $c = 0$ can be, but is not shown inside the circles, since clock pulses are used to probe signals only during their presence, and z is always 0 during $c = 0$.

Table 8.1.1 Analysis of the Clocked Network in Fig. 8.1.7

a. **Transition Table**

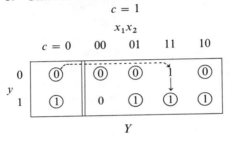

		$c=0$					$c=1$		
		00	01	11	10	00	01	11	10
y	0	0	0	0	0	0	0	1	0
	1	1	1	1	1	0	1	1	1

Y

b. **Condensed Transition table**

		$c=0$	$c=1$			
			00	01	11	10
y	0	⓪	⓪	⓪	1	⓪
	1	①	0	①	①	①

Y

c. **Condensed output Table**

		$c=0$	$c=1$			
			00	01	11	10
y	0	0	0	1	0	1
	1	0	1	0	1	0

z

The performance of the network in Fig. 8.1.7 can also be described by the following switching expressions, which can be easily obtained from Table 8.1.1:

$$Y = cx_1x_2 \vee x_1y \vee x_2y \vee \bar{c}y \quad \text{and} \quad z = c(x_1 \oplus x_2 \oplus y).$$

Sometimes switching expressions may be convenient to see functional relationships. For example, z is the parity function of x_1, x_2, and y during $c = 1$, a relationship that is easier to grasp conceptually than in tables.

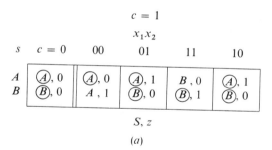

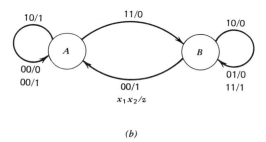

Fig. 8.1.8 Description of the clocked network in Fig. 8.1.7. **(a)** *State-output table.* **(b)** *State diagram.*

The analysis of a network where all the gates are synchronized can be treated similarly.

Synthesis of Clocked Sequential Networks in Fundamental Mode

Clocked sequential networks in fundamental mode can be synthesized in the same way as the asynchronous sequential networks in fundamental mode discussed in Chapter 7, treating clock input c as a variable in the same way as input variables $x_1, x_2, \ldots, x_n$. This is simply a reversal of the analysis procedure explained with respect to Table 8.1.1 and Fig. 8.1.8.

8.2 Raceless Flip-Flops

Raceless flip-flops are flip-flops that have complex structures but eliminate network malfunctions due to hazards.

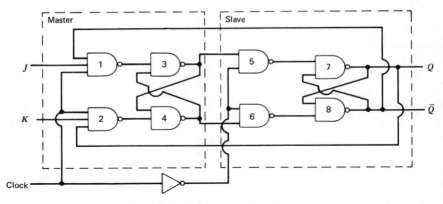

Fig. 8.2.1 **J-K** *master-slave flip-flop.*

J-K Master-Slave Flip-Flop

A **master-slave flip-flop** (or, simply, an MS flip-flop) consists of a pair of flip-flops called a master flip-flop and a slave flip-flop. The *J-K* master-slave flip-flop, for example, is shown in Fig. 8.2.1.* (Other types of master-slave flip-flops will be discussed later.) Its symbol is shown in Fig. 8.2.2, where the letters *MS* are shown inside the rectangle. Each action of the master-slave flip-flop is controlled by the leading and trailing edges of a clock pulse, as explained in the following paragraphs. Briefly speaking, the leading edge of a clock pulse isolates the slave from the master and reads in the input information to the master. The trailing edge isolates the *J*, *K* inputs from the master and then transfers the information from the master to the slave.

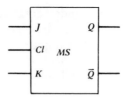

Fig. 8.2.2 Symbol for **J-K** *master-slave flip-flop.*

---

* For the sake of simplicity, all the gates in these flip-flops are assumed to have equal switching times. But in actual electronic implementations, this may not be true.

Operation of a *J-K* Master-Slave Flip-Flop

The *J-K* master-slave flip-flop in Fig. 8.2.1 works with the clock pulse in Fig. 8.2.3 as follows, where the rise and fall of the pulse are exaggerated for illustration. When the clock has the value 0, NAND gates 1 and 2 have outputs 1, and then the flip-flop consisting of gates 3 and 4 does not change its state. As long as the clock stays at 0, gates 5 and 6 force the flip-flop consisting of gates 7 and 8 to assume the same outputs as those of the flip-flop consisting of gates 3 and 4 (i.e., the former is slaved to the latter, which is the master).

When the clock starts to rise to the lower threshold at time t_1, gates 5 and 6 are disabled—in other words, the slave is cut off from the master. (The lower threshold value of the clock pulse still presents the logic value 0 to gates 1 and 2. The clock waveform is inverted to gates 5 and 6 through the inverter. The inputs to gates 5 and 6 from the inverter present the logic value 0 also to gates 5 and 6, because the inverted waveform is still close to logic value 1 but is not large enough to let gates 5 and 6 work. The inverter is actually implemented by a diode which is forward-biased, so that the network works in this manner.)

When the clock pulse reaches the upper threshold at t_2, gates 1 and 2 are enabled, and the information at J or K is read into the master flip-flop through gate 1 or 2. Since the slave is cut off from the master by disabled gates 5 and 6, the slave does not change its state, maintaining the previous output values.

When the clock pulse falls to the upper threshold at t_3 after its peak, gates 1 and 2 are disabled, cutting off J and K from the master. In other words, the outputs of 1 and 2 become 1, and the master maintains the current output values.

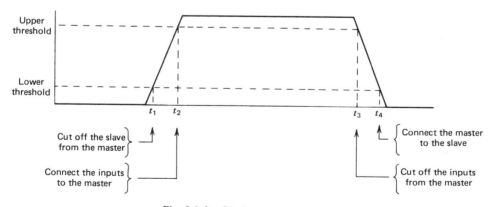

Fig. 8.2.3   Clock pulse waveform.

When the clock pulse falls further to the lower threshold at t_4, gates 5 and 6 are enabled and the information stored at the master is transferred to the slave, gates 7 and 8.

The relationship between the combination of J and K, and the value of Q after the trailing edge of the clock pulse, is not different from that in Fig. 8.1.3b.

The use of J-K master-slave flip-flops (and also other types of flip-flops) can be facilitated by the addition of a **direct reset-input terminal R_D** and a **direct set-input terminal S_D**, though these input terminals are omitted in Fig. 8.2.1 and in the figures for other flip-flops for the sake of simplicity. Input $\bar{S}_D$, that is, the complement of S_D, is connected to gates 3 and 7. Also, $\bar{R}_D$ is connected to gates 4 and 8. These input terminals are convenient for initial setting or resetting. If we want to set the J-K master-slave flip-flop to $Q = 1$, S_D is set to 1. If we want to reset to $Q = 0$, R_D is set to 1. For normal operation, S_D and R_D are kept at 0. When the J-K master-slave flip-flop has direct set- and reset-input terminals, these input terminals are added to the symbol in Fig. 8.2.2.

Features of Master-Slave Flip-Flops

The important feature of the master-slave flip-flop is that the reading of information into the flip-flop and the establishment of new output values are done at different times; in other words, the outputs of the flip-flop can be completely prevented from feeding back to the inputs while the network that contains the flip-flop is still in transition. Now we have a J-K flip-flop that works reliably regardless of how long a clock pulse or signal 1 at terminal J or K lasts, since input gates 1 and 2 in Fig. 8.2.1 are gated by output gates 7 and 8, which do not assume new values before gates 1 and 2 are disconnected from J and K. (In the case of the clocked J-K flip-flop in Fig. 8.1.3a, new output values are immediately fed back to the input gates, and this makes this flip-flop unreliable.) As we will see later, sequential networks that work reliably can be constructed compactly with master-slave flip-flops, without worrying about hazards.

Notice that, like the S-R latch in Fig. 7.1.1a, **outputs Q and $\bar{Q}$ of the J-K master-slave flip-flop** (and also other types of flip-flops to be discussed later) **briefly assume identical values during transition**, even if the flip-flop works properly. But all J-K master-slave flip-flops (also other types of flip-flops) in a network do not respond to their input changes, during this transition, and consequently this spurious signal (i.e., the identical values of Q and $\bar{Q}$) has no chance of causing hazards (even if the outputs of a particular flip-flop under consideration are fed back to its inputs, or supplied to other flip-flop's inputs). Similarly, possible extrinsic or intrinsic hazards from the gates without a clock in Fig. 8.1.5 cannot reach master-slave flip-flops in the next level. Thus design of networks is greatly simplified with the use of master-slave flip-flops, since we do not need to worry about hazards.

The use of a pair of flip-flops in the master-slave flip-flop appears wasteful, but it is essential for elimination of hazards.

If the master-slave flip-flop is implemented as a single IC package, its cost is almost equal to that of a latch, since the cost of extra gates is much smaller than other costs, such as those of a container and labor, in the case of such a small IC package. However, if a large number of flip-flops are contained in the chip, the use of latches may occasionally be desirable for economy if we can use them without hazards, since master-slave flip-flops occupy a larger chip area than latches. Also, when the outputs of flip-flops are connected to displays (e.g., light-emitting diodes) or clocked networks, latches are completely satisfactory. In these cases we do not need to use master-slave flip-flops; latches are sufficient.

T (Type) Master-Slave Flip-Flop

The T (type) master-slave flip-flop (or toggle flip-flop, trigger flip-flop, or T flip-flop) has only a single input, labeled T, as shown in Fig. 8.2.4a. Whenever we have $T = 1$ during the clock pulse, the outputs of the flip-flop change

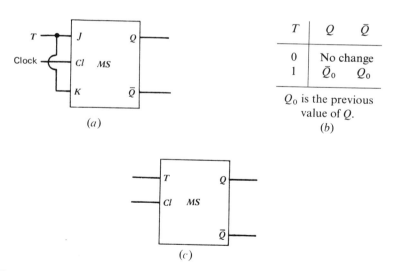

Fig. 8.2.4 T-type master-slave flip-flop. (a) Implementation of a T-type master-slave flip-flop based on J-K master-slave flip-flop. (b) Input-output relationship. (c) Symbol for T-type master-slave flip-flop.

456 Synthesis of Synchronous Sequential Networks

(if $Q = 1$, Q becomes 0, and if $Q = 0$, Q becomes 1), as shown in Fig. 8.2.4b. The flip-flop in Fig. 8.2.4a is essentially the J-K master-slave flip-flop in Fig. 8.2.2 with J and K tied together as T. The T-type flip-flop, denoted as Fig. 8.2.4c, is often used in counters.

D (Type) Master-Slave Flip-Flop

The D (type) master-slave flip-flop (D implies "delay")* has only a single input D, and an implementation is shown in Fig. 8.2.5a, using an extra inverter. As shown in Fig. 8.2.5b, no matter what value Q has, Q is set to the value of D that is present during the clock pulse. The D-type flip-flop is used for delay of a signal or data storage and is denoted by the symbol shown in Fig. 8.2.5c.

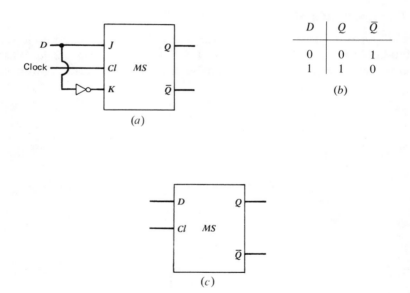

Fig. 8.2.5 *D-type master-slave flip-flop.* **(a)** *Implementation of a* **D**-*type master-slave flip-flop based on* **J-K** *master-slave flip-flop.* **(b)** *Input-output relationship.* **(c)** *Symbol for* **D**-*type master-slave flip-flop.*

---

* The word "latch" often means a D-type flip-flop that is clocked but not a master-slave (see catalogs of Texas Instruments, Motorola, and Fairchild). If we want to differentiate this type from an S-R latch, we need to call it a D latch.

Edge-Triggered Flip-Flops

Edge-triggered flip-flops are another type of raceless flip-flops. When either the leading or the trailing edge of a clock pulse (not both*) causes a flip-flop to respond to an input and then immediately disengage the input from the flip-flop by the feedback of some gate output to its input gates, the flip-flop is called an **edge-triggered flip-flop**. If networks are designed with edge-triggered flip-flops, no network malfunction occurs as in the case of the master-slave flip-flop, because the inputs are disengaged from the edge-triggered flip-flop shortly after the flip-flop accepts its new input value at the edge of a clock pulse, and they are not engaged until the edge of the next clock pulse.

Edge-triggered flip-flops that respond at leading edges of a clock are called **positive edge-triggered flip-flops**, and those that respond at trailing edges are termed **negative edge-triggered flip-flops**. One is obtained from the other simply by inverting a clock, so usually they need not be differentiated.

As an example of this type of flip-flop, an edge-triggered J-K flip-flop is shown in Fig. 8.2.6. Once the flip-flop consisting of gates 5 and 6 assumes new output values, gates 1 and 2 immediately become insusceptible to input changes. (Detailed analysis of Fig. 8.2.6 is left to readers as Exercise 8.2.4.)

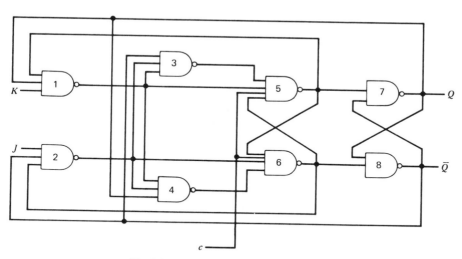

Fig. 8.2.6 Edge-triggered J-K flip-fllop.

---

* The master-slave flip-flops may be interpreted to be edge-triggered for both leading and trailing edges of clock pulses.

458 Synthesis of Synchronous Sequential Networks

As another example of this type of flip-flop, an edge-triggered D-type flip-flop is shown in Fig. 8.2.7. Each of the three dotted-line rectangles is the dotted-line rectangle in the S-R latch in Fig. 7.1.1d (i.e., an $\bar{S}$-$\bar{R}$ latch). As long as all inputs to each dotted-line rectangle are kept as 1, the outputs of the two NAND gates in each dotted-line rectangle maintain the same values. When the clock is 0, the dotted-line rectangle F_3 has inputs of 1, and Q and $\bar{Q}$ continue to maintain the previous output values. (F_3 is a slave flip-flop, and F_1 and F_2 represent a kind of master flip-flop.) When the clock changes from 0 to 1 (at the so-called positive edge), the value of the input D is transferred to Q after a small multiple of τ, where τ is the switching time of each NAND gate. Then Q continues to maintain the value during the clock pulse and also after the clock falls to 0. (Detailed analysis of Fig. 8.2.7 is left to readers as Exercise 8.2.5.)

Notice that, **unlike edge-triggered flip-flops, which become insusceptible to inputs immediately after the flip-flops accept new input values, master-slave**

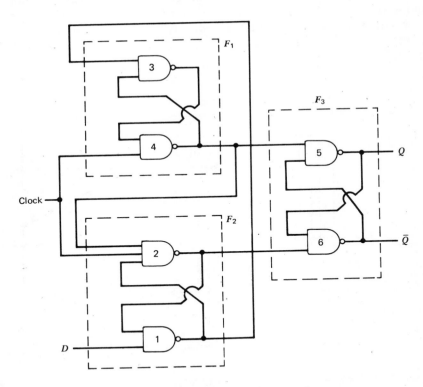

*Fig. 8.2.7 Edge-triggered **D**-type flip-flop.*

flip-flops are susceptible to new input values roughly throughout the pulse durations, as compared in Fig. 8.2.8. For example, a J-K master-slave flip-flop is set to 1 (or 0) if J (or K) maintains value 1 for at least 2τ time at any time from t_2 to t_3 in Fig. 8.2.3. The value of J immediately before t_3 is not particularly important. Thus master-slave flip-flops and edge-triggered flip-flops have completely different periods of susceptibility to input changes. But D master-slave flip-flops (Fig. E8.2.7) are not different from D edge-triggered flip-flops (Fig. 8.2.7) in this respect. Both D edge-triggered flip-flops and D master-slave flip-flops are set to new values by the input values that are present immediately before the trailing edges of pulses.

Edge-triggered flip-flops are particularly useful when we want a network responsive to inputs only during a short, prespecified period (e.g., inputs from keyboards). In other cases, use of edge-triggered flip-flops or master-slave flip-flops usually does not make much difference.

Different Flip-Flops

Master-slave flip-flops and edge-triggered flip-flops are collectively called **raceless flip-flops** or simply **flip-flops**. The S-R flip-flop discussed in Section 7.1 is called an **S-R latch**, and a J-K flip-flop and a D flip-flop with similar structures (i.e., without masters and slaves), such as those shown in Fig. 8.1.3a, are called a **J-K latch** and a **D latch**, respectively. A D latch is sometimes called simply a **latch**. There are other varieties of flip-flops. (See, for example, Lenk [1972] and catalogs published by semiconductor manufacturers.)

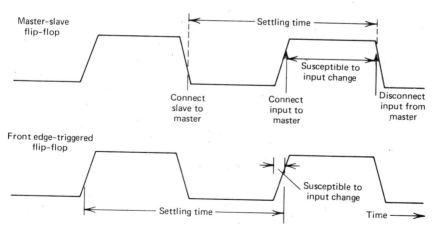

Fig. 8.2.8 Clock waveforms for comparison of susceptibility of master-slave and edge-triggered flip-flops.

460 Synthesis of Synchronous Sequential Networks

Remark 8.2.1: For detailed discussion of flip-flops, see, for example, Taub and Schilling [1977], Millman and Halkias [1972], Strauss [1970], Barna and Porat [1973], and IC Applications Staff of TI [1971]. Different ways to control inputs to flip-flops are described in Delhom [1968]. Electronic circuit design of S-R latches is discussed, for example, in Sifferlen and Vartanian [1970] and Delhom [1968]. Criticisms of IC flip-flops are given in Forbes [1971] and Percival and Gray [1972]. The network malfunction that may occur if a clock pulse appears when signal 1 disappears (i.e., the transition of a clock pulse shown by the slow slope in Fig. 8.2.3 occurs when signal 1 has the transition to 0) is discussed in Chaney, Ornstein, and Littlefield [1972], Chaney and Molnar [1973], Mayne and Moore [1974], Pečhouček [1976], and Catt [1966]. ∎

Exercises

8.2.1 (M) A student wants to use the network in Fig. E8.2.1 as a T-type flip-flop, where an S-R latch and two AND gates constitute the network. Describe how the network behaves, in terms of the duration of the input x in relation to the switching time of the gates (including those in the latch), and also in terms of the interval between two consecutive pulses coming in x. Can the network work as a T-type flip-flop under an appropriate assumption regarding the duration of the input x? If we replace the latch by a J-K master-slave flip-flop (clock c must be supplied to the flip-flop), does the network operate properly?

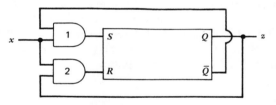

Fig. E8.2.1

8.2.2 (M) (i) Unless the duration of a clock pulse, the interval between adjacent clock pulses, and the switching time of gates have an appropriate ratio, the network in Fig. E8.2.2(i) does not work as a J-K latch. Discuss the performance of the network, and determine how long or short (describe in multiples of τ) each of the above should be so that the network works as a J-K latch, where τ is the switching time of each gate.

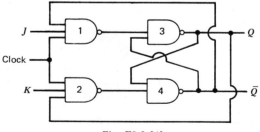

Fig. E8.2.2(i)

(ii) Discuss whether the network in Fig. E8.2.2(ii) can be used as a clocked *J-K* flip-flop without malfunctioning, if we assume that every gate has equal switching time τ. [Inabe and Kataoka 1977]

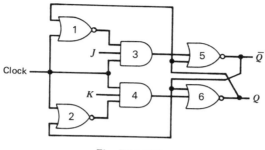

Fig. E8.2.2(ii)

8.2.3 (M) Let us analyze the performance of the *J-K* master-slave flip-flop in Fig 8.2.1.

(i) How long should the input maintain the same value in order to let the flip-flop assume the new output value without malfunction? Describe the duration as a multiple of τ, where τ is the switching time of each gate.

(ii) During what time interval on the clock waveform in Fig. 8.2.3 is the flip-flop susceptible to the inputs of the flip-flop? Describe the interval as a multiple of τ.

8.2.4 (MT) Let us analyze the edge-triggered *J-K* flip-flop shown in Fig. 8.2.6. Assuming that every gate has switching time τ, answer the following questions, expressing durations and other times as multiples of τ.

(i) When and for how long must *J* and/or *K* assume a new value in order to make the flip-flop work correctly? When can *J* and/or *K* return to the original value without making the network malfunction? How long must *c* maintain the value 1 in order to make the network operate properly?

(ii) Find the delay time, until the outputs *Q* and *Q̄* establish new values, corresponding to each combination of values of *J* and *K*. (Count it from the instant when *c* changes.)

(iii) When *c* and *J* and/or *K* change from 0 to 1 simultaneously, can the network operate correctly? If not, find an appropriate timing of the changes of *c* and *J* and/or *K*. Also determine when the network settles down.

(iv) Do *Q* and *Q̄* assume the same value at any time?

8.2.5 (M) Let us analyze the edge-triggered *D*-type flip-flop in Fig. 8.2.7. Assuming that every gate has switching time τ, answer the following questions, expressing durations and other times as multiples of τ.

(i) When and for how long must *D* assume a new value in order to make the flip-flop work correctly? Assume that each clock pulse has a sufficiently long duration. When can *D* return to the original value without making the network malfunction? How long must *c* maintain the value 1 in order to make the network operate properly?

462 Synthesis of Synchronous Sequential Networks

(ii) Find the delay time until the outputs of F_3 settle to the values corresponding to a new value of D. (Count it from the instant when c changes.)

(iii) When D and c change from 0 to 1 simultaneously, can the network operate correctly? If not, find an appropriate relative timing of the changes of D and c. Also determine when the network settles down.

(iv) Do Q and $\bar{Q}$ assume the same value at any time?

8.2.6 (M) Design a *J-K* master-slave flip-flop with one of the following types of gates, using as few gates as possible.

A. MOS gates. (*Hint*: Since an MOS gate can express a negative function which is more general than NAND or NOR, Fig. 8.2.1 is not an appropriate starting point. Therefore first design a *J-K* master-slave flip-flop with other types of gates (e.g., a mixture of AND gates, NOR gates, and an inverter). Then convert it into an MOS network.)

B. IIL gates (described in Exercise 6.10).

8.2.7 (M) Solve Exercise 8.2.5 for the *D* master-slave flip-flop in Fig. E8.2.7. For the sake of simplicity, assume that the inverter has the same switching time as the other gates. (In actual electronic implementations this assumption may not be true.)

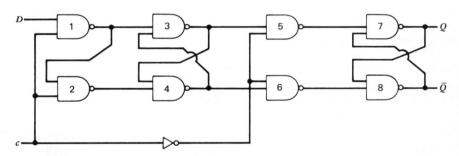

Fig. E8.2.7 **D** *master-slave flip-flop.*

8.2.8 (M) Let us analyze the edge-triggered *S-R* flip-flop in Fig. E8.2.8. Assuming that every gate has switching time τ, answer the following questions, expressing durations and other times as multiples of τ.

(i) When and for how long must S or R assume a new value in order to make the flip-flop work correctly? When can S or R return to the original value without making the network malfunction? How long must c maintain value 1 in order to make the network operate properly?

(ii) Find the delay time until the outputs Q and $\bar{Q}$ establish new values corresponding to each combination of values of S and R. (Count it from the instant when c changes.)

(iii) Suppose that S returns from 1 to 0 while $R = 0$ without making the flip-flop malfunction as we found in (i). Then, if R changes to 1 before the trailing edge of a clock pulse, will the flip-flop still be set to $Q = 1$ and $\bar{Q} = 0$, which correspond to the combination $S = 1$ and $R = 0$?

Sequential Networks with Master-Slave Flip-Flops in Skew Mode

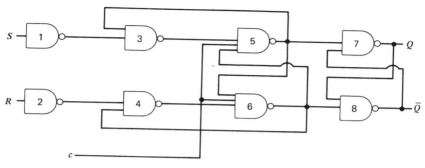

Fig. E8.2.8   Edge-triggered S-R flip-flop.

(iv) Can S and R assume value 1 simultaneously, without making the network malfunction?

(v) Do Q and $\bar{Q}$ assume the same value at any time?

8.2.9 (M) Discuss how the outputs Q and $\bar{Q}$ of the J-K master-slave flip-flop in Fig. 8.2.1 change when signal 1 appears alternately at inputs J and K during one clock pulse. [Although Q and $\bar{Q}$ change to the complements of the current values when J and K are kept at 1 (according to Fig. 8.1.3b), this is another possibility to change Q and $\bar{Q}$, if the signals at J and K are not steady.]

8.3 Sequential Networks with Master-Slave Flip-Flops in Skew Mode

Clocked sequential networks with S-R latches were analyzed and synthesized as networks in fundamental mode in Section 8.1, in the same manner as those in Chapter 7. In this section we analyze clocked networks with master-slave flip-flops. (Those with edge-triggered flip-flops can be similarly treated, as we mention later.)

Each master-slave flip-flop consists of two latches, master and slave. If these two latches represented two separate internal variables, a clocked sequential network with master-slave flip-flops could be analyzed in fundamental mode, as we did in Section 8.1. But if only the output of each master-slave flip-flop, that is, the output of the slave, is treated as an internal variable, ignoring the output of the master (i.e., not treating the output of the master as another internal variable), the transition-output table, or state-output table, for the network works in a manner different from a corresponding table in fundamental mode, as we see in the following. The tables in this different mode are said to be in "skew mode."

464 Synthesis of Synchronous Sequential Networks

Suppose that we substitute a *J-K* master-slave flip-flop for the *S-R* latch in the network in Fig. 8.1.7, as shown in Fig. 8.3.1. (We can substitute an *S-R* master-slave flip-flop as well. But let us substitute a *J-K* master-slave flip-flop, since we explained it in detail in Section 8.2, and this flip-flop works as an *S-R* master-slave flip-flop because $J = K = 1$ does not occur with this particular network.) A clock is connected to the flip-flop, deleting the clock inputs to the AND gates in Fig. 8.1.7.

The master in the *J-K* master-slave flip-flop in the network in Fig. 8.3.1 is connected to its inputs until time t_3 in the waveform in Fig. 8.3.2 (as explained with respect to Fig. 8.2.3 in Section 8.2). Hence the flip-flop is susceptible to input changes until* t_3 in Fig. 8.3.2, but after this instant the flip-flop does not pick up any new input value, even if the input changes. Then, at time t_4, the master is connected to the slave (as explained in Section 8.2), and the flip-flop establishes its new output values of y and $\bar{y}$. Therefore, at the trailing edge of each clock pulse, the *J-K* master-slave flip-flop can

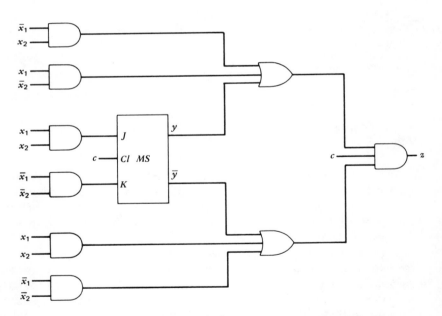

Fig. 8.3.1 A clocked sequential network in skew mode (a serial adder).

---

* Strictly speaking, the flip-flop is susceptible to input changes until "slightly" before t_3, because the input value must be kept at the same value for a while in order to set the master to its new value, as explained with Fig. 7.1.2*b*.

Sequential Networks with Master-Slave Flip-Flops in Skew Mode 465

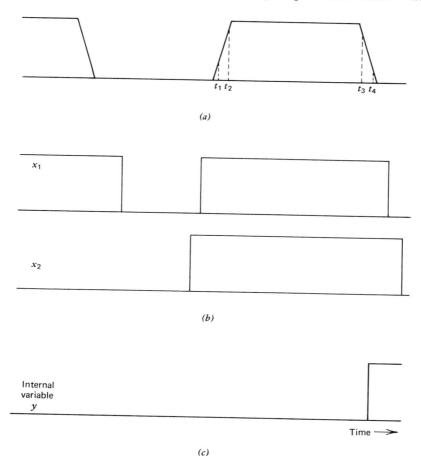

Fig. 8.3.2 *Relation between clock, inputs, and an internal variable.* (a) *Clock* c. (b) *Inputs* x_1 *and* x_2. (c) *Internal variable* y.

change its outputs. Notice that **internal variable y, that is, the output of the flip-flop, does not change until then**.

To analyze this network, let us consider the transition-output table in Table 8.3.1a. Here the letter y is used instead of c, in order to differentiate networks in skew mode from those in fundamental mode, as in Fig. 8.1.7 and Table 8.1.1b, where internal variable y can change immediately when a clock pulse appears as well as when it disappears. Since the internal state and the output of the network do not change during the absence of clock pulses, the columns for $y = 0$ can be condensed into a single column in

Table 8.3.1 Analysis of Fig. 8.3.1

a. **Condensed Transition-Output Table in Skew Mode**

$\gamma = 1$

$x_1 x_2$

y	$\gamma = 0$	00	01	11	10
0	0, 0	0, 0	0, 1	1, 0	0, 1
1	1, 0	0, 1	1, 0	1, 1	1, 0

Y, z

b. **State-Output Table in Skew Mode**

$\gamma = 1$

$x_1 x_2$

s	$\gamma = 0$	00	01	11	10
A	$A, 0$	$A, 0$	$A, 1$	$B, 0$	$A, 1$
B	$B, 0$	$A, 1$	$B, 0$	$B, 1$	$B, 0$

S, z

c. **Condensed Transition Table in Skew Mode**

$\gamma = 1$

$x_1 x_2$

y	$\gamma = 0$	00	01	11	10
0	0	0	0	1	0
1	1	0	1	1	1

Y

d. **Compact Transition Table in Skew Mode**

$\gamma = 1$

$x_1 x_2$

y	00	01	11	10
0	0	0	1	0
1	0	1	1	1

Y

Sequential Networks with Master-Slave Flip-Flops in Skew Mode 467

Tables 8.3.1a and 8.3.1b, like the case of Tables 8.1.1b and 8.1.1c. (As explained with respect to Table 8.1.1, this is not always possible for some other networks if, for example, the output gate has no clock at its input and changes its output value depending on the values of x_1 and x_2 during the absence of clock pulses.) For simplicity assume* that inputs x_1 and x_2 can change only during the absence of clock pulses, as illustrated in Fig. 8.3.2. (The output z appears at the output gate being probed by the clock pulse.)

Suppose that during the absence of a clock pulse the network is in state $(\gamma, x_1, x_2, y) = (0\ 0\ 0\ 0)$ in the condensed transition-output table in Table 8.3.1a. Suppose that inputs (x_1, x_2) change from $(0\ 0)$ to $(1\ 1)$ sometime during the absence of a clock pulse and $(x_1, x_2) = (1\ 1)$ lasts at least until the trailing edge of the clock pulse. If this transition is interpreted on Table 8.3.1a, the network moves from $(\gamma, x_1, x_2, y) = (0\ 0\ 0\ 0)$ to $(1\ 1\ 1\ 0)$, as shown by the dotted-line arrow, at the leading edge of the clock pulse. Then the network must stay in this state, $(1\ 1\ 1\ 0)$, until the trailing edge of the clock pulse, because the internal variable y does not change yet. (This is different from the fundamental mode, in which the network does not stay in this state unless the state is a stable one, and vertically moves to a stable state in a different row in the same column.) Then the network moves from $(\gamma, x_1, x_2, y) = (1\ 1\ 1\ 0)$ to $(0\ 1\ 1\ 1)$, as shown by the solid-line arrow in Table 8.3.1a, when y assumes the new value at the trailing edge of the clock pulse. Thus the transition occurs horizontally and then **diagonally** in Table 8.3.1a.

Notice, however, that in the new state $(1\ 1\ 1\ 0)$ at the leading edge of the clock pulse in the above transition from $(\gamma, x_1, x_2, y) = (0\ 0\ 0\ 0)$, the network z assumes the new output value. Consequently, **in this new stable state during $\gamma = 1$ in Table 8.3.1a, the new current value of the network output, z, is shown, while the value of the internal variable in this new stable state shows the next value, Y.** In fundamental mode, when the network moves horizontally to a new unstable state in the same row in the transition-output table, the current value of the network output is shown and the value of the internal variable represents the next value, Y (in this sense the situation is not different), but the network output lasts only during a short, transient period, unless the new state is stable. In contrast, in skew mode the output value for the new state is not transient (because the network stays in this state during the clock pulse) and is essential in the description of the network performance.

Thus the clock pulses in Table 8.3.1 in skew mode must be interpreted in a manner completely different from clock c in the network in Table 8.1.1b, which operates in fundamental mode. Hence we use the different symbol γ, and γ might be called a **virtual clock**.

---
* This can be relaxed as follows: the inputs can change at any time except in the neighborhood prior to the trailing edges of clock pulses. Still the network works in the same manner except different nature of new output values during clock pulses.

468 Synthesis of Synchronous Sequential Networks

The transition in a sequential network with master-slave flip-flops, such as the one in Fig. 8.3.1, occurs horizontally and then diagonally in its transition table (unlike the transition in the sequential networks discussed in Chapter 7 and Section 8.1). Such a network is said to be in **skew mode**.* (This mode could be also called the diagonal transition mode.)

The network in Fig. 8.3.1 is actually a **serial adder**. (Often, in practice, in Fig. 8.3.1 a D master-slave flip-flop is connected to the output gate for convenience.) The output y of the J-K master-slave flip-flop represents a carry from the previous digit position in addition. The state-output table is shown in Table 8.3.1b, where states A and B correspond to carries 0 and 1, respectively. Before starting addition, the network must be placed in state A, that is, y (and z if a D master-slave flip-flop is added) must be set to 0 (by the direct reset-input terminal mentioned in Section 8.2). Otherwise, the adder shows erroneous outputs.

Features of Skew Mode

Extending the above observation, we find that the transition-output tables in skew mode (such as Table 8.3.1a) show the following important differences from those in fundamental mode (such as Tables 8.1.1b and 8.1.1c).

1. Whether input variables change or not, the network moves to a new column (for $\gamma = 1$) in the *same row* at the leading edge of every clock pulse.
2. The network moves to the first column (for $\gamma = 0$) in the *same row* at the trailing edge of a clock pulse, if internal variable y does not change, whether input variables x_1 and x_2 change or not.
3. The network moves *diagonally* to the first column in a different row at the trailing edge of a clock pulse, if y changes, whether x_1 and x_2 change or not. (In the case of fundamental mode, y cannot change if the network stays in the same row during the clock pulse.)

---

* The skew mode is similar to the **pulse mode**, which has been defined by some authors as a way of operation of an asynchronous network where (1) the input pulses last long enough to cause latches to change a state and (2) the input pulses are short enough so that they are no longer present when the change in latch outputs is fed back to the inputs of the network. (For example, networks in pulse mode may be asynchronous sequential networks that are connected to the output of synchronous networks which are controlled by clocks in this manner.) Such an intricate control of the duration of input pulses is usually impractical, as at least some industry circles agree with the author. The use of clock pulses whose duration is not critically short, and also the use of level signals for input variables, are very common. Also, **even if clock "pulses" are supplied, a network can be in fundamental mode**, as discussed in regard to Fig. 8.1.7.

For these reasons the term "pulse mode," which occurs often in the literature, is not used in this book, even though the transition of a network in a transition table in pulse mode is accidentally not different from that in skew mode.

Sequential Networks with Master-Slave Flip-Flops in Skew Mode 469

4. The network stays in every state for a long period of time (i.e., the entire pulse repetition period during the absence or presence of a clock pulse), so every state is not transitory. (Thus every state looks like a stable state at first thought, but is not a stable state in the sense of a stable state in fundamental mode, for the following reason.) The Y entry in each state in the columns for $\gamma = 1$ shows the next state, not the current state, whereas the z entry in these columns shows the current value of the network output. In the case of synchronous networks in fundamental mode, however, a state in a column for $c = 1$ may be unstable; and, if so, the z entry, which shows the current value of the network output, is present only during the transition, having no importance except for hazard consideration.

Simplification of Transition-Output Tables in Skew Mode

We form the condensed transition table in Table 8.3.1c from Table 8.3.1a by considering only Y. Then this condensed transition table can be further condensed. The first column for $\gamma = 0$ can be deleted, as shown in Table 8.3.1d, because Y in this column is identical to coordinate y in each row (since the flip-flop does not change its state y during the absence of clock pulses, unlike a network in fundamental mode which may change its state y, if nonclocked gates are connected to latches). Table 8.3.1d works as follows. For example, suppose that the network is in state $(\gamma, x_1, x_2, y) = (1\ 1\ 1\ 0)$ in Table 8.3.1d. Since the entry in this state is $Y = 1$ in Table 8.8.1d, the network must go to state $(0\ 1\ 1\ 1)$ when γ changes to 0 from 1. Thus the network will be in the row for $y = 1$ in the first column for $\gamma = 0$ in Table 8.3.1c. In this case, even if the first column for $\gamma = 0$ in Table 8.3.1c is deleted in Table 8.3.1d, we can easily determine that the network must be in the row for $y = 1$ in the deleted first column for $\gamma = 0$, since this information is shown in the entry of the state $(1\ 1\ 1\ 0)$ in the remaining columns for $\gamma = 1$ in Table 8.3.1d. Thus, even if the first column for $\gamma = 0$ in Table 8.3.1c is deleted in Table 8.3.1d, all the information about the network performance is retained. This unique property of the transition table (also state table) is due to the fact that only at the trailing edge of a clock pulse does next-state value Y, which is stored in the master, emerge as a new value of an internal variable (since Y is transferred to the slave from the master). This compact transition table in skew mode (also, compact state or flow tables, which can be similarly obtained) will be conveniently used in the network synthesis in Section 8.5.

In an output table, however, the columns for $\gamma = 0$ generally cannot be deleted, because, unlike the case of transition tables, the output values in these columns for $\gamma = 0$ generally have no relationship to those in the columns for $\gamma = 1$. For example, in the output table that can be formed from Table 8.3.1a, $z = 1$ in the state $(\gamma, x_1, x_2, y) = (1\ 1\ 1\ 1)$ is different from the value of z, 0, in the row $y = 1$ in the first column for $y = 0$, into which the network goes from state (1 1 1 1). This property of z's values is due to the fact that the next value of z is not stored anywhere in the network, unlike the next value, Y.

Networks with Edge-Triggered Flip-Flops

Networks that contain edge-triggered flip-flops can be similarly analyzed with transition tables in skew mode, because positive and negative edge-triggered flip-flops respond at leading and trailing edges of a clock pulse, respectively.

Advantages of Skew Mode

Master-slave or edge-triggered flip-flops can completely eliminate the hazard problem of sequential networks, as we already see. Also, the transition- (or state-) output tables for networks with these flip-flops are expressed in skew mode. As we will see in Section 8.5, the preparation of tables in skew mode is usually easier than that of tables in fundamental mode. Also, the tables tend to be simpler, and consequently the designed networks also are likely to be simpler.

Interpretation of Skew Mode in Terms of Fundamental Mode

As a matter of fact, the skew mode is simply another way to look at the fundamental-mode operation of a network with master-slave (or edge-triggered) flip-flops. The J-K master-slave flip-flop contains two flip-flops: the master and the slave. If the outputs of the master and slave are treated as two independent internal variables, y_m and y_s, respectively, the network can be expressed with a transition table in fundamental mode. But when we consider only one internal variable for the entire master-slave flip-flop, a succession of horizontal-vertical transitions in a transition table in fundamental mode is interpreted as a horizontal transition (from A to B) and then a direct diagonal transition (from B to E, shown in the dotted line) on the transition table in skew mode, as illustrated in Table 8.3.2. (The first transition, shown in the solid line from A to C through B, takes place at the leading edge of a clock pulse, changing y_m. The second transition, in the solid line from C to E through D, takes place at the trailing edge of the clock pulse, changing y_s.) Therefore the skew mode is simply a concise expression of the fundamental mode for a network with master-slave (or edge-triggered) flip-flops.

Sequential Networks with Master-Slave Flip-Flops in Skew Mode

Table 8.3.2 Interpretation of Skew Mode with an Extra Internal Variable in Fundamental Mode

		$c=0$	$c=1$ $x_1 x_2$ 00	01	11	10
	00	A			B	
$y_m y_s$	01					
	11	E				
	10	D			C	

Notice that, even if a clock is provided, a network cannot be in skew mode unless master-slave or edge-triggered flip-flops are used (change of y_m has no influence on the operations of other gates in the network, unlike internal variables in networks with latches).

Networks Not in Fundamental or Skew Mode

Notice that **a sequential network in fundamental or skew mode must settle down** (i.e., there can be no transition inside the network) **before the network inputs change**. Otherwise the analysis is generally too complicated [an example is Exercise 8.1(iii)]. Of course, networks that do not settle down before input changes may be useful for special purposes. Also, if a sequential network in neither fundamental nor skew mode is designed, the number of gates or levels may be fewer than the number in fundamental or skew mode. But such a network is usually difficult to design. Also notice that, even with the use of master-slave or edge-triggered flip-flops, sequential networks that are neither in fundamental nor in skew mode can be realized. One example is a sequential network, part of which is in fundamental mode and the rest in skew mode. [Another example is Exercise 8.1(iii).] Also, networks whose transitions are more complicated than those in fundamental or skew mode may be realized. This is true without master-slave or edge-triggered flip-flops. Anyway, fundamental and skew modes are simply operation types of sequential networks that are now conveniently and widely used.

8.4 Translation of Word Statements into Flow-Output Tables

In this section we translate word statements for given problems into flow-output tables in skew mode. Flow-output tables in skew mode are usually simpler and can be obtained more quickly than those in fundamental mode. This is due to the fact that, in preparing skew-mode tables, we have more freedom because, for each transition, we need to consider only one state, unlike the case of fundamental mode, where generally a pair of unstable and stable states must be considered. Let us compare the two modes, with examples.

Example 8.4.1—A clocked network with a clocked output: Suppose that a network with two inputs, x_1 and x_2, and an output z is to be designed.

1. Inputs do not change during the presence of clock pulses. Inputs x_1 and x_2 cannot assume value 1 simultaneously during the presence of clock pulses. (We do not care whether the simultaneous appearance or no appearance of 1 occurs during the absence of clock pulses.)
2. The output z can assume value 1 or 0 during the presence of the clock pulses, but is always 0 during the absence of clock pulses, as illustrated in Fig. 8.4.1.
    (a) The value of z is 1 only when the value 1 appears during the clock pulse at the same input at which the last value 1 appeared.

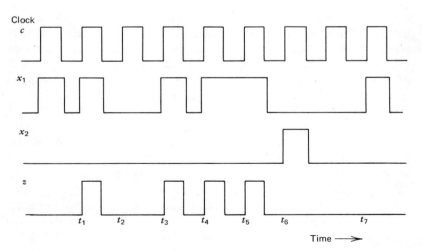

Fig. 8.4.1 Waveforms for Example 8.4.1.

Translation of Word Statements into Flow-Output Tables 473

(If clock pulses during which the value 1 appears at neither x_1 nor x_2 follow the clock pulse during which the value 1 appears at x_i, $z = 1$ only when the value 1 appears at the same x_i again, but $z = 0$ during the intermediate clock pulses. For example, $z = 1$ during the fourth clock pulse, which starts at t_3, but $z = 0$ during the third clock pulse, which starts at t_2, in Fig. 8.4.1.)

(b) Otherwise the value of z is 0. (In other words, the value of z is 0 when the value 1 appears at neither input, or alternates between inputs x_1 and x_2, during clock pulses.)

Let us prepare a flow-output table for this network.

1. The network must have at least two states, A and B, because if we look at the network during the absence of clock pulses, the network must have had the last 1 either at x_1 (then the network is in the corresponding internal state A) or at x_2 (the network is in the corresponding state B). The network may need more than two states, but let us proceed with two for the moment. If two states are found insufficient, we will try with more states later. Therefore suppose that we have two states, A and B, in the flow-output table in Table 8.4.1. The first column must have A and B, since the network must stay in each state during the absence of a clock.

2. Assume that the network was in state A during the absence of a clock pulse (i.e., the last 1 appeared at x_1). If the value 1 appears at x_1 during the current clock pulse, then $z = 1$, since the value 1 is repeated at the same input. The network must stay in state A since, after the current clock pulse disappears, current 1 at x_1 will become the last 1 at x_1. Therefore enter A in the last column in the first row as the next state, and enter the corresponding output value, 1. If the value 1 appears at x_2 (instead of x_1) during the current clock pulse, then $z = 0$,

Table 8.4.1 Flow-Output Table for Example 8.4.1

			$c = 1$			
			$x_1 x_2$			
	s	$c = 0$	00	01	11	10
Last 1 at						
x_1	A	$A, 0$	$A, 0$	$B, 0$	—	$A, 1$
x_2	B	$B, 0$	$B, 0$	$B, 1$	—	$A, 0$
			S, z			

474 *Synthesis of Synchronous Sequential Networks*

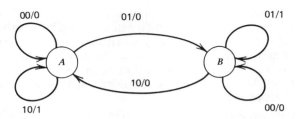

Fig. 8.4.2 State diagram for Example 8.4.1.

since the value 1 alternates between x_1 and x_2. The network must go to state B because, after the current pulse disappears, current 1 at x_2 will become the last 1 at x_2. Therefore enter the next state B in the third column in the first row and the corresponding output value, 0. If the value 1 appears at neither x_1 nor x_2, the network must stay in state A. Therefore enter A in the second column in the first row and the corresponding output value, 0. Thus the first row of Table 8.4.1 has been obtained.

When the network is assumed to be in state B, we can fill in the second row in the flow-output table in a similar manner. All the entries in column $(x_1, x_2) = (1\ 1)$ are dashes, since this column is assumed not to occur. Thus Table 8.4.1 has been obtained. The corresponding state diagram is shown in Fig. 8.4.2. A state diagram such as Fig. 8.4.2 is convenient for preparing a flow-output table, or verifying that a prepared table really meets a given network performance requirement.

3. Derivation of a network in skew mode: If the flow-output table is assumed to be in skew mode, using master-slave (or edge-triggered) flip-flops, we have completed the flow-output table by changing c to y, since the entries in columns for $y = 1$ represent next states and the current output values, and we have diagonal transitions to the column for $y = 0$ at the trailing edges of clock pulses. By assigning internal variables $y = 0$ to state A and $y = 1$ to state B, we can get the transition-output table in skew mode, as shown in Table 8.4.2.

Let us implement a network for this Table 8.4.2 according to the general model of sequential networks in Fig. 7.4.3. If we choose to use a *J-K* master-slave flip-flop, this general model will be the network in Fig. 8.4.3. (When we choose other types of flip-flops, we can design in a similar manner.) Since we need to find new input values of J and K for next output values Y of the flip-flop, we need to prepare an excitation table, as discussed in Section 7.4. Table 8.4.3a shows how the current output value y of a *J-K* master-slave flip-flop changes into the next output value Y for each combination of the values of the flip-flop inputs J and K. If we reverse this input-output relationship, we get

Table 8.4.2 Transition-Output Table in Skew Mode for Example 8.4.1

	$\gamma = 0$	$\gamma = 1$ $x_1 x_2$			
		00	01	11	10
y 0	0, 0	0, 0	1, 0	d, d	0, 1
y 1	1, 0	1, 0	1, 1	d, d	0, 0
			Y, z		

Table 8.4.3b. This table shows what values inputs J and K must take for each change of internal variable y to its next value Y. Using this table, we can convert the transition-output table in Table 8.4.2 into the excitation and output tables in Table 8.4.4. The excitation table in Table 8.4.4a shows what values the flip-flop inputs J and K must take next for each change of the network inputs x_1 and x_2. (Notice that, as explained in Section 8.3, the first column for $\gamma = 0$ in Table 8.4.2 is ignored in Table 8.4.4a, but the first column for $\gamma = 0$ in Table 8.4.2 cannot be ignored in Table 8.4.4b.) Preparing J and K in Table 8.4.4a separately into the two Karnaugh maps in Tables 8.4.5a and 8.4.5b, and preparing z in Table 8.4.4b into the Karnaugh map in Table 8.4.5c, we can find a minimal sum for each of J, K, and z. On the basis of these switching expressions, we have designed the sequential network shown

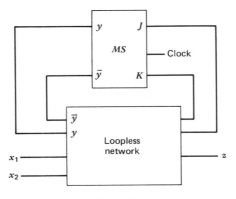

Fig. 8.4.3

Table 8.4.3 Input-Output Relationship of *J-K* Master-Slave Flip-Flop

a.

INPUTS		OUTPUTS	
J	K	y	Y
0	0	0	0
		1	1
0	1	0	0
		1	0
1	0	0	1
		1	1
1	1	0	1
		1	0

b.

OUTPUTS		INPUTS	
y	Y	J	K
0	0	0	d
0	1	1	d
1	0	d	1
1	1	d	0

in Fig. 8.4.4. Since the output z can appear only during the presence of clock pulses, not only the flip-flop but also the AND gates are clocked. Notice that this network has the minimum number of internal variables (i.e., only one), since the given problem cannot be implemented by a loopless network alone and consequently requires at least one internal variable.

  4. **Feasibility of a network in fundamental mode:** Next let us check whether Table 8.4.1 can be in fundamental mode. Now latches instead of master-slave flip-flops must be used, and the internal variable can change at the leading edges and also the trailing edges of clock pulses, instead of at the trailing edges only. If the network in fundamental mode is in state A in the column for $c = 0$ and signal value 1 appears at x_2, the network must move to $(c, x_1, x_2, s) = (1\ 0\ 1\ B)$ through

Table 8.4.4 Excitation-Output Table

a. **Excitation Table**

		$x_1 x_2$		
	00	01	11	10
y 0	0, d	1, d	d, d	0, d
1	d, 0	d, 0	d, d	d, 1

J, K

b. **Output Table**

			$y = 1$	
		$x_1 x_2$		
$y = 0$	00	01	11	10
y 0	0	0	0	d	1
1	0	0	1	d	0

z

Table 8.4.5 Karnaugh Maps

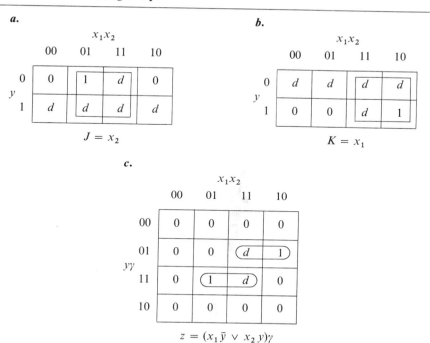

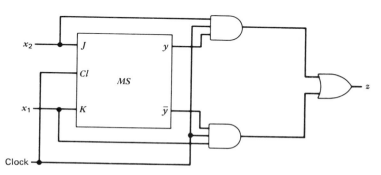

Fig. 8.4.4 Network for Example 8.4.1 in skew mode using a J-K master-slave flip-flop.

(1 0 1 A) at the leading edge of a clock pulse; and once the network is in state B, $z = 0$ must be shown at the output because signal value 1 alternates between x_1 and x_2. But we have already written $z = 1$ in the state $(c, x_1, x_2, s) = $ (1 0 1 B) because, if this state is entered from state B in the column for $c = 0$, signal 1 appears at the same input, x_2. This is a contradiction.

In conclusion, the flow-output table cannot be in fundamental mode if we use two states, A and B, which are interpreted to correspond to the last 1 at x_1 and x_2, respectively.

Remark 8.4.1: Notice that the network in Fig. 8.4.4 can be interpreted to work in fundamental mode, if the master and the slave are assigned two internal variables as we explained with respect to Table 8.3.2. This means that there must exist a flow-output table for Example 8.4.1 with four states in fundamental mode, and Fig. 8.4.4 must be one of the corresponding networks. ■

Example 8.4.2—A clocked network where the output z is a level signal (i.e., the value of z is maintained even during the absence of clock pulses): The performance of the network is the same as in Example 8.4.1 except for z being a level signal. During clock pulses, an input signal of value 1 appears at exactly one of two inputs, x_1 and x_2, of the network, or does not appear at all. As long as signal 1 continues to appear at the same input during the presence of clock pulse, we have $z = 1$. (This includes the following case. Suppose that we have $z = 0$, when signal 1 appears at one of the inputs during a clock pulse. Then signal 0 follows at x_1 and x_2 during the succeeding clock pulses. If signal 1 comes back to the same input, z becomes 1.) Regardless of the presence or absence of the clock pulses, z remains 1 until signal 1 starts to alternate between the two inputs during clock pulses. Once z becomes 0, z remains 0 regardless of the presence or absence of clock pulses, until signal 1 starts to appear at the same input during clock pulses.

1. An output value must depend on an internal state only, because z is a level signal and the input signal 1 may not be present during the absence of clock pulses. Let us assume that there are exactly two states, A and B, such that $z = 0$ when the network is in state A and $z = 1$ when the network is in state B. (We will try more than two states if two are found insufficient.)

2. Assume that during the absence of clock pulses the network is in state A. Since the network must stay in this state as long as $c = 0$, the next state, S, in the column for $c = 0$ must be A, as shown in Table 8.4.6a. Similarly, the next state for the second row in the column $c = 0$ must be B.

Table 8.4.6 Two States Are Not Sufficient for Example 8.4.2

a.

z	s	$c = 0$	$c = 1$ $x_1 x_2$ 00	01	11	10
0	A	$A, 0$			—	A
1	B	$B, 1$			—	
			S, z			

b.

s	$c = 0$	$c = 1$ $x_1 x_2$ 00	01	11	10
A	$A, 0$			—	B
B	$B, 1$			—	
		S, z			

3. When the network is in state A during $c = 0$, suppose that $x_1 = 1$ appears at the next clock pulse. Let us choose A as the next state, S, as shown in Table 8.4.6a. But this choice means that, if we apply value 1 repeatedly at x_1, the network goes back and forth between the two states in the first and last columns in the first row in Table 8.4.6a. Thus $z = 1$ must result from the requirement on the network performance, but since the network stays in the first row, we must have $z = 0$. This is a contradiction. Thus the next state for $(x_1, x_2, s) = (1\ 0\ A)$ cannot be A.

4. Next assume that the next state for $(x_1, x_2, s) = (1\ 0\ A)$ is B, as shown in Table 8.4.6b. Suppose that the value 1 has been alternating between x_1 and x_2. If we had the last 1 at x_2, the network must be currently in A because of $z = 0$ for alternating 1's. When the next 1 appears at x_1, $z = 0$ must still hold because the value 1 is still alternating. But the network will produce $z = 1$ for state $(1\ 0\ A)$, because B is assumed to correspond to $z = 1$. This is a contradiction, and the choice of B is also wrong.

5. In conclusion, two states are not sufficient, so we try again with more states. For this example, other interpretations of the two states are not possible because the values of z must correspond to the two states, as mentioned in Step 1.

6. In the above we had contradictions by assuming only two states, A and B, corresponding to $z = 0$ and 1; we did not know which input's 1 led to each state. Thus, in addition to the values of z, let us consider which input had the last 1. In other words, we have four states corresponding to the combinations of z and the last 1, as shown in Table 8.4.7. (At this stage we do not know whether or not three states are sufficient. But Procedure 7.6.1, which can be applied to flow-output tables in skew mode also, can eliminate any unnecessary states, though the elimination may sometimes be time-consuming.) Let us assume four

Table 8.4.7 Flow-Output Table in Skew Mode for Example 8.4.2

			$y = 0$	$y = 1$ $x_1 x_2$			
Last 1 at	z	s		00	01	11	10
x_1	0	A	A, 0	A, 0	C, 0	—	B, 1
x_1	1	B	B, 1	B, 1	C, 0	—	B, 1
x_2	0	C	C, 0	C, 0	D, 1	—	A, 0
x_2	1	D	D, 1	D, 1	D, 1	—	A, 0
				S, z			

states for the moment. As a matter of fact, both three states and four states require two internal variables. Hence, in terms of the number of internal variables, it does not matter whether we have three or four states, though the two cases may lead to different networks.

7. **Derivation of a flow-output table in skew mode:** Let us form a flow-output table in skew mode. In the column of $y = 0$ in Table 8.4.7, the network must stay in each state during the absence of a clock pulse. Thus all the next states S must be identical to s in each row. Suppose that the network is in state $(y, x_1, x_2, s) = (0\ 0\ 0\ A)$ with output $z = 0$ after having the last 1 at x_1. When the value 1 appears at x_1 and y becomes 1, the next state S must be B for the following reason. When the current clock pulse disappears, this 1 at x_1 will be "the last 1" at x_1 (so S must be A or B) and z will have to be 1 because of the repeated occurrence of 1's at x_1. (This contradicts $z = 0$, which we will have if A is entered as S.) Hence the possibility of S being A is ruled out, and S must be B. Since value 1 is repeated at x_1, we have $z = 1$.

The next states and the values of output z in all other cells in Table 8.4.7 can be found in a similar manner.

Therefore the network obviously can work in skew mode. A network can be constructed in a manner similar to Example 8.4.1, but this will be done later, in Section 8.5.

8. **Derivation of a flow-output table in fundamental mode:** Next let us interpret this table in fundamental mode. Table 8.4.7 with c in place of y shows that, when the network placed in state $(c, x_1, x_2, s) = (0\ 0\ 0\ A)$ receives $x_2 = 1$, the next state will be C. But the network goes to unstable state $(c, x_1, x_2, s) = (1\ 0\ 1\ C)$ that has entry D, since if

Translation of Word Statements into Flow-Output Tables

we assume fundamental mode it must move vertically, instead of the diagonal transition in skew mode. Hence the network must go further to stable state (1 0 1 D), without settling in the desired stable state, C, if the network still keeps $x_2 = 1$ and $c = 1$. Therefore the network cannot be in fundamental mode. (Recall that the next state entries in a flow-output table do not show intermediate unstable states but do show the destination stable states, as discussed in Section 7.3.)

The above difficulty can be avoided by adding two new rows, E and F, as shown in Table 8.4.8. When the network placed in state (0 0 0 A) receives $x_2 = 1$, the network goes to the new stable state F in column $(x_1, x_2) = (0\ 1)$. For this state F, $z = 0$, without causing contradiction. When the clock pulse disappears, the network goes to stable state (0 0 0 C) after passing through unstable state (0 0 0 F). The problem with the other states is similarly eliminated. All stable states are encircled as stable states.

The values of z for stable states are easily entered. The values of z for unstable states can be entered with certain freedom. For example, suppose that the network placed in state $(c, x_1, x_2, s) = (0\ 0\ 0\ A)$ receives $x_1 = 1$. Then the next state S is B. Since this 1 at x_1 may be interpreted as not yet influencing the network output, 0 may be entered as z for (1 1 0 A). But we could enter $z = 1$ corresponding to the next stable state. This does not make much difference as far as the external behavior of the network is concerned, because it simply means that

Table 8.4.8 Flow-Output Table in Fundamental Mode for Example 8.4.2

			$c = 0$	\multicolumn{4}{c}{$c = 1$ x_1x_2}
Last 1 at	z	s		00	01	11	10
x_1	0	A	Ⓐ, 0	Ⓐ, 0	F, 0	—	B, d
x_1	0	E	A, 0	—	—	—	Ⓔ, 0
x_1	1	B	Ⓑ, 1	Ⓑ, 1	F, d	—	Ⓑ, 1
x_2	0	C	Ⓒ, 0	Ⓒ, 0	D, d	—	E, 0
x_2	0	F	C, 0	—	Ⓕ, 0	—	—
x_2	1	D	Ⓓ, 1	Ⓓ, 1	Ⓓ, 1	—	E, d
			\multicolumn{5}{c}{S, z}

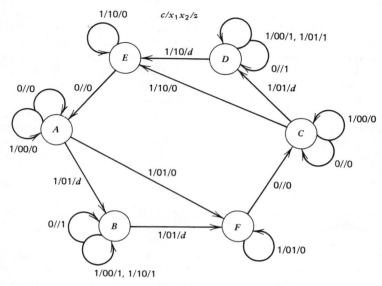

Fig. 8.4.5   State diagram for Table 8.4.8.

$z = 1$ appears a little bit earlier or later. The network that results, however, may be different. Accordingly, $z = d$ (d denotes "don't-care") would be the best assignment, since this gives flexibility in designing the network, as we will see in Section 8.5.

The state diagram is shown in Fig. 8.4.5.

Appropriate Use of Clock in Network Simplification

When Examples 8.4.1 and 8.4.2 are compared, the required network performances are found to be identical except that the output is a level signal in the latter example and a clocked signal in the former. But even in skew mode, the flow-output table for Example 8.4.1 is much simpler than that for Example 8.4.2. In general, the number of states in a flow table can be reduced by appropriate use of clock pulses.

Advantages of Flow-Output Tables in Skew Mode

As seen above in the preparation of a flow-output table for a given problem, we need to prepare a next stable state only, for each transition, in the case of skew mode. On the contrary, we need to prepare both an unstable state and a next stable state, for each transition, under the restriction to avoid multiple changes of internal variables, in the case of fundamental mode. Consequently, there are greater chances of successful realization of flow-output tables in skew mode with fewer states than those in fundamental mode.*

Remark 8.4.1: Notice that every state in a flow-output table in skew mode is not transitory, unlike unstable states in the case of fundamental mode. Consequently, the use of dashes (or don't-care d's) for outputs in skew mode is different from that in fundamental mode, in the sense that output values last for short transient periods in the case of fundamental mode (e.g., d's in Table 8.4.8), though in the case of skew mode dashes (or don't-care d's) may be entered if some cells cannot be entered by the network because of restrictions or are not important. ∎

8.5 Design of Sequential Networks in Skew Mode

After preparing a flow-output table for a given problem, we can reduce the number of states by the state reduction method, Procedure 7.6.1. Although the method was explained for fundamental mode in Section 7.6, we need simply to compare states in which a network will settle down and also output values, which it will assume, for two states under comparison. Hence there is no essential difference from the case for fundamental mode.

Now let us design a network based on a flow-output table in skew mode, using master-slave flip-flops.

The flow-output table in skew mode for Example 8.4.2 in Table 8.4.7 is reproduced here as Table 8.5.1. When we form a flow table for our synthesis in Table 8.5.2, the first column for $y = 0$ in Table 8.5.1 can be deleted, as discussed in Section 8.3.

As pointed out in Section 8.2, when master-slave flip-flops are used, the network does not have hazards even if internal variables make multiple changes, because the flip-flops do not respond to any input changes during the transition of the slaves. Thus **we need not worry about hazards due to multiple changes of internal variables in finding a state assignment**. But the number of gates, connections, or levels in a network to be designed can differ, depending on how binary numbers are assigned to states. If we want to obtain a network with a minimum number of gates, connections, or levels, we need to try different state assignments, though there are too many possible assignments to consider. In many cases, however, this is not really necessary (it is of secondary importance compared with the hazard problem, which makes

---

* The advantages of a flow-output table with diagonal transitions were probably first discussed by McCluskey [1965] (without raceless flip-flops), with good foresight before the advent of raceless flip-flops. With examples he showed that, if we use diagonal transitions, simpler flow-output tables are usually obtained, leading to simpler networks; also, the networks are free of malfunctions due to hazards.

484 Synthesis of Synchronous Sequential Networks

Table 8.5.1 Flow-Output Table in Skew Mode for Example 8.4.2

	$\gamma = 0$	$\gamma = 1$ $x_1 x_2$			
		00	01	11	10
S A	A, 0	A, 0	C, 0	—	B, 1
B	B, 1	B, 1	C, 0	—	B, 1
C	C, 0	C, 0	D, 1	—	A, 0
D	D, 1	D, 1	D, 1	—	A, 0

S, z

networks useless if they malfunction). Making state assignments without considering multiple changes of internal variables is much easier than having to take these changes into account.

Let us derive the transition table shown in Table 8.5.3 from Table 8.5.2, using a state assignment as shown.

Let us use *J-K* master-slave flip-flops. Using the output-input relationship shown in Table 8.4.3b (other master-slave flip-flop types can be treated in a

Table 8.5.2 Compact Flow Table in Skew Mode

	$\gamma = 1$ $x_1 x_2$			
S	00	01	11	10
A	A	C	—	B
B	B	C	—	B
C	C	D	—	A
D	D	D	—	A

S

Design of Sequential Networks in Skew Mode 485

Table 8.5.3 Transition Table in Skew Mode

s	$y_1 y_2$	\$x_1 x_2\$ 00	01	11	10
A	00	00	11	dd	01
B	01	01	11	dd	01
C	11	11	10	dd	00
D	10	10	10	dd	00

$Y_1 Y_2$

similar manner), we form the excitation table in Table 8.5.4. From this, Karnaugh maps* are formed in Table 8.5.5. The output table for z is formed in Table 8.5.6, directly from Table 8.5.1. In this case notice that, as explained in Section 8.3, the first column in Table 8.5.1 cannot be deleted, since in each row the value of z is different in different columns.

From Tables 8.5.5 and 8.5.6 a network has been synthesized in Fig. 8.5.1. Here notice that, in deriving switching expressions for J_i and K_i from Table

Table 8.5.4 Excitation Table Derived from Table 8.5.3

		$x_1 x_2$ 00	01	11	10
	00	0d, 0d	1d, 1d	dd, dd	0d, 1d
$y_1 y_2$	01	0d, d0	1d, d0	dd, dd	0d, d0
	11	d0, d0	d0, d1	dd, dd	d1, d1
	10	d0, 0d	d0, 0d	dd, dd	d1, 0d

$J_1 K_1, J_2 K_2$

---

* For the sake of simplicity, Karnaugh maps for individual functions are used instead of deriving a minimal multiple-output network (e.g., those in Section 5.2, if we want to have a network in two levels).

Table 8.5.5 Karnaugh Maps for J_1, K_1, J_2, K_2

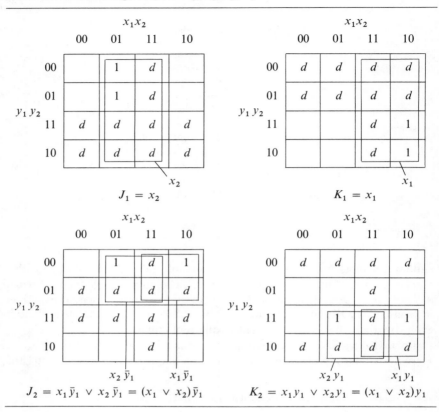

Table 8.5.6 Output Table

$$z = (x_2 \bar{y}_1 \vee x_1 y_1 \vee \bar{x}_2 y_1 y_2 \vee \bar{x}_1 \bar{y}_1 \bar{y}_2)\gamma \vee \bar{\gamma}(y_1 y_2 \vee \bar{y}_1 \bar{y}_2)$$

8.5.5 (or 8.5.4), $\gamma = 1$ need not be considered, that is, conjunction of γ with the switching expressions need not be considered, since the connection of the clock to the clock terminal of the flip-flop is doing this (e.g., J_1 in Fig. 8.5.1 receives x_1 only when the clock is on, and this is essentially $J_1 = x_1\gamma$).

Master-slave flip-flops do not respond to changes in their inputs when and after their outputs change until the leading edges of next clock pulses. Thus no network malfunction due to hazards or races occurs, and no post-analysis of designed networks is necessary. This is the advantage of clocked networks with master-slave or edge-triggered flip-flops.

Different Output Specifications

When the flow-output table in Table 8.5.1 (i.e., Table 8.4.7) was prepared in Example 8.4.2, the network was specified to assume new output values in some columns for $\gamma = 1$ [e.g., $z = 1$ for the column (1 0) for $\gamma = 1$ in row

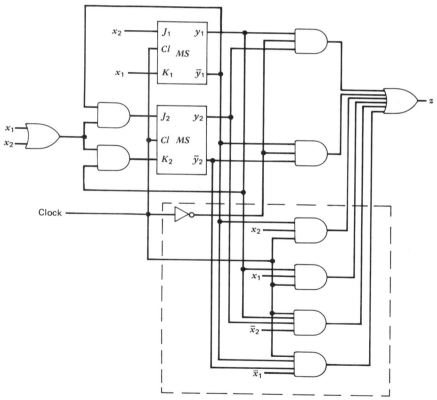

Fig. 8.5.1 Synthesized network based on Table 8.5.1 (for Example 8.4.2).

Table 8.5.7 Flow-Output Table in Skew Mode

	$y = 0$	$y = 1$ $x_1 x_2$			
s		00	01	11	10
A	$A, 0$	$A, 0$	$C, 0$	—	$B, 0$
B	$B, 1$	$B, 1$	$C, 1$	—	$B, 1$
C	$C, 0$	$C, 0$	$D, 0$	—	$A, 0$
D	$D, 1$	$D, 1$	$D, 1$	—	$A, 1$

S, z

A]—in other words, at the leading edges of clock pulses—and then maintain the new output values during these clock pulses.

When the given requirements on network performances allow us certain freedom in specifying when the network outputs can assume new values, however, we can often derive simpler networks. For example, consider Table 8.5.7. In this table z can assume a new output value only at the trailing edge of a clock pulse, while the transition among states is identical to that in Table 8.5.1. (In practice, this is often preferable to Table 8.5.1.) Table 8.5.7 can be written as Table 8.5.8a, placing the values of z outside the rectangle,

Table 8.5.8

a. Flow-Output Table

	$y = 0$	$y = 1$ $x_1 x_2$				z
s		00	01	11	10	
A	A	A	C	—	B	0
B	B	B	C	—	B	1
C	C	C	D	—	A	0
D	D	D	D	—	A	1

b. Karnaugh Map for z

	$y_1 = 0$	$y_1 = 1$
$y_2 = 0$		①
$y_2 = 1$	①	

z

$z = y_1 \bar{y}_2 \lor \bar{y}_1 y_2$

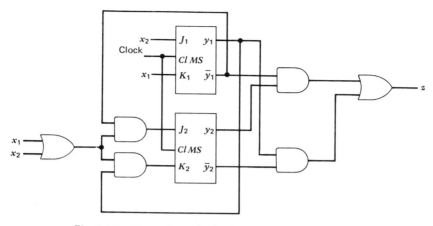

Fig. 8.5.2 Network synthesized on the basis of Table 8.5.7.

since the value of z in each row is the same in different columns. Then the Karnaugh map for z is obtained in Table 8.5.8b, if we use the state assignment shown in Table 8.5.3. (Since the values of z are independent of x_1 and x_2, z can be expressed in the Karnaugh map with y_1 and y_2 as the only variables, in Table 8.5.8b.) The excitation table in this case is identical to Table 8.5.4. Thus we have obtained the network in Fig. 8.5.2 from the Karnaugh maps in Tables 8.5.8b and 8.5.5. The network in Fig. 8.5.2 is simpler than that in Fig. 8.5.1, which contains the extra gates shown inside the dotted line.

Many other output specifications are possible. For example, the network can assume new output values at both the leading and trailing edges of clock pulses.

Advantages of Skew Mode

The advantages of skew-mode operation with master-slave or edge-triggered flip-flops may be summarized as follows:

1. We can often use simpler flow-output tables (or state-output tables) than are required in fundamental mode, making design easier. In other words:

   (a) Flow-output tables in skew mode require no more, and often fewer, states than those in fundamental mode (because we need not consider both unstable and stable states for each input change, and need not consider entering dashes or extra states, which are to avoid multiple changes of internal variables).

490 *Synthesis of Synchronous Sequential Networks*

(b) The column for $\gamma = 0$ in a flow (or state or transition) table can be omitted, though **the columns for $\gamma = 0$ in an output table cannot always be ignored**. (Notice that, unlike the next-state values, output values in the columns for $\gamma = 0$ in the output table have no relationship to those in the columns for $\gamma = 1$.)

2. State assignments are greatly simplified, because we need not worry about hazard due to multiple changes of internal variables. (If we want to minimize the number of gates, connections, or levels, we need to try different state assignments. This is not important, since the networks to be synthesized will work reliably anyway.)

3. Networks synthesized in skew mode usually require fewer internal variables than those in fundamental mode. (Since master-slave flip-flops contain many gates, if we count these gates the total number of gates in a network may not be fewer. We cannot rule out the possibility of getting simpler networks when we do not use master-slave flip-flops.)

4. After the network synthesis we do not need to check whether the networks contain hazards or not. This is probably the greatest of all the advantages of skew mode, since checking hazards and finding remedies is usually very cumbersome and time-consuming.

8.6 General Comments on Sequential Networks

In Chapters 7 and 8 we have discussed the analysis and synthesis of sequential networks in general. To synchronize the operations of gates, every computer in current use, probably without exception, has a clock. (In smaller digital systems there may be the possibility of no clock.) By this synchronization of the operation of gates, extrinsic hazards can be easily eliminated without adding extra gates, unlike the remedies in Section 6.7.

All the gates in digital systems are not necessarily connected to clocks (except in electronic circuit technology, which requires clock connection of all gates), since all the gates can be essentially synchronized. Then a digital system generally consists of three types of sequential networks: asynchronous networks (i.e., the networks without clocks in Chapter 7), clocked networks in fundamental mode (Section 8.1), and clocked networks in skew mode (Sections 8.3 through 8.5). Network malfunctions due to hazards which the probing of signals by clocks (e.g., Fig. 8.1.4) cannot eliminate (e.g., hazards that simultaneous, identical output values of flip-flops cause) can be eliminated by clocked networks in skew mode. We need not worry about network malfunctions due to hazards (including races). Also, design procedures are simplified by the skew mode. All three types of networks, however, are used in practice, depending on situations. For example, if speed is the most important factor, asynchronous networks are still preferred. If only spurious outputs are a problem, clocked networks in fundamental mode will be sufficient.

General Comments on Sequential Networks 491

Sequential Networks Not Discussed in Chapters 7 and 8

Not every aspect of sequential networks is discussed in Chapters 7 and 8. Although the clock terminals of flip-flops are always connected to clocks in Chapter 8, the flip-flops can be more freely used by connecting the clocked terminals to other gates or flip-flops in networks (also to control signals supplied from some other part of networks). Although latches and flip-flops are discussed in Chapters 7 and 8, there are many other variations of flip-flops. (As new IC logic families are developed, different flip-flops pertinent to these families, particularly MOS logic families, become available.) Also, if sequential networks are to contain many internal variables, they cannot be handled by the procedures discussed in Chapters 7 and 8. We need to design such large networks heuristically by decomposing them into smaller networks of manageable size, or by modifying known networks. Some of these problems not discussed in Chapters 7 and 8 will be examined in Chapter 9.

Realization of Combinational Networks with Networks Having Loops

The interesting question arises of whether loops have any significance for combinational networks, though throughout the preceding chapters we synthesized combinational networks by means of loopless networks. Loops are an absolute necessity for sequential networks, however, if no other kinds of memory devices are used. More specifically, let us consider the following question: can combinational networks with loops have fewer gates than loopless networks for certain functions? There has been no systematic research on this subject, though a few casual observations have been published (Remark 8.6.1). But in practice, when we need to design large networks, we sometimes should consider networks with loops or memories as alternative ways of designing combinational networks, even though they can be realized with loopless networks. For example, a combinational network that detects whether exactly one out of n variables assumes the value 1 can be realized with a counter and a shift register (both are sequential networks), using far fewer gates than a loopless network, as shown in Section 9.4.

Remark 8.6.1: There are very few reports dealing with the question of whether networks with loops can have fewer gates than loopless networks when we want to design combinational networks for certain functions. Short [1960] provided an example* of a combinational network of relay contacts with loops, which has fewer contacts than a loopless network for a certain function. An interesting example from McCaw [1963] is quoted on p. 65 of Wood [1968] and p. 169 of

---

* His example also establishes the fact, observed by Lee [1959], that a computer program that consists entirely of binary decisions must generally contain loops if the total number of such decisions is to be minimized.

Brzozowski and Yoeli [1976]. Another example is discussed in Kautz [1970] and Nozaki [1973]. These examples are interesting, since all minimization problems for combinational networks are usually approached with an implicit or explicit assumption that only loopless networks are to be considered. (The minimization of the number of NOT gates in combinational networks by Huffman [1971] also uses a network with loops whose output values of gates inside oscillate. This is a related but different problem, since the number of NOT's, instead of the number of gates, is to be minimized.) But we know very little of how to design combinational networks with loops (in particular, no work on single-output networks with loops has been reported). Whether or not such networks are useful in practice remains to be seen (the internal oscillation reported in Huffman [1971] might be a desirable feature for diagnosis, for example), and more systematic knowledge would be desirable. ■

Exercises

8.1 (MT)   In the network in Fig. E8.1, input x is a level signal, c is a clock, input x does not change during the presence of clock pulses, every gate has switching time τ, and the S-R latches of Fig. 7.1.1a are used and have response time 2τ (i.e., time until their outputs assume the values corresponding to a new input value).

(i) Assume that the duration of a clock pulse is much longer than τ and the duty cycle of the clock is 50 percent.
   Derive a transition-output table.
   Derive a word statement describing the network performance, if you can.
   A state that a network cannot enter unless the network is placed in that state is called a **transient state**. Check whether there are such states; if there are, identify them.

(ii) When the S-R latches are replaced by S-R master-slave flip-flops with clock connected, how does the network behave in the case of (i)? Derive a transition-output table. When the network works stably with S-R latches, does the replacement with S-R master-slave flip-flops cause any change? (Some networks work reliably even with non-master-slave-type flip-flops.)

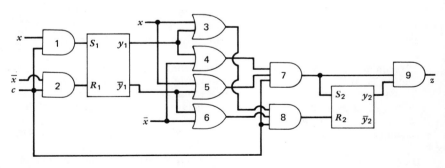

Fig. E8.1

(iii) Assume that the duration of a clock pulse is 2τ, and the interval between an adjacent pair of clock pulses is 2τ (i.e., the duty cycle is 50 percent). Also assume that the S-R latches of Fig. 7.1.1a are used again. Then do either of the following:
(a) Discuss the general difficulty in analyzing this type of network.
(b) Analyze the network, that is, derive waveforms that show the responses of the output z and internal variables y_1 and y_2 against the change of input x. [Since this network is in neither fundamental nor skew mode, the analysis would be complex and time-consuming. If readers prefer, they work (a). This is the reason why networks not in fundamental or skew mode are seldom used.]

(iv) If all the gates are clocked in (iii), can we avoid the difficulty encountered in this case?

8.2 (M) Let us analyze the network in Fig. E8.2 with J-K master-slave flip-flops, which has the single input x and two outputs, z_1 and z_2. Assume that input x changes only when the network settles down.

(i) Derive a transition-output table in skew mode for the network.

(ii) Draw a state diagram, and then make a word statement that describes the performance of the network.

(iii) If the J-K master-slave flip-flops are replaced by J-K latches (i.e., the flip-flops shown in Fig. 8.1.3a), can the network still work as before? Discuss what will happen. Assume that the duration of a clock pulse and the value 1 of J and K are kept to appropriate time lengths.

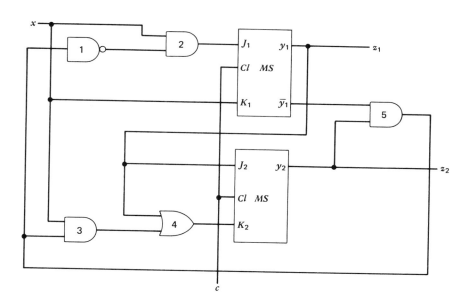

Fig. E8.2

494 Synthesis of Synchronous Sequential Networks

8.3 (E) Prepare a transition-output table in skew mode for the network with J-K master-slave flip-flops in Fig. E8.3.

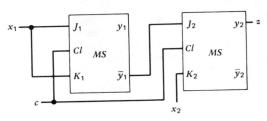

Fig. E8.3

8.4 (M) Excess-3 code is a binary expression of decimal numbers, as shown in Table E8.4, where each decimal number I is assigned the binary number that represents decimal number $I + 3$.

We want to design a network to detect errors in excess-3 code which is serially sent to network input x, starting with the rightmost bit in Table E8.4. If the network receives any binary number other than the excess-3 code, the network output z is set to 1, and otherwise to 0. Input and output signals appear only during clock pulses.

(i) Draw a state diagram, using as few states as possible. Explain concisely what each state represents physically. (For example, state A in Table 8.4.1 represents physically "last 1 at x_1.")

(ii) Show a flow-output table in skew mode.

Table E8.4

DECIMAL NUMBER	EXCESS-3 CODE
0	0011
1	0100
2	0101
3	0110
4	0111
5	1000
6	1001
7	1010
8	1011
9	1100

8.5 (M) We want to design a network to convert a serially sent excess-3 code back into the corresponding binary number (i.e., from the binary numbers in the right-hand side in Table E8.4 to those that the decimal numbers in the left-hand side represent).

The excess-3 code is sent to the network input x in synchronization with the clock, starting with the rightmost bit in Table E8.4. At every fourth bit the corresponding binary number is to be displayed at the network outputs.

(i) Draw a state diagram, using as few states as possible. Explain concisely what each state represents physically.
(ii) Show a flow-output table in skew mode.

8.6 (M) Suppose that binary numbers, each two bits long, are serially sent to the network input x, with the least significant bit first and in synchronization with the clock. The network adds these binary numbers, but the carry from the most significant bit is ignored. At every two bits a new sum is to be displayed at the network outputs; until then, the previous sum is kept at the outputs.

(i) Draw a state diagram with as few states as possible.
(ii) Show a flow-output table in skew mode.

8.7 (MT) We want to design a network in fundamental mode for Example 8.4.1, using more than two states, under either condition A or B, as specified below.

(i) Derive a flow-output table in fundamental mode with as few states as possible.
(ii) Derive a state assignment free of multiple changes of internal variables.
(iii) Design a network in double-rail input logic.
A. Use AND and OR gates only, without any flip-flops.
B. Use S-R latches and then, if necessary, as few AND and OR gates as possible.

8.8 (MT) Do the following for Exercise 7.14:

(i) Write a flow-output table in skew mode, with as few states as possible.
(ii) Design a network in double-rail input logic, using AND and OR gates and clocked J-K master-slave flip-flops. Assume that the inputs from the buttons do not change during the presence of clock pulses (possibly by providing a simple network to take care of this, between the two buttons and the sequential network to be designed).

8.9 (M) We want to design a binary counter of bit length two. The input x does not change during the presence of a clock pulse. If $x = 1$ lasts for m consecutive clock pulses, where m is restricted to be at most 3, then the counter is interpreted to receive m 1's. Whenever we have $x = 0$ during any clock pulse, the counter is set to 0. Draw a flow-output table in skew mode. Design a network with T master-slave flip-flops and any types of gates. Assume that x, clock c, and their complements are available as the network inputs. Use a minimum number of flip-flops.

8.10 (M) We want to design a network with a single input x and a single output z. The value at x can change only during the absence of a clock pulse. Whenever $x = 1$ continues for three consecutive clock pulses (i.e., 1 1 1), output z becomes 1 only during the presence of a clock pulse at the next pulse, as illustrated in Fig. E8.10, where c is the clock. Otherwise z is 0.

(i) Draw a flow-output table in skew mode with as few states as possible. Draw a state diagram.
(ii) Design a network with J-K master-slave flip-flops and any types of gates. Assume that x, c, and their complements are available as the network inputs. Use as few flip-flops as possible.

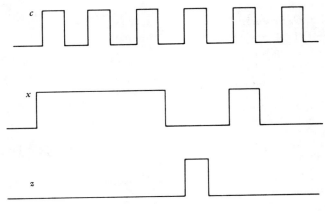

Fig. E8.10

8.11 (M) We want to design a binary counter of bit length two, where input x is 0 or 1 only during the presence of a clock pulse. Whenever $x = 1$ lasts for consecutive clock pulses, the count must be reset to 0. Assume that the counter outputs can change only at the trailing edges of clock pulses.

(i) Draw a flow-output table in skew mode with as few states as possible. Also draw a state diagram.

(ii) Design a network with J-K master-slave flip-flops and any types of gates. Use as few flip-flops as possible.

8.12 (M) We want to design a counter that counts up for each occurrence of a clock pulse when $x = 1$, in the count sequence 0, 1, 4, 7, and then returns to 0. When $x = 0$, the counter does not count at each occurrence of a clock pulse. Assume that the counter input x can change only during the absence of clock pulses.

(i) Show a flow-output table in skew mode, with a minimum number of states.

(ii) Design such a counter, using the minimum number of J-K master-slave flip-flops and, if any extra gates are necessary, as few AND, OR, and NOT gates as possible, as the secondary objective.

8.13 (M) We want to design a divide-by-3 counter* whose count increases by 1 whenever signal 1 appears at the counter input x during a clock pulse. The signal at x can change only during the absence of a clock pulse.

(i) Show a flow-output table in skew mode, using a minimum number of states.

(ii) Design such a counter with the minimum number of J-K master-slave flip-flops. If any extra gates are necessary, use as few AND, OR, and NOT gates as possible.

8.14 (M) We want to design a network to compare two binary numbers, $X = (x_1, x_2, \ldots, x_n)$ and $Y = (y_1, y_2, \ldots, y_n)$, which are serially sent to the network inputs x and y, in synchronization with the clock, starting with the least significant bits x_1 and

---

* See Exercise 7.17 for the definition of "divide-by-N counter."

y_1. The network has two outputs, z_X and z_Y. When $X > Y$, z_X is set to 1, keeping $z_Y = 0$; when $X < Y$, z_Y is set to 1, keeping $z_X = 0$. When $X = Y$, both z_X and z_Y are set to 0.

(i) Show a flow-output table in skew mode, using a minimum number of states. Explain concisely what each state represents physically.

(ii) Design such a network in double-rail input logic, with the minimum number of *J-K* master-slave flip-flops. If any extra gates are necessary, use as few AND and OR gates as possible.

8.15 (M) We want to design an accumulator that can add a binary number X to a binary number Z stored in the accumulator itself, as illustrated in Fig. E8.15. (An **accumulator**, in general, is a register in which the result of an arithmetic or logic operation is formed and retained.)

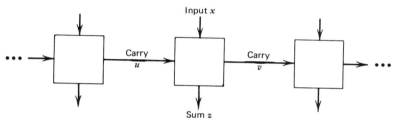

Fig. E8.15   Accumulator.

The network for one bit position of the accumulator has input x, which receives one bit of the number to be added, and input u, which receives the carry from the next lower bit position. The outputs of the network are carry v, to the next higher bit position, and sum z for this bit position. (In other words, x and u are added to the current value of z to form carry v and sum z.) Each of x, u, v, and z is 0 or 1, and changes in synchronization with the clock. A binary number X appears only during one clock pulse for an addition.

(i) Show a flow-output table in skew mode, using a minimum number of states.

(ii) Design such a network with the minimum number of *J-K* master-slave flip-flops. If extra gates are necessary, use as few AND, OR, and NOT gates as possible.

8.16 (MT) We want to design an accumulator with control terminal t, as illustrated in Fig. E8.16. When $t = 0$, the accumulator adds a binary number X to a binary number

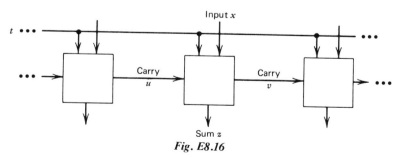

Fig. E8.16

498 *Synthesis of Synchronous Sequential Networks*

Z stored in the accumulator, as described in Exercise 8.15. When $t = 1$, the current value z of the output of the network for one bit position of the accumulator is replaced by $x \oplus z$, setting v to 0 at the same time.

(i) Show a flow-output table in skew mode, using a minimum number of states.

(ii) Design such a network with the minimum number of *J-K* master-slave flip-flops. If any extra gates are necessary, use as few AND, OR, and NOT gates as possible.

8.17 (M) Solve Exercise 7.15 for the following:

(i) Draw a flow-output table in skew mode, using as few states as possible and assuming that a clock is available. (Can you obtain fewer states than in Exercise 7.15?)

(ii) Design a network in double-rail input logic, with edge-triggered *J-K* flip-flops and any types of gates. Use as few flip-flops as possible. Assume that the inputs from the keys can change only during the absence of clock pulses. (Unless carefully designed, the network may change its output at each clock pulse.)

8.18 (M) The states in a completely specified flow-output table can be classified into indistinguishability equivalence classes by applying input sequences of different lengths (e.g., 0 1 1 0 1 is called an input sequence of length 5). This procedure is called the **Moore reduction procedure**. The unique feature of this procedure is that only the process of finding distinguishable pairs of states is repeated, and we need no process for finding indistinguishable pair of states. The procedure works as follows.

(i) (E) Derive the indistinguishability classes of one of the flow-output tables in out what states are distinguishable from others, and partition the states into subsets that are mutually distinguishable.

Let us illustrate Step 1 with the flow-output table in skew mode in Table 8.18a.

Table E8.18a

	$y = 0$	$y = 1$	
		x	
		0	1
A, 0	C, 0	B, 0
B, 0	E, 0	C, 1
C, 0	A, 0	D, 0
D, 0	A, 0	C, 0
E, 0	B, 0	A, 1
F, 0	E, 0	D, 1

S, z

Table E8.18b

$P_0 = (ABCDEF)$
$P_1 = (ACD)(BEF)$
$P_2 = (A)(CD)(BEF)$
$P_3 = (A)(CD)(BF)(E)$
$P_4 = (A)(CD)(BF)(E)$

First let us apply input sequences of length 1. Suppose we apply $x = 1$ when the network is in A, C, or D. When clock pulse $\gamma = 1$ arrives, the network will have the same output 0 along with next state B, D, or C, respectively. When the network is in B, E, or F, the network will have the same output 1 along with next states C, A, or D, respectively. Next, suppose we apply $x = 0$. When $\gamma = 1$ arrives, the network will have the same output 0, no matter which state the network was initially placed. Thus all the states shown in P_0 are partitioned into two subsets, (ACD) and (BEF), as shown in P_1.

2. Repeat Step 1, by increasing R by $R + 1$, until no new distinguishable pairs of states are found.
3. Then the partitioned subsets represent indistinguishability (equivalence) classes of states.

Next, P_1 is partitioned into (A), (CD), and (BEF) as shown in P_2, by comparing output values for input sequences of length 2. By applying input sequences of length 3, the states are further partitioned as P_3. But P_4 corresponding to input sequences of length 4 is identical to P_3. Then the procedure terminates and P_4 expresses the indistinguishability classes.

(i) (E) Derive the indistinguishability classes of one of the flow-output tables in Table E8.18c by the Moore reduction procedure.

Table E8.18c

A		$\gamma = 1$		
		$x_1 x_2$		
$\gamma = 0$	00	01	11	10
A, 0	B, 1	A, 1	A, 1	G, d
B, 0	B, 1	B, 0	A, d	G, d
C, 0	C, 1	C, 0	A, d	G, d
D, 0	D, 1	E, 1	D, 0	H, 1
E, 0	C, 1	E, 1	E, 1	G, d
F, 0	F, 1	A, 1	D, d	F, 1
G, 0	F, 1	G, 1	G, 1	G, 0
H, 0	D, 1	E, d	H, 0	H, 1

S, z

Table E8.18c (Continued)

B		$x_1 x_2$ $\gamma = 1$		
$\gamma = 0$	00	01	11	10
A, 0	A, 1	D, 1	E, d	A, 0
B, 0	B, 0	F, 1	B, 1	B, 1
C, 0	B, d	C, 1	C, 0	H, d
D, 0	A, 1	D, 1	E, 1	D, 0
E, 0	B, d	C, 1	E, 1	E, 1
F, 0	F, 1	F, 1	H, 1	D, d
G, 0	B, d	G, 1	G, 0	H, d
H, 0	B, d	G, 1	H, 1	H, 1

S, z

(ii) (M) Prove the following property: If there exist two states, s_i and s_j, such that they are in the same subset in P_{k-1} but not in the same subset in P_k, then $N(P_{t-1}) < N(P_t)$ holds for every $t = 1, 2, \ldots, k$, where $N(P_t)$ is the number of different subsets in P_t.

(iii) (M) On the basis of the property in (ii), prove that, if two states, s_i and s_j, are distinguishable, they are distinguishable for input sequences of length $R - 1$ or less, where R is the number of states in the given flow-output table.

(iv) (M) On the basis of the property in (ii), prove that, when partition P_k corresponding to input sequences of length k is identical to partition P_{k-1} corresponding to input sequences of length $k - 1$, the procedure terminates, yielding all indistinguishability classes of states. [Obviously, $k \leq R$ from (iii).]

8.19 (E) Minimize the number of states in the flow-output table* in skew mode in Table E8.19 by the Moore reduction procedure described in Exercise 8.18.

---

* This table may be interpreted as a network whose output z appears only during the absence of clock pulses.

Table E8.19

	$y = 1$	
	x	
$y = 0$	0	1
A, 0	C, 0	B, 0
B, 1	E, 0	C, 0
C, 0	A, 0	D, 0
D, 0	A, 0	C, 0
E, 1	B, 0	A, 0
F, 1	E, 0	D, 0
	S, z	

8.20 (E) Minimize the number of states in the flow-output table shown in Table E8.19 by Procedure 7.6.1.

CHAPTER 9

Practical Considerations in Logic Design

In the preceding chapters we discussed mostly the design of small networks and switching algebra. In logic design practice, however, many other things must be considered.

First, as digital systems (including computers) become widely used in our society, we have to design them under very diversified motivations.

Many networks that digital systems contain are often too large to be designed by the procedures in the preceding chapters, and somewhat different approaches are required. As semiconductor memories become compact and inexpensive, memories are becoming widely used to implement networks, being mixed with gates. Also, software is often implemented with ROMs (read-only memories), since ROMs are cheaper than RAMs (random-access memories), and firmware, that is, software implemented by ROMs, can be executed faster than software stored in RAMs. When digital systems are too large to be designed by the procedures in the preceding chapters, or are to be designed with a mixture of gates, software, and memories, it is an important but not easy problem to determine how to decompose digital systems into subnetworks of manageable sizes, software, and memories. Flow charting is helpful in this connection.

It is becoming very important to design without errors before hardware implementation, and also to find faulty gates after hardware implementation.

These problems are discussed in this chapter. Since space is limited, only aspects related to the preceding chapters are considered, leaving out some other aspects such as microprogramming and microcomputers.

9.1 Diversified Design Motivations and Design Approaches

As digital systems become widely used in many items of merchandise, logic design is being done under diversified motivations. Consequently, we need to take different design approaches. Also, in logic design practice, we have to consider many aspects.

Diversified Design Motivations

Recently, digital systems have been incorporated in many items of merchandise other than computers. (In some cases the digital systems incorporated are so small that they might be more appropriately called digital networks.) Examples are digital systems for cameras (e.g., Polaroid's SX-70), electronic watches, electronic weight scales, electronic locks, calculators, video games, personal computers, electronic controls of automobiles, sewing machines, and electronic measurement instruments. As integrated circuits become more inexpensive and compact, many items of merchandise that never existed before will be introduced, based on digital systems.

Consequently, logic design must be done under many different motivations. Since each case is different, we have different problems. For example, we have to choose an appropriate IC logic family, since each case has different performance requirements. (Scientific computers require fast speed. Wristwatches require very low power consumption.) Not only performance requirements but also compactness is very important in some cases such as cameras, so digital systems must accordingly be implemented in LSI packages rather than in many MSI packages. In short, there are many things to be considered. Here, however, let us consider two important cases that lead to two sharply contrasting logic design approaches—in other words, quick design and high-performance design.

Often quick introduction of new computers or new merchandise with digital systems is very important for a manufacturer in terms of profits. (In some cases, introduction of a new product one year earlier than a competitor's generates about twice the total income that the competitor gets [Davidow 1974]). This is so because the firm that introduces the product can capture all initial potential customers, at highest prices, and latecomers are left with only the remaining, fewer customers, selling at lower prices. This difference in timing often means a big difference in profits. In other words, **the profits due to faster introduction of new products on a market often far exceed the profits due to careful design.** The use of off-the-shelf IC packages, including off-the-shelf microcomputers, is often essential for shortening design time. **By shortening design time, the required manpower is reduced, and thus the design cost contained in the product cost is reduced.**

The other case we consider here is logic design to attain highest performance, utilizing most advanced technology. Designers usually try to improve the economic aspect, that is, performance per cost, at the same time. In this case design and development usually take many years, because new technology often must be explored at the same time. Hence this is the other extreme to the above quick design in terms of design time.

Of course, when manufacturers introduce new products or computers, it is ideal to introduce them with the highest performance in the shortest time. But it is usually very difficult to attain both, so one aspect must be emphasized, based on the firms' marketing strategy against competitors.

Quick Design

When logic design needs to be completed in a short time, it is usually done with off-the-shelf packages such as standard networks in MSI packages or microcomputers in LSI packages. Most networks that are commonly used in digital systems are available in MSI or LSI packages. By using them, logic designers need not redesign these networks, except possibly for some interface networks, thereby eliminating a major part of design efforts and time. (These IC packages can be assembled into systems by using only their input-output relationships, which can usually be described in word statements like those in Chapters 7 and 8 [Blakeslee 1975]. Even the procedures and switching algebra in the preceding chapters may not be needed, except for the design of interface networks.) In some cases a 20-to-1 reduction in manpower due to the use of off-the-shelf IC packages is reported [Noyce 1971]. With this approach the design of a minicomputer probably can be accomplished by a few people in a few months.

If off-the-shelf microcomputers are used, design time is even further reduced, since designers do programming instead of logic design in order to let the microcomputers do what logic networks are supposed to do [McDermott 1975b].

By emphasizing quick design, however, we usually sacrifice performance. Also, when production volume is high, quick design does not yield the lowest manufacturing cost. Thus, when production volume is high and longer design time can be justified, custom LSI designed by the library approach is often used. (As discussed in Section 2.4, there are different approaches in custom LSI design. In the library approach the layouts for networks, subnetworks, or gates that are commonly used in digital systems are stored in a library. Designers can use these layouts without redesigning from scratch each time. Thus the layouts in the library are used somewhat like the off-the-shelf IC packages mentioned above.) Computer-aided design (CAD) is extensively used not only for layouts (the library approach itself is a CAD) but also in other stages of design. With the library approach of custom LSI design, both performance and cost are improved, although design time is often longer than is required with the use of off-the-shelf microcomputers. (See, e.g., Gold [1976].)

When compactness of logic networks (e.g., control networks of cameras) is mandatory, we usually need to use custom LSI.

High-Performance Design

When we want to develop computers or digital systems of high performance, using the most advanced technology, much greater manpower and design time are required than are needed with the quick design approaches described above. The actual time requirement depends on how ambitious the designers are. The design process typically takes a few years, whether a large computer or a microcomputer (advanced) is being designed. In a typical case 3 to 5 years is required to design a large, high-speed computer.* If the computer is not drastically different from the previous models, this time can be shorter, but if the computer is based on many new ideas, it may be longer. If gates are implemented with completely new technology, networks probably need to be designed from scratch, so a long time is spent on logic design, as well as on other problems.†

Complete design of a large computer (including logic design) requires a large number of people, ranging perhaps from fifty to a few hundreds. These people are divided into groups for the designs of the central processing unit, control unit, input-output unit, memories, and software. Representatives of these groups, who are often called computer architects (usually several people), coordinate the joint work of all the groups.

As we become able to pack an increasingly large number of networks in a single VLSI chip every year, the design of custom LSI chips (including microcomputers by custom design) of high performance with the most advanced technology is beginning to require greater effort and longer time. It is not uncommon for the development of a new microcomputer chip with the most advanced technology to take a few years although such design perhaps requires much less manpower than is needed in the case of large computers. At least key networks (i.e., some standard networks, like adders) in a microcomputer must be designed from scratch, considering the best layouts‡ for them on a VLSI chip so that parasitic capacitance is reduced to

---
* Mason [1974], Stein [1975], McDermott [1975a].
† Bell, Kotok, Hastings, and Hill [1978], Russell [1978], Case and Padegs [1978], Borgerson, Hanson, and Hartley [1978], Brooks [1975], Wise [1966], Mack [1974].
‡ In this case, the best layouts are usually found by hand (possibly aided by CAD for simple tasks) without using the library approach, since layouts by the library approach yield at least 15 percent larger chips than do hand-prepared layouts. (The library approach itself offers different approaches. The library approach in which the layouts for MOS networks or subnetworks are stored in a library requires shorter time to use but yields less compact layouts than the library approach in which the layouts for the components constituting the networks or subnetworks are stored in a library.) Hand-prepared layouts developed for custom VLSI chips of high performance will be used for the library approach in the future, when the development of such custom VLSI chips proves successful.

We are packing more gates on a single chip every year, and will eventually face the following

(Continued)

improve speed. (Logic network design, discussed in the preceding chapters, and layout design are highly interactive. For example, if the parallel connections of two MOSFETs for x and y and the MOSFET for z are exchanged in the left-most gate in Fig. 2.3.8, the parasitic capacitance of that gate will be changed, with the different speed of the gate. In this sense, layout design is an important part of logic design.) Such high-performance design requires a large initial investment, but when production volume is very high,* it can greatly reduce manufacturing cost (despite very high design cost) for the highest performance, compared with all other approaches discussed so far.

Design Stages of a Digital System

In the case of the high-performance design discussed above, every effort is made to realize digital systems with the best performance (usually speed), while simultaneously considering the reduction of cost per performance.†

First, the performance or cost of the entire system is predominantly determined by architectural design, which must be done based on good knowledge of all other aspects of the system, including logic design (also software). If an inappropriate architecture is chosen, the best performance or lowest cost of the system cannot be achieved, even if logic networks, or other aspects like software, are designed to yield the best results. For example, if microprogramming is chosen for the control logic of a microcomputer based on MOS, it occupies too much of the precious chip area, sacrificing performance and cost, though we have an advantage of short design time and design flexibility. Thus, if performance or manufacturing cost is important, implementation by logic networks is preferred (e.g., microprocessors 8080 and Z80 [Shima and Faggin 1974, Shima, Faggin, and Ungermann 1976]).

Next, appropriate IC logic families and the corresponding electronic circuit technology are chosen for each network in the system. (Other aspects such as memories are simultaneously determined in greater detail.) We do not use expensive, high-speed IC logic families where speed is not required.

---

dilemma: the use of CAD is desirable because of the extensive design time required for very large, complex networks and the possibility of design errors, but CAD tends to yield larger networks and chips than does design by hand, as the integration size increases.

* When production volume is low (this is usually the case with large computers), gate arrays (or masterslices) are often used for high performance computers. [*Electronics* Apr. 28, 1978; May 25, 1978]. [Nakano et al. 1978]. This is another custom LSI design approach which economizes low production volume, though best economy cannot be attained because gates are prearranged with sufficient spacing for possible connections. Also, the number of different LSI chips is often minimized, instead of minimizing the entire system, in order to reduce total development cost. In this case, however, minimization of the size of each LSI chip is still desirable to improve the performance (and also to reduce cost).

† For cost analysis, see, for example, [Phister, 1976, 1978].

Diversified Design Motivations and Design Approaches 507

Architecture and electronic circuits are outside the scope of this book, so they are not discussed further.

Finally, each network is designed to give the highest performance, considering cost reduction, as explained in the following paragraphs.

The above design stages are highly interactive and iterative, since if it is found impossible to attain the design goals in a certain stage, adjustments are needed in other design stages.

Performance Improvement

Although the performance of the entire system (i.e., speed in this case) is predominantly determined by the architectural design, it is strongly influenced also by the key networks of the system such as adders. Therefore we need to design fast key networks.

Network Size Reduction

Size reduction of the entire system is closely related to cost reduction, other things such as electronic circuits being the same. But it is also related to speed improvement (or improvement of other aspects of performance). For example, if a system is too large to be put in a single chip (chips), we need more chips. The more frequently the signals propagate on the connections on a pc board, the slower the speed is. Hence size reduction is vitally important not only for cost reduction but also for speed improvement.

The network minimization procedures in the preceding chapters have been limited to small networks. The performance and the cost of the entire digital system, however, are the final criteria for the evaluation of the system. Therefore minimization of small networks is meaningless if these networks are only a small part of the entire system. But the minimization of small networks is still meaningful in the following sense. Since network size reduction is very difficult as a direct design objective (as discussed in Section 4.1), we need to use the minimization procedures discussed in the preceding chapters on subnetworks. (The results may be satisfactory. If not, they can serve as a basis for further improvement.)

Large networks are usually designed using the following approaches, since no procedure to directly design large minimal networks is known:

1. Some important standard networks are designed by cascading small networks (as discussed in Section 9.2).
2. Large networks are decomposed into small networks of manageable sizes (probably with the help of the flow charts in Section 9.5), which can be designed by the procedures in the preceding chapters or by heuristic procedures based on their functional relationships (word statement type).

3. Large networks are designed based on switching expressions that express functional relationships between their inputs and outputs (e.g., the look-ahead adders in Section 9.2).
4. Large networks are designed by modifying known networks (which in turn were designed by the above approaches).

Thus, if compact, small subnetworks are obtained, the large networks are certainly improved, though it is not guaranteed that the large networks will be the most compact (except the few cases mentioned in Section 9.2).

Also we need to minimize networks under restrictions due to architectural considerations or implimentation constraints. For example, suppose that many processors must be distributed in different locations from architectural consideration, or that the same subnetwork which has many connections with the rest of the entire network must be duplicated in separate LSI chips because of pin-number limitation of LSI packages. In these cases, the size of the entire system or the entire network is not minimized but we still need to minimize the size of each processor (in the former case) or each LSI chip (in the latter case), under such restrictions, in order to improve performance or economy.

Minimization Time of Networks

As seen in the preceding chapters, network minimization is very time-consuming for some functions and is practically impossible for networks with a large number of gates, if the number of gates or connections is to be absolutely minimized. In logic design practice, "absolute" minimization is usually not necessary, and near-minimal (or subminimal) networks are satisfactory. But what does "near-minimal" mean in this context? Nobody knows how close near-minimal networks are to absolutely minimal networks, since the latter cannot be known. In the literature heuristic design procedures alleged to yield near-minimal networks are presented, but the authors simply hope that their procedures achieve this. Two heuristic design procedures that do not guarantee absolutely minimal networks often yield networks with drastically different numbers of gates. (For example, the network of 25 NOR gates derived by a heuristic procedure for some functions, published in *IEEE Trans. Comput.*, is improved by another heuristic procedure into a network of 15 NOR gates, as mentioned in Section 6.6. Another example is the carry-ripple adders, which have been implicitly considered the minimal network for addition since the beginning of computer history, and were not further simplified until recently, as discussed in Section 9.2.) When designers or authors are tired, they tend to say that their networks are near-minimal.

Diversified Design Motivations and Design Approaches 509

Designers must consider a trade-off between processing time and the advantages of network minimization. Then, if minimization is too time-consuming, what is a reasonable compromise? Although the final criterion is whether competitors can do better in regard to the performance or cost of the entire system (not the individual networks inside), a difference of 10 or 15 percent in the entire system or chip usually appears to be tolerable. At least in the case of small networks, experienced logic designers appear to know intuitively, from their experience, that their networks are reasonably good, and they well know the best networks for functions that they encounter often, as long as they continue to use the same IC logic families and the same standard networks. (Therefore experienced logic designers often do not need minimization procedures such as those in the preceding chapters, including Karnaugh maps or switching expressions. Since they usually work under very strong pressure to complete a logic design in the shortest time to compete against other firms, they do not have the time to design minimal networks or to explore the possibilities of finding networks better than the standard networks already known.)

But a difference of 10 or 15 percent is sometimes significantly large. When the LSI chip size is critical (beyond the critical size, the chip cost rapidly increases), the chip cost easily doubles with an area increase of 10 or 15 percent. Therefore, when a VLSI chip of high performance is to be developed, the layout is usually done by hand, since the library approach tends to produce layouts larger by at least 15 percent than those done by hand.* Thus size reduction to the full extent is very important if the chip itself is to be sold as an individual commercial product. As mentioned in Section 2.4, however, if the chip is part of a much more expensive product, such size reduction is not important for cost reduction (but vitally important for speed improvement).

Frequently, in the case of large networks, probably nobody knows how much different the designed network is from the best one. Minimization of large networks is impossible, so is not attempted. Hence designed networks often contain a large number of redundant gates or connections, and the difference in network size achieved by different designers can be very great in some cases. Therefore any efforts to reduce the size (possibly by heuristic approaches) may pay off. Also, novice designers frequently obtain greatly oversized systems. [For students 50 percent or even 100 percent is not rare. Novice designers do not have complete repertories of networks, as experienced designers have. They must build up their experience by designing good networks by themselves (using the procedures in the preceding chapters, or other approaches), or by looking for good networks in the literature. Of

---
* Zilog microcomputers Z80 and Z8000 are such examples. Also see the third footnote in this section.

course, when new digital systems with high performance are to be developed with new technology, or new layout rules, even experienced logic designers may not be much different from novice designers, since they need to design even standard networks from scratch, though they can borrow some ideas from standard networks of old technology. (The layouts stored in the library in the library approach for custom LSI design must be updated, designing again from scratch whenever new technology is developed.)]

Some minimization effort, not necessarily for absolute minimization, is always desirable if it can be done within the time constraints.

Other Design Considerations

In the preceding chapters, minimal networks (combinational or sequential) in terms of economy or performance have been designed under restrictions, if we have any. (As discussed in Section 4.1, if restrictions are simple such as maximum fan-in, networks are minimized, simultaneously considering the restrictions. If restrictions are complex, minimal networks derived without the restrictions are later modified so that the restrictions are met.) In the practice of logic design, however, many other problems must be considered, such as design time, error-free design, ease in making design changes (when design errors are discovered), testability,[*] diagnosability, and use of ROM, PLA, or RAM, in addition to cost reduction and performance improvement due to minimization, as discussed so far. For most of these problems we do not have clear-cut solutions, even though some design procedures may be helpful, as we will point out in the following sections. A practical approach is to first design networks without considering them and then modify the designed networks to take care of these problems. For very complex logic design practice **experience and the intuition based on it are usually very important**.

9.2 Design of Standard and Large Networks

Certain networks such as adders and counters are used very often in digital systems. Let us call them **standard networks**. Some standard networks are very large and cannot be directly designed by the procedures in the preceding chapters. Many of these standard networks, however, whether combinational or sequential, have the unique features described below and can be

---

[*] Even when functional facilities such as testability (i.e., by adding extra gates and terminals, a network can be conveniently tested) are to be added as an extra restriction, performance improvement or cost reduction by minimization procedures, architectural considerations, or possibly other means is still desirable.

Design of Standard and Large Networks 511

designed based on these features. For other large networks the design can be based on the functional relationships of their inputs and outputs, or on modifications of other networks.

Features of Standard Networks

1. Their performances can be conveniently described in word statements. To express them in switching expressions or truth tables is very tedious, requiring complex expressions or large tables. Examples of important standard networks are as follows:

   (a) **Counter.** The number of occurrences of signal 1 at its single input terminal is shown at its multiple outputs. This is an ordinary counter. An **up-and-down counter** is a special counter with an input terminal x and a control terminal t. When $t = 0$, signal 1 at x is added to the count; when $t = 1$, signal 1 at x is subtracted from the count. If an up-and-down counter, say, with 10 outputs (i.e., the count is expressed in a binary number 10 bits long), is expressed in a flow table, the table requires 1024 states for stable states alone.

   (b) **Adder.** A ripple adder is discussed in Section 1.3. Again, if this is to be expressed in a flow table, a huge number of states must be filled in, since the adder may have about 30 to 100 inputs and about the same number of outputs.

   (c) **Shift register.** A shift register is a register in which a stored number can be shifted to the right or left by a controlling signal.

   (d) **Decoders** (in Exercise 2.10) and **multiplexers** (in Exercise 4.17) are other examples of standard networks.

2. Some standard networks such as adders and counters have a **modular structure**. This is convenient, because as different number lengths are required on different occasions (though the basic performance requirement is not changed), an entire network can be implemented by cascading a desired number of modules. For example, when we need an adder for binary numbers of different bit lengths, a ripple adder can be easily constructed by cascading the networks of the full adders discussed in Section 1.3. The modular structure is also convenient for compact layout on IC chips. This is also convenient for diagnosis, particularly of sequential networks, which are usually extremely time-consuming to diagnose. The design practice for modular networks is as follows:

   (a) A good network for each module is designed, possibly by making time-consuming efforts. For example, a ripple adder can be designed with modules as discussed in Section 1.3. Another example is the parity function of n variables, for which a minimal network for $x_1 \oplus x_2$ can be used as a module. Such good networks are usually published. What a logic designer must do is simply to find good networks for modules in books, magazines, or catalogs, and to cascade basic module networks as many times as required. This approach greatly facilitates logic design. We need not worry about complex switching expressions or flow-output tables; what we need to consider is simply the relationship between the performance of the module network under consideration and that of the entire network to be designed.

(b) When a logic designer needs a network with a slightly different function, the designer modifies an existing good network according to the new assumptions or restrictions.

3. Some standard networks such as multiplexers, input-output interface networks, and serial communication interface networks (see e.g., L. Smith [1976]) have a certain flexibility (as illustrated in Exercises 4.17 and 4.20). They can be used in many different cases, by supplying control signals or simple programs.

Design of Ripple Adders (or Carry-Ripple Adders)

Minimal full adders are designed under various conditions in Liu, Hohulin, Shiau, and Muroga [1974], using the integer programming logic design method mentioned in Sections 5.4 and 6.6. For example, a full adder in double-rail input logic with a minimal number of AND and OR gates as the primary objective and a minimal number of connections as the secondary objective obtained in Liu, Hohulin, Shiau, and Muroga [1974] is shown in Fig. 9.2.1. By cascading this network, a ripple adder of arbitrary bit length can be constructed. Fig. 9.2.2 shows, as a more sophisticated modular approach, an adder in double-rail input logic with AND and OR gates for an arbitrary bit length, where three different modules designed by the integer

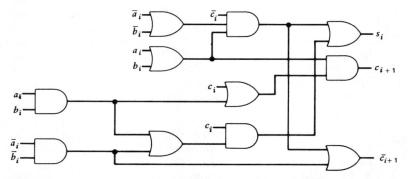

Fig. 9.2.1 Minimal AND-OR networks in double-rail input logic for full adder (*11 gates and 23 connections*).

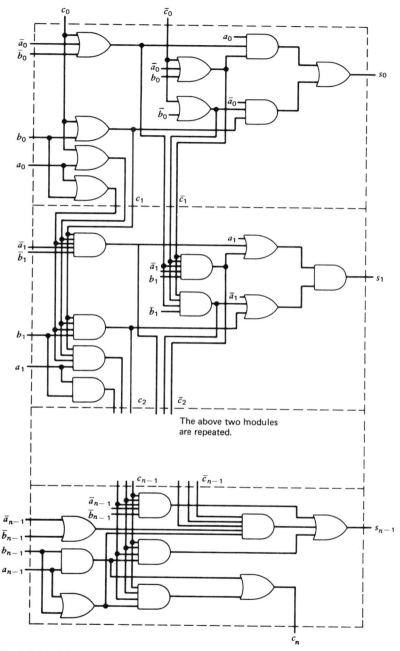

Fig. 9.2.2 Adder in double-rail input logic with one net gate delay per bit position.

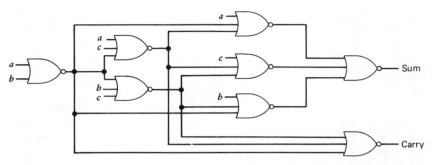

Fig. 9.2.3 Minimal NOR network in single-rail logic for full adder (8 NOR gates and 23 connections).

programming logic design method are used. (The first two modules are repeated in this adder; when the first module is to be repeated as the third module for s_2, the three lines for each of c_2 and $\bar{c}_2$ must be connected to the OR gates to which c_0 and $\bar{c}_0$ are connected in the first module for s_0, respectively.) This adder has one net gate delay per bit position, so its carry propagation time is half that of the ripple adder implemented by cascading the full adder in Fig. 9.2.1. Also, the number of gates per bit position is only 9, instead of the 11 gates in the latter. (For more detail and the variations of this adder, see Yu and Muroga [to be published].)

As another example, Fig. 9.2.3 shows the full adder in Liu, Hohulin, Shiau, and Muroga [1974], which has the minimal number of NOR gates as the primary objective and the minimal number of connections as the secondary objective, under the assumption that only noncomplemented variables a, b, and c are available as network inputs (c is the carry from the previous bit position) and there is no fan-in or fan-out restriction. This is the only minimal network under this assumption. Cascading the full adder in Fig. 9.2.3, we can construct a ripple adder of arbitrary bit length. (As will be discussed later in this section, a ripple adder constructed in this way does not have a minimal number of gates, though it is a reasonably good network.) By replacing NOR gates by NAND gates in Fig. 9.2.3, the minimal NAND network for the full adder can be obtained under the same assumption, since the sum and carry are self-dual.

The ripple adder designed by cascading the full adder in Fig. 9.2.3 will be compared to adders obtained by a more sophisticated modular approach later in this section.

Shortcomings of Modular Networks

The convenient modular approach described in (a) and (b) of feature 2 in the preceding list has the following shortcomings:

1. The performance, particularly speed, is usually not very good.
2. Even if each module is minimal in terms of the number of gates, connections, or levels, the entire network may not be minimal.

Large Minimal Networks

Usually we can do nothing about the second shortcoming above, since the entire network is often so large that it is usually very difficult to reduce the number of gates or connections. But for the adder and parity functions, the following modular approaches, which guarantee the minimality of the entire networks for any bit length are known.

An adder in single-rail input logic with a minimum number of NOR gates for an arbitrary bit length can be realized by cascading the set of three modules shown in Fig. 9.2.4, where a carry (e.g., c_i) and its complement each are transmitted over two or more lines. In this case, for the least significant bit position (i.e., $i = 0$), the two lines for c_0 should be replaced by a single line, and the module for the most significant bit position must be modified so that the carry is obtained on a single line. (If the most significant bit position is i or $i + 1$ in Fig. 9.2.4, a NOR gate is connected to the lines shown for $\bar{c}_{i+1}$ or $\bar{c}_{i+2}$ to obtain the carry for the most significant bit position. If it is $i + 2$, the entire module for the bit position $i + 2$ must be modified.) The adder has net gate delay 5 per three stages, though the number of levels from c_i to c_{i+3} in Fig. 9.2.4 is larger than five (Exercise 9.3.9), so the net gate delay per stage is 5/3 (though an adjustment calculation is necessary for the most significant bit positions). This is roughly 17 percent faster than the conventional ripple adder derived by cascading the full adder shown in Fig. 9.2.3, assuming that each gate has the same switching time. (For more details and the variations of this adder, see Lai and Muroga [to be published in 1979] and also for delay time, see Fukushima and Muroga [to be published].)

When both complemented and noncomplemented variables are available as network inputs, an adder with a minimum number of NOR gates and a single net gate delay per bit position has also been obtained. This adder is based on the repetition of two basic modules which consist of six NOR gates each and have 15 and 16 connections, respectively (thus 15.5 connections for each bit position, on the average). The adders mentioned in this and the preceding paragraph (and their variations) always have a minimal number of NOR gates, no matter what bit length they have. They have fewer gates than the ripple adders based on the full adders in Fig. 9.2.3. If NOR gates are replaced by NAND gates, we have adders with a minimal number of NAND gates. All the adders of NOR gates discussed above were derived, and their minimality was theoretically proved, on the basis of the minimal networks for modules designed by the integer programming logic design method [Lai and Muroga, to be published in 1979].

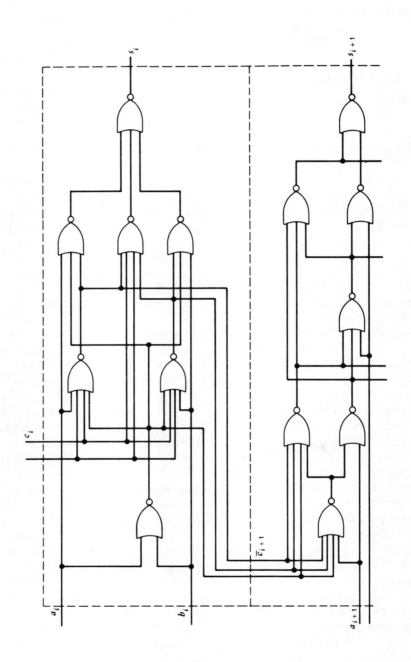

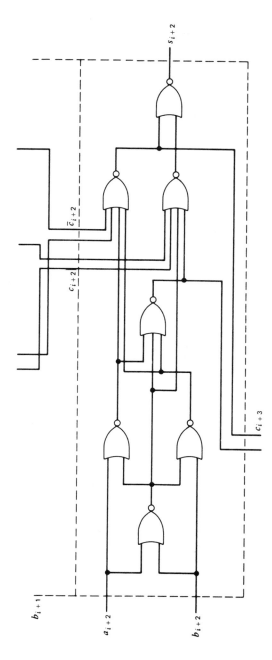

Fig. 9.2.4 Modules for an adder that has the minimal number of NOR gates for an arbitrary bit length. (This set of modules is repeated for an arbitrary bit length.)

518 Practical Considerations in Logic Design

Multipliers that require fewer gates than those based on full adders and half adders can be designed on the basis of the adders in Figs. 9.2.2 and 9.2.4.

NOR networks for parity functions

$$x_1 \oplus x_2 \oplus \cdots \oplus x_n \quad \text{and} \quad \overline{x_1 \oplus x_2 \oplus \cdots \oplus x_n}$$

were designed under two different minimality criteria, minimization of the number of gates and minimization of the number of connections, based on minimal networks for small values of n, by the integer programming logic design method. Their minimality was proved for any value of n by a long, sophisticated argument [Lai 1976].

In general, no procedure is known for the design of large networks with a minimum number of gates or connections. Hence the above networks with a minimum number of gates or connections for any number of variables (or bit length) are very exceptional. Also, these networks are minimal, or very close to minimum, even if the minimization of the number of gates and the minimization of connections are exchanged* [e.g., a network with a minimum number of gates as the primary objective and a minimum number of connections as the secondary objective is also a network with a minimum number of connections as the primary (not secondary) objective and a minimum number of gates as the secondary objective]. Roughly speaking, these networks are most compact, according to the argument in Muroga and Lai [1976].

Fast Adders

The performance problem (shortcoming 1 in the preceding list) is often essential. To improve performance we may have to abandon the modular approach, depending on the functions that we want to design. For example, the speed of the adder is vitally important in high-speed computers. We want to have a faster adder even if it requires many more gates.

Since the adders based on the modules explained above (including ripple adders) are slow because the maximum carry propagation time is linearly proportional to bit length, an adder called the **look-ahead adder** (or carry-look-ahead adder) is used for high-speed addition.† As discussed in Section

---

* These networks are interesting examples of the discussion in Section 4.1 about whether minimization of the number of gates minimizes the number of connections at the same time.

† When all gates have identical switching times, the look-ahead adder is faster than the ripple adder. But when gates have different switching times, the look-ahead adder may not be faster. For example, in the case of MOS LSI, MOS gates with many fan-out connections have longer switching times than those with few fan-out connections, because of parasitic capacitance. In the Intel microcomputer 8080, the ripple adder is used because the look-ahead adder is too slow (some MOS gates with many fan-out connections greatly slow down the speed of the look-ahead adder) and the variations of the look-ahead adder (such as those in MacSorley [1961]) occupy too much chip area.

1.3, the addition of two numbers $(a_n \cdots a_2 a_1)$ and $(b_n \cdots b_2 b_1)$ generates the following sum s_i and carry c_{i+1} at the ith position ($i = 1, 2, \ldots, n$):

$$s_i = a_i \oplus b_i \oplus c_i, \tag{9.2.1}$$

$$c_{i+1} = a_i b_i \vee a_i c_i \vee b_i c_i. \tag{9.2.2}$$

The c_{i+1} may be written as follows:

$$c_{i+1} = v_i \vee w_i c_i$$

where

$$v_i = a_i b_i \quad \text{and} \quad w_i = a_i \vee b_i.$$

Thus

$$\left.\begin{aligned}
c_2 &= v_1 \vee w_1 c_1, \\
c_3 &= v_2 \vee w_2 c_2 = v_2 \vee w_2(v_1 \vee w_1 c_1) \\
    &= v_2 \vee w_2 v_1 \vee w_2 w_1 c_1, \\
c_4 &= v_3 \vee w_3 v_2 \vee w_3 w_2 v_1 \vee w_3 w_2 w_1 c_1, \\
c_5 &= v_4 \vee w_4 v_3 \vee w_4 w_3 v_2 \vee w_4 w_3 w_2 v_1 \vee w_4 w_3 w_2 w_1 c_1, \\
&\vdots
\end{aligned}\right\} \tag{9.2.3}$$

On the basis of (9.2.3) for carries* and (9.2.1) for sums, we get the look-ahead adder shown in Fig. 9.2.5. This adder does not have identical modules for all bits. If every gate is assumed to have the same switching time τ, the look-ahead adder is faster than the ripple adder, at the expense of a great increase in the number of gates. If there is no maximum fan-in or fan-out restriction, the maximum delay time for the carry is 3τ, regardless of the bit length. But if there is a restriction, we have to cascade a look-ahead adder of small bit length, as a module, at the expense of an increase in carry propagation time. (The increase in carry propagation time can be mitigated by connecting the modules using extra gates, instead of connecting them straightforwardly without extra gates, as in the case of the ripple adder [MacSorley 1961].)

Since gates are becoming cheaper because of LSI technology, more complex but faster adders are conceived of (e.g., see Singh and Waxman [1972, 1973]).

There are many variations of adders (Remark 9.2.1).

---

* By using different expressions for the carries (but still along the same line of thinking), we can get different versions of the look-ahead adder. An example is the look-ahead adder of H. Takahashi and E. Goto (see p. 378 of Muroga [1971]).

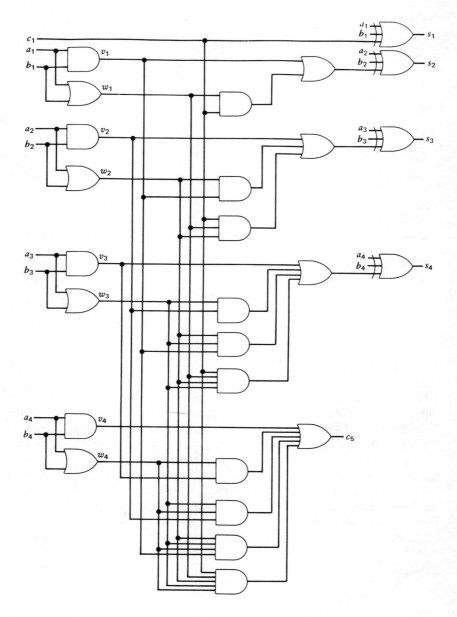

Fig. 9.2.5 Look-ahead adder.

Design of Large Networks in General

Large networks that are difficult to design by the procedures in the preceding chapters can be designed by considering the functional relationships between inputs and outputs. This can be done by using appropriate switching expressions like those for the look-ahead adder above, or by combining known networks with appropriate modification. The design of such networks is not a simple task, however, and often requires ingenuity.

Other Standard Networks

For other types of standard networks such as counters, shift registers, and multipliers, see the references in Remark 9.2.1. (Counters are discussed in the next section also.)

> *Remark 9.2.1:* The following lists of references on standard networks are not exhaustive. Books: Maley [1970], Morris and Miller [1971], Flores [1963], Blaauw [1976], Richards [1955], Hill and Peterson [1973]. For counters: Loui [1971], Langdon [1971], Carlow [1972], Oberman [1973], Meahl [1973]. For shift registers: Blair [1972], Eimbinder [1971], Golomb [1967], Percival [1971]. For adders: MacSorley [1961], Singh and Waxman [1972, 1973], Svoboda [1970], Lucas [1969], Weinberger and Smith [1956], Quatse and Keir [1967], Hendrickson [1960], Unger [1977], Lehman and Burla [1961], Gilchrist, Promerene, and Wong [1955]. For multipliers: Partridge [1973], Parasuraman [1977], Stenzel, Kubitz, and Garcia [1977], Baugh and Wooley [1973, 1976], Pezaris [1971], Ueda, [1976]. ■

▲ 9.3 Design of Sequential Networks with Given Building Blocks

In this section we discuss how to design a sequential network, using pre-designed subnetworks as building blocks, by modifications of the design procedure discussed in Section 8.5. The subnetworks in this case can be the modules or standard networks discussed in Section 9.2. But here let us use flip-flops as building blocks, since many important sequential networks can be synthesized with flip-flops with extra gates.

Sequential networks that cannot be derived by the design procedure in Section 8.5 can be obtained by the procedures in this section. The clock terminals of flip-flops in networks designed by the procedure in Section 8.5 are always connected to clocks, although in some cases they need not be. In the networks designed in this section, the clock terminals* are not necessarily

---

* The direct set- and reset-input terminals of the flip-flops mentioned in Section 8.2 also can be liberally used, though their use is not discussed in this book for the sake of simplicity.

522 Practical Considerations in Logic Design

connected to clocks but are connected to gates or flip-flops. Thus asynchronous networks (i.e., without clocks) can be designed with the master-slave flip-flops which have clock terminals. In this sense the procedure in Section 8.5 can yield only some of all possible sequential networks with flip-flops, and we may miss better networks.

To convey some idea of how differently a network can be constructed with flip-flops, as opposed to the procedure in Section 8.5, let us explain some simple examples. A binary counter can be implemented by cascading *J-K* master-slave flip-flops as shown in Fig. 9.3.1. (When this is to be differentiated from more complex but faster counters, it is called a **ripple counter**.) In Fig. 9.3.1 the clock terminal of each flip-flop is treated in the same manner as the other two terminals, J and K (i.e., a clock is not necessarily connected to the terminal Cl), and is supplied with the data signal. Constant value 1 is supplied to J and K of all the flip-flops. Whenever a data signal of value 1 appears at x, one is added to the counter. (The data signal 1 at x of flip-flop F_1 must be kept at a new value for a sufficiently long time interval if we want to let the counter work reliably.) When the counter receives $x = 1$ three times, for example, its outputs (z_4, z_3, z_2, z_1) display the count (0 0 1 1). Suppose that we want a binary counter which returns to count (0 0 0 0) when the counter receives $x = 1$ ten times. This decade counter is realized in Fig. 9.3.2 by connecting the flip-flops used in Fig. 9.3.1 in a different manner and adding one extra AND gate. (The explanation of why this counter works in that way is left to the reader as Exercise 9.3.11.)

Both counters in Figs. 9.3.1 and 9.3.2 are now available as MSI packages. Notice that these counters are asynchronous, though the flip-flops used have clock terminals.

Let us introduce the new procedures, illustrating them with a **divide-by-N counter**, that is, a counter that returns to the initial count when the counter receives the signal value 1 N times.

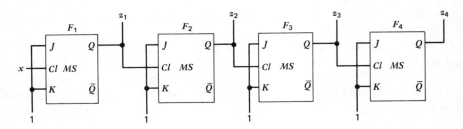

Fig. 9.3.1 Binary counter with J-K master-slave flip-flops (more precisely speaking, a binary ripple counter).

Design of Sequential Networks with Given Building Blocks 523

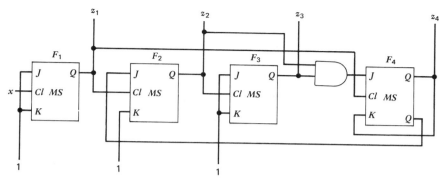

Fig. 9.3.2 Decade counter (or divide-by-10-counter) with J-K master-slave flip-flops (more precisely speaking, a decade ripple counter).

(a) Design of a Counter with J-K Master-Slave Flip-Flops, by a Straightforward Modification of the Procedure in Section 8.5

Instead of using the trial-and-error approach by which networks such as the one in Fig. 9.3.2 can be obtained, let us design a counter that has an arbitrary sequence of counts, by a straightforward modification of the procedure discussed in Section 8.5. As an example, let us design a divide-by-5 counter which counts 0, 1, 2, 3, 4 and then returns to 0. By using the clock terminal of each J-K master-slave flip-flop as the input for signal x, the counter has the transition table shown in Table 9.3.1, where q_i is the current output value of the

Table 9.3.1 Transition Table for a Divide-by-5 Counter

	x	
	0	1
000	000	001
001	001	010
010	010	011
011	011	100
100	100	000

$q_3 q_2 q_1$ (row labels); $Q_3 Q_2 Q_1$ (output)

Table 9.3.2 Output-Input Relationship of J-K Flip-Flop

OUTPUTS		INPUTS	
q	Q	J	K
0	0	0	d
0	1	1	d
1	0	d	1
1	1	d	0

ith bit position of the counter, and Q_i is the next output value (i.e., the outputs of the flip-flops are assumed to be the counter outputs*). Unlike the clock, x may not appear periodically with a fixed period. All flip-flops can assume their new output values when x returns to 0 from 1 (as they do when a clock returns to 0 from 1). Hence Table 9.3.1 may be interpreted to be in skew mode, though no clock is used, and we need only the right-hand column of Table 9.3.1, as explained in Sections 8.3 and 8.5.

Using the output-input relationship of a master-slave flip-flop in Table 9.3.2, which is obtained by reversing the input-output relationship as we did in Table 8.4.3, we derive the excitation table in Table 9.3.3, which is rewritten

Table 9.3.3 Excitation Table

		$x = 1$	
	000	$0d$, $0d$,	$1d$
	001	$0d$, $1d$,	$d1$
$q_3 q_2 q_1$	010	$0d$, $d0$,	$1d$
	011	$1d$, $d1$,	$d1$
	100	$d1$, $0d$,	$0d$
		$J_3 K_3, J_2 K_2, J_1 K_1$	

* If the flip-flop outputs (i.e., internal variables) are assumed not necessarily to be the network outputs, the procedure here must be correspondingly modified.

Table 9.3.4 Karnaugh Maps

q_3 \ $q_2 q_1$	00	01	11	10
0			1	
1	d	d	d	d

$$J_3 = Q_2 Q_1$$

q_3 \ $q_2 q_1$	00	01	11	10
0	d	d	d	d
1	1	d	d	d

$$K_3 = 1$$

	00	01	11	10
0		1	d	d
1		d	d	d

$$J_2 = Q_1$$

	00	01	11	10
0	d	d	1	
1	d	d	d	d

$$K_2 = Q_1$$

	00	01	11	10
0	1	d	d	1
1		d	d	d

$$J_1 = \bar{Q}_3$$

	00	01	11	10
0	d	1	1	d
1	d	d	d	d

$$K_1 = 1$$

in Karnaugh maps in Table 9.3.4. (Since x works in the same manner as a clock, the column for $x = 0$ in Table 9.3.1 can be ignored and x can be ignored in Table 9.3.4, as explained in Section 8.5.) Then we have the counter shown in Fig. 9.3.3.

This approach, however, will not yield some networks, that is, those in which the clock terminal of a flip-flop is connected not to x, but to other flip-flops or gates, as in Fig. 9.3.2. Therefore let us develop a general approach as follows.

(b) Design of a Sequential Network Using Given Building Blocks

The above approach can be extended to a more general case, that is, the design of a sequential network with given building blocks. As an example, let us design a divide-by-5 counter that counts from 0 to 4 consecutively, using J-K master-slave flip-flops as building blocks in a manner different from case (a).

526 *Practical Considerations in Logic Design*

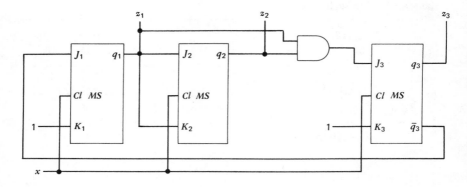

Fig. 9.3.3 A divide-by-5 counter.

Let x denote the input to this counter, whose current output values* are q_1, q_2, q_3 and whose next output values are Q_1, Q_2, Q_3. Now suppose that x need not be connected to the clock terminals of the flip-flops but may be connected to other terminals of the flip-flops or gates, and also that the clock terminals of the flip-flops may be connected to the outputs of gates or flip-flops. The transitions of all flip-flops in the network may not simultaneously occur. (For example, suppose that the clock terminal of flip-flop F_1 is connected to x, and the clock terminal of another flip-flop F_2 is connected to a gate or the output of a third flip-flop F_3. Then the outputs of flip-flop F_1 can change only at the trailing edge of $x = 1$ because F_1 is a master-slave flip-flop. But the outputs of F_2 may not change at the trailing edge of $x = 1$ because the gate or the output of F_3, which is connected to the clock terminal of F_2, may change from 0 to 1 when x changes from 1 to 0.) Therefore, unlike the case of a flow (or transition) table in skew mode, where a clock is connected to the clock terminals of all the flip-flops and consequently the leftmost column for $y = 0$ may be ignored, **we cannot ignore the column for $x = 0$**. (Notice that our table cannot look like Table 8.3.1*d* since all the flip-flops no longer receive identical signals at their clock terminals, and consequently all the flip-flops do not make diagonal transitions at the same time.)

The input-output relationship of a *J-K* master-slave flip-flop is shown in Table 9.3.5, considering the clock terminal *Cl* as an independent variable, unlike Table 9.3.2, since the clock terminal *Cl* is no longer necessarily supplied with a clock and consequently can be used as an independent terminal.

---

* See the second footnote in this section.

Design of Sequential Networks with Given Building Blocks 527

Table 9.3.5 Input-Output Relationship of a J-K Master-Slave Flip-Flop

cJK	qQ
000	$\begin{cases}00\\11\end{cases}$
001	$\begin{cases}00\\11\end{cases}$
010	$\begin{cases}00\\11\end{cases}$
011	$\begin{cases}00\\11\end{cases}$
100	$\begin{cases}00\\11\end{cases}$
101	$\begin{cases}00\\10\end{cases}$
110	$\begin{cases}01\\11\end{cases}$
111	$\begin{cases}01\\10\end{cases}$

Reversing the input-output relationship in Table 9.3.5, we obtain Table 9.3.6. For example, let us try to find what values (c, J, K) must assume in order to have $(q, Q) = (0\ 0)$ in Table 9.3.6. From Table 9.3.5 we find that $(q, Q) = (0\ 0)$ holds for any of $(c, J, K) = (0\ 0\ 0), (0\ 0\ 1), (0\ 1\ 0), (0\ 1\ 1), (1\ 0\ 0), (1\ 0\ 1)$. Notice that $(0\ 0\ 0), (0\ 0\ 1), (0\ 1\ 0), (0\ 1\ 1)$ can be expressed as $(0\ d\ d)$ (i.e., if $c = 0$, J and K can be 0 or 1) and that $(1\ 0\ 0)$ and $(1\ 0\ 1)$ can be expressed as $(1\ 0\ d)$. Thus the first row in Table 9.3.6 has been obtained, and we proceed similarly for other rows.

Then, using Table 9.3.6, we convert Table 9.3.1 into the excitation table in Table 9.3.7. Since Table 9.3.1 contains 22 entries where $q_i = Q_i$ holds [i.e., $(q_1, Q_1) = (0\ 0)$ or $(1\ 1)$, $(q_2, Q_2) = (0\ 0)$ or $(1\ 1)$, and $(q_3, Q_3) = (0\ 0)$ or $(1\ 1)$], and since we can use $(c, J, K) = (0\ d\ d)$ or $(1\ 0\ d)$ for $(q, Q) = (0\ 0)$ and $(c, J, K) = (0\ d\ d)$ or $(1\ d\ 0)$ for $(q, Q) = (1\ 1)$, we have 2^{22} possible excitation tables from Table 9.3.1 by way of Table 9.3.6. (There are more, as explained later.) Of course these 2^{22} excitation tables generally do not produce 2^{22} different networks, but far fewer. (It is usually not easy to find which

Table 9.3.6 Output-Input Relationship of a J-K Master-Slave Flip-Flop, Considering c as an Additional Input

qQ	cJK
00	$\begin{cases} 0dd \\ \text{or} \\ 10d \end{cases}$
01	$11d$
10	$1d1$
11	$\begin{cases} 0dd \\ \text{or} \\ 1d0 \end{cases}$

is the best choice. Probably the best strategy is to choose appropriate entries such that the entries in an excitation table to be obtained later facilitate the derivation of a simpler network, but this is hard to do.)

Note that **constant input 1 cannot be connected to the CI terminals under any circumstance** because the J-K master-slave flip-flops will not change their outputs. Hence we must avoid excitation tables where c is identically 1.

Table 9.3.7 Excitation Table

		x	
		0	1
	000	$0dd, 0dd, 0dd$	$0dd, 0dd, 11d$
	001	$0dd, 0dd, 0dd$	$0dd, 11d, 1d1$
$q_3 q_2 q_1$	010	$0dd, 0dd, 0dd$	$0dd, 0dd, 11d$
	011	$0dd, 0dd, 0dd$	$11d, 1d1, 1d1$
	100	$0dd, 0dd, 0dd$	$1d1, 0dd, 0dd$

$$c_3 J_3 K_3, c_2 J_2 K_2, c_1 J_1 K_1$$

Design of Sequential Networks with Given Building Blocks 529

Table 9.3.7 is obtained by using $(c, J, K) = (0\ d\ d)$ for every $(q_i, Q_i) = (0\ 0)$ or $(1\ 1)$, where $i = 1, 2, 3$. By drawing Karnaugh maps in Table 9.3.8, we obtain the counter shown in Fig. 9.3.4a.

As a second example, let us use $(c, J, K) = (0\ d\ d)$ for $(q_i, Q_i) = (0\ 0)$ and $(1\ 1)$ in the column for $x = 0$ in Table 9.3.1, and $(c, J, K) = (1\ 0\ d)$ for $(q_i, Q_i) = (0\ 0)$ and $(c, J, K) = (1\ d\ 0)$ for $(q_i, Q_i) = (1\ 1)$ in the column

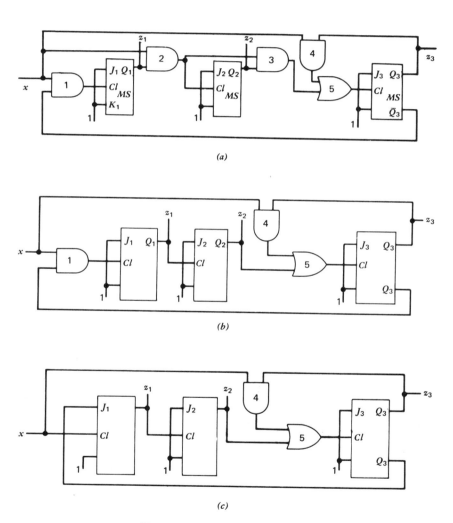

Fig. 9.3.4 Divide-by-5 counters.

Table 9.3.8 Karnaugh Maps

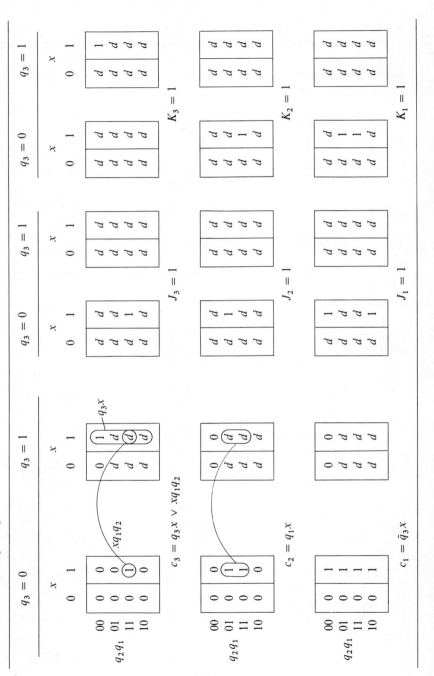

Design of Sequential Networks with Given Building Blocks 531

Table 9.3.9 Excitation Table

		x	
		0	1
	000	0dd, 0dd, 0dd	10d, 10d, 11d
	001	0dd, 0dd, 0dd	10d, 11d, 1d1
$q_3 q_2 q_1$	010	0dd, 0dd, 0dd	10d, 1d0, 11d
	011	0dd, 0dd, 0dd	11d, 1d1, 1d1
	100	0dd, 0dd, 0dd	1d1, 10d, 10d

$$c_3 J_3 K_3, \; c_2 J_2 K_2, \; c_1 J_1 K_1$$

for $x = 1$ in Table 9.3.1. Then we obtain the excitation table in Table 9.3.9. This leads to the counter in Fig. 9.3.3, where x is supplied to the clock terminals of all the flip-flops like those derived by the approach in case (a). (In other words, the design approach here includes the one in case (a) as a special situation.) As further examples, counters derived by some other conversions are shown in Fig. 9.3.5.

After having obtained these networks, we need to check whether they work as counters, because Table 9.3.1 is constructed under the assumption that a network moves horizontally from the left column to the right column when x changes from 0 to 1, and diagonally from the right column to the left column when x changes from 1 to 0. In other words, we have assumed that the transition of the network to the next state can occur at the trailing edges of $x = 1$. But this may not be true of the network obtained, since all the clock terminals of flip-flops are not necessarily connected to x, as explained in the second paragraph of this case, (b). Another possible cause for deviation of the performance of the obtained network from the desired performance is that in Table 9.3.5 the transition of q to Q in a flip-flop can take place when c changes from 0 to 1 and then to 0 (in the first four rows in Table 9.3.5), or when c changes from 1 to 0 (in the last four rows in Table 9.3.5). But in some flip-flop in the obtained network, J or K may recive a new value, before q changes to Q by the change of c, corresponding to the current value of J and K. Hence, after obtaining a network, we have to check whether it works as intended. (We can do this by simply continuing to change x between 0 and 1 and finding out whether the network shows a correct count for each change.)

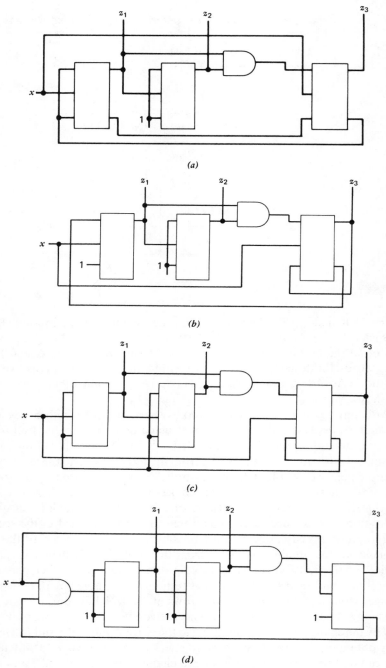

Fig. 9.3.5 *Divide-by-5 counters.*

Design of Sequential Networks with Given Building Blocks 533

When we check whether the designed networks work as intended, by continuing to change their inputs, we may find that the networks can sometimes be further simplified or modified. For example, by checking the counter in Fig. 9.3.4a, we can find that AND gates 2 and 3 can be eliminated, thus deriving the simpler counter in Fig. 9.3.4b. By further checking this counter carefully, we can find that another gate, 1, can be eliminated by changing a connection, deriving the still simpler counter in Fig. 9.3.4c. The counter in Fig. 9.3.4a has more gates than that in 9.3.4c but works faster (i.e., for the changes of x, outputs z_1, z_2, and z_3 assume new values sooner).

Notice, however, that the counter in Fig. 9.3.4c cannot be obtained from any excitation table that can be obtained from Table 9.3.1 by way of Table 9.3.6, using the conversion described above. This means that there are other excitation tables, derived by complex conversions other than the 2^{22} conversions mentioned already (Remark 9.3.1). But since the derivation of such excitation tables appears complex, simplification of the networks derived by the 2^{22} possible conversions would be more practical.

In the above design approach, flip-flops are used. As their clock terminals are treated as independent inputs, these flip-flops are treated as building blocks which are more general than the flip-flops discussed in Chapter 8. Possibly we can design some networks with some other building blocks in this manner, starting with transition tables in skew or fundamental mode.

Remark 9.3.1: The excitation table for the counter in Fig. 9.3.4c is shown in Table 9.3.10. The underlined entries (1 1 1) cannot be derived from Table 9.3.1 by way of Table 9.3.6. But these (1 1 1)s can be present in Table 9.3.10, since

Table 9.3.10 Excitation Table for Fig. 9.3.4c

		x	
		0	1
	000	0dd, 0dd, 0dd	0dd, 0dd, 11d
	001	0dd, <u>111</u>, 0dd	0dd, 11d, 1d1
$q_3 q_2 q_1$	010	<u>111</u>, 0dd, 0dd	<u>111</u>, 0dd, 11d
	011	<u>111</u>, <u>111</u>, 0dd	11d, 1d1, 1d1
	100	0dd, 0dd, 0dd	1d1, 0dd, 10d

$c_3 J_3 K_3, c_2 J_2 K_2, c_1 J_1 K_1$

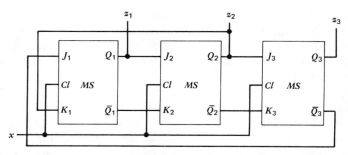

Fig. 9.3.6 *A divide-by-5 counter with nonconsecutive counts, using no extra gates.*

(c_3, J_3, K_3), for example, can be (1 1 1) for (q_3, q_2, q_1, x) = (0 1 1 0) because of no change of c_3 until (q_3, q_2, q_1, x) becomes (1 0 0 0). Such excitation tables can be derived by considering entries different from those in the conversion [explained in case (b)] for some states, when Table 9.3.1 is converted by way of Table 9.3.6. But this generally appears complex. ∎

Minimization of the Number of Extra Gates in Counter Design

When a divide-by-N counter is to be synthesized, the number of extra gates can be minimized by using counting sequences different from the natural consecutive counting sequence (i.e., 0, 1, 2, 3, . . .).

By using arbitrary counting sequences, the divide-by-N counter shown in Fig. 9.3.6 is derived without any extra gates. Counters for other values of N are also discussed in Oberman [1973].

Exercises

9.3.1 (M−) A **shift register** can be realized as a cascaded network of flip-flops, each of which stores one bit of a binary number (i.e., the far left flip-flop stores the first bit, the second flip-flop stores the second bit, and so on), and at each clock pulse these bits move to the right (or left) by one bit position. A shift register that shifts right only is

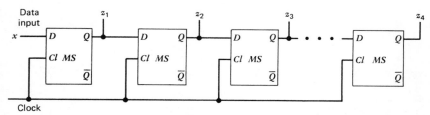

Fig. E9.3.1 *Shift register with D-type master-slave flip-flops.*

Design of Sequential Networks with Given Building Blocks 535

called a **right-shift register**. As an example, a right-shift register with D-type master-slave flip-flops is shown in Fig. E9.3.1. Assume that the register stores the binary number $(0\ 1\ 1\ 0\ \cdots)$ during the absence of the clock pulse, where $z_1 = 0$, $z_2 = 1$, $z_3 = 1$, $z_4 = 0$, and so on. When the first clock pulse arrives, the register shifts the stored binary number right by one bit position and enters x at the leftmost position. In other words, the register stores $(1\ 0\ 1\ 1\ 0\ \cdots)$ or $(0\ 0\ 1\ 1\ 0\ \cdots)$, according as $x = 1$ or 0.

Design a right-shift register, using J-K master-slave flip-flops and inverters only.

9.3.2 (M−) By adding some gates of any types, modify the shift register in Fig. E9.3.1 so that all the n bits constituting a binary number can be simultaneously loaded into the shift register. This **loadable shift register** needs a mode-control terminal t such that, if $t = 1$, the binary number is loaded, and if $t = 0$, the register works as a right-shift register.

9.3.3 (M−) Adding some gates of any types, modify the shift register in Fig. E9.3.1 so that it can shift right or left. We need two mode-control terminals, r and l, for right or left shift, such that, if $r = 1$ and $l = 0$, the network works as a right-shift register, and if $r = 0$ and $l = 1$, the network works as a left-shift register. (The combinations $r = l = 0$ and $r = l = 1$ are assumed not to occur.)

9.3.4 (M) A **comparator** is a network to determine which of two given binary numbers, $X = (x_1, x_2, \ldots, x_n)$ and $Y = (y_1, \ldots, y_n)$, is greater, where x_1 and y_1 are the least significant bits. The output z_n is 1 only when X is greater than Y. As shown in Fig. E9.3.4, this network can be designed as a **ripple comparator**, that is, a cascade of modules $M_1, M_2, \ldots M_n$, whose outputs are $z_1, z_2, \ldots, z_n$, respectively, with the following property:

$$z_1 = 1 \text{ only when } x_1 > y_1,$$
$$z_2 = 1 \text{ only when } (x_1, x_2) > (y_1, y_2),$$
$$z_3 = 1 \text{ only when } (x_1, x_2, x_3) > (y_1, y_2, y_3),$$
$$\cdots$$

Here $(x_1, \ldots, x_k) > (y_1, \ldots, y_k)$ means that the binary number $(x_1, \ldots, x_k)$ is greater than the binary number $(y_1, \ldots, y_k)$.

(i) Assuming that all modules $M_2, \ldots, M_n$ except M_1 have an identical network configuration, design M_1 and M_2, using as few AND, OR, and NOT gates as possible.
(ii) Explain concisely why this network works as a comparator.

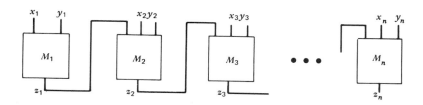

Fig. E9.3.4 *Ripple comparator.*

9.3.5 (M) (i) Design the fastest network for the comparator to compare two binary numbers, X and Y, each three bits long. The network output is 1 only when $X > Y$. (This comparator is called a **look-ahead comparator**, in contrast to the ripple comparator in Exercise 9.3.4.) Use as few AND, OR, and NOT gates as possible.

(ii) Explain why the network output is 1 only when $X > Y$.

9.3.6 (M) The **look-ahead counters** shown in Figs. E9.3.6a and E9.3.6b* are fast counters, in contrast to **ripple counters**, such as in Fig. 9.3.1, which are a cascade of flip-flops without extra gates.

(i) Explain concisely why the look-ahead counter in Fig. E9.3.6a shows a new count faster than the ripple counter in Fig. 9.3.1, and also why the look-ahead counter in Fig. E9.3.6b shows a new count faster than the counter in Fig. E9.3.6a. Compare their response times, assuming that all gates have equal switching times and that the flip-flops are those shown in Fig. 8.2.1.

(ii) The AND gate 3 in Fig. E9.3.6b exceeds the fan-in of 3. Redesign this counter, using extra AND gates and/or OR gates, so that every gate satisfies the maximum fan-in restriction of 3 and, under this condition, the counter has the fastest response time As a secondary objective, use as few gates as possible.

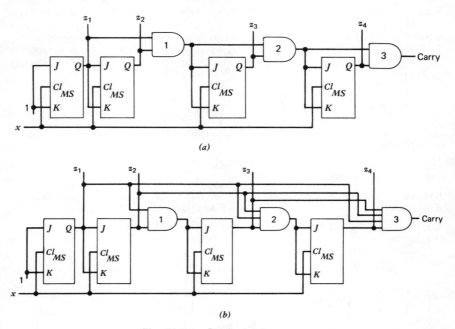

Fig. E9.3.6 Look-ahead counters.

* Strictly speaking, x must be connected to AND gate 3 in Figs. E9.3.6a and E9.3.6b. But if we do so, when we cascade many of the networks in either of these figures, the networks in the higher level positions do not work properly. Therefore the catalogs of semiconductor manufacturers present look-ahead counters as they are shown here.

Design of Sequential Networks with Given Building Blocks   537

9.3.7 (MT)   When each digit of a decimal number is represented by a four-bit binary number with bit weights 8, 4, 2, and 1, it is called the **8-4-2-1 binary coded decimal** (or the 8-4-2-1 BCD, as explained in Exercise 2.12).

Design a network for the addition of two 8-4-2-1 BCD numbers such that the addition in any bit length can be done by cascading a corresponding number of module networks. Use any gate types.

9.3.8 (MT)   Design a counter in which the count is in two decimal digits expressed in the 8-4-2-1 BCD (explained in Exercise 9.3.7). Use any gate types.

9.3.9 (MT)   Find the effective and net gate delays from c_i to c_{i+3} in the adder in Fig. 9.2.4.

▲ **9.3.10** (E)   Discuss whether the counter in Fig. 9.3.1 does not work correctly when the duration of pulse $x = 1$ is short. If so, find the minimal duration of $x = 1$, as a multiple of τ, such that the counter works correctly. Assume that the flip-flop in Fig. 8.2.1 is used here and that τ is the switching time of each gate.

▲ **9.3.11** (ET)   Explain why the counter in Fig. 9.3.2 works, showing waveforms for x, z_1, z_2, z_3, and z_4.

▲ **9.3.12** (M − T)   Using the approach in case (a) of Section 9.3, design a binary counter that returns to count 0 after showing counts 0, 1, 2, 3, 4, 5 in this order, using J-K master-slave flip-flops. If necessary, use extra gates of any types, but limit the number as much as possible.

▲ **9.3.13** (M)   Using J-K master-slave flip-flops, design a divide-by-4 counter that counts (0 0), (0 1), (1 1), (1 0) in this order and then returns to (0 0) (i.e., counts 0, 1, 3, 2, and then 0 are shown at the counter outputs, in this order). Input signal x must be connected to the clock terminals of all the flip-flops used.

If the flip-flops are appropriately connected, no gate is required. But if finding such a connection configuration proves time-consuming, you may use gates of any types, but use as few gates as possible.

▲ **9.3.14** (M)   Design an up-down counter for three bits, using J-K master-slave flip-flops. (For the definition of up-down counter, see Exercise 7.5.4.) Assume that no clock is available (i.e., the clock terminals of the flip-flops are not connected to a clock). Use as few extra gates of any types as possible.

▲ **9.3.15** (RT)   Using J-K master-slave flip-flops, design a divide-by-5 counter that counts (0 0 0), (0 0 1), (0 1 1), (1 1 1), (1 1 0) in this order and then returns to 0 (i.e., counts 0, 1, 3, 7, 6, and then 0). Use as few extra gates of any types as possible, and assume that the clock terminals of the flip-flops may be connected to any outputs of gates or flip-flops or to the input signal x. Use the approach in case (b) of Section 9.3.

▲ **9.3.16** (RT)   Using J-K master-slave flip-flops, design a divide-by-6 counter that counts (0 0 0), (0 0 1), (0 1 1), (1 1 1), (1 1 0), (1 0 1) in this order and then returns to 0 (i.e., counts 0, 1, 3, 7, 6, 5, and then 0). Use as few extra gates of any types as possible, and assume that the clock terminals of the flip-flops may be connected to any outputs of gates or flip-flops or to the input signal x. Use the approach in case (b) of Section 9.3.

9.4 Design of Networks with a Mixture of Memories and Gates

Random-access memories are memories into which new information can be written and also from which stored information can be read out. Traditionally, RAMs are used to store computer programs, data, and intermediate computational results in computers. Read-only-memories are memories in which information is permanently stored and from which information can be read out, but into which new information cannot be written. Traditionally, ROMs are used to store programs and data permanently. Particularly when the control units of the computer are implemented with ROMs by microprogramming, the ROMs are called **firmware**, since they are halfway between software and hardware.

In this section we will discuss usage of semiconductor memories that is different from the above traditional usage. In some cases, memories, particularly ROMs, can be used to replace some gates in a network, reducing the cost and space of the entire network, though their speed is slower than that of gates. The extensive use of semiconductor memories as substitutes for gates in networks is due to the fact that electronic peripheral circuits to operate a memory, such as a decoder (a decoder, a network to access an address of the memory, requires high output power and considerable space in the case of core memory) and a memory register (this serves to temporarily store a word during writing in or reading out information), are now so compact that they are conveniently encased in a tiny IC package along with a memory. Also, this increased use is due to drastic cost reduction.*

If ROMs are manufactured in large quantities, they are cheaper than RAMs. Also, ROMs, unlike RAMs, do not present the danger that stored information will be damaged when the power supply is cut off. Moreover, ROMS are generally faster. Therefore ROMs, rather than RAMs, are usually used to replace gates. As we see in this section, **ROMs are a new and very important dimension in logic design**.

Implementation of ROMs

Implementation examples† of ROMs are shown in Fig. 9.4.1. Fig. 9.4.1a shows a ROM based on diodes. One of the horizontal lines is chosen corresponding to each set of values of the address inputs (i.e., when the decoder receives a particular set of values of address inputs, only one horizontal line

---

* For example, memories are extensively used in the minicomputer PDP-11/70 to reduce cost [*Electronics* Oct. 27, 1977].
† For electronic details see, for example, Taub and Schilling [1977] and Millman and Halkias [1972].

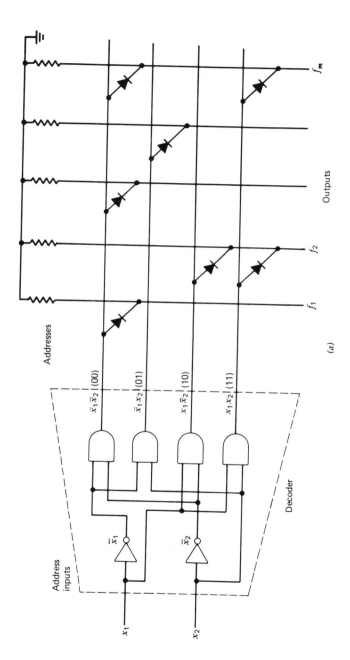

Fig. 9.4.1 Implementation example of ROMs. (a) ROM based on diodes.

540 Practical Considerations in Logic Design

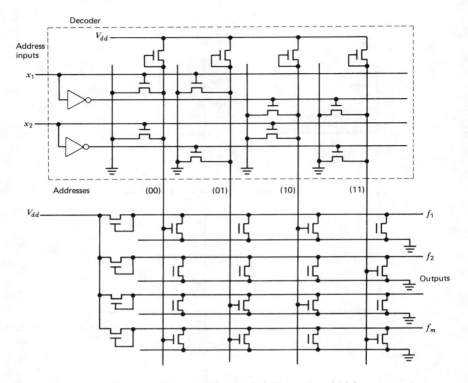

Fig. 9.4.1. (Continued) **(b) ROM based on MOS.**

of the corresponding address receives a high positive voltage, while other lines continue to have zero voltages). Then only the vertical lines which are connected to that horizontal line by diodes receive high positive voltages representing signal 1 (if positive logic is used), and all other vertical lines have zero voltages representing signal 0. Thus information is stored by placing diodes at necessary intersections in the matrix of lines. (Because of long lines which have parasitic capacitance, ROMs are generally slower than gates.)

Fig. 9.4.1b shows an example* of a ROM based on MOSFETs. For each set of values of the address inputs, only one of the vertical lines has the power source voltage; all other vertical lines have zero voltages. For example,

---

* There are other memory configurations. See, for example, Wilson [1978] and Holdt and Yu [1978]. (Fig. 9.4.1b, which is shown here for the sake of simplicity, is not appropriate in the case of a large number of addresses, since the fan-out connections of inputs in the decoder become too large. Therefore each address needs to be accessed, for example, by an X-decoder and a Y-decoder instead of by a single decoder.)

suppose that *n*-channel MOSFETs are used and are connected to the power supplies of a positive voltage. If address inputs (1 1) are received, i.e., x_1 and x_2 are positive voltages, the rightmost vertical line has the positive voltage and all other vertical lines have zero voltages. Then only the horizontal lines in the lower matrix of lines (i.e., the matrix outside the decoder shown in the dotted line) which are connected to that vertical line through MOSFETs have zero voltages; all other horizontal lines in the lower matrix have the positive voltage. In this case, information is stored by the connections of the gates of appropriate MOSFETs to the vertical lines in the lower matrix. (The connections of the gates of the MOSFETs inside the decoder have a fixed pattern such that only one vertical line has the positive voltage for any combination of the values of the address inputs. Thus their connection pattern has no relation to the stored information.)

ROMs as Substitutes for Gates in a Network

A straightforward approach in using ROMs to realize given functions is to store truth tables for the functions, such as Table 9.4.1, in a ROM. Input variables $x_1, \ldots, x_n$ are used as addresses of the ROM, and the values of $f_1, f_2, \ldots, f_m$ are stored in the memory register in each address. By this simple approach, any set of functions can be realized, but an excessively large memory is usually required.

If we do not use this approach, there are many other possibilities. It requires ingenuity and trial-and-error efforts to find a good way to implement networks with a mixture of gates and ROMs, since there is no simple general procedure. As an example of the use of ROMs to replace some gates in a network, let us consider the following problem.

An **m-out-of-n detector** is a network whose output is 1 when exactly *m* out of the *n* inputs to this network assume value 1, and whose output is 0 otherwise. This property is used to detect whether or not *n* transmission lines carry erroneous signals (if the number of 1's is more or less than *m*, the lines carry erroneous signals).

Table 9.4.1 Truth Table

$x_1 \, x_2 \cdots x_n$	$f_1 \, f_2 \cdots f_m$
0 0 $\cdots$ 0 0	0 1 $\cdots$ 1
0 0 $\cdots$ 0 1	1 0 $\cdots$ 0
0 0 $\cdots$ 1 0	0 0 $\cdots$ 0
0 0 $\cdots$ 1 1	1 1 $\cdots$ 1
$\vdots$	$\vdots$

542   Practical Considerations in Logic Design

As the simplest example, let us consider a 1-out-of-n detector.* Let us design this using four different approaches and compare the results, as follows.

First, let us design a 1-out-of-n detector using a loopless gate configuration, as shown in Fig. 9.4.2. This network works fast, but it requires n inverters, at least n AND gates, and at least one OR gate. When n increases, the numbers of levels and gates also increase because of maximum fan-in restriction.

The second approach is based on the combination of a loadable shift register (see Exercise 9.3.2) and a two-bit counter, as shown in Fig. 9.4.3. Inputs $x_1, \ldots, x_n$ are loaded into the shift register. Then, in synchronization with a clock, the bits in the shift register are serially fed into the counter, which counts the number of 1's in $x_1, \ldots, x_n$. When the counter shows count 1, the detector output shows 1; otherwise 0. [In other words, if the counter shows output (0 1), the detector output is 1. If the counter shows (0 0), the detector output is 0. Also, if the counter shows (1 0) or (1 1), the detector output is 0, and gate 1 is inhibited to transmit the shift register output to the counter in order to prevent counter overflow. In other words, even when the number of 1's is more than three, the detector output is always 0.] This approach requires far fewer gates than the first approach as n increases

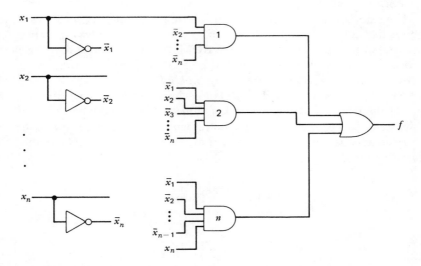

Fig. 9.4.2   1-out-of-n detector by a loopless gate network.

---

* Kobylar, Lindsay, and Pitroda [1973] with a modification.

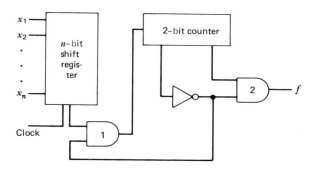

Fig. 9.4.3   1-out-of-n detector with a shift register and counter.

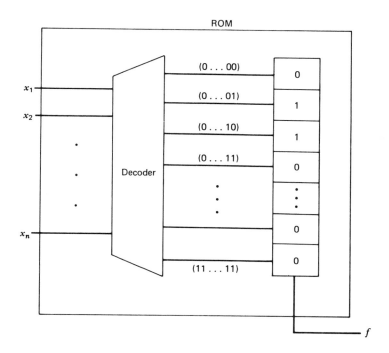

Fig. 9.4.4   1-out-of-n detector with a single ROM.

(particularly for an m-out-of-n detector). Because of the serial counting operation, the response time becomes slower as n increases, if the clock rate is fixed (or the clock rate must increase as n increases, requiring expensive, high-speed gates).

The third approach is based on the use of ROMs. A simple-minded approach is to store the output values of the detector in a single ROM, using input values $(x_1, \ldots, x_n)$ as addresses, as shown in Fig. 9.4.4 (i.e., the approach discussed in connection with Table 9.4.1). The ROM in this case needs to have a memory capacity of $2^n \times 1$ bits. The detector's output is read out of the ROM.

When n is large, this approach requires a ROM with a large capacity, and the following approach, with many ROMs of smaller capacity, is preferred because a ROM with a large capacity is either not available or very expensive. The detector is constructed with ROMs having t inputs each, as illustrated in Fig. 9.4.5 with $t = 8$ as an example. At the outputs of each ROM with a memory capacity of 2^t bits, where $t < n$, three states in t inputs are distinguished: all 0's, a single 1, and two or more 1's. Since the state of all 0's can be identified from knowledge of the other two states, each ROM has only two outputs for the other two states, that is, one output which is 1 only when the number of 1's in the t inputs is two or more, and another output* which is 1 only when the number of 1's is exactly one. (In the case of $t = 8$, each ROM has a memory capacity of 256×2 bits, where two bits are for two outputs.)

The first stage in Fig. 9.4.5 consists of $\lceil n/t \rceil$ ROMs. The first outputs of ROMs which represent the state of "the number of 1's is two or more" are collected by OR gates in groups of t in the second stage. The second outputs of ROMs are connected to ROMs in the second stage, in groups of t, each of which determines whether the number of 1's in its t inputs is exactly one, or two or more.

In the succeeding stages, this is repeated. In the second, third, ... stages, we need $\lceil n/t^2 \rceil$, $\lceil n/t^3 \rceil$, ..., ROMs, respectively. The AND gate in the last stage receives an inhibiting signal through the chain of OR gates when any ROM in the first stage receives two or more 1's in its inputs.

The total number of t-bit ROMs required for the 1-out-of-n detector is

$$\sum_{i=1}^{\lceil \log_t n \rceil} \lceil n/t^i \rceil.$$

(Note that all the OR gates in Fig. 9.4.5 also can be replaced by fewer

---

* If each ROM has only this output, the network does not work correctly. For example, the network shows a faulty output value when one ROM has exactly one 1 in its inputs and another ROM has two or more 1's in its inputs.

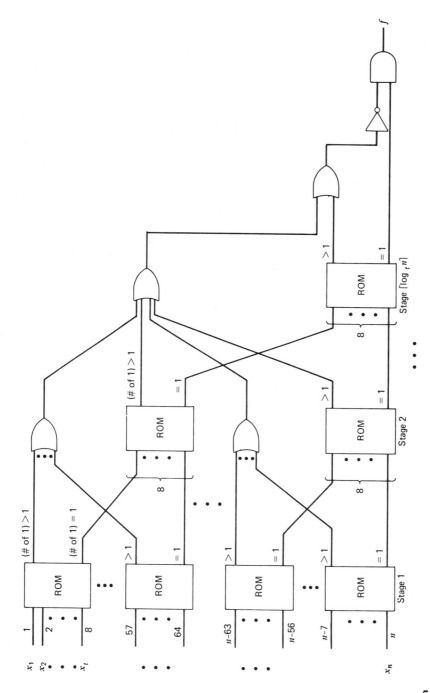

Fig. 9.4.5 1-out-of-n detector with ROMs.

546 *Practical Considerations in Logic Design*

ROMs, or the corresponding information can be stored in unused addresses in other memories if there are any.)

The fourth approach is based on PLAs, which are a special type of ROM, as discussed later in this section. In the case of the 1-out-of-n detector, by using PLAs in Fig. 9.4.5 instead of ROMs, we can obtain a detector with fewer IC packages and a faster response time. (The design of a detector with PLAs is left to readers as Exercise 9.4.1 [Cavlan and Cline 1976].)

A comparison of the number of IC packages and the response times for the four approaches described above is shown in Table 9.4.2 for different values of n. The first approach, with a loopless network of gates, is the fastest, but it requires the largest number of IC packages and accordingly is the most expensive. The second approach is the most economical but the slowest.

Table 9.4.2 Comparison of Four Approaches to the Implementation of the 1-out-of-n Detector

APPROACH	n				
	10	50	80	800	5000
Gate network (Fig. 9.4.2)					
Number of levels	3	5	5	7	9
Response time (ns)[a]	30	50	50	70	90
Number of IC packages[b]	3	311	737	71,369	2,781,056
Shift register and counter (Fig. 9.4.3)					
Number of levels	13	53	83	803	5,003
Response time (ns)[a]	130	530	830	8,803	50,003
Number of IC packages[b]	5	9	12	84	504
ROMs (Fig. 9.4.5 with ROMs)					
Number of levels	5	5	6	6	7
Response time (ns)[c]	200	200	240	240	280
Number of IC packages[b]	7	12	19	133	810
PLAs (Fig. 9.4.5 with PLAs)					
Number of levels	3	5	5	6	7
Response time (ns)[c]	120	200	200	240	280
Number of IC packages[b]	3	9	10	62	385

[a] A switching time of 10 nanoseconds (ns) per level of gate or shift is assumed.

[b] An AND gate with 10 inputs, an OR gate with 10 inputs, 10 NOTs, a shift register of 10 bits, a single counter, a ROM with 8 inputs and 2 outputs [memory capacity of 512 bits and matrix size of $(8 \times 2 + 2) \times 256 = 4608$], or a PLA with 16 inputs and 2 outputs [matrix size of $(16 \times 2 + 2)(16 + (16 \times 15(2)) = 4624$] is assumed for each IC package.

[c] An access time of 40 nanoseconds is assumed for a ROM and a PLA.

Design of Networks with a Mixture of Memories and Gates

The third approach, with ROMs, ranks between the first two approaches. It is slower than the first approach, but not by much, and is much faster than the second approach. In terms of the number of IC packages, it is more expensive than the second approach, but not by much (each ROM of 512 bits costs slightly more than an IC package of a gate with 10 inputs), and is much less expensive than the first approach.

Additional advantages of the third approach, with ROMs, and the fourth approach, with PLAs, are obvious. When the number of IC packages is reduced, the labor, time, and material (fewer pc boards may be used) for assembling are also reduced, so cost savings are further enhanced and development and manufacturing time is shortened. Another important advantage of the reduction in the number of IC packages is greater reliability. Since IC packages replace the soldering of many connections, the reliability is greatly improved (badly soldered connections are one of the major causes of faulty networks). Testing the network is also easier because of the reduction in the number of connections and the systematic configuration. The fourth approach, the detector with PLAs, is more economical than the detector with ROMs but is not always better in the case of other design problems. One reason why a network with PLAs is better for some functions will be clear after the discussion of PLAs. It is important to note that the choice between ROMs and PLAs depends on the functions to be implemented. In the particular example of the detector above, PLAs yield a better network because of the simplicity of the functions involved.

We have explained the use of ROMs to substitute for gates, based on the 1-out-of-n detector. But ROMs are used in many different ways. (See, e.g., Barna and Porat [1973].)

Advantages and Disadvantages of the Use of ROMs

As seen in Fig. 9.4.1, ROMs have a simpler iterative structure than gates in random-logic networks. Ordinary logic networks, unlike ROMs, do not have fixed configurations; the network configurations differ, depending upon the output functions to be realized. Therefore ordinary logic networks are called **random-logic networks**, in contrast to networks with iterative structures, fixed independently of the functions to be realized, such as memories.

Because of their simple iterative structure, ROMs as substitutes for gates have the following advantages, depending on how appropriately a network is designed with a mixture of gates and ROMs. If a network is not appropriately designed, it may become more expensive and bulkier with the use of ROMs. It requires ingenuity to find a best way of using ROMs, since there is no simple systematic procedure for general design problems. (In the above example of the 1-out-of-n detector, use of a single ROM is too expensive.

548 *Practical Considerations in Logic Design*

Thus it is essential to find an appropriate network configuration with ROMs of appropriate memory capacity, storing appropriate information in each ROM. Possibly there may be approaches other than those explained.)

Because of the systematic structure of a ROM, more components (such as diodes or MOSFETs) can be packed in the same area than is possible with the gates of random-logic networks. Thus the number of IC packages can be reduced if a network is appropriately designed with a mixture of gates and ROMs. Also, if a network is manufactured in large quantities, so that the initial cost per package for setting data in ROMs becomes negligibly small, compared with the manufacturing cost of each ROM package, then the cost is reduced, sometimes drastically.

Also, because of the simple, iterative structure, design with ROMs is more immune to errors, design changes are easier, and testability (i.e., ease in testing ROMs) is improved. Moreover, logic information to be stored in ROMs need not be changed, even if an old IC technology is replaced by a new one (in the case of gate networks, the networks must be redesigned if an IC logic family is changed).

The use of ROMs to replace gates has the following disadvantages. First, the speed of the network is reduced. Second, the data storage patterns of ROMs can be set only by semiconductor manufacturers during IC fabrication. The fabrication of IC chips for random-logic networks in general requires many different masks (usually five to seven) to set the layout of the networks on the chips, but in the case of ROMs only one of these masks (which is for connections) must be custom-made. Hence, if a ROM with the same pattern is manufactured in large quantities, it is much cheaper than a random-logic network consisting exclusively of gates with a comparable number of transistors (whether custom-made or off-the-shelf). Otherwise ROMs are not cheap, since the preparation of a mask for connections requires a fixed initial investment.

To circumvent this inconvenience of setting information into ROMs by semiconductor manufacturers only, we often use the following types of ROMs. In contrast to them, the ROMs discussed above are often called **mask-programmable ROMs**. Data storage patterns, however, can be easily set up in **EAROM** (electrically alterable read-only memory) or **PROM** (programmable read-only memory) by users (i.e., not necessarily by semiconductor manufacturers). The EAROM is a ROM that is used when information is required to be written only occasionally. (Information is written with a higher voltage than a normal operating voltage. **EPROMs** (erasable PROMs) are similar to the EAROMs, but require that the previous information be erased by ultraviolet light exposure before writing new information.) The PROM is a ROM that is used when a user wants to set information permanently (e.g., by blowing up thin Nichrome wires by excessively large currents),

Design of Networks with a Mixture of Memories and Gates 549

but, once set, the information cannot be changed. Although EAROMs and PROMs are convenient, since users can set information in a very short time, the cost reduction advantage of ROMs over gates in the case of large production volume is usually significantly lost.

Programmable Logic Array (PLA)

A PLA is illustrated in Fig. 9.4.6a. MOSFETs are prearranged in a matrix on an IC chip, along with horizontal and vertical lines. Connections between the MOSFET gates and vertical or horizontal lines are set up by the semiconductor manufacturer during IC fabrication, according to customers' specifications. (Only one mask, for the connections, is custom-made.) When the same networks are manufactured in large quantities, PLAs are inexpensive. They are very versatile because of low cost and design flexibility. PLAs are also often put on chips for some calculators, microcomputers, or intelligent terminals.

A ROM with n address inputs has a decoder with 2^n outputs (i.e., 2^n vertical lines in Fig. 9.4.1b), whereas a PLA does not, though a PLA is a special type of ROM. A PLA with n address inputs has a decoder counterpart, which is different from a decoder and usually has fewer than 2^n outputs (i.e., far fewer than 2^n vertical lines, in Fig. 9.4.6a). Thus, if the matrix size (i.e., the product of the numbers of vertical and horizontal lines) is kept the same on an IC chip, a PLA can have many more address inputs than a ROM of the same size.

As shown in Fig. 9.4.6a, MOSFETs are arranged in two matrices. When MOSFET gates are connected as denoted by dots, we get $\overline{(x\bar{y}\bar{z})}, \overline{(\bar{x}z)}, \overline{(xyz)}$ at the outputs P_1, P_2, P_3 of the upper matrix, respectively, since P_1, P_2, P_3 express NAND gates, if negative logic* is used with n-MOS. Then the outputs f_1, f_2, f_3 of the lower matrix are again NAND gates, with P_1, P_2, P_3 as their inputs. Thus

$$f_1 = \overline{P_1 P_3} = \bar{P}_1 \vee \bar{P}_3 = x\bar{y}\bar{z} \vee xyz,$$
$$f_2 = \bar{P}_2 \phantom{= \bar{P}_1 \vee \bar{P}_3} = \bar{x}z,$$
$$f_3 = \overline{P_1 P_2} = \bar{P}_1 \vee \bar{P}_2 = x\bar{y}\bar{z} \vee \bar{x}z.$$

(Notice that the lower matrix does not contain inverters, though the upper matrix does, right after inputs x, y, z.) Therefore the two matrices of MOSFETs in Fig. 9.4.6a represent a network of NAND gates in two levels. This is interpreted as a network of AND gates in the first level and OR gates

---

* If positive logic is used, as we did in Section 2.3, then P_1, P_2, and P_3 express NOR gates, and f_1, f_2, and f_3 are expressed in conjunctive forms.

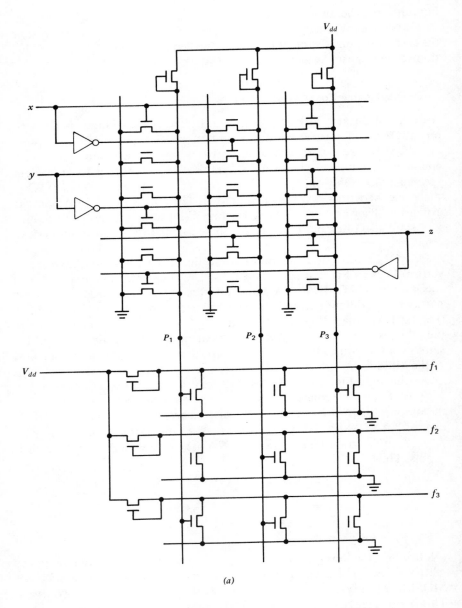

Fig. 9.4.6 *MOS PLA for combinational networks.* (a) *MOS PLA.*

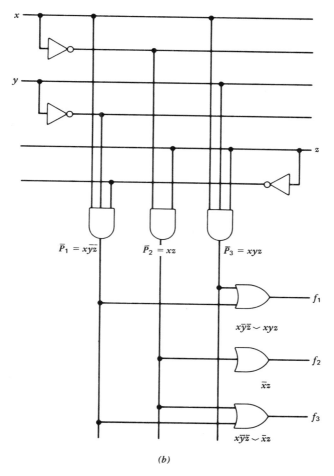

Fig. 9.4.6 *(Continued)* **(b)** *AND-OR network that is equivalent to Fig. 9.4.6a.*

in the second (output) level, as discussed in Section 6.1. The outputs of these AND gates represent $\bar{P}_1$, $\bar{P}_2$, and $\bar{P}_3$. This is shown in Fig. 9.4.6b, where AND and OR gates replace the actual NAND gates realized by MOSFETs. Thus the upper and lower matrices in Fig. 9.4.6a are often called **AND and OR matrices** (or AND and OR arrays), respectively. To facilitate our discussion, the vertical lines that run through the two matrices in Fig. 9.4.6a will be called **product lines**, since they correspond to the products in disjunctive forms for the network output functions.

552 Practical Considerations in Logic Design

Any combinational network with AND and OR gates in two levels is realized by determining only connections. (The connections between MOSFET gates and horizontal or vertical lines are usually denoted by dots, as shown in Fig. 9.4.7.) The network has multiple outputs. Also, a PLA can realize many independent networks with single or multiple outputs.

If the number of different multiple-output prime implicants (more specifically, paramount prime implicants) in disjunctive forms for the output functions is minimized, as we did in Section 5.2, we can minimize the number of product lines, that is, the number of P_i's, as can easily be seen. Thus **the required matrix size, $(2n \times m)t$, of a PLA is minimized** when the PLA has n inputs, m outputs, and t product lines. Also, if the total number of connections in a two-level AND-OR network is minimized as the secondary objective, as we did in Section 5.2, then the number of connections (i.e., dots) in the corresponding PLA is minimized. Then the chances of faults due to bad connections have been minimized, since the connections are often causes of faults. Therefore the **derivation of a minimal two-level network with AND and OR gates in Section 5.2 is very important for the minimal and reliable design of PLAs.**

However, **minimization of the number of connections in a minimal two-level AND-OR network is not as important as minimization of the number of different multiple-output prime implicants**, because the chance of faults can be greatly lessened by careful manufacturing, but the required size of the PLA is determined by the number of different multiple-output prime implicants and cannot be changed by any other factors. Also, instead of making connections as they become necessary on a PLA, a PLA is sometimes prepared by disconnecting unnecessary connections after it has been manufactured with the gates of all MOSFETs connected to the lines. In this case the chance of faults may be lessened by increasing the number of connections in the two-level AND-OR network.

When minimal two-level AND-OR networks cannot be derived by the procedures discussed in Section 5.2 (if the number of connections need not be minimized, we can use a more effective procedure such as Procedure 5.2.2 or its variations), we must be content with switching expressions that appear reasonably simple* (e.g., when functions have many variables, say 20 or 30).

Sequential networks also can be easily implemented on a PLA. Some outputs of the lower matrix are connected to master-slave flip-flops, as shown in Fig. 9.4.7. (Commercially available PLAs often contain a small fixed number of master-slave flip-flops, particularly of the J-K type.) Then the outputs of the flip-flops are fed back to the inputs of the upper matrix of MOSFETs. More than one sequential network can be realized on a single

---
* See, for example, Hong, Cain, and Ostapko [1974].

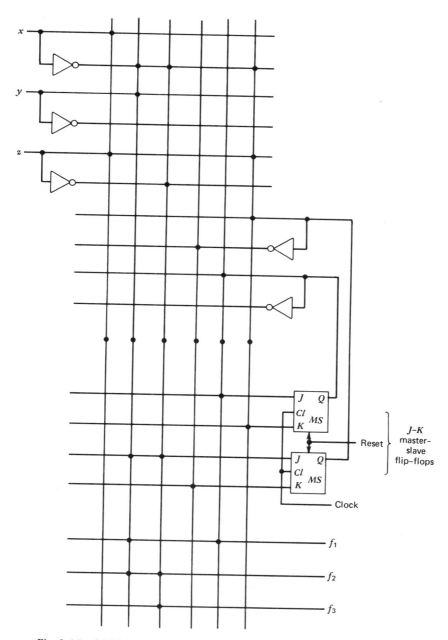

Fig. 9.4.7 MOS PLA for sequential networks (MOSFETs are not shown).

554 *Practical Considerations in Logic Design*

PLA, along with many combinational networks. Minimization of the number of flip-flops is important, since flip-flops occupy a chip area of significant size because of the horizontal lines and MOSFETs (loads and drivers) required, and consequently not many flip-flops are usually provided in a PLA chip. (The number of flip-flops can be reduced by Procedure 7.6.1. But, as mentioned in Section 8.5, there is no efficient method to find a state assignment that minimizes the size of a PLA.)

Notice that, if switching functions that can be implemented with a PLA are implemented with a ROM by storing their truth table, the ROM usually requires a larger size. [For example, if the three functions implemented by the PLA in Fig. 9.4.6a are implemented by their truth table (Table 9.4.1), instead of by switching expressions, using the ROM in Fig. 9.4.1b, the ROM has a matrix size of $(6 + 3) \times 8 = 72$ versus $(6 + 3) \times 3 = 27$ in Fig. 9.4.6a, since we need 8 vertical lines in Fig. 9.4.1b.] However, ROMs with a large number of addresses do not use the decoder implementation in Fig. 9.4.1b, so comparison of ROMs and PLAs is more complicated.

In commercially available PLA packages the number of inputs (i.e., $x_1, \ldots, x_n$) is roughly somewhere between 10 and 30, the number of outputs is less than 20, and the number of product lines is between 50 and 150. But PLAs of larger sizes are also used.

A **field-programmable PLA**, usually abbreviated as a FPLA, is also available. In this, the user can set up a dot pattern by burning short Nichrome wires connected in series with diodes in the PLA (made by modifying the ROM in Fig. 9.4.1a), by temporarily feeding excessive currents. (When an FPLA is not encased in a package, undesired cross points can be cut by laser beam, though this approach requires special equipment.) In large volume production FPLAs are slightly more expensive than PLAs, but they greatly facilitate the implementation of networks. Even if a logic designer needs only one PLA according to his specification (or in very small quantities), he can make it with FPLA by himself inexpensively and quickly. In contrast to FPLAs, the PLAs discussed so far are called **mask-programmable**.

There are variations of PLA (Remark 9.4.1).

Applications of PLAs

Computers and digital systems are becoming increasingly large and complex, because of the cost reduction, performance improvement, and size reduction of integrated circuits. Consequently, design is becoming more time-consuming, or more prone to errors. At the same time the race to introduce new products before competitors do is becoming more intense. For these reasons PLAs and FPLAs are becoming indispensable tools in some cases of logic design. Since only dot patterns such as those in Fig. 9.4.7 need to be prepared,

Design of Networks with a Mixture of Memories and Gates 555

the checking, redesigning, or testing of networks is easy and fast if PLAs are used. In particular, FPLAs (also PROMs) are convenient when prototypes of networks are prepared to find design errors, because FPLAs (or PROMs) can be easily replaced by users, instead of having to rely upon semiconductor manufacturers to prepare new masks for PLAs (or ROMs).

Usually PLAs are used for networks that require design changes, even though any networks in a digital system can be implemented with PLAs. Networks that do not require design changes are preferably designed with random-logic network configurations, because any network implemented in a PLA can be replaced by a network in a freer configuration (i.e., a random-logic network) than the matrix configuration of a PLA, thereby reducing the size and improving the speed. (By a freer arrangement of MOSFETs and diodes, lines can be shorter, or the network can be in multilevels, reducing the size* and parasitic capacitance. Thus the speed can be improved and, in the case of large volume production, the cost also can be reduced.)

Some microcomputer chips contain PLAs as part of them, so that users can easily change the functions of the microcomputers by changing the PLAs.

In addition, PLAs are extensively used in microinstruction decoders to yield starting addresses of microsubroutines (Exercise 9.4.6. Also see Reyling [1974a, 1974b].)

Another application of PLAs is to change part of the information stored in ROMs. When data or parts of programs stored in some addresses in a ROM (i.e., firmware) need to be changed or are found erroneous, a PLA can be used as a patch instead of throwing the ROM away. When the PLA receives these particular addresses, it inactivates the ROM and then either provides new data or turns on another ROM. Moreover, PLAs are also used as the decision tables mentioned in Section 3.7. (PLAs work much faster than software implementation, though decision tables are usually implemented in software.)

As a simple example of network implementation by a PLA, Fig. 9.4.8 shows a code converter, which converts code words at its inputs x and y into object code words at its outputs f_1 and f_2, as shown in Table 9.4.3. If a product line is provided for each row in Table 9.4.3, we need four product lines. By finding switching expressions with a minimum number of products, however:

$$f_1 = \bar{x}\bar{y} \vee xy \quad \text{and} \quad f_2 = \bar{x},$$

three product lines are sufficient, as shown in Fig. 9.4.8. Code conversion in general (e.g., between any two of the following: binary numbers, Gray code,

---

* Roughly speaking, for each gate in a random-logic network it is estimated that a PLA implementing the same functions and appropriately used will require 10 to 40 bits ("bits" here means the product of the numbers of vertical and horizontal lines in a PLA) [Donath 1974].

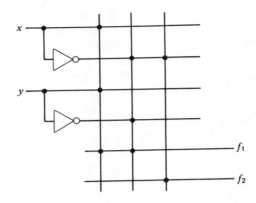

Fig. 9.4.8   PLA as a code converter.

binary-coded decimal numbers, and excess-3 code) is very time-consuming, even for the fastest computers, by software processing. But with a PLA like the one in Fig. 9.4.8 code conversion can be processed with high speed (probably less than 100 nanoseconds from inputs to outputs of a PLA; this is roughly 1000 times faster than code conversion by software).

For further reading about PLAs, see, for example, Texas Instruments [1970], National Semiconductor [1977], Fleisher and Maissel [1975], Logue, Brickman, Howley, Jones, and Wu [1975], Maggiore [1974], Mitchell [1976], Sherwood [1977], Roth [1978].

Remark 9.4.1—Variations of PLA: There are some variations of PLAs. In one version, an inverter can be added to each of the outputs $f_1, f_2, \ldots, f_m$ in Fig. 9.4.6a as an option. Thus functions that have too many multiple-output prime implicants in their disjunctive forms to be realized by a single PLA but whose complements (i.e., $\bar{f}_i$'s instead of f_i's) have fewer multiple-output prime implicants can be implemented by a single PLA, based on the multiple-output prime implicants for $\bar{f}_i$'s. Then, by the inverters, we obtain f_i's as the outputs [Hemel 1976].

Table 9.4.3   Conversion Table

x	y	f_1	f_2
0	0	1	1
0	1	0	1
1	0	0	0
1	1	1	0

In another version of PLA, the entire set of input variables is partitioned into subsets of variables, and then a decoder is provided for each subset of variables. (For example, inputs x_1, x_2, x_3, and x_4 are partitioned into the set of x_1 and x_2 and the set of x_3 and x_4. Then x_1 and x_2 have a decoder whose outputs represent functions $x_1 x_2, x_1 \bar{x}_2, \bar{x}_1 x_2$, and $\bar{x}_1 \bar{x}_2$; also, x_3 and x_4 have a separate decoder whose outputs represent $x_3 x_4, x_3 \bar{x}_4, \bar{x}_3 x_4$, and $\bar{x}_3 \bar{x}_4$. These decoder outputs are the inputs to the upper matrix of a PLA.) This adds the possibility of reducing the matrix size of the PLA, though the decoders add complexity [Wood 1975].

Combining many PLAs in a matrix form is discussed in [Cox, Devine, and Kelly 1976], in order to improve the packing density.

In another version of PLA, which is called an **associative logic array**, more than two matrices are vertically cascaded, by letting some vertical lines go through two or more matrices, but not letting all vertical lines go through all the matrices. With this PLA a NAND (or NOR) network in more than two levels can be implemented, so the size of the entire PLA, consisting of many matrices, can be reduced [Greer 1976], since a multilevel network implementation generally requires fewer gates than a two-level AND/OR network implementation, as discussed in Section 5.4. ∎

Integration of Logic Gates, Software, Firmware, and Hardware in Network Design

As we see in this section, gates can be replaced by ROMs or PLAs, software can be replaced by ROMs as firmware,* firmware can be replaced by gates, and some gates can be replaced by transistors (and/or diodes) without the explicit concept of "gate."† In these replacements, cost, performance, compactness, ease of change, design time, and convenience are traded. When hardware was expensive, only simple but important functions (or jobs), such as addition and multiplication, which were encountered most often during computation, were implemented by hardware, as adders and multipliers. Now integrated circuits are so inexpensive that many more complex functions which were traditionally implemented by software can be economically implemented by hardware (in fact, are being so implemented) in appropriate forms of gates, memories, or transistors. Since RAMs with very large capacity are available in packages as small as IC packages, logic networks can contain computer programs stored in these RAMs, in addition to gates, and memories used as substitutes for some gates. (Consequently, logic networks in some cases essentially become microcomputers, or their simplified versions. As a matter of fact, some digital systems (or even microcomputers) consist of

---
* ROMs as firmware are explained in Exercise 9.4.3. For details see, for example, Gshwind and McCluskey [1975] and Mano [1972].
† As explained in Section 5.5, the "gate" concept is sometimes not important and we can derive better logic networks by transistor-level logic design.

many microcomputers.) This means that **logic design is now based on software, firmware, hardware, and gates (often gates are also considered as hardware), without clear border lines.** We can say that, in designing digital systems or computers, **there is now no distinction among software, firmware, hardware, and gates, and we need integrated knowledge of all of them.** Integrated use of all of them in network design can result in computers and digital systems of improved performance, convenience, and economy.*

For simple job functions such as binary addition, switching expressions can be immediately found, and then we can design networks based on them. For more complex job functions such as decimal addition, design can still be done by modifying the binary addition. But for still more complex job functions such as calculations of $\sin \theta$ or $\cos \theta$, switching expressions cannot immediately be found, so the procedures or algorithms that are most suitable for each case must be determined before deriving switching expressions for designing networks. Also, there is no simple general procedure to determine where we should use gates, ROMs, RAMs, firmware, software, or transistors in networks (as seen in the 1-out-of-n detector in this section). For each job function we have to compare different implementations.† If there are too many products in disjunctive forms for functions to be implemented, PLAs cannot be used, and ROMs can be a choice (Remark 9.4.2). If networks are to be manufactured in large quantities, gates are preferred to ROMs in terms of cost and speed, as mentioned earlier. Microcomputers are sometimes used in a digital system or a computer, but PLAs are preferred to microcomputers if job functions are simple (Remark 9.4.3). Shift registers which are traditionally implemented by gates can be implemented by RAMs, more economically and compactly, if a large number of them are necessary and parallel loading is not important [Wyland 1974].

In conclusion, logic design is becoming much more complex than conventional design with gates only, requiring a great deal of ingenuity and a wide spectrum of knowledge.

---

* For example, Conklin and Rodgers [1978].

† Hardware is usually harder to change than software, so software is preferred when frequent changes are expected. But when software is very complex, as in operating systems, hardware is preferred. For some job functions, software development may be easier if special computers dedicated to these job functions are designed, since operating systems need not be developed. Emulation is more economical than extensive reprogramming. Also, when microprocessors of high performance are used, operating systems are simpler because there is no need for multiprogramming [Wagner 1976]. For the advantages of compilers in ROM, see Langer and Dugan [1978].

Since gate networks and solid-state memories (e.g., Fig. 9.4.1b) use transistors in the same manner, only difference between them is configuration, i.e., whether transistors (or more generally solid-state devices) are arranged in iterative structure. So programming with ROMs (or microprogramming) may be interpreted as logic design that connections among transistors are determined.

Design of Networks with a Mixture of Memories and Gates 559

Remark 9.4.2: When the number of products required for switching expressions for output functions $f_1, f_2, \ldots, f_m$ in the PLA in Fig. 9.4.6a increases, many product lines are required for the PLA, and its advantage over a ROM diminishes. When a PLA and a ROM have about the same matrix size [i.e., the product of the numbers of vertical and horizontal lines; for a PLA it is $(2n + m)t$ and for a ROM, $(2n + m)2^n$, when both have n inputs and m outputs and the PLA has t product lines, assuming that the ROM implementation in Fig. 9.4.1b is to be compared], which makes the two comparable in chip size (thus cost), the PLA has more address inputs, as a trade-off with fewer vertical lines. Therefore, if the functions to be implemented can be expressed by switching expressions with a small number of terms, PLAs can be useful, as in the case of the 1-out-of-n detector in Fig. 9.4.5. However, PLAs with a large number of product lines are not feasible, because the fan-out connections of each input become too large. In the case of a ROM, this fan-out problem is solved by using decoders of more sophisticated structure than the one in Fig. 9.4.1b. ∎

Remark 9.4.3: Microcomputers are extensively used in logic design. Although they are far more versatile than PLAs, the latter can be used alone for simple jobs or functions and can process much faster. For example, code conversions, mentioned earlier in this section, can be done by a PLA alone nearly 1,000 times faster than by microcomputers [Derman 1976]. When we use a microcomputer, it can be either software-intensive or hardware-intensive with many auxiliary IC packages. Hence a trade-off between software development cost (and time) and hardware cost must be considered in each case of production volume [*Electronic Design* January 1977, Pokoski 1978].

Also, notice that random-logic networks offer better performance and economy than microcomputers if they are designed for simple job functions and manufactured in very large quantities. ∎

Exercises

9.4.1 (M) Discuss what information should be stored in each ROM in the 1-out-of-n detector in Fig. 9.4.5. Also, design the 1-out-of-n detector, using PLAs and gates, instead of ROMs and gates. (*Hint*: Straightforward replacement of ROMs by PLAs may not work. If so, consider different network configurations or different output functions for PLAs.) Also, discuss how many address inputs a PLA can have if the PLA is to have a matrix size* roughly equal to that of a ROM† with 8 address inputs, assuming that the ROM implementation in Fig. 9.4.1b is to be compared. (If the matrix size is comparable, a ROM and a PLA may be considered to have roughly equal costs.)

9.4.2 (M) Design an m-out-of-n detector, using ROMs and OR gates in a manner similar to that shown in Fig. 9.4.5. Each ROM may have at most $m + 1$ outputs. Discuss what information should be stored in each ROM. Calculate the number of ROMs and the response time for $m = 3$ and $n = 10$, 80, and 800, assuming that each ROM has a memory capacity of 1024 bits and an access time of 40 nanoseconds. If necessary, gates of any types may be used in addition to ROMs. Use as few ROMs and gates as possible.

---

* The matrix size of a PLA is the product of the numbers of vertical and horizontal lines, as defined in the text. The matrix size in Fig. 9.4.6a is $(3 \times 2 + 3)3 = 27$.
† The matrix size of the ROM in Fig. 9.4.1b is $(2 \times 2 + 4)4 = 32$.

560 **Practical Considerations in Logic Design**

9.4.3 (M)   The control network of a computer can be designed with a ROM, as illustrated in Fig. E9.4.3a. The ROM can be divided into a section for next address control and another section for data path control. Inputs $x_1, \ldots, x_n$, which show status information of other parts of the computer, arrive at each clock pulse. When they arrive, they are combined with the information from the ROM section for next address control in the branch logic, to determine a next address of the ROM. The next address is decoded, and the information in that address is read out. Then outputs $z_1, \ldots, z_m$

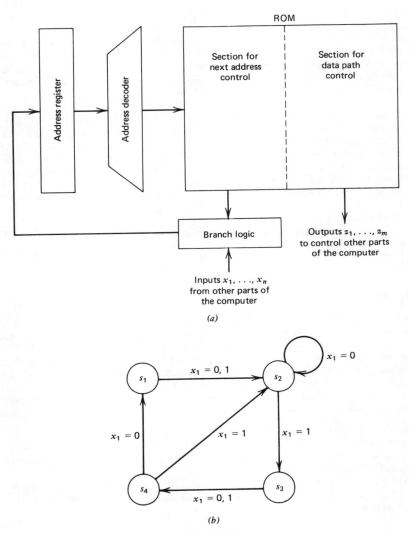

Fig. E9.4.3

Design of Networks with a Mixture of Memories and Gates   561

are sent to other parts of the computer to direct the information flow among various registers by activating appropriate gates (during transmission the information may be complemented or shifted by $z_1, \ldots, z_m$ also). The ROM used in this way for controlling a computer is called **firmware**, and the information in the ROM is termed **microinstructions**. (Note that the control unit of a computer can also be implemented with gates only. In this case it is usually called a **hard-wired control unit**.

(i) As a special case, suppose that we have $n = 1$ and $m = 4$ with the state diagram shown in Fig. E9.4.3b. When the ROM section for next address control shows the current state s_i, the output z_i is 1, while the other three outputs are 0, for $i = 1, 2, 3,$ and 4.

Design this control network. (The address register and decoder need not be designed, but show the detail of the branch logic and the information pattern in the ROM.) Assuming that each ROM bit is much cheaper than a gate and also that, if two ROMS have equal memory bit sizes (excluding the decoder), a ROM with fewer addresses is cheaper, implement a network which is the most economical even for large n.

(ii) The above ROM essentially implements a transition-output table, and the above network is one way of implementing a sequential network (though it may not be the best way). Discuss the advantages and disadvantages of the above implementation of a sequential network with a ROM, compared with the conventional approach discussed in Chapter 8.

9.4.4 (R) Implement the multiple-output network shown in Fig. 5.2.5 with a PLA. It is sufficient to show only the dot pattern.

9.4.5 (R) Implement one of the following sequential networks, using the PLA shown in Fig. 9.4.7. It is sufficient to show only the dot pattern. Use as few product lines as possible. (We can change the numbers of inputs, outputs, product lines, and flip-flops in the PLA in Fig. 9.4.7, but no other extra connection can be made, though, if necessary, each clock terminal of the flip-flops can be connected to a newly provided horizontal line.)

A. The sequential network in Fig. 8.5.2.
B. The divide-by-5 counter in Fig. 9.3.4a.
C. The divide-by-5 counter in Fig. 9.3.4b.

9.4.6 (M−) Suppose that a calculator has instructions (0 1 0 0), (1 1 0 1), (0 0 − 1), (1 0 1 1), and (1 1 1 0) (e.g., corresponding to "add," "subtract," "square-root," etc.), where the dash means "don't-care." The corresponding microsubroutines start at addresses (0 1 1), (0 1 0), (1 0 1), (1 1 0), and (1 1 1), respectively. (To be realistic, these addresses may have a longer bit length, but three bits are assumed here for simplicity.) We want to design a PLA that converts the former set of code words into the latter.

Showing only dots, derive a PLA that has a minimum number of product lines as the primary objective and a minimum number of dots as the secondary objective.

9.4.7 (M−) The correspondence between Gray code and 8-4-2-1 binary coded decimal (BCD) is shown in Table E9.4.7. Design a converter from Gray code to 8-4-2-1 BCD, economically, under each of the following conditions. Assume that only noncomplemented variables are available as inputs in each case.

(i) Design with a ROM only.
(ii) Design with four 8-input multiplexers. (For multiplexers, see Exercise 4.17.)

562 *Practical Considerations in Logic Design*

(iii) Design with a PLA only.

(iv) Design with a decoder and gates (any of AND, OR, NOT, NOR, or NAND gates may be used) only.

(v) Design with NAND gates only.

(vi) Compare approaches (i) through (v) in terms of economy and speed. (If necessary, make appropriate, reasonable assumptions in regard to the production volume, design time, and speed of each ROM, PLA, or gate.)

Table E9.4.7

GRAY CODE				8-4-2-1 BCD			
0	0	0	0	0	0	0	0
0	0	0	1	0	0	0	1
0	0	1	1	0	0	1	0
0	0	1	0	0	0	1	1
0	1	1	0	0	1	0	0
0	1	1	1	0	1	0	1
0	1	0	1	0	1	1	0
0	1	0	0	0	1	1	1
1	1	0	0	1	0	0	0
1	1	0	1	1	0	0	1

9.4.8 (E) Suppose that a PLA which has n inputs, m outputs, and t product lines is given and does not contain flip-flops inside. Thus the matrix size is $(2n + m)t$. Find a formula for the number p of address inputs of a ROM whose matrix size, $(2p + m)2^p$, is roughly equal to $(2n + m)t$. Assume that the ROM implementation in Fig. 9.4.1b is to be compared, and that m is the same for the PLA and the ROM.

Calculate the number p of address inputs of the ROM for the following combinations of values: $n = 17$, $m = 18$, $t = 60$; $n = 25$, $m = 25$, $t = 40$.

9.4.9 (M) Suppose that we need to realize a network for a function $f(x_1, \ldots, x_n)$. But we have only PLAs each of which has k inputs, where $k < n < 2k$. Also, we have a sufficient number of AND, OR, and NOT gates.

Devise and compare all possible network configurations from the viewpoints of speed and economy (check two PLAs in parallel, two PLAs in cascade, and possibly other configurations; gates may be used to connect PLAs), discussing advantages and disadvantages.

9.4.10 (M) Design a sequencer (i.e., a network to generate a sequence of control signals) for the following "addition" instruction, using the PLA shown in Fig. 9.4.7, as illustrated in Fig. E9.4.10. Assume that the PLA has sufficient numbers of inputs, outputs, product lines, and J-K master-slave flip-flops. Only dots need to be shown on the PLA.

By the "addition" instruction, a number A stored in a memory address K is to be added to a number B in an accumulator in the computer, using the following sequence of steps. (For "accumulator," see Exercise 8.15.)

Design of Networks with a Mixture of Memories and Gates 563

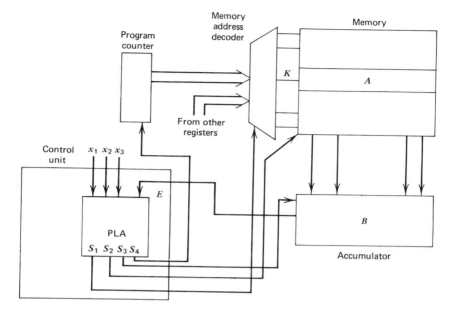

Fig. E9.4.10

(i) When the PLA receives the "addition" instruction code (1 0 1) at its inputs (x_1, x_2, x_3), which lasts only during one clock pulse, the PLA sends signal 1 at its output S_1 to let the memory address decoder accept address K. Output S_1 is 1 for only one clock pulse.

(ii) One clock pulse later, the PLA sends signal 1 at its output S_2 to read out the number A, stored at address K of the memory. Output S_2 is 1 for only one clock pulse.

(iii) Two clock pulses later (after sending S_2), the PLA sends signal 1 at its output S_3 to open the gates to the accumulator in order to add B to A in the accumulator. Output S_3 is 1 for only one clock pulse.

(iv) When the PLA receives the end signal 1 of the addition from the accumulator (the carry at the most significant bit position of the accumulator can be used as the end signal 1) at its input E, all the flip-flops used for Steps (i) through (iii) are reset, and then the PLA sends signal 1 at its output S_4 to the program counter to read the next instruction from the memory.

[Thus the PLA must have, at least, inputs x_1, x_2, x_3, E, and outputs S_1, S_2, S_3, S_4. This PLA implements only the sequencer for "addition" instruction. But a PLA (or PLAs) in the control unit of a computer generally must implement sequencers for all instructions. Input E may become 1 for other instructions. Of course, the control unit in general are not always realized with PLAs.]

9.5 Logic Design with Flow Charts

If a digital system is to perform many complex job functions, nobody can design the entire system at once. The system must be decomposed by the designer into subsystems of manageable size, so that each subsystem can be easily designed and the entire system will have good performance and low cost. (If the subsystems are standard networks with good performance and low cost, the design will be simpler and the entire system will probably have reasonably good performance and low cost.) This is a major concern of every logic designer of digital systems, and achieving this decomposition is highly dependent on the logic designer's skill.

For example, suppose that we want to design a control network for an elevator for a 100-story building. If we try to design a sequential network by a design procedure discussed in Chapters 7 and 8, by treating every call signal for "up" or "down" from each floor or the elevator as an independent variable, then it is impossible to use the design procedure because of an excessively large number of input variables (i.e., we have a huge transition table). But if "up" and "down" signals from all the floors and the elevator are merged into a single "up" signal and a single "down" signal, respectively (though we need flip-flops for each floor to light up "up" and "down" indication lamps), and if these two merged signals are used as inputs to a sequential network, then we have a sequential network with a greatly reduced number of variables, and we can design it without difficulty (i.e., we have a much smaller transition table). As can be seen from this example, how to decompose an entire system is a vitally important branch point, leading to a good or bad system. By carefully examining and preprocessing the performance requirements and variables before the design of subsystems, we can have each subsystem with simple performance requirements and few variables; otherwise it will have too complex requirements and too many variables to be successfully designed. Flow charts, to be discussed in this section, are a very convenient tool for doing this.

The design of large digital systems is time-consuming and cumbersome, and it is easy to make design errors or to overlook some of the required functions. But in many cases it is—probably much more important than good performance or low cost—to have no mistake in design. Once errors sneak into a large digital system, it is difficult to find out where they are. Careful design without mistakes from the beginning leads to a much less time-consuming and consequently much less costly overall development. For example, in designing networks to be put on an LSI chip, it is not possible to make partial corrections, unlike the situation for networks with discrete components, in which design errors can be easily corrected by changing some components. In the case of LSI, the entire chip must be designed and manufactured again, discarding the faulty chip. Some systems are so complex that all required functions cannot be tested by manufacturers, and customers may find mistakes after using the systems for a while. The history is abundant with such horror stories (e.g., design mistakes in a fancy calculator were found accidentally by a customer, and all calculators already on the market had to be called back to the manufacturer). This situation is similar to the debugging problem with large software, but is much more serious, since corrections are expensive and time-consuming. Flow charts, to be discussed in the following, are aids in avoiding design mistakes.

The performance that a network or a digital system is required to have is conveniently expressed with the flow chart used for computer programming, although there is a subtle difference from the flow chart in programming, as we will see later. In Chapters 7 and 8 we found state diagrams useful for designing complex sequential networks, where it is easy to overlook part of the required functions, though simple networks can be designed directly without state diagrams. Flow charts are essentially a generalization of state diagrams. Flow charts are also convenient for later modification, testing, or partial computer program implementation.

Flow charts may be used at different levels. In other words, they are used to show relationships between a digital system and its networks, or between a network and its gates. Also, it is arbitrary what is expressed by flow charts, as we will see in later examples. A large system is usually designed by a "top-down" approach [Caplener and Janku 1974]. In other words, after the overall organization is designed, the finer details are designed. Flow charting is convenient in such an approach.

Example 1: To illustrate how flow charts are used, let us consider a simple example. Suppose that a factory wants to design a network to control the inspection process of manufactured products. If a product is short in length or width, it is rejected. Every 10 products that pass this test are put into a shipping box. There are two sensors, which have outputs x and y. If a product is short in length, $x = 0$; otherwise $x = 1$. If a product is short in width, $y = 0$; otherwise $y = 1$. Whenever a product passes the inspection, the network sends a pulse signal p (i.e., $p = 1$ for a short period when the product passes the inspection, and $p = 0$ at any other time) and the product is dropped into a shipping box. Then 1 is added to a counter A. When the counter A counts 10 products, the network sends signal q to move a belt conveyor so that the box loaded with 10 products moves away. When the next box on the belt conveyor is underneath the hopper through which products are dropped from the inspection table, a sensor to detect the box position sends a

566 *Practical Considerations in Logic Design*

pulse signal r to the network. Then the next product is put on the inspection table. Also, when a product is rejected, the next product is put on the inspection table, while 1 is added to a counter B, which counts the number of rejected products. The entire performance required is shown in the flow chart in Fig. 9.5.1, using diamonds and rectangles. A diamond is used for the decision for a certain question and has two outgoing branches, for "yes" and "no." A rectangle denotes an event to be reached, the state, or the output of a subnetwork or the entire network.

For this simple network performance requirement, a network can be immediately drawn, as shown in Fig. 9.5.2. Notice that the cascade of decision diamonds encircled in the dotted line in Fig. 9.5.1 has three outputs, which can be easily expressed as xy, $\bar{x}$, and $x\bar{y}$, and that this is realized as the subnetwork encircled in the dotted line in Fig. 9.5.2. (This subnetwork can be easily simplified to a subnetwork of two gates. Readers should try to do this.)

Notice that here, **unlike the situation in the flow charts used in programming for a single processor,*** many subsystems or networks in a

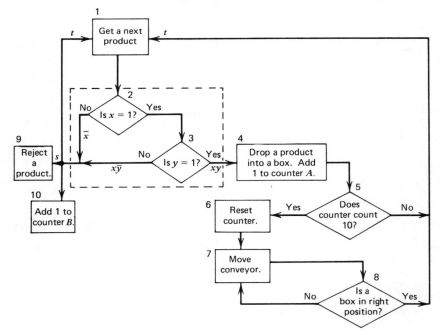

Fig. 9.5.1   Flow chart for a network to control the inspection process.

---

* The flow charts in programming for multiple processors are closer to those for logic design, but the flow charts for logic design can usually have much greater concurrency.

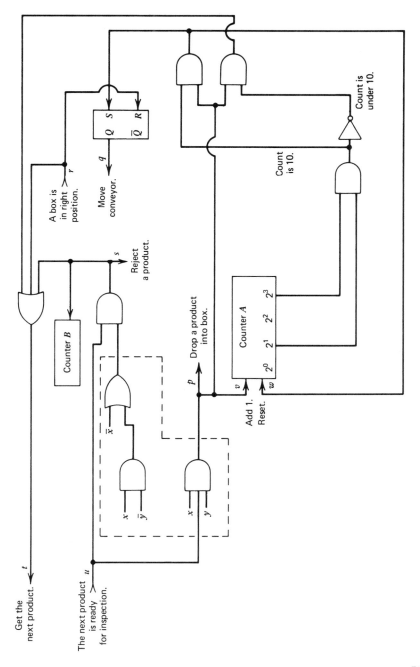

Fig. 9.5.2 Network for Fig. 9.5.1.

flow chart can work simultaneously. This is a subtle but important difference between the flow charts for logic design and those for programming. (Because of this, networks in concurrent operations can be designed and can work faster than software.) In the flow chart in Fig. 9.5.1, each of diamonds 2 and 3 has three branchings, to squares 1, 9, and 10. Then the entire network in Fig. 9.5.2 has three simultaneous operations on all the branchings: adding 1 to counter B, rejecting a product by signal s, and getting a next product by signal t.

The entire network has many inputs (four) and outputs (four). But notice that the network has the following special property. When an output such as q becomes 1, the entire network does not change at all until the input r becomes 1 (another such pair is t and u). Because of this straightforward functional relationship between a particular input-output pair, designing the network based on the flow chart by breaking the performance requirement into functional relationships facilitates inspection, repair, or modification of the network. Networks that have special functional relationships among inputs and outputs, like this network, which is quiescent after some outputs become 1 until the corresponding inputs receive return signals, are special cases of the general sequential network discussed in Chapters 7 and 8. Such networks are often encountered in practice, rather than general sequential networks which have a much more intricate relationship between inputs and outputs. Such networks are often much easier to design by flow charts than by the design procedures in Chapters 7 and 8.

Some blocks are not shown in detail in a flow chart, such as the counters in Fig. 9.5.1. This is not necessary when these blocks are available as standard networks in MSI or LSI packages. Therefore, when we want to design networks or digital systems with MSI or LSI packages, or when we want to design modular computer systems such as PDP-16,* flow charts are a convenient means to show the functional relationships among blocks.

Example 2: Some blocks that are not shown in detail in a flow chart for an entire digital system can be further expressed in other flow charts. For example, let us design a clocked J-K flip-flop which has the same performance as that of the J-K master-slave flip-flop described in Section 8.2. The performance requirement of this network is as follows:

1. The network has two inputs J and K, a clock input c, and two outputs Q and $\bar{Q}$, where $\bar{Q}$ is the complement of the first output Q.
2. The network is clocked.
3. If $J = 1$ and $K = 0$ during the presence of a clock pulse, that is, during $c = 1$, then Q and $\bar{Q}$ are set to 1 and 0, respectively, at the

---
* For details, see Bell, Carson and Newell [1972] and Grason, Bell and Eggert [1973].

moment of c's change from 1 to 0, while Q and $\bar{Q}$ do not change before then. Similarly if $J = 0$ and $K = 1$ during $c = 1$, then Q and $\bar{Q}$ are set to 0 and 1, respectively, at the moment of c's change from 1 to 0. Similarly also, if $J = K = 1$ during $c = 1$, the current values of Q and $\bar{Q}$ change to their complements at the moment of c's change from 1 to 0. If $J = K = 0$ during $c = 1$, Q and $\bar{Q}$ do not change.

A flow chart is drawn in Fig. 9.5.3. Four rectangles show stable states of the network, with the corresponding output values given inside these rectangles. Let us assign binary numbers to these states as shown outside the rectangles, using two internal variables, y_1 and y_2. [If we use only two states, it is difficult to draw a flow chart. As a matter of fact, as we see with the J-K flip-flop in Fig. 8.1.3a, it is difficult to obtain a reliable J-K flip-flop if we use only two states. Thus let us try with four states, that is, two internal variables.]

In Fig. 9.5.3 the network is in rectangle 0 0 or 1 1, when $c = 0$. Suppose that the network is in the rectangle 0 0 [i.e., the rectangle that corresponds to the state $(y_1, y_2) = (0\ 0)$] with the network outputs

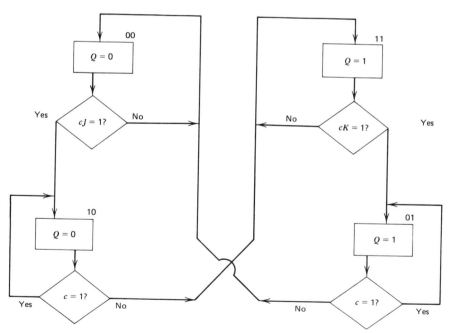

Fig. 9.5.3 Flow chart for clocked J-K flip-flop. [A binary number outside each rectangle shows the values of two internal variables, (y_1, y_2).]

$Q = 0$ and $\bar{Q} = 1$. If $c = 0$ or $J = 0$ (regardless of whether $K = 0$ or 1), the network remains in rectangle 0 0. When $c = 1$ and $J = 1$, the network moves to rectangle 1 0, whether $K = 0$ or 1. When the network moves into rectangle 1 0, it stays there as long as c continues to be 1. When c changes to 0 (i.e., the clock pulse disappears), the network moves to rectangle 1 1, where the outputs change to $Q = 1$ and $\bar{Q} = 0$. If $c = 1$ and $K = 1$, the network moves to rectangle 0 1; otherwise the network stays in rectangle 1 1; and so on. Notice that, when $c = J = K = 1$, the network always makes a transition from rectangle 0 0 to 1 1 or from rectangle 1 1 to 0 0. A flow chart such as Fig. 9.5.3 is a generalization, in a much freer style, of the state diagram discussed in Chapters 7 and 8.

A transition table is obtained from Fig. 9.5.3 as shown in Table 9.5.1a. Notice that Table 9.5.1a is in fundamental mode, because every internal variable contained in the network is not ignored. [Only if a network is designed with raceless flip-flops and some internal variables (i.e., those for masters) are ignored, do we have a transition table in skew mode, as discussed in Section 8.3.] Then Table 9.5.1a is split into each of Y_1 and Y_2, as shown in Tables 9.5.1b and 9.5.1c, which consist of two Karnaugh maps. In Table 9.5.1b Y_1 is expressed as

$$Y_1 = y_1\bar{c} \lor y_1\bar{y}_2 c \lor y_1 c\bar{K} \lor \bar{y}_2 cJ$$

as the disjunction of the loops on two maps. This is simplified as

$$Y_1 = y_1\bar{c} \lor y_1\bar{K} \lor y_1\bar{y}_2 \lor \bar{y}_2 cJ = y_1(\bar{c} \lor \bar{K} \lor \bar{y}_2) \lor \bar{y}_2 cJ$$
$$= y_1(\overline{cKy_2}) \lor \bar{y}_2 cJ.$$

Similarly, from Table 9.5.1c,

$$Y_2 = y_1\bar{c} \lor y_2 c.$$

In Table 9.5.2 we have an output table, and we obtain

$$Q = y_2 \text{ and } \bar{Q} = \bar{y}_2.$$

Based on these Y_1, Y_2, Q, and $\bar{Q}$, a network is designed as shown in Fig. 9.5.4. This is simply one of the possible networks that can be derived from these Y_1, Y_2, Q, and $\bar{Q}$.

The network in Fig. 9.5.4 is identical to the J-K master-slave flip-flop shown in Fig. 8.2.1, except for the presence of gates 10 and 11. But Fig. 8.2.1 can easily be obtained from Fig. 9.5.4 by deleting gates 10 and 11 and changing their output connections to gates 3 and 7, respectively, without changing the network performance.

Significance of Flow Charts

As shown in the above examples, flow charts can be used in many different ways to facilitate the design of complex digital systems.

Table 9.5.1 Transition Table in Fundamental Mode for Fig. 9.5.3

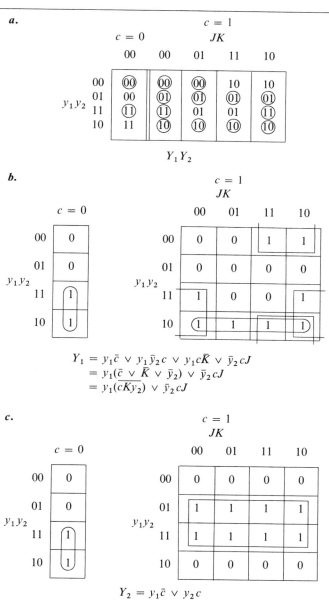

$$Y_1 = y_1 \bar{c} \lor y_1 \bar{y}_2 c \lor y_1 c \bar{K} \lor \bar{y}_2 c J$$
$$= y_1 (\bar{c} \lor \bar{K} \lor \bar{y}_2) \lor \bar{y}_2 c J$$
$$= y_1 (\overline{cKy_2}) \lor \bar{y}_2 c J$$

$$Y_2 = y_1 \bar{c} \lor y_2 c$$

571

572 *Practical Considerations in Logic Design*

Table 9.5.2 Output Table for Fig. 9.5.3

	$c = 0$
00	0
01	1
$y_1 y_2$ 11	1
10	0

$Q = y_2$

As integrated circuits become cheaper, increasingly complex and large digital systems are being designed. When the design of such large digital systems are beyond the capability of a single person and must be done by a group of designers, it is a formidable task for anyone inside or outside this group to find out how the systems work. Also, it is rare to have such a digital system work correctly at first trial, and it is not easy to find out why the system does not work. In such cases the availability of the corresponding flow charts greatly helps the designers. Thus **it is becoming an absolute necessity to prepare flow charts along with switching expressions, logic network drawings, and, ultimately the electronic circuit drawings**. Thus logic design is becoming an iterative procedure of drawing flow charts and logic networks. After drawing and redrawing them and merging common functions or simplifying flow charts or networks, logic designers can reach the final version of flow charts and networks.

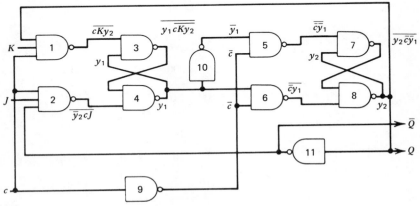

Fig. 9.5.4 Clocked J-K flip-flop.

Also, as mentioned at the end of Section 9.4, digital systems (or computers) are increasingly being designed by integrated approaches involving gates, memories, software, firmware, and hardware. For some functions (jobs or subsystems) we need to find the sophisticated procedures (or algorithms) which are most suitable for such integrated approaches. Flow charts are convenient for complex and cumbersome thinking in such integrated approaches, and for breaking the entire systems into subsystems of manageable size, so that each subsystem can be conveniently implemented by one of the following: gates, ROMs to substitute for gates, ROMs for firmware, software (programs stored in RAMs), PLAs, and hardware (transistor circuits to be designed without explicit use of the gate concept).

For further study of flow charts, see Clare [1973] (this booklet shows many examples of the use of flow charts; according to T. E. Osborne, the use of flow charts started in the early 1960s and has successfully continued in the Hewlett-Packard Co.), Lee [1976], Engineering Staff of AMI [1972], Richards [1973], Hardy [1973], Criscimagna [1971], Eggert [1973], Gladstone [1973] (flow charting in the design of a system with a microprocessor), Riley [1974] (a list of symbols used in flow charting), Jaeger [1974] (flow charting in microprogramming), Johnson [1976], or Peterson [1977] (Petri nets are sometimes useful for describing and analyzing the flow of information and control in digital systems).

Exercises

9.5.1 Draw a flow chart for a network to control the cash dispenser of a candy vending machine with the following performance requirement, and then design a network. A candy costs 15 cents, and 5, 10, or 25 cent coins may be inserted into the slot on the vending machine. Each coin must be put into one of three hoppers for 5, 10, and 25 cent coins, by opening or closing the lids of the hoppers, which are electrically controlled by binary electric signals. Change must be released from the hoppers. If no change with an appropriate combination of coins is available or the candies are all sold out, the inserted coins must be returned.

9.5.2 Draw a flow chart for a four-function calculator that handles eight digits, detailed to the level of standard networks whose functions are known. (For example, an adder need not be designed. Simply show a rectangle indicating an adder.) There are keys for digits 0 to 9, $+$, $-$, $\times$, $\div$, $=$, and clearing. Assume that all calculations are limited to integers (i.e., no floating point operations are needed).

9.5.3 Draw a flow chart for a special calculator, a "check checker" for keeping track of a checking account balance, which has the following specifications. There are keys for digits from 0 to 9. By depressing these keys, digits can be entered into a register, shifting the previous digits left. A number in the register is displayed on LEDs. By pressing a "clear" key, an erroneous entry in the register can be cleared without disturbing a balance. By pressing a "balance" key, the current balance is entered into the register from a memory. By pressing a "check key," the amount in the register is deducted from the stored balance. By pressing a "deposit" key, a new deposit entered in the register is added to the current balance and is stored in the memory.

9.5.4 Draw a flow chart for floating-point operation for one of the following:

A. Addition-subtraction.
B. Multiplication.

9.5.5 Draw a flow chart for a network to control the elevator described in Exercise 7.5.9, for a four-story building (instead of a flow-output table for a three-story building).

9.6 Simulation, Test, and Diagnosis

After networks are designed, it is important to find design errors before hardware implementation, since correction thereafter is expensive and time-consuming. Logic simulation by computer programs is a common practice for checking logic design.

When logic designers are convinced that there are no design mistakes (as mentioned at the beginning of Section 9.5, a complete check is impossible because of the excessive time that would be required for large systems, computers, or even calculators), networks are implemented in hardware (in discrete components or IC chips). When IC packages are produced for the networks by semiconductor manufacturers, the packages must be tested. When IC packages or discrete components are assembled on pc boards, the pc board must be tested. When pc boards are mounted on a subassembly board, the subassembly board must be tested. When final products are assembled, the products must be tested before shipment.

At each stage of this manufacturing process, we need a test because a faulty component (discrete component or IC package) that slips past one manufacturing stage usually costs much more to catch at the next stage because it is then more time-consuming to find and fix the faulty component. (It is ten times more costly in some cases. Suppose that inspection of components costs 30 cents per component. The cost rises to $3 to detect each faulty component that sneaks into a pc board. In the next stage the cost will be $30 per component, and so on.*) During assembling, bad connections are introduced as new sources of faults, in addition to faulty components. Usually components can be quickly and economically tested by automatic testers, but once components are assembled, automatic tests become increasingly difficult.

The importance of finding faulty components in early stages is indicated also by the following discussion. Suppose that each IC package has an

---

* The industry average cost of finding and removing a defective IC package from some finished product is approximately $100, whereas the high-volume cost of completely testing each SSI package individually before assembling is well under 10 cents. If the finished product consists of 100 IC packages, the total cost of the test is $10, which is only one tenth of the cost incurred by testing later [Hubbs 1974].

average yield of p percent (i.e., p out of 100 components is good). If we mount N IC packages on a pc board, the assembled pc board has an average yield of p^N. This is illustrated in Fig. 9.6.1. For example, when $p = 97$ percent and $N = 100$, the yield of the assembled pc board is only 4.8 percent. Therefore we have to try to eliminate faulty IC packages as completely as possible before mounting them on a pc board.

When the integration size of an IC package is large, testing each package is time-consuming, expensive, and often impossible. For example, if an IC package contains 50 flip-flops that cannot be directly accessible from the package pins, we need 2^{50} different input sequences to test these flip-flops alone (2^{50} is roughly 10^{15}). Even if testing with each input sequence takes only 10 nanoseconds, a few months would be required!

Suppose that all faulty components which can be found by short tests (a "short" time such that manufacturing is still profitable) are eliminated and networks are assembled into a digital system or a computer. A manufacturer tests it with sample problems for a while; then it is delivered to a customer. But at the customer's site many errors in design are usually found after successful operation for several months or years. For certain combinations of states or instructions that the manufacturer did not test (the manufacturer cannot test all combinations, since this would take months or even years), the system does not work as expected. It is not unusual that a large computer

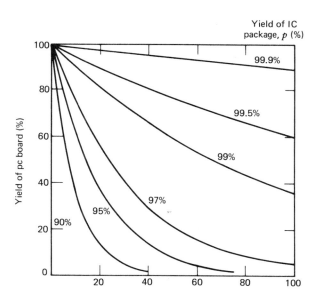

Fig. 9.6.1 Yield of an assembled pc board.

is found to require corrections of a few thousand connections, at the customer's site, because of design errors, even if logic design is checked by simulation and, during the manufacturing, every step from components to the system is tested very carefully. Only after many years of operation can large digital systems work without new design errors being discovered. (Precisely speaking, nobody can be sure whether all the design mistakes are ever found. But it is known, at least, that all design errors which are easily detected by customers can be found only after many years.)

During the operation of a system, some components (or gates) become faulty. It is important to quickly find faulty gates by diagnosis, whenever the system malfunctions, and then to resume its normal operation. Because a system now contains so many gates, such diagnosis is becoming expensive.

The simulation of logic networks, the testing of components or IC packages, and the system diagnosis have somewhat different motivations under different situations. All, however, have the same objective: to find gates that behave differently from the original design [Mattera 1975, Allan 1976]. In this section let us discuss these three processes. For all of them, computer programs have been extensively developed, and switching algebra is very useful in some aspects.

Simulation

When a network with a large number of gates or a digital system is designed, many design errors are usually involved. It is desirable to eliminate such errors from the network before implementing it with hardware (i.e., electronic circuits), because once the hardware implementation is done, it is costly and time-consuming to find mistakes and correct them. Therefore the logic operations of the designed network are usually simulated by computer programs* before the implementation with hardware. (In some of these CAD programs appropriate use of switching algebra reduces the memory space requirement and improves speed.) In other words, combinations of input variable values are supplied to the designed network, and a check is done to see whether or not the corresponding values of the network outputs are correct. When the network has many inputs, it is too time-consuming to test all combinations of values of input variables (particularly for sequential networks). Thus it is very important to find the fewest possible combinations of input variable values (for a combinational network) or the fewest possible shortest sequences of input variable values (for a sequential network), so

* Chang and Chappell [1975], Armstrong [1972], Szygenda and Thompson [1975], Breuer [1964], Case et al [1964], Seshu [1965], Hardie and Suhocki [1967], Biancomano [1977], Lake [1970] (this discusses simulation cost), Teets [1972], Scheff and Young [1972], Lewin [1977].

that as many mistakes as possible can be found. No efficient, systematic method to find such minimal combinations or sequences is known, but it is important to simulate the network without excessive computation time.

In addition to the incomplete test due to time limitation, simulation has another shortcoming. Some hazards of a network are difficult to detect, since they are usually due to faulty performance of the electronic circuits that implement gates, and do not explicitly show up in the logic operations simulated by computer programs. Hence, in addition to the **logic simulation** described so far, we need **circuit simulation**. Unlike the logic simulation, which simulates the logic operations of gates, circuit simulation simulates the performance of the electronic circuits that constitute gates, by modeling electronic circuits with differential equations or other mathematical expressions and then solving these equations by numerical analysis.* Circuit simulation often lacks desired accuracy. If an entire network is to be implemented on an IC chip, we have no good means to test other than circuit simulation. But if a network is to be manufactured with discrete components, test of the designed network by a prototype of hardware implementation is the best option. A hardware prototype of a designed network is called a **breadboard**.† Unfortunately, to construct a breadboard is expensive and time-consuming. (If prototypes with slower speed than the designed one are used, the expenses can be reduced, but complete tests are often not achieved.) Hence, in many cases, simulation is preferred in spite of its shortcomings.

   Architectural simulation simulates the architecture of a digital system, using a programming language such as GPSS. The aim here is for an overall evaluation of the system from the viewpoint of statistical measurement of throughput, resource utilization, queue lengths, busy times, system bottlenecks, and so on. Thus architectural simulation has a different objective from that of logic simulation.

Test

Semiconductor manufacturers usually test IC packages before shipment to customers. They reject faulty packages by visual inspection of package damage and by automatic measurement of electronic performance (i.e., measurement of currents, resistances, response time, logic operations, etc.) at room temperature. When a stricter test is necessary (e.g., at a customer's request) for further elimination of faulty packages, they do the above testing in combination with heat cycling, mechanical shock, and burn-in. The burn-in test calls for the operation of IC packages for many hours or

---
* See, for example, Herskowitz and Schilling [1972] and Herskowitz [1968].
† See, for example, Lyman [1976].

days, often under higher voltages than the normal operating ones. This test, although time-consuming and expensive, is very effective, because potentially faulty IC packages usually become faulty during the burn-in test.*

From the standpoint of the nature of testing, tests can be divided into parametric and functional types. In **parametric testing**, electronic performance such as delay time, rise time, fall time, voltages, currents, and pulse distortions is examined. **Functional testing** tries to find out whether a package has incorrect output values for any combinations of input values. This is essentially logic simulation by electronic means. We need not locate faulty gates in an IC package, since the entire package must be discarded if any output value is faulty for even a single input value combination.

The memory space required to store correct output values in the above test becomes very large as integration size increases. (Hence, instead of storing a truth table, the output of a faultless package is often used as a reference.) Also, semiconductor manufacturers produce many IC packages in large quantities. Thus automation of testing is becoming an absolute necessity from the viewpoints of economy and test time.

Dedicated computer systems for automatic testing† (rather than automatic test systems based on minicomputers) have been developed and widely used. Development of software for testing usually represents a greater expenditure than the hardware cost of an automatic testing system. Testing procedures and parameters to be tested are programmed, being tailored to the particular characteristics of each product that should be tested. In a typical testing procedure, IC packages are inserted one at a time into a testing system, and random signals or fixed test patterns are supplied to the input pins of the package. Then the signal values at the output pins and possibly those at extra pins provided for testing (if flip-flops are contained, extra pins can be provided for testing these flip-flops) are compared with the signal values from the corresponding pins of a package known to be faultless.

Testing of assembled pc boards requires the locating of faulty components or IC packages. This is essentially the problem of diagnosis, to be discussed in the next subsection.

For further study of testing, see the following: Hnatek [1977a], Tawfiq [1974], Jackson [1974], Chrones [1978] (this discusses the cost of testing LSI chips), or Rubinstein [1977].

Diagnosis

Suppose that a computer or digital system is successfully completed, after correcting all design mistakes and eliminating the hazards of its electronic

---
* See, for example, Loranger [1975].
† See, for example, To and Tulloss [1974], Grossman [1974], Torrero [1977], Knowles [1976].

circuits and all faulty gates. During the operation of the digital system, however, some gates become faulty (e.g., transistors or other components may deteriorate; some connections may be short- or open-circuited). Then faulty gates must be found and replaced in a short time. This is important in many cases. For example, if repair of a computer takes a long time for each trouble, customers may switch to the computers of another manufacturer. But it is not a simple task to find faulty gates among many. Computer manufacturers are making considerable efforts to improve the diagnosis facilities of computers.

The test of a computer is divided into the **functional test** and the **structural test**. Both are done by computer programs, usually aided by special electronic circuits* (i.e., hardware monitors). First a small part of the system, including a small memory space (this part is called a hard core), is tested, and then a computer program for testing is stored in that memory space in order to test the entire system. This is called a **hard-core test**. Then the functional test verifies whether each network or each block of networks in the system performs its task properly. For example, it checks whether an adder adds correctly, and a shift register shifts correctly. Computer programs for the functional tests are made, being tailored to specific tasks of networks. After pinpointing which network is faulty by the functional test, we proceed to the structural test discussed in the rest of this section, that is, a test to find which gate is faulty in the network. The diagnosis can be done only after a small faulty network has been located by the functional test; then the network is scrutinized more carefully by the structural test. If we skip the functional test, a structural test of the entire system will be impossible because of excessively long processing time. [The approach in this paragraph was introduced for the IBM 360 and called the Fault Locating Test (FLT).]

In the following, let us describe the **path-sensitizing method**, which is widely used for the diagnosis of combinational networks. Suppose that we want to detect whether one of the inputs of gate 2, x_2, is permanently fixed to the value y (either 1 or 0), by a fault in the NAND gate network in Fig. 9.6.2. (When y is fixed to 0, it is customarily said to be **stuck-at-0**. When y is fixed to 1, it is said to be **stuck-at-1**. The path-sensitizing method and most other methods are effective only when a single fault that is stuck-at-1 or 0 is to be detected.) Suppose that all other inputs to the gates through which signal y goes to one of the network outputs are set to 1 (if we have a network of NOR gates, they are set to 0) by setting inputs x_1, x_3, and x_4 to appropriate values, as shown in Fig. 9.6.2. We can do this by setting $x_1 = 0$ and $x_3 = x_4 = 1$ (i.e., gates 2, 4, and 6 have inputs of 1 except the bold-lined inputs in

---

* IBM 370/115 and 125, for example, have a diagnostic processor in addition to arithmetic/logic and I/O processors. The VAX-11 780 computer of Digital Equipment Corp. has a microcomputer for diagnosis.

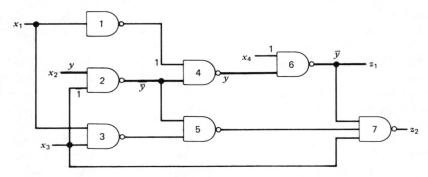

Fig. 9.6.2 *A sensitized path (shown in the bold lines).*

Fig. 9.6.2). Then the network output z_1 becomes $\bar{y}$ (in the case of a general network, y or $\bar{y}$). This path is called a **sensitized path** from input x_2 of gate 2 to network output z_1. Thus, when the value of input x_2 to gate 2 is set to a value different from the stuck-at value, y, we can find out whether y is stuck at 1 or 0. In other words, when we set $x_2 = 1$ along with $x_1 = 0$ and $x_3 = x_4 = 1$, we can find out that y is stuck at 0 if z_1 shows output value 1, because z_1 must be 0 for $x_1 = 0$ and $x_2 = x_3 = x_4 = 1$ if the network is not faulty. Similarly, when we set $x_2 = 0$ along with $x_1 = 0$ and $x_3 = x_4 = 1$, we can find out that y is stuck at 1 if $z_1 = 0$.

Furthermore, a fault at any point on this path from input x_2 of gate 2 to z_1 can be found similarly. For example, a fault on the input to gate 6 from gate 4 can be found. If the input is stuck at 1, the fault can be found by setting $x_2 = 0$ along with $x_1 = 0$ and $x_3 = x_4 = 1$. Also, by setting $x_2 = 1$, we can detect whether the input to gate 6 from gate 4 is stuck at 0.

If we determine the input vectors corresponding to the values 1 of the gate inputs set in order to form the above sensitized path, we have obtained test patterns to determine whether any point on the path is faulty or not. For the above case, $(x_1, x_2, x_3, x_4) = (0\ 1\ 1\ 1)$ is obtained for determining whether the input to gate 2 from x_2 is stuck at 0, the input to gate 4 from gate 2 is stuck at 1, or the input to gate 6 from gate 4 is stuck at 0. Similarly, $(0\ 0\ 1\ 1)$ is obtained for determining whether the inputs are stuck at 1, 0, or 1, respectively. However, in order to find which input on this bold-lined path is faulty, we must sensitize different paths which partly share the bold-lined path in all different ways. For example, if we find the value of output z_1 faulty for the bold-lined path and we also find value z_2 faulty by sensitizing the path of gates 1, 4, 6, 7 to output z_2, we can conclude that the fault must be somewhere from gate 4 to the output of gate 6, if only one fault is expected.

However, in general we have no guarantee that there can exist different paths which can be sensitized so that all gates can be separated individually. (For example, if a network contains a cascade of two gates between which there is only one connection, we cannot determine which one of these two gates contains a fault.) The maximum number of gates in a set of gates in which a fault cannot be located in a particular gate is called the **resolution**.

Therefore, if we obtain test patterns, that is, a set of input vectors that sensitizes a set of paths containing all the connections in each network in a given system and separating gates as much as possible, we can detect faults. The entire test patterns are stored in a memory (e.g., disks); and whenever a system malfunctions, the network under suspicion in the system, after being located by the functional test, is tested with the test patterns for that network.

A procedure based on switching expressions to find the test pattern for a given network that is more formal than the procedure discussed above is known, though it does not necessarily yield a minimal set of input vectors for the test. (See Armstrong [1966].)

In the case of Fig. 9.6.2, if we use network output z_2, we cannot determine whether y is faulty or not, for the following reasons. The inputs from gate 3 to gate 5 and from x_3 to gate 7 become 1 for the above test pattern. But two other inputs to gate 7 are y and $\bar{y}$. This means that network output z_2 is always 1, not showing y or $\bar{y}$. In other words, the path from input x_2 of gate 2 to output z_2 through gates 2, 4, 6, 7 cannot be sensitized. If, however, the path from input x_4 of gate 6 to output z_2 through gates 6 and 7 is sensitized, a fault on the input from gate 6 to 7 can be detected. Therefore, if we use an inappropriate path, a fault of some gate cannot be detected.

Another problem is the difficulty in determining the test patterns for some sensitized paths. For example, let us consider the sensitized path from y of gate 2 to the network output in Fig. 9.6.3. Two inputs to gate 6 must be set

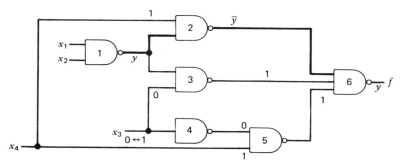

Fig. 9.6.3   Contradiction for some sensitized path.

to 1. Also, x_4 must be set to 1. Then one input to gate 3 must be 0. Also, one input to gate 5 must be 0; consequently, at the input of gate 4, x_3 must be 1— but must be 0 from the previous consideration. This is a contradiction. Hence we cannot determine a test pattern to sensitize this path.

These two problems are not unique to the path-sensitizing method, but the path-sensitizing method has one unique problem, as follows. Let us consider the network in Fig. 9.6.4. If the fault at × is stuck-at-1, we can easily find a test pattern that sensitizes a single path (Exercise 9.6.4). Next suppose that the fault is stuck-at-0. The fault stuck-at-0 indicated by × in Fig. 9.6.4 does not have a test pattern if we consider a single path (either 2-5-8 or 2-6-8). Since the fault at × is stuck-at-0, $x_1 = x_2 = 0$ obviously must hold. If we want to sensitize path 2-5-8, then the output of gate 6 must be 0, requiring $x_4 = 1$, and also the output of gate 7 must be 0, requiring that the output of gate 3 be 1 and x_4 be 0. Thus we have a contradiction regarding the value of x_4, and similarly with path 2-6-8 by symmetry of the network. Therefore, if we want to sensitize only a single path, we have no test pattern.

As diagnosis methods for combinational networks, the Boolean difference method, the *D*-algorithm, and others have also been proposed [Susskind 1973]. In particular, the *D*-algorithm suggested by Roth, Bouricius, and Schneider [1967] is a sophisticated version of the path-sensitizing method. They proved that, by considering multiple sensitized paths, the *D*-algorithm can find a test pattern if one exists, unlike the path-sensitizing method. For example, the test pattern $(x_1 x_2, x_3, x_4) = (0\ 0\ 0\ 0)$ for Fig. 9.6.4 can be found by considering two paths, 2-5-8 and 2-6-8, simultaneously (Exercise 9.6.11). In this sense the *D*-algorithm is the first test generation method that

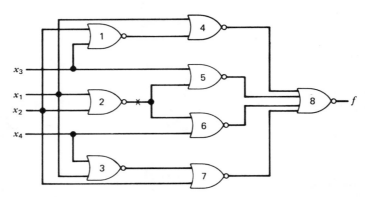

Fig. 9.6.4 The fault stuck-at-0 shown by × does not have a test pattern if only a single sensitized path is considered in the path-sensitizing method.

has been proved to be algorithmic. But the *D*-algorithm appears to require longer test time and more memory space than the path-sensitizing method. The path-sensitizing method and the *D*-algorithm are effective only when networks contain single faults that are stuck at 1 or 0, and may detect or locate only some of multiple faults (not all of them). This shortcoming may not be so bad from the practical viewpoint, since the probability of multiple faults is small (the probability is practically 0 when a computer system is tested or diagnosed sufficiently often). But there are many types of faults other than the stuck-at-0 and stuck-at-1 type. The others include intermittent faults (which occur more often in memories than in logic networks), input shorts, and "wrong" components faults. In spite of the existence of other types of faults, computer programs for diagnosis are usually written under the assumption of a stuck-at-0 or stuck-at-1 single fault. For some multiple faults the path-sensitizing method sometimes cannot precisely locate faults (one report says that the single-fault assumption is valid only in 60 to 70 per cent of all cases) but can detect practically all faults (about 80 to 90 per cent of all cases), including multiple stuck-at type faults.

In spite of its occasional inability to produce a test pattern even when one exists (e.g., Fig. 9.6.4), probably the path-sensitizing method, along with the *D*-algorithm and its extension (sequential networks also can be treated [Putzolu and Roth 1971]), are currently the most widely used for many computers, because essentially all faults (including the non-stuck-at type) can be found, the memory space for test patterns is acceptable (though large), and computer time for diagnosis is reasonably short. The path-sensitizing method, the *D*-algorithm, and its extension are often collectively called the **D-algorithm**. For some networks other methods (e.g., testing the outputs of IC packages according to truth tables; see Exercises 9.6.8) are used differently, depending on the networks. When a faulty network is located, it is partitioned into subnetworks of manageable size (up to about 1000 gates; diagnosis of networks of more gates is too time-consuming with any of the above methods), and then the above methods are applied to each subnetwork.

If the above methods take too much time to locate faulty IC packages (or gates), a frequent practice is to locate faulty pc boards. This is much less time-consuming than locating faulty gates. Then the faulty pc boards are replaced by good ones and are sent to a manufacturing plant, where they are tested with lowered power supply voltages, or suspected gates or IC packages are examined with an oscilloscope.

The diagnosis of sequential networks is much more difficult than that of combinational networks. One approach used in practice is the extended *D*-algorithm developed by Putzolu and Roth in 1971 (after Kubo [1968]), whereby a sequential network is tested as a combinational network by temporarily disconnecting the feedback loops. Since this approach can locate only about 50 to 60 per cent of all faults, other test patterns must be inserted as auxiliary means. Another approach is to activate an extra clock solely provided for diagnosis, whereby all flip-flops in a network are connected as a shift register and all information in the flip-flops is shifted into extra terminals for diagnosis (when the diagnosis clock is deactivated, these flip-flops are disconnected and work as ordinary flip-flops again). This approach is often used in practice but is somewhat expensive (networks with the clock are about 15 to 30 percent more expensive than ones without it), and it is still hard to precisely locate faulty gates.

The preceding discussion is limited to the case where a digital system is implemented with discrete components, SSI, or MSI. When a digital system is implemented with many LSI packages, however, diagnosis based on the functional relationships among networks can be more efficient than diagnosis by the method described above, since we do not need to locate faulty gates inside packages but must find only which package is faulty. In this case we can handle more than a thousand gates. (But when a large network cannot be put into a single LSI package and part of the network goes to another LSI package, along with other networks, the diagnosis will be much harder, because these networks contain many gates and a faulty gate must be located among them.) Small networks for failure monitoring also can be incorporated in LSI chips.

The diagnosis programs, test patterns, and correct network outputs require considerably large memory space and are usually stored in many disks. In some small digital systems they are stored in ROMs.

Diagnosis by software and hardware monitors (including microcomputers) is often supported by microprogramming (so-called microdiagnosis).

Usually intermittent faults are very difficult to eliminate. Hence, when such faults are discovered by hardware monitors, computation is usually retried and continued. The intermittent faults are recorded and later utilized for diagnosis.

Usually, memories are much less reliable than logic networks, so faults in memories are found by error-checking codes such as the Hamming code and parity check. Because memories have a simpler structure than logic networks, diagnosis of memories is much easier [Goldblant 1976, Rickard 1976, Levine and Meyers 1976].

Quick diagnosis of computers is very important for computer manufacturers to satisfy their customers. Therefore the manufacturers are making considerable efforts in regard to logic design, software, and hardware monitors, for diagnosis.

There are numerous publications about diagnosis. Books: Chang, Manning, and Metze [1970], Friedman and Menon [1971], Sellers, Hsiao, and Bearnson [1968b], Breuer and Friedman [1976], Lee [1976]. Introductory surveys:

Susskind [1973], Preiss [1972], Su [1974], Maling and Allen [1963], Motorola [1972a]. Bibliographies: Scola [1972], Bennets and Lewin [1971], Breuer [1972b], McCluskey [1971]. Discussions of practical aspects: Hackl and Shirk [1965], Carter, Montgomery, Preiss, and Reinheimer [1964], Bartow and McGuire [1970], Andrews [1973], Ulrich and Baker [1974], Schneider [1974], Jhu [1975], Kitamura and Tashiro [1974], Gillow [1975], Guffin [1971], Johnson [1971]. Original contributions; Armstrong [1961, 1966], Roth [1966], Roth, Bouricius, and Schneider [1967], Poage [1963], Eichelberger [1964], Sellers, Hsiao, and Bearnson [1968a], Kubo [1968], Putzolu and Roth [1971], Bossen and Hong [1971], Kohavi and Spires [1971], Akers [1973].

Exercises

9.6.1 (R) Suppose that 50 IC packages, each of which has 99.9 percent yield, are mounted on a pc board. Then 50 such pc boards are assembled into a final product. Find the yield of the final product, using Fig. 9.6.1 (a rough estimation is sufficient).

9.6.2 (R) Determine a test pattern(s) to sensitize one of the following paths in Fig. E9.6.2:

A. Path x_1-1-3-5-z_1.
B. Path x_3-2-4-5-z_1.

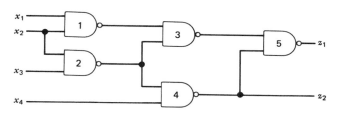

Fig. E9.6.2

9.6.3 (E) We want to find a fault in each gate in one of the following sets in Fig. 9.6.2. What path(s) should we sensitize? Also show a test pattern(s) for each gate.

A. Gates 2, 4, 6.
B. Gates 3, 5, 7.

9.6.4 (E) When the fault at × is stuck-at-1 in Fig. 9.6.4, find a test pattern to sensitize path 2-5-8. (Notice that, if the fault is stuck-at-1, we need not consider two or more paths, unlike the case of stuck-at-0 discussed in the text.)

9.6.5 (M) By adding an extra gate (or gates), modify the network in Fig. 9.6.3 so that the sensitized path from y of gate 2 to the network output can have a test pattern. (Do not change the network output f by this addition.)

Then, if you can, develop an appropriate procedure to find a gate addition which can be applied to similar cases. (If your procedure is valid only when a network is under a certain condition, state that condition.)

9.6.6 (M) Suppose that a two-level network that has AND gates in the first level and one OR gate in the second (output) level is designed on the basis of an irredundant disjunctive form such that each prime implicant in the irredundant disjunctive form is expressed by the output of an AND gate. The network contains a single fault stuck at 0 or 1 at an input or the output of a certain AND gate. (A fault at the output of gate 1 is shown by × in Fig. E9.6.6.)

Discuss a procedure to detect a single fault at an input or the output of an AND gate, by observing the output of the OR gate. Prove why your procedure works. Also discuss whether we can locate an AND gate with a faulty input or output, in each case.

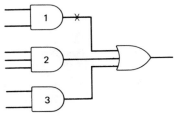

Fig. E9.6.6

9.6.7 (E) Suppose that we have a network in which no gate or connection can be removed without changing the output function of the network (this is called an **irredundant network**). Prove or disprove each of the following two statements:

(i) A single fault at any connection can be detected by the path-sensitizing method.
(ii) It can also be located by the path-sensitizing method.

9.6.8 (E) The simplest approach to find faults in a network is the use of a **fault table**, which shows the relationship between input vectors and detectable faults.

Consider the network in Fig. E9.6.8a and the corresponding fault table in Table E9.6.8. The fourth row $(x_1, x_2, x_3) = (0\ 1\ 1)$, for example, in the fault table can detect the faulty connections B stuck at 0, C stuck at 0, and E stuck at 1 (they are denoted by B_0, C_0, and E_1, respectively), where the detectability is indicated by ×. These faults

Table E9.6.8 Fault Table

x_1	x_2	x_3	A_0	B_0	C_0	D_0	E_0	A_1	B_1	C_1	D_1	E_1
0	0	0				×	×					
0	0	1				×	×	×	×			
0	1	0				×	×	×		×		
0	1	1		×	×							×
1	0	0	×								×	
1	0	1	×								×	
1	1	0	×								×	
1	1	1										

can be detected by observing the network output to be different from that of the faultless network. The procedure is similar with other rows. The last row does not detect any fault.

If we find a set of a minimal number of rows such that every column contains × in these rows, we can detect in the shortest time whether the network is faulty or not. (This is essentially the covering problem mentioned in Procedure 4.5.2. The table reduction techniques discussed there can be used here.)

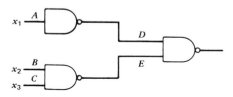

Fig. E9.6.8a *A network.*

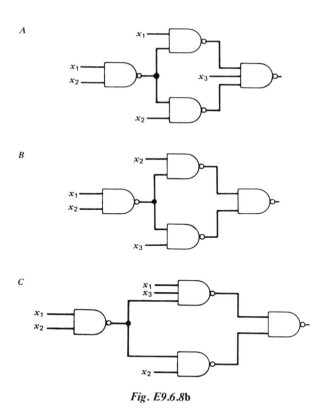

Fig. E9.6.8b

588   *Practical Considerations in Logic Design*

For one of the networks in Fig. E9.6.8b do the following:

(i) Derive a fault table.
(ii) Find a set of a minimum number of rows.

Remark: A set of a minimum number of input vectors obtained from a fault table could be used for logic simulation of a network before hardware implementation. As mentioned at the beginning of Section 9.6, logic simulation serves to test whether a designed network performs an intended task or the logic design is correct (e.g., the network may contain incorrect connections), instead of trying to find faulty gates or connections in a network whose logic is known to be correct.

For large networks the preparation of fault tables is very time-consuming because of the large sizes of the tables. Storage of switching expressions derived from fault tables is preferable to storage of the tables, because of smaller memory space requirement. Note that, unlike the situation in the path-sensitizing method, each row may sensitize more than one path. ■

9.6.9 (E) The **Boolean difference** is a convenient concept for network diagnosis. It is defined as

$$\frac{df}{dx_i} = f(x_1, \ldots, x_i, \ldots, x_n) \oplus f(x_1, \ldots, \bar{x}_i, \ldots, x_n),$$

where $f(x_1, \ldots, x_i, \ldots, x_n)$ denotes the output of a network with inputs $x_1, \ldots, x_i, \ldots, x_n$. In other words, it is the modulo 2 sum of the original function and the value of the function with x_i complemented.

(i) Prove the following calculation rules of the Boolean difference:

$$\frac{df}{dx_i} = \frac{d\bar{f}}{dx_i}, \qquad (1)$$

$$\frac{dfg}{dx_i} = f\frac{dg}{dx_i} \oplus g\frac{df}{dx_i} \oplus \frac{df}{dx_i}\frac{dg}{dx_i}, \qquad (2)$$

$$\frac{d(f \vee g)}{dx_i} = \bar{f}\frac{dg}{dx_i} \oplus \bar{g}\frac{df}{dx_i} \oplus \frac{df}{dx_i}\frac{dg}{dx_i}. \qquad (3)$$

(ii) On the basis of the Boolean difference, we can diagnose a network. When a Boolean difference is identically equal to 0, a fault at input x_i does not cause an error in f. When a Boolean difference is identically equal to 1, however, a fault at input x_i always causes an error in f. When a Boolean difference becomes a function g, a fault at x_i causes an error in f if and only if $g = 1$.

For example, the network shown in Fig. E9.6.9a has the output $f = x_1x_2 \vee x_3$. Let us calculate the Boolean difference, df/dx_1. We can calculate this directly, but if we use the calculation rules derived in (i), we can find it quickly as follows:

$$\frac{df}{dx_1} = \bar{x}_3 \frac{d(x_1 x_2)}{dx_1} \qquad \text{[by (3) above]}$$

$$= \bar{x}_3 x_2 \frac{dx_1}{dx_1} \qquad \text{[by (2) above]}$$

$$= \bar{x}_3 x_2.$$

This means that a fault in x_1 causes the output f to be in error if $\bar{x}_3 x_2 = 1$, that is, $x_3 = 0$ and $x_2 = 1$.

Find out, for one of the networks in Fig. E9.6.9b, under what conditions an error in input x_1 causes the output to be in error.

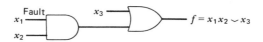

Fig. E9.6.9a

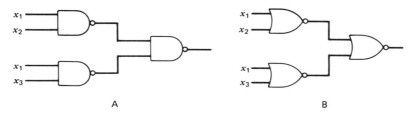

Fig. E9.6.9b

9.6.10 (E) Prove the statements in the first paragraph of Exercise 9.6.9(ii).

9.6.11 (E) Suppose that an AND gate has only two inputs, x and y, as shown in Fig. E9.6.11a. If one of them is fixed to 1 and a value D, which is 0 or 1, is assigned to the other input, the gate output is D. Thus we get Table E9.6.11a for the AND gate.

Table E9.6.11

a. For AND			b. For OR		
x	y	f	x	y	f
1	D	D	0	D	D
D	1	D	D	0	D

c.

1	2	3	4	5	6
D	1	0	D		
			D	0	D
D	1	0	D	0	D

Fig. E9.6.11a   Fig. E9.6.11b

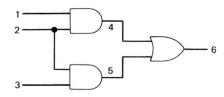

Fig. E9.6.11c

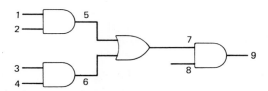

Fig. E9.6.11d

Similarly, we get Table E9.6.11b for the OR gate shown in Fig. E9.6.11b, by setting one of the inputs to 0. These tables are called **primitive D cubes**.

When we have a network as shown in Fig. E9.6.11c, suppose that we want to test path 1-4-6. By choosing and arranging appropriate rows of the above tables, we can get Table E9.6.11c. Thus, if we set $x_2 = 1$, $x_5 = 0$, and, accordingly, $x_3 = 0$, we can detect a fault on path 1-4-6, where x_i denotes the signal value of input i.

(i) Using this approach, prepare a table to detect a fault on path 1-5-7-9 in the network in Fig. E9.6.11d, and show the input values that should be set.

(ii) Show that, if we sensitize paths 2-5-8 and 2-6-8 simultaneously, we can detect the fault stuck-at-0 at $\times$ in Fig. 9.6.4 (though we cannot detect it if we sensitize only one of these paths, as explained in the text).

[The **D-algorithm** works something like the above approach (simplified though), that is, first we derive primitive D-cubes in order to characterize a fault in a gate or a building block. Then we generate all possible paths from the point of failure to all network outputs by cascading primitive D-cubes.]

9.6.12 (R) Intermittent faults in a computer system are hard to find. Discuss how to eliminate them. (Try to stretch your imagination beyond the discussion in the text.)

9.6.13 (E) Design an *S-R* latch that has an extra input t for testing, such that, if $t = 0$, the network works as an ordinary latch, and if $t = 1$, it works as a combinational network, breaking the feedback loops.

9.6.14 (M−) Suppose that there are two separate *S-R* latches. Using them, design a network that has an extra clock t for testing, so that the network works as a shift register which shifts right at each clock pulse $t = 1$, and if $t = 0$, the network becomes two independent *S-R* latches.

References

Abd-Alla, A. M., and A. C. Meltzer, *Principles of Digital Computer Design*, vol. 1, Prentice-Hall, 1976, 318 pp.

Akers, S. B., Jr. (GE), "On maximum inversion with minimum inverters," *IEEE Trans. Electron. Comput.*, February 1968, pp. 134–135.

Akers, S. B., Jr., "Universal test sets for logic networks," *IEEE Trans. Comput.*, September 1973, pp. 835–839.

Allan, R., "The failure tracers," *IEEE Spectrum*, October 1976, pp. 33–39.

Amarel, S., G. Cooke, and R. O. Winder (RCA), "Majority gate networks," *IEEE Trans. Electron. Comput.*, February 1964, pp. 4–13.

Anderson, J. L. (Fairchild), "Multiplexers double as logic circuits," *Electronics*, Oct. 27, 1969, pp. 100–105.

Anderson, R. E., "Digital testing glossary reflects industry usage," *Electronics*, May 27, 1976, pp. 119, 121.

Andrews, M., "Simplify minicomputer maintenance with microdiagnostics. The programs provide fault-isolation to single functional elements," *Electron Des.*, Aug. 2, 1973, pp. 86–87.

Appels, J. Th., and B. H. Geels, *Handbook of Relay Switching Technique*, Springer, 1966, 321 pp. (Reviewed in *Comput. Rev.*, June 1969, p. 293.)

Arden, B. W., B. A. Galler, and R. M. Graham, "An algorithm for translating Boolean expressions," *J. ACM*, April 1962, pp. 222–239.

Arden, B. W., and R. M. Graham, "On GAT and the construction of translators," *Commun. ACM*, July 1959, pp. 24–26.

Armstrong, D. B. (Bell Laboratories), "A general method of applying error correction to synchronous digital systems," *Bell Syst. Tech. Jr.*, March 1961, pp. 577–593.

Armstrong, D. B., "On finding a nearly minimal set of fault detection tests for combinational logic nets," *IEEE Trans. Electron. Comput.*, February 1966, pp. 66–73.

Armstrong, D. B., "A deductive method for simulating faults in logic circuits," *IEEE Trans. Comput.*, May 1972, pp. 464–471.

Armstrong, D. B., A. D. Friedman, and P. R. Menon, "Realization of asynchronous sequential circuits without inserted delay elements," *IEEE Trans. Comput.*, February 1968, pp. 129–134.

Barna, A., and D. J. Porat, *Integrated Circuits in Digital Electronics*, John Wiley, 1973, 482 pp.

Bartee, T. C., "Computer designs of multiple-output logical networks," *IRE Trans. Electron. Comput.*, March 1961, pp. 21–30.

Bartee, T. C., I. L. Lebow, and I. S. Reed, *Theory and Design of Digital Machines*, McGraw-Hill, 1962.

References

Bartow, N., and R. McGuire, "System/360 Model 85 Microdiagnostics." *AFIPS Spring Joint Computer Conference Proceedings*, 1970, pp. 191–197.

Baugh, C. R., C. S. Chandersekaran, R. S. Swee, and S. Muroga, "Optimal networks of NOR-OR gates for functions of three variables," *IEEE Trans. Comput.*, February 1972, pp. 153–160.

Baugh, C. R., T. Ibaraki, T. K. Liu, and S. Muroga, "Optimum Network Design Using NOR and NOR-AND Gates by Integer Programming," Report 293, Dept. of Computer Science, Univ. of Illinois, January 1969.

Baugh, C. R., and B. A. Wooley," A two's complement parallel array multiplication algorithm," *IEEE Trans. Comput.*, December 1973, pp. 1045–1047.

Baugh, C. R. and B. A. Wooley, U.S. Patent 3,978,329, Aug. 31, 1976.

Bell, C. G., J. Carson, and A. Newell, *Designing Computers and Digital Systems* (*Using PDP-16RTMs*), Digital Press, 1972.

Bell, C. G., and A. Newell, *Computer Structure: Readings and Examples*, McGraw-Hill, 1971, 668 pp.

Bell, C. G., A. Kotok, T. N. Hastings, and R. Hill, "The evolution of the DEC system 10," *Commun. ACM*, January 1978, pp. 44–63.

Bell, E. T., *Men of Mathematics*, Simon and Schuster, 1937, 592 pp.

Bell, E. T., *The Development of Mathematics*, 2nd ed., McGraw-Hill, 1945, 637 pp.

Belzer, J., "Can present methods for library and information retrieval service survive?" (panel), *Proceedings of ACM Annual Conference*, 1971, pp. 564–577.

Bennets, R. G., and D. W. Lewin, "Fault diagnosis of digital systems—a review," *Computer*, July/August 1971, pp. 12–21.

Berger, H. H., and S. K. Wiedman, "Merged-transistor logic—a low cost bipolar logic concept," *IEEE J. Solid-State Circuits*, October 1972, pp. 340–346.

Biancomano, V., "Logic-simulator programs set pace of computer-aided design," *Electronics*, Oct. 13, 1977, pp. 98–101.

Birkhoff, G., *Lattice Theory*, rev. ed., American Mathematical Society, 1948, 283 pp.

Blaauw, G. A., *Digital System Implementation*, Prentice-Hall, 1976, 384 pp.

Blair, K. A., "Penetrate the mystique of MOS specs," *Electron Des.*, Apr. 13, 1972, pp. 64–69.

Blakeslee, T. R., *Digital Design with Standard MSI and LSI*, John Wiley, 1975, 357 pp.

Booth, T. L., *Digital Networks and Computer Systems*, John Wiley, 1971, 451 pp.

Borgerson, B. R., M. L. Hanson, and P. A. Hartley, "The evolution of the Sperry Univac 1100 Series: A history, analysis, and projection," *Commun. ACM*, January 1978, pp. 25–43.

Bossen, D. C., and S. J. Hong, "Cause-effect analysis for multiple fault detection in combinational networks," *IEEE Trans. Comput.*, November 1971, pp. 1252–1257.

Bouchet, A., "An algebraic method for minimizing the number of states in an incomplete sequential machine," *IEEE Trans. Electron. Comput.*, August 1968, pp. 795–799.

Brandhorst, W. T., "Simulation of Boolean logic constraints through the use of term weights," *American Documentation*, vol. 17, July 1966, pp. 145–146.

Bredeson, J. G., and P. T. Hulina, "Generation of a clock pulse for asynchronous sequential machines to eliminate critical races," *IEEE Trans. Comput.*, February 1971a, pp. 225–226.

Bredeson, J. G., and P. T. Hulina, "Generation of prime implicants by direct multiplications," *IEEE Trans. Comput.*, April 1971b, pp. 475–476.

Breuer, M. A., "Techniques for the simulation of computer logic," *Commun. ACM*, July 1964, pp. 443–446.

Breuer, M. A. (Ed.), *Design Automation of Digital Systems*, vol. 1: Theory and Techniques, Prentice-Hall, 1972a, 420 pp.

Breuer, M. A., "Recent development in the automated design and analysis of digital systems," *Proc. IEEE*, January 1972b, pp. 12–27.

Breuer, M. A. (Ed.), *Digital System Design Automation: Languages, Simulation and Data Base*, Computer Science Press, 1975, 417 pp.

Breuer, M. A., and A. D. Friedman, *Diagnosis and Reliable Design of Digital Systems*, Computer Science Press, 1976, 308 pp.

Brooks, F. P., Jr., *The Mythical Man-Month*, Addison-Wesley, 1975, 195 pp.

Brown, F. M., "The origin of the method of iterated consensus," *IEEE Trans. Comput.*, August 1968, p. 802.

Brzozowski, J. A., and M. Yoeli, *Digital Networks*, Prentice-Hall, 1976, 398 pp.

Burgoon, R., "Improve your Karnaugh mapping skills," *Electron. Des.*, Dec. 21, 1972, pp. 54–56.

Caldwell, S. H., *Switching Circuits and Logical Design*, John Wiley, 1958, 686 pp.

Capell, A., D. Knoblock, L. Mather, and L. Lopp (Hewlett–Packard), "Process refinements bring C-MOS on sapphire into commercial use," *Electronics*, May 26, 1977, pp. 99–105.

Caplener, H. D., and J. A. Janku, "Top-down approach to LSI system design," *Comput. Des.*, August 1974, pp. 143–148.

Carlow, E. F., "Design of shift-counters," *Electron. Des.*, June 22, 1972, pp. 70–74.

Carr, W. N., and J. P. Mize, *MOS/LSI Design and Application*, McGraw-Hill, 1972, 331 pp.

Carter, W. C., H. C. Montgomery, R. J. Preiss, and H. J. Reinheimer, "Design of serviceability features for IBM System/360," *IBM J. Res. Dev.*, April 1964, pp. 115–126.

Case, P. W., H. H. Graff, L. E. Griffith, A. R. Leclereq, W. B. Murley, and T. M. Spence, "Solid logic design automation," *IBM J. Res. Dev.*, April 1964, pp. 127–140.

Case, R. P., and A. Padegs, "Architecture of the IBM System/370," *Commun. ACM*, January 1978, pp. 72–96.

Catt, I., "Time loss through gating of asynchronous logic signal pulses," *IEEE Trans. Comput.*, February 1966, pp. 108–111.

Cavlan, N., and R. Cline, "FPLA applications—exploring design problems and solutions," *EDN* Apr. 5, 1976, pp. 63–69.

Chaney, T. J., and C. E. Molnar, "Anomalous behavior of synchronizer and arbiter circuits," *IEEE Trans. Comput.*, April 1973, pp. 421–422.

Chaney, T. J., S. M. Ornstein, and W. M. Littlefield, "Beware the synchronizer," *Compcon 72*, September 1972.

Chang, D. M. Y., and T. H. Mott, Jr., "Computing irredundant normal forms abbreviated presence functions," *IEEE Trans. Electron. Comput.*, June 1965, pp. 335–342.

References

Chang, H. Y., and S. G. Chappell (Bell Laboratories), "Deductive techniques for simulating logic circuits," *Computer*, March 1975, pp. 52–59.

Chang, H. Y., E. Manning, and G. Metze, *Fault Diagnosis of Digital Systems*, John Wiley, 1970, 159 pp.

Chrones, C. (DCA Reliability Laboratory), "Calculating the cost of testing LSI chips," *Electronics*, Jan. 5, 1978, pp. 171–173.

Chu, Y., *Computer Organization and Microprogramming*, Prentice-Hall, 1972, 533 pp.

Clare, C. R., *Designing Logic Systems Using State Machines*, McGraw-Hill, 1973, pp. 41–42.

Cobbold, R. S. C., *Theory and Application of Field-Effect Transistors*, John Wiley, 1970, 534 pp.

Cobham, A., R. Fridshal, and J. H. North, "An application of linear programming to the minimization of Boolean functions," *Proceedings of AIEE 2nd Annual Symposium on Switching Circuit Theory*, October 1961, pp. 3–9.

Cobham, A., R. Fridshal, and J. H. North, "A Statistical Study of the Minimization of Boolean Functions Using Integer Programming," IBM Research Report RC-756, 1962.

Cobham, A., and J. H. North, "Extensions of the Integer Programming Approach to the Minimization of Boolean Functions," IBM Research Report RC-915, Apr. 2, 1963.

Conklin, P. F., and D. P. Rodgers (Digital Equipment Corp.), "Advanced minicomputer designed by team evaluation of hardware/software tradeoffs," *Comput. Des.*, April 1978, pp. 129–137.

Cox, D. T., W. T. Devine, and G. J. Kelly (IBM), "High density logic array," U.S. Patent 3,987,287, Oct. 19, 1976.

Criscimagna, T. N. (IBM), "Start logic design with flow diagrams," *Electron. Des.*, Dec. 9, 1971, pp. 56–59.

Culliney, J. N., M. H. Young, T. Nakagawa, and S. Muroga, "Results of the Synthesis of Optimal networks of AND and OR gates for Four-Variable Switching Functions by a Branch-and-Bound Computer Program," UIUCDCS-R-76-789, March 1976. (Also to appear in *IEEE Trans. Comput.* in January 1979.)

Dao, T. T., "Threshold I^2L and its applications to binary symmetric functions and multivalued logic," *IEEE J. Solid-State Circuits*, October 1977, pp. 463–472.

Dao, T. T., E. J. McCluskey, and L. K. Russell, "Multivalued integrated injection logic," *IEEE Trans. Comput.*, December 1977, pp. 1233–1241.

Dao, T. T., L. K. Russell, D. R. Preedy, and E. J. McCluskey, "Multilevel I^2L with threshold gates," *IEEE International Solid-State Circuits Conference*, 1977, pp. 110–111.

Das, S. R., "A new algorithm for generating prime implicants," *IEEE Trans. Comput.*, December 1971, pp. 1614–1615.

Date, C. J., *An Introduction to Data Base Systems*, 2nd ed., Addison-Wesley, 1977, 536 pp.

Davidow, W., "How microprocessors boost profits," *Electronics*, July 11, 1974, pp. 105–108.

Davidson, E. S., "An algorithm for NAND decomposition under network configuration," *IEEE Trans. Comput.*, December 1969, pp. 1098–1109.

Davies, P. M., "Reading in microprogramming," *IBM Syst. J.*, No. 1, 1972, pp. 16–40.

Delhom, L., *Design and Application of Transistor Switching Circuits*, McGraw-Hill, 1968, 278 pp.

Delobel, C., and R. G. Casey, "Decomposition of a data base and the theory of Boolean switching functions," *IBM J. Res. Dev.*, September 1973, pp. 374–386.

Derman, S., "PLAs or microprocessors? At times they compete and at other times they cooperate," *Electron. Des.*, Sept. 1, 1976, pp. 24, 26, 28, 30.

De Vries, R. C., "Comment on Lawler's multi-level Boolean minimization," *Commun. ACM*, April 1970, pp. 265–266.

Dietmeyer, D. L., *Logical Design of Digital Systems*, Allyn and Bacon, 1971.

Dinsmore, S. H., "Clean up your logic schematics," *Electron. Des.*, June 21, 1974, pp. 80–84.

Dlugatch, I., "Use statistics in your logic design," *Electron. Des.*, Dec. 23, 1971, pp. 60–62.

Donath, W. E., "Equivalence of memory to 'random logic,'" *IBM J. Res. Dev.*, September 1974, pp. 401–407.

Dunham, B., and R. Fridshal, "The problem of simplifying logical expressions," *J. Symb. Logic*, March 1959, pp. 17–19.

Dwinger, P., *Introduction to Boolean Algebra*, Physica-Verlag, 1971, 71 pp.

Eggert, J. L. (Digital Equipment Corp.), "A flow chart technique for computer aided design of small computers," *Comput. Des.*, November 1973, pp. 73–81.

Eichelberger, E. B., "Hazard detection in combinational and sequential switching circuits," *IEEE International Conference Record on Switching Circuits, Theory, and Logic Design*, 1964, pp. 111–121.

Eimbinder, J. (Ed.), *Semiconductor Memories*, John Wiley, Chapter 18, 1971, 214 pp.

Electronics, "Emitter-follower logic pushes bipolar designs into LSI realm," Jan. 24, 1974, pp. 29–30.

Electronics, "Architect of the 11/60," Oct. 27, 1977, p. 110.

Electronics, "Gate arrays have marketers raring to go," Apr. 27, 1978, pp. 83–84.

Electronics, "Signetics readying 2,000-gate arrays," May 25, 1978, pp. 41–42.

Electronic Design, "The economics of software vs. hardware," Jan. 18, 1977, p. 56.

Elineau, G., and W. Wiesbeck, "A new J-K flip-flop for synchronizers," *IEEE Trans. Comput.* December 1977, pp. 1277–1279.

Engineering Staff of AMI, *MOS Integrated Circuits*, Van Nostrand, 1972, 474 pp.

Ercoli, P., and L. Mercurio, "Threshold logic with one or more than one threshold," *Proceedings of IFIP Congress*, 1962, North-Holland, pp. 341-345.

Ettinger, M. A., and G. W. Jacob, "An algorithm for sequential circuit design," *Comput. Des.*, May 1968, pp. 46–53.

Even, S., I. Kohavi, and A. Paz, "On minimal modulo 2 sums of products for switching functions," *IEEE Trans. Comput.*, October 1967, pp. 671–674.

Fagin, F., "Functional dependencies in a relational database and propositional logic," *IBM J. Res. Dev.*, November 1977, pp. 534–544.

Falk, H., "Hard-soft tradeoff," *IEEE Spectrum*, February 1974, pp. 34–39.

Falk, H., "Computers," *IEEE Spectrum*, April 1975, p. 47.

Fleisher, H., and L. I. Maissel, "Introduction to array logic," *IBM J. Res. Dev.*, March 1975, pp. 98–109.

Flores, I., *The Logic of Computer Arithmetic*, Prentice-Hall, 1963, 493 pp.

Forbes, B., "A system designer's lament: give us better digital ICs," *Electronics*, Dec. 6, 1971, pp. 70–75.

Fox, P. E., and W. J. Nestork, "Design of logic circuit technology for IBM System/370 Models 145 and 155," *IBM J. Res. Dev.*, September 1971, pp. 384–390.

Friedman, A. D., and P. R. Menon, *Fault Detection in Digital Circuits*, Prentice-Hall, 1971, 220 pp.

Friedman, T. D., and S. C. Yang, "Quality designs from automatic logic generator—ALERT," *Proceedings of Design Automation Workshop* 1970, pp. 71–82.

Fujioka, A., K. Taniguchi, A. Hayasaka, K. Matsui, O. Yumoto, and K. Sakai (Hitachi), "500 MHz clock-rate logic LSI," *International Solid-State Circuits Conference*, 1973, pp. 164–165. [Also, *Nikkei Electron.* (in Japanese), 5–7, 1973, pp. 84–102.]

Garrett, L. S., "Integrated circuit digital logic families, III," *IEEE Spectrum*, December 1970, pp. 30–42.

Garrett, L. S., "C-MOS may help majority logic win designers' vote," *Electronics*, July 19, 1973, pp. 107–112.

Ghazala, M. J., "Irredundant disjunctive and conjunctive forms of a Boolean function," *IBM J. Res. Dev.*, April 1957, pp. 171–176.

Gilchrist, B. J., B. J. H. Promerene, and S. Y. Wong, "Fast carry logic for digital computers," *IRE Trans. Electron. Comput.*, December 1955, pp. 133–136.

Gildersleeve, T. R., *Decision Tables and Their Practical Application in Data Processing*, Prentice-Hall, 1970.

Gillow, G. (NCR), "Simplifying processor maintenance with a carefully designed maintenance panel," *Comput. Des.*, July 1975, pp. 95–99.

Gimpel, J. F., "The minimization of TANT networks," *IEEE Trans. Electron. Comput.*, February 1967, pp. 18–38.

Ginsburg, S., *An Introduction to Mathematical Machine Theory*, Addison-Wesley, 1962, 148 pp.

Gladstone, B., "Design with microprocessors instead of wired logic asks more of designers," *Electronics*, Oct. 11, 1973, pp. 91–104.

Goetz, M. A., "What's good for IBM," *Datamation*, April 1975, p. 103.

Gold, J., "To use custom LSI or μPs? That is often the question . . . ," *Electron. Des.*, July 19, 1976, pp. 26–32.

Goldblant, R. C. (Univac), "How computers can test their own memories," *Comput. Des.*, July 1976, pp. 69–73.

Golomb, S. W., *Shift Register Sequences*, Holden-Day, 1967, 224 pp.

Goto, E., and H. Takahashi, "Some theorems in threshold logic for enumerating Boolean functions," *Proceedings of IFIP Congress*, 1962, North-Holland, pp. 747–752.

Grason, J., C. G. Bell, and J. Eggert, "The commercialization of register transfer modules," *Computer*, October 1973, pp. 23–27.

Grasselli, A., and F. Luccio, "A method for minimizing the number of internal states in incompletely specified sequential networks," *IEEE Trans. Electron. Comput.*, June 1965, pp. 350–359.

Grasselli, A., and F. Luccio, "A method for the combined row-column reduction of flow tables," *IEEE Conference Record, Seventh Annual Symposium on Switching Theory and Automata Theory*, 1966, pp. 136–147.

Grasselli, A., and F. Luccio, "Some covering problems in switching theory," in *Network and Switching Theory* (A NATO Advanced Study Institute), edited by G. Biorci, Academic Press, 1968, pp. 536–557.

Greer, D. L., "An associative logic matrix," *IEEE J. Solid-State Circuits*, October 1976, pp. 679–691.

Grossman, S. E., "Automated testing pays off for electronic system makers," *Electronics*, Sept. 19, 1974, pp. 95–109.

Gschwind, H., and E. J. McCluskey, *Design of Digital Computers*, 2nd ed., Springer-Verlag, 1975, 548 pp.

Guffin, R. M., "Microdiagnostics for the Standard Computer MLP-900 Processor," *IEEE Trans. Comput.*, July 1971, pp. 803–808.

Hackl, F. J., and R. W. Shirk, "An integrated approach to automated computer maintenance," *IEEE Conference Record on Switching Circuit Theory and Logical Design*, 1965, pp. 289–302.

Halmos, P. R., *Lectures on Boolean Algebra*, Van Nostrand, 1963, 147 pp.

Hamilton, D. J., and W. G. Howard, *Basic Integrated Circuit Engineering*, McGraw-Hill, 1975, 587 pp.

Hampel, D., and R. O. Winder, "Threshold logic," *IEEE Spectrum*, May 1971, pp. 32–39.

Hardie, F. H., and R. J. Suhocki, "Design and use of fault simulation for Saturn computer design," *IEEE Trans. Comput.*, August 1967, pp. 412–429.

Hardy, R. M. (General Electric), "A flow chart hardware correlation," *Digital Des.*, June 1973, pp. 46–47.

Harper, C. A. (Ed.), *Handbook of Components for Electronics*, McGraw-Hill, 1977.

Harrison, M. A., *Introduction to Switching and Automata Theory*, McGraw-Hill, 1965, 499 pp.

Harrison, M. A., "Counting theorems and their application to classifications of switching functions," Chapter 4 of *Recent Developments in Switching Theory*, edited by A. Mukhopadhyay, Academic Press, 1971, pp. 85–120.

Hart, H., and A. Slob, "Integrated injection logic—a new approach to LSI," *IEEE J. Solid-State Circuits*, October 1972, pp. 346–351.

Hazeltine, B., "Encoding of asynchronous sequential circuits," *IEEE Trans. Electron. Comput.*, October 1965, pp. 727–729.

Hellerman, L., "A catalog of three-variable OR-Invert and AND-Invert logical circuits," *IEEE Trans. Electron. Comput.*, June 1963, pp. 198–223.

Hemel, A. (Monolithic Memories), "The PLA: a 'different kind' of ROM," *Electron. Des.*, Jan. 5, 1976, pp. 78–84.

Hendrickson, H. C., "Fast high-accuracy binary parallel addition," *IRE Trans. Electron. Comput.*, December 1960, pp. 469–479.

Herskowitz, G. J. (Ed.), *Computer-Aided Integrated Circuit Design*, McGraw-Hill, 1968, 432 pp.

Herskowitz, G. J., and R. B. Schilling (Eds.), *Semiconductor Device Modeling for Computer-Aided Design*, McGraw-Hill, 1972, 360 pp.

References

Higonnet, R. A., and R. A. Grea, *Logical Design of Electrical Circuits*, McGraw-Hill, 1958, 220 pp.

Hilbert, D., and W. Ackermann, *Principles of Mathematical Logic*, Chelsea Publishing Co., 1950, 172 pp.

Hilburn, J. L., and P. M. Julich, *Microcomputers/Microprocessors: Hardware, Software and Applications*, Prentice-Hall, 1976, 368 pp.

Hill, F. J., and G. R. Peterson, *Digital System: Hardware Organization and Design*, John Wiley, 1973, 481 pp.

Hill, F. J., and G. R. Peterson, *Introduction to Switching Theory and Logical Design*, 2nd ed., John Wiley, 1974, 590 pp.

Hittinger, W. C., "Metal-oxide-semiconductor technology," *Sci. Amer.*, August 1973, pp. 48–57.

Hlavička, J., "Essential hazard correction without the use of delay elements," *IEEE Trans. Comput.*, March 1970, pp. 232–238.

Hnatek, E. R., *A User's Handbook of Integrated Circuits*, John Wiley, 1973, 449 pp.

Hnatek, E. R., "High reliability semiconductors: paying more doesn't always pay off," *Electronics*, Feb. 3, 1977, pp. 101–105.

Hnatek, E. R., *A User's Handbook of Semiconductor Memories*, John Wiley, 1977b, 635 pp.

Ho, I. T., and T. C. Chen, "Multiple addition by residue threshold functions," *Compcon 72*, pp. 283–286.

Hohulin, K., and S. Muroga, "Alternative Methods for Solving the cc-Table in Gimpel's Algorithm for Synthesizing Optimal Three Level NAND networks," Report 720, Dept. of Computer Science, Univ. of Illinois, 1975.

Holdt, T., and R. Yu, "V-MOS configuration packs 64 kilobits into 175-mil's chip," *Electronics*, Mar. 30, 1978, pp. 99–104.

Hong, S. J., R. G. Cain, and D. L. Ostapko, "MINI: a heuristic approach for logic minimization," *IBM J. Res. Dev.*, September 1974, pp. 443–458.

Hubbs, J. C., "Why dc-only testing is out for IC's," *EDN*, July 5, 1974, pp. 55–57.

Huffman, D. A., "Synthesis of sequential switching circuits," *J. Franklin Inst.*, vol. 25, March 1954, pp. 161–190, and April 1954, pp. 275–303. (Also in the book *Sequential Machines—Selected Papers*, edited by E. F. Moore, Addison-Wesley, 1964, pp. 3–62.)

Huffman, D. A., "A Study of the Memory Requirements of Sequential Switching Circuits," Technical Report 293, Research Laboratory of Electronics, MIT, April 1955.

Huffman, D. A., "Combinational circuits with feedback," in *Recent Advances in Switching Theory*, edited by A. Mukhopadhyay, Academic Press, 1971, pp. 28–55.

Hulme, B. L., and R. B. Worrell, "A prime implicant algorithm with factoring," *IEEE Trans. Comput.*, November 1975, pp. 1129–1131.

Humby, E., *Programs from Decision Tables*, American Elsevier, 1973, 91 pp.

Huntington, E. V., "Sets of independent postulates for the algebra of logic," *Trans. Amer. Math. Soc.*, July 1904, pp. 288–309.

Huntington, E. V., "New sets of independent postulates for the algebra of logic, with special reference to Whitehead and Russell's *Principia Mathematica*," *Trans. Amer. Math. Soc.*, January 1933, pp. 274–304. (Correction, pp. 557–558, 971.)

Huskey, H. D., and W. H. Wattenburg, "Compiling techniques for Boolean expressions and conditional statements in ALGOL 60," *Comm. ACM*, January 1961, pp. 70–75.

Husson, S. S., *Microprogramming: Principles and Practice*, Prentice-Hall, 1970, 614 pp.

Ibuki, K., K. Naemura, and S. Nozaki, "General theory of complete sets of logical functions," *Electron. Commun. Jap.* (IEEE translation), vol. 46, No. 7, July 1963, pp. 55–56.

IC Applications Staff of TI, *Designing with TTL Integrated Circuits*, McGraw-Hill, 1971, 322 pp.

IEEE Trans. Comput., Special Issue on Multiple-Valued Logic, December 1977.

Iker, H. P., "Solution of Boolean equations through use of term weights to the base two," *American Documentation*, January 1967, p. 47.

Ikran, M., and D. A. Roy, "A simple technique to improve the pi algorithm for prime implicant determination," *IEEE Trans. Comput.*, November 1976, pp. 1184–1187.

Inabe, Y., and K. Kataoka, "An integrated J-K flip-flop circuit," *IEEE J. Solid-State Circuits*, August 1977, pp. 403–406.

Jackson, P. C., "What the user should know about computer-controlled testing," *Comput. Des.*, October 1974, pp. 97–102.

Jaeger, R. (Signetics), "Microprogramming: A general design tool," *Comput. Des.*, August 1974, pp. 150, 152, 155–157.

Jesse, J. E., "A more efficient use of Karnaugh maps," *Comput. Des.*, February 1972, pp. 80, 82.

Jhu, J. H. (Plessy), "Design for fault isolation: Increasingly the man who creates the circuit must ensure that it can be easily tested," *Electron. Des.*, Nov. 8, 1975, pp. 86–90.

Johnson, A. M., "The microdiagnostics for the IBM System/360 model 30," *IEEE Trans. Comput.*, July 1971, pp. 798–803.

Johnson, D. W. (Control Data Corp.), "Go from flow chart to hardware," *Electron. Des.*, Sept. 1, 1976, pp. 90–95.

Kantorovich, L. V., "On a mathematical symbolism convenient for performing machine calculators," *Trans. USSR Acad. Sci.*, vol. 113, No. 4, 1957, pp. 738–741 (in Russian).

Karnaugh, M., "The map method for synthesis of combinational logic circuits," *Commun. Electron.*, November 1953, pp. 593–599.

Kautz, W. H., "The necessity of closed circuit loops in minimal combinational circuits," *IEEE Trans. Comput.*, February 1970, pp. 162–164.

Kella, J., "State minimization of incompletely specified sequential machines," *IEEE Trans. Comput.*, April 1970, pp. 342–348.

Kitamura, T., and S. Tashiro (Nippon Electric Corp.), "Techniques and methods for fault diagnosis," *J. Inst. Elec. Commun. Eng. Jap.*, October 1974, pp. 1160–1167 (in Japanese).

Knowles, R., *Automatic Testing: Systems and Applications*, McGraw-Hill, 1976, 254 pp.

Kobylar, A. W., R. L. Lindsay, and S. G. Pitroda, "ROMs cut cost, response time of m-out-of-n detectors," *Electronics*, Feb. 15, 1973, pp. 112–114.

Kodandapani, K. L., and R. V. Setlur, "A note on minimum Reed–Muller canonical forms of switching functions," *IEEE Trans. Comput.*, March 1977, pp. 310–313.

Kohavi, A., and D. A. Spires, "Designing sets of fault-detection tests for combinational logic circuits," *IEEE Trans. Comput.*, December 1971, pp. 1463–1469.

Kohavi, Z., *Switching and Finite Automata Theory*, McGraw-Hill, 1970, 592 pp.

Kubo, H. (Nippon Electric Corp.), "A procedure for generating test sequences to detect sequential circuits failure," *NEC Res. Dev.*, No. 12, October 1968, pp. 69–78.

Kuntzman, J., *Algèbre de Boole*, Dunod, Paris, 1965, 319 pp.

Lai, H. C., Ph.D. Thesis, Dept. of Computer Science, Univ. of Illinois, 1976.

Lai, H. C., and S. Muroga, "Minimal parallel binary adders with NOR (NAND) gates," to appear in *IEEE Trans. Comput.* in 1979.

Lai, H. C., T. Nakagawa, and S. Muroga, "Redundancy check technique for designing optimal networks by branch-and-bound method," *Int. J. Comput. Inf. Sci.*, September 1974, pp. 251–271.

Lake, D. W., "Logic simulation in digital systems," *Comput. Des.*, May 1970, pp. 77–83.

Langdon, G. G., Jr., "A survey of counter design technique," *Comput. Des.*, October 1971, pp. 83–93.

Langdon, G. G., Jr., *Logic Design—A Review of Theory and Practice*, Academic Press, 1974, 179 pp.

Langer, R., and T. Dugan, "Say it in a high-level language with 64K read-only memories," *Electronics*, Apr. 13, 1978, pp. 119–124. ("Readers' comments," June 22, 1978, p. 6.)

Lawler, E. L., "An approach to multi-level Boolean minimization," *J. ACM*, July 1964, pp. 283–295.

Lazăr, I. S., and D. Almer, "Algoritm si program pentru determinarea claselor de echi valentă ale functiilor logice," *Autom. Electron.*, vol. 16, No. 6, 1972, pp 251–257.

Lazăr, I. S., D. Almer, and C. Donciu, "Algoritm si catalog pentru scheme logice optime ale functiilor logice de trei variable binare," *Lucrările ICPE*, No. 27, 1972, pp. 103–112.

Lee, C. Y., "Representation of switching functions by binary decision programs," *Bell Syst. Tech. J.*, 1959, pp. 985–999.

Lee, S. C., *Digital Circuits and Logic Design*, Prentice-Hall, 1976, 594 pp.

Lehman, M., and N. Burla, "Skip techniques for high-speed carry propagation in binary arithmetic units," *IRE Trans. Electron. Comput.*, December 1961, pp. 69–698.

Lemke, C. E., H. M. Salkin, and K. Spielberg, "Set covering by single branch enumeration with linear programming subproblems," *Oper. Res.* July–August 1971, pp. 998–1022.

Lenk, J. D., *Handbook of Logic Circuits*, Reston Publishing Co., 1972, 307 pp.

Levine, L. (Xerox), and W. Meyers, "Semiconductor memory reliability with error detecting and correcting codes," *Computer*, October 1976, pp. 43–50.

Levine, R. I., "Logic minimization beyond the Karnaugh map," *Comput. Des.*, March 1967, pp. 40–43.

Lewin, D., *Computer-Aided Design of Digital Systems*, Crane, Russak, 1977, 313 pp.

Lin, W. C. *Microprocessors: Fundamentals and Applications*, IEEE Press, 1977, 335 pp.

Liu, C. L., *Elements of Discrete Mathematics*, McGraw-Hill, 1977, 294 pp.

Liu, C. N., "State variable assignment method for asynchronous sequential switching circuits," *J. ACM*, April 1963, pp. 209–216.

Liu, T. K., K. Hohulin, L. E. Shiau, and S. Muroga, "Optimal one-bit full adders with different types of gates," *IEEE Trans. Comput.*, January 1974, pp. 63–70.

Logue, J. C., N. F. Brickman, F. Howley, J. W. Jones, and W. W. Wu, "Hardware implementation of a small system in programmable logic arrays," *IBM J. Res. Dev.*, March 1975, pp. 110–119.

London, K. R., *Decision Tables*, Auerbach Publishers, 1972, 205 pp.

Loranger, J. A., Jr., "The case for component burn-in. The gain is well worth the price," *Electronics*, Jan. 23, 1975, pp. 73–78.

Loui, J. S., "A one-map approach to hybrid counter design," *Comput. Des.*, December 1971, pp. 58–61.

Lucas, P., "An accumulator chip," *IEEE Trans. Comput.*, February 1969, pp. 105–114.

Luccio, F., "Reduction of the number of columns in flow table minimization," *IEEE Trans. Electron. Comput.*, October 1966, pp. 803–805.

Luecke, G., J. P. Mize, and W. N. Carr, *Semiconductor Memory Design and Application*, McGraw-Hill, 1973, 320 pp.

Lyman, J., "Handy breadboard systems speed the development of prototypes," *Electronics*, Feb. 19, 1976, pp. 97–104.

Lyman, J., "Growing pin count is forcing LSI package changes," *Electronics*, Mar. 17, 1977a, pp. 81–91.

Lyman, J., "Chip carriers are making inroads," *Electronics*, Nov. 24, 1977b, pp. 86, 88, 89.

Lyman, J., "New methods and materials stir up printed wiring," *Electronics*, Apr. 27, 1978, pp. 114–122.

Mack, R. P., "Planning for the IBM System/360," in *Systems Planning and Design*, edited by R. DeNeufville and D. Masks, Prentice-Hall, 1974, pp. 370–380.

MacSorley, O. L., "High-speed arithmetic in binary computers," *Proc. IRE*, January 1961, pp. 67–91.

Maggiore, J. "PLA—a Universal logic element," *Electron. Prod. Mag.*, Apr. 15, 1974, pp. 67–75.

Maley, G. A., *Manual of Logic Circuits*, Prentice-Hall, 1970, 256 pp.

Maley, G. A., and J. Earle, *The Logic Design of Transistor Digital Computers*, Prentice-Hall, 1963, 322 pp.

Maling, K., and E. L. Allen, "A computer organization and programming system for automated maintenance," *IEEE Trans. Electron. Comput.*, December 1963, pp. 887–895.

Mano, M. M., *Computer Logic Design*, Prentice-Hall, 1972, 450 pp.

Mano, M. M., *Computer System Architecture*, Prentice-Hall, 1976, 478 pp.

Marcovitz, A. B., and C. M. Shub, "An improved algorithm for the simplification of switching functions using identifiers on a Karnaugh map," *IEEE Trans. Comput.*, April 1969, pp. 376–378.

Marcus, M. P., "Derivation of maximal compatibles using Boolean algebra," *IBM J. Res. Dev.*, November 1964, pp. 537–538.

Marinković, S. B., and Ž. Tošić, "Algorithm for minimal polarized polynomial form determination," *IEEE Trans. Comput.*, December 1974, pp. 1313–1315.

Mason, J. F., "How to be a top designer and remain a designer, despite corporate lures," *Electron. Des.*, Apr. 26, 1974, pp. 100–103.

Mattera, L., "Component reliability, Part 1: Failure data bears watching," *Electronics*, Oct. 27, 1975, pp. 9–98; "Part 2: Hearing from users," Oct. 30, 1975, pp. 87–94; "Reliability revisited: failure-rate comparisons are given a second look," Dec. 25, 1975, pp. 83–85. (Readers' comments: Dec. 11, 1975, pp. 6, 8.)

Mayne, D., and R. Moore, "Minimize computer 'crashes,'" *Electron. Des.*, Apr. 26, 1974, pp. 168–172.

McCaw, C. R., "Loops in Directed Combinational Switching networks," Technical Report 6208-1, Stanford Electronics Laboratory, April 1963.

McCluskey, E. J., "Minimum-state sequential circuits for a restricted class of incompletely specified flow tables," *Bell Syst. Tech. J.*, November 1962, pp. 1759–1768.

McCluskey, E. J., "Logical design theory of NOR gate networks with no complemented inputs," *Proceedings of Fourth Annual IEEE Symposium on Switching Circuit Theory and Logical Design*, September 1963, pp. 137–148.

McCluskey, E. J., *Introduction to the Theory of Switching Circuits*, McGraw-Hill, 1965, 318 pp.

McCluskey, E. J., "Test and diagnosis procedure for digital networks," *Computer*, January/February 1971, pp. 17–20.

McCluskey, E. J., Jr., and S. H. Unger, "A note on the number of internal variable assignments for sequential switching circuits," *IRE Trans. Electron. Comput.*, December 1959, pp. 439–440.

McDaniel, H., *An Introduction to Decision Logic Tables*, John Wiley, 1968, 96 pp.

McDaniel, H., *Decision Table Software*, Brandon/System Press, 1970, 84 pp.

McDermott, J., "Focus on consumer IC's," *Electron. Des.*, Aug. 3, 1972, pp. 42–48.

McDermott, J., "Tomorrow's big computers are taking shape today; the changes are radical," *Electron. Des.*, Oct. 25, 1975a, p. 37.

McDermott, J., "Experts tell how to hold down high cost of processor programs," *Electron. Des.*, Dec. 20, 1975b, pp. 20–22, 24, 26.

McDermott, J., "IC packages are changing shape to handle growing LSI chips," *Electron. Des.*, Dec. 20, 1977, pp. 30, 32.

McGlynn, D. R., *Microprocessors: Technology, Architecture, Software, and Applications*, John Wiley, 1976, 222 pp.

Meahl, E., "Counter look-ahead technique," *Electron. Des.*, Dec. 6, 1973, pp. 104–108 (Part 1), and Dec. 20, 1973, pp. 68–75 (Part 2).

Mealy, G. H., "A method for synthesizing sequential circuits," *Bell Syst. Tech. J.*, September 1955, pp. 1045–1079.

Mei, K. C-Y., "Multiple output logic circuits," U.S. Patent 3,965,367, June 22, 1976.

Meisel, W. S., "A note on internal state minimization in incompletely specified sequential networks," *IEEE Trans. Electron. Comput.*, August 1967, pp. 508–509.

Michalski, R. S., and Z. Kulpa, "A system of programs for the synthesis of switching circuits using the method of disjoint stars," *Proceedings of IFIP Congress*, 1971, pp. 61–65.

Miiler, H. S., and R. O. Winder, "Majority logic synthesis by geometric methods," *IRE Trans. Electron. Comput.*, February 1962, pp. 89–90.

Miller, E. F., Jr., and J. H. Pugsley, "Sequential machines (Bibliography 22)," *Comput. Rev.*, May 1970, pp. 303–325.

Miller, R. E., *Switching Theory*, vols. I and II, John Wiley, 1965, 351 pp. and 250 pp.
Millman, J., and C. C. Halkias, *Integrated Electronics*, McGraw-Hill, 1972, 911 pp.
Minnick, R. C., P. T. Bailey, R. M. Sanfort, and W. L. Semon, "Magnetic bubble computer system," *Fall Joint Computer Conference*, 1972, pp. 1279–1298.
Mitani, N., "On transmission of numbers in a sequential computer," Convention of the Institute of Electrical Engineers of Japan, Tokyo, November 1951 (in Japanese).
Mitchell, T. W., "Programmable logic arrays," *Electron. Des.*, July 19, 1976, pp. 98–101.
Moisil, G. R. C., *The Algebraic Theory of Switching Circuits*, Pergamon Press, 1969, 719 pp.
Montalbano, M., "Tables, flow charts and program logic," *IBM Syst. J.*, September 1962, pp. 51–63.
Moore, E. F., "Gedanken-experiments on sequential machines," in *Automata Studies*, edited by C. E. Shannon and J. McCarthy, Princeton University Press, 1956, pp. 129–153.
Moore, E. F., *Sequential Machines: Selected Papers*, Addison-Wesley, 1964, 266 pp.
Moriwaki, Y., "Synthesis of Minimum Contact Networks based on Boolean Polynomials and Its Programming on a Digital Computer," Report 138, Inst. of Industrial Science, Univ. of Tokyo, March 1972, 55 pp.
Morrealle, E., "Partitioned list algorithms for prime implicant determination from canonical forms," *IEEE Trans. Electron. Comput.*, October 1967, pp. 611–620.
Morrealle, E., "Computational complexity of partitioned list algorithms," *IEEE Trans. Comput.*, May 1970a, pp. 42–428.
Morrealle, E., "Recursive operators for prime implicant and irredundant normal form determination," *IEEE Trans. Comput.*, June 1970b, pp. 504–509.
Morrealle, E., "Computer experience on partitioned list algorithms," *IEEE Trans. Comput.*, November 1970c, pp. 1099–1105.
Morris, R. L., and J. R. Miller (Eds.), *Designing with TTL Integrated Circuits*, McGraw-Hill, 1971, 322 pp.
Motorola, Inc., "LSI with 112 Gate TTL Array," Brochure, 1972a.
Motorola, Inc., *MECL System Design Handbook*, new rev. ed., December 1972b, 237 pp.
Mott, T. H., Jr., "Determination of irredundant normal forms of a truth function by iterated consensus of the prime implicants," *IRE Trans. Electron. Comput.*, June 1960, pp. 245–252.
Mukhopadhyay, A., "Complete sets of logic primitives," in *Recent Developments in Switching Theory*, edited by the same author, Academic Press, 1971, pp. 1–26.
Mukhopadhyay, A., and G. Schmitz, "Minimization of EXCLUSIVE-OR and LOGICAL EQUIVALENCE switching circuits," *IEEE Trans. Comput.*, February 1970, pp. 132–140.
Muller, D. E., "Metric Properties of Boolean Algebra and Their Application to Switching Circuits," Internal Report 46, Univ. of Illinois, Digital Computer Laboratory, 1953.
Muller, D. E., "Applications of Boolean algebra to switching circuit design and error detection," *Trans. IRE*, September 1954, pp. 6–12.
Muller, D. E., "A Theory of Asychronous Circuits," Report 66, Univ. of Illinois, Digital Computer Laboratory, December 1955.

606 References

Muller, D. E., "Infinite sequences and finite machines," *Proceedings of Fourth Annual IEEE Symposium on Switching Circuit Theory and Logical Design*, vol. S-156, September 1963, pp. 9–16.

Muller, D. E., and W. S. Bartky, "A Theory of Asychronous Circuits," Univ. of Illinois, Digital Computer Laboratory, I, Report 75, November 1956, and II, Report 78, March 1957.

Muller, D. E., and W. S. Bartky, "A theory of asynchronous circuits," in "Proceedings of International Symposium on Theory of Switching, April 1957," in *Annals of Computational Laboratory of Harvard University*, vol. 29, pp. 204–243, Harvard University Press, 1959.

Muroga, S., "Logical design of optimal digital networks by integer programming," Chapter 5 in *Advances in Information System Science*, vol. 3, edited by J. T. Tou, Plenum Press, 1970, pp. 283–348.

Muroga, S., *Threshold Logic and Its Applications*, John Wiley, 1971, 478 pp.

Muroga, S., and T. Ibaraki, "Design of optimal switching networks by integer programming," *IEEE Trans. Comput.*, June 1972, pp. 573–582.

Muroga, S., and H. C. Lai, "Minimization of logic networks under a generalized cost function," *IEEE Trans. Comput.*, September 1976, pp. 893–907.

Nakagawa, T., H. C. Lai, and S. Muroga, "Pruning and Branching Methods for Designing Optimal Networks by the Branch-and-Bound Method," Report 471, Dept. of Computer Science, Univ. of Illinois, August 1971.

Nakano, T., O. Tomisawa, K. Anami, M. Nakaya, M. Ohmori, and I. Ohkura, "A 920 gate DSA MOS masterslice," *IEEE J. Solid-State Circuits*, October 1978, pp. 536–541.

National Association of Relay Manufacturers, *Engineers' Relay Handbook*, Hayden Book Co., 1966, 300 pp.

National Semiconductor Corp., "How to design with programmable logic arrays," in *Memory Data Book*, 1977.

Nelson, R. J., "Simplest normal truth functions," *J. Symb. Logic*, June 1954, pp. 105–108.

Nelson, R. J., *Introduction to Automata*, John Wiley, 1968, 400 pp.

Noyce, R., "Bob Noyce of Intel speaks out on the integrated circuit industry," *EDN/EE*, Sept. 15, 1971, p. 30.

Nozaki, A., "On the circuits given by W. Kautz, D. Huffman, and D. Muller," *J. Inf. Process. Jap.*, March 1973, pp. 182–187.

Oberman, R. M. M., *Electronic Counters*, Barnes and Noble, 1973, 232 pp.

Ogdin, C. A., "μC design course," *EDN*, Nov. 20, 1976, pp. 126–316.

Ogdin, C. A., *Microcomputer Design*, Prentice-Hall, 1978a, 208 pp.

Ogdin, C. A., *Software Design for Microcomputers*, Prentice-Hall, 1978b, 200 pp.

Oliver, F. J., *Practical Relay Circuits*, Hayden Book Co., 1971, 363 pp. (Reviewed in *IEEE Spectrum*, September 1971, pp. 100–101.)

Parasuraman, B., "Hardware multiplication techniques for microprocessor systems," *Comput. Des.*, April 1977, pp. 75–82.

Partridge, J. E., "Cascade adder improves system speed for high-speed multiply operations," *EDN*, Apr. 20, 1973, pp. 74–77.

Paull, M. C., and S. H. Unger, "Minimizing the number of state incompletely specified switching functions," *IRE Trans. Electron. Comp.*, September 1959, pp. 356–367.

Pearson, H. R., and F. B. Allen, *Modern Algebra*, Ginn, 1970, 621 pp.

Peatman, J. B., *Microcomputer-Based Design*, McGraw-Hill, 1977, 540 pp.

Pěchouček, M., "Anomalous response times of input synchronizers," *IEEE Trans. Comput.*, February 1976, pp. 133–139.

Percival, R., "Dynamic MOS shift registers can also simulate stack and silo memories," *Electronics*, Nov. 8, 1971, pp. 85–89.

Percival, R., and J. Gray, "Functional digital ICs: device to delight the systems man." *Electronics*, Mar. 13, 1972, pp. 78–81.

Perrin, J. P., M. Denouette, and E. Daclin, *Switching Machines*, vol. 2, Springer-Verlag, 1972, 421 pp.

Peterson, J. L., "Petri nets," *ACM Comput. Surv.*, September 1977, pp. 223–252.

Petrick, S. R., "On the minimization of Boolean functions," *Proceedings of the International Conference on Information Processing*, 1959, published by Oldenbourg, Germany, 1960, pp. 422–423.

Petritz, R. L., "The pervasive microprocessors: trends and prospects," *IEEE Spectrum*, July 1977, pp. 18–24.

Pezaris, S. D., "A 40-ns 17-bit by 17-bit array multiplier," *IEEE Trans. Comput.*, April 1971, pp. 442–447.

Phister, M., Jr., *Data processing technology and economics*, Santa Monica Publishing Co., 1976, 573 pp.

Phister, M., Jr., "Analyzing computer technology costs," *Comput. Des.*, Part 1, September 1978, pp. 91–98; Part 2, October 1978, pp. 109–118.

Poage, J. F., "Derivation of optimum tests to detect faults in combinational circuits," *Proceedings of Symposium on Mathematical Theory of Automata*, Polytechnic Press, 1963, pp. 483–528.

Pollack, S. L., H. T. Hicks, Jr., and W. J. Harrison, *Decision Tables: Theory and Practice*, John Wiley, 1971, 179 pp.

Pokoski, J. L., "Software analyses for combinatorial logic," *Computer Des.*, June 1978, pp. 113–118.

Pooch, V. W., "Translation of decision tables," *ACM Comput. Surv.*, June 1974, pp. 125–151.

Post, E. L., *The Two-Valued Iterative Systems of Mathematical Logic*, Princeton University Press, 1941, 122 pp.

Prather, R. E., and H. T. Casstevens II, "Realization of Boolean expressions by atomic digraphs," *IEEE Trans. Comput.*, August 1978, pp. 681–688.

Preiss, R. J., "Fault test generation," Chapter 7, pp. 335–410, in *Design Automation of Digital Systems*, vol. 1, edited by M. A. Breuer, Prentice-Hall, 1972.

Preparata, F. P., and R. T. Yeh, *Introduction to Discrete Structures*, Addison-Wesley, 1973, 354 pp.

Puckett, G., J. Marley, and J. Gragg, "Automotive electronics, II: The Microprocessor is in," *IEEE Spectrum*, November 1977, pp. 37–45.

Putzolu, G. R., and J. P. Roth, "A heuristic algorithm for the testing of asynchronous circuits," *IEEE Trans. Comput.*, June 1971, pp. 639–647.

608 References

Quatse, J. T., and R. A. Keir, "A parallel accumulator for a general purpose computer," *IEEE Trans. Electron. Comput.*, April 1967, pp. 165–171.

Quine, W. V., "Two theorems about truth functions," *Bol. Sol. Math. Mex.*, vol. 10, Nos. 1 and 2, 1953, pp. 64–70.

Quine, W. V., "A way to simplify truth functions," *Amer. Math. Mon.*, November 1955, pp. 627–631.

RCA, *COS/MOS Integrated Circuits Manual*, 1972, 224 pp.

Reusch, B., "Generation of prime implicants from subfunctions and a unifying approach to the covering problem," *IEEE Trans. Comput.*, September 1975, pp. 924–930.

Reyling, G., Jr., "PLAs enhance digital processor speed and cut component count," *Electronics*, Aug. 8, 1974a, pp. 109–114.

Reyling, G., Jr., "Extend LSI-processor capabilities," *Electron. Des.*, Oct. 25, 1974b, pp. 90–95.

Rhyne, V. T., and P. S. Noe, "On the number of distinct state assignments for a sequential machine," *IEEE Trans. Comput.*, January 1977, pp. 73–75.

Richards, C. L., "An easy way to design complex program controllers," *Electronics*, Feb. 1, 1973, pp. 107–113.

Richards, R. K., *Arithmetic Operations in Digital Computers*, Van Nostrand, 1955, 397 pp.

Richards, R. K., *Digital Design*, John Wiley, 1971, 577 pp.

Rickard, B., "Automatic error correction in memory systems," *Comput. Des.*, May 1976, pp. 179–182. (Corrections, November 1976, p. 10.)

Riley, W. B., "Design blending hardware and software," *Electronics*, July 11, 1974, pp. 103–104.

Rodgers, T. J., and J. E. Meindl, "VMOS: High-speed TTL compatible MOS logic," *IEEE J. Solid-State Circuits*, October 1974, pp. 239–250.

Rosin, R. F., "An introductory problem in symbol manipulation for the student," *Commun. ACM*, September 1960, pp. 448–489.

Roth, J. P., "Algebraic topological methods for the synthesis of switching systems, I," *Trans. Amer. Math. Soc.*, July 1958, pp. 301–326.

Roth, J. P., "Algebraic topological methods in synthesis," in "Proceedings of International Symposium on Theory of Switching, April 1957," in *Annals of Computational Laboratory of Harvard University*, vol. 29, pp. 57–73, Harvard University Press, 1959.

Roth, J. P., "Diagnosis of automata failures: A calculus and a method," *IBM J. Res. Dev.*, July 1966, pp. 278–291.

Roth, J. P., "Programmed logic array optimization," *IEEE Trans. Comput.*, February 1978, pp. 174–176.

Roth, J. P., W. G. Bouricius, and P. R. Schneider, "Program algorithms to compute tests to detect and distinguish between failures in logic circuits," *IEEE Trans. Electron. Circuits*, October 1967, pp. 567–579.

Roth, J. P., and E. G. Wagner, "Algebraic topological methods for the synthesis of switching systems, III: Minimization of non-singular Boolean trees," *IBM J. Res. Dev.*, October 1959, pp. 326–344.

Rubinstein, E., "Independent test labs: caveat emptor," *IEEE Spectrum*, June 1977, pp. 44–50.

Rudeanu, S., *Boolean Functions and Equations*, North-Holland, 1974, 442 pp.
Russell, R. M., "The CRAY-1 computer system," *Commun. ACM*, January 1978, pp. 63–72.
St. Clair, P. R., Jr., "Decision tables clear the way for sharp selection," *Comput. Decis*, February 1970, pp. 14–18.
Salkin, H. M., *Integer Programming*, Addison-Wesley, 1975, 537 pp.
Saucier, G., "Encoding of asynchronous sequential networks," *IEEE Trans. Comput.*, June 1967, pp. 365–369.
Scheff, B. H., and S. P. Young, "Gate-level logic simulation," in *Design Automation of Digital Systems*, vol. 1, edited by M. A. Breur, Prentice-Hall, 1972, Chapter 3, pp. 101–172.
Schindler, M. J., "Computer-aided design beats trial-and-error—just find the software," *Electron. Des.*, June 7, 1977, pp. 44–49.
Schmid, H. (General Electric), *Decimal Computation*, John Wiley, 1974, 266 pp.
Schneider, D. (General Radio), "Designing logic boards for automatic testing," *Electronics*, July 25, 1974, pp. 100–104.
Schultz, G. W., "An algorithm for the synthesis of complex sequential networks," *Comput. Des.*, March 1969, pp. 49–55.
Scola, P., "An annotated bibliography of test and diagnostics," *Honeywell Comput. J.*, vol. 6, No. 2, 1972, pp. 97–102.
Sellers, F. F., Jr., M. Y. Hsiao, and W. L. Bearnson, "Analyzing errors with the Boolean differences," *IEEE Trans. Comput.*, July 1968a, pp. 676–683.
Sellers, F. F., Jr., H. Y. Hsiao, and L. W. Bearnson, *Error Detecting Logic for Digital Computers*, McGraw-Hill, 1968b, 295 pp.
Senders, S. L., and J. R. Lucchesi, "Design of sequential switching circuits with the cube logic technique," *Comput. Des.*, April 1971, pp. 59–64.
Seshu, S., "On an improved diagnosis program," *IEEE Trans. Electron. Comput.*, February 1965, pp. 76–79.
Shannon, C. E., "A symbolic analysis of relay and switching circuits," *Trans. AIEE*, vol. 57, 1938, pp. 713–723.
Sherwood, W., "PLATO-PLA Translator/Optimizer," *Proceedings of Symposium on Design Automation and Microprocessors*, 1977, pp. 28–35.
Shima, M., and F. Faggin, "In switch to n-MOS microprocessor gets on 2 μs cycle time," *Electronics*, Apr. 18, 1974, pp. 95–100.
Shima, M., F. Faggin, and R. Ungermann, "Z-80 chip set heralds third microprocessor generation," *Electronics*, Aug. 19, 1976, pp. 89–93.
Short, R. A., "A Theory of Relations between Sequential and Combinational Realizations of Switching Functions," Technical Report 098-1, Stanford Electronics Laboratory, 1960, pp. 33–34, 102–104.
Sifferlen, T. S., and V. Vartanian, *Digital Electronics with Engineering Applications*, Prentice-Hall, 1970, 307 pp.
Sikorski, R., *Boolean Algebra*, 2nd ed., Academic Press, 1964, 237 pp.
Singh, S., and R. Waxman, "Multiple operand addition and multiplication," *Fall Joint Computer Conference*, 1972, pp. 367–373.
Singh, S., and R. Waxman, "Multiple operand addition and multiplication," *IEEE Trans. Comput.*, February 1973, pp. 113–120.

References

Skelly, P. G., "A beginning for hardware documentation standard," *EDN*, July 5, 1973, pp. 70–76.

Skokan, Z. E., "EFL logic family for LSI," *International Solid-State Circuits Conference*, 1973, pp. 162, 163, 218.

Slagle, J. R., C. L. Chang, and R. C. T. Lee, "A new algorithm for generating prime implicants," *IEEE Trans. Comput.*, April 1970, pp. 304–310.

Smith, L., "USART—a universal µp interface for serial data communications," *EDN*, Sept. 5, 1976, pp. 81–86.

Smith, R. A., "Minimal three-variable NOR and NAND logic circuits," *IEEE Trans. Electron. Comput.*, February 1965, pp. 79–81.

Smith, R. J., *Circuits, Devices and Systems*, 3rd ed., John Wiley, 1976, 767 pp.

Smother, C. L., "A composite mapping technique to streamline the use of Karnaugh maps," *Comput. Des.*, November 1970, pp. 128–132.

Sobel, A., "Electronic numbers," *Sci. Amer.*, June 1973, pp. 64–73.

Sommar, H. G., and D. E. Dennis, "A new method of weighted term searching with a highly structured thesaurus," *Proc. Amer. Soc. Inf. Sci.*, vol. 6, 1969, pp. 193–198.

Stein, P. "What's in a word?" *Comput. Decis*, January 1975, pp. 16–17.

Stenzel, W. J., W. J. Kubitz, and G. H. Garcia, "A compact high-speed parallel multiplication scheme," *IEEE Trans. Comput.*, October, 1977, pp. 948–957.

Strauss, L., *Wave Generation and Shaping*, 2nd ed., McGraw-Hill, 1970, 775 pp.

Streetman, B. G., *Solid State Electronic Devices*, Prentice-Hall, 1972, 463 pp.

Su, S. Y. H., "Logical design and its recent development, Part 5: Fault diagnosis in digital networks," *Comput. Des.*, January 1974, pp. 87–92.

Susskind, A. K., "Diagnostics for logic networks," *IEEE Spectrum*, October 1973, pp. 40–47.

Svoboda, A., "Adder with distributed control," *IEEE Trans. Comput.*, August 1970, pp. 749–751.

Szygenda, S. A., and E. W. Thompson, "Digital logic simulation in a time-based table-driven environment," *Computer*, March 1975, pp. 24–49.

Tabloski, T. F., and F. J. Mowle, "A numerical expansion technique and its application to minimal multiplexer logic circuits," *IEEE Trans. Comput.*, July 1976, pp. 684–702.

Taub, H., and D. Schilling, *Digital Integrated Electronics*, McGraw-Hill, 1977, 650 pp.

Tawfiq, T. K., "Production logic tester checks a variety of ICs," *Electronics*, Apr. 18, 1974, pp. 116–117.

Teets, J. J. (IBM), "The role of simulation in LSI design," *Spring Joint Computer Conference*, 1972, pp. 1065–1070.

Texas Instruments, Inc., "MOS Programmable Logic Arrays," Application Bulletin CA-158, 1970.

Tison, P., Ph.D. Thesis, Faculty of Science, Univ. of Genoble, France, 1965.

Tison, P., "Generalization of consensus theory and application to the minimization of Boolean functions," *IEEE Trans. Electron. Comput.*, August, 1967, pp. 446–456.

To, K., and R. E. Tulloss (Western Electric), "Automatic test systems," *IEEE Spectrum*, September 1974, pp. 44–51.

Todd, C. D., "An annotated bibliography on NOR and NAND logic," *IEEE Trans. Electron. Comput.*, October 1963, pp. 462–464.

Torng, H. C., "An algorithm for finding secondary assignments for synchronous sequential machines," *IEEE Trans. Comput.*, May 1968, pp. 461–469.

Torng, H. C., *Switching Circuits: Theory and Logic Design*, Addison-Wesley, 1972, 414 pp.

Torrero, E. A., "Don't write off the reed relay; it's faster, tinier, and cheaper," *Electron. Des.*, Oct. 14, 1971, pp. 28–30.

Torrero, E. A., "Automatic test equipment: not so easy," *IEEE Spectrum*, April 1977, pp. 29–34.

Tracey, J. H., "Internal state assignments for asynchronous sequential machines," *IEEE Trans. Electron. Comput.*, August 1966, pp. 551–560.

Ueda, T., Japanese Patent 848,503 (Sho51-22779, July 12, 1976).

Ulrich, E. G., and T. Baker (GTE), "Concurrent simulation of nearly identical digital systems," *Computer*, April 1974, pp. 39–44.

Unger, S. H., "Simplifications of state tables," in *A Survey of Switching Circuit Theory*, edited by E. J. McCluskey, Jr., and T. C. Bartee, McGraw-Hill, 1962, pp. 145–170.

Unger, S. H., "Flow table simplification—some useful aids," *IEEE Trans. Electron. Comput.*, June 1965, pp. 472–475.

Unger, S. H., *Asynchronous Sequential Switching Circuits*, John Wiley, 1969, 290 pp.

Unger, S. H., "Tree realizations of iterative circuits," *IEEE Trans. Comput.*, April 1977, pp. 365–383.

Van Cleemput, W. M., *Computer Aided Design of Digital Systems—A Bibliography*, Computer Science Press, vol. I, 1960–1974, 1976a, 374 pp.; vol. II, 1975–1976, 1976b, 278 pp.; vol. III, 1976–1977, 1977.

Van Holten, C., "Double multiplexer logic capability," *Electron. Des.*, Aug. 16, 1974, pp. 86–89.

Van Tuyl, R., and C. Liechti, "Gallium arsenide spawns speed," *IEEE Spectrum*, March 1977, pp. 41–47.

Veitch, E. W., "A chart method for simplifying truth functions," *Proceedings of ACM Conference*, May 2–3, 1952, pp. 127–133.

Vikas, O., "Combine multichart Karnaugh maps into single, easy-to-handle versions," *Electron. Des.*, July 19, 1975, p. 70.

Wagner, F. V., "Is decentralization inevitable?" *Datamation*, November 1976, p. 93.

Warner, R. M., Jr. (Ed.), *Integrated Circuit-Design Principles and Fabrication*, McGraw-Hill, 1965, 385 pp.

Weinberger, A., and J. L. Smith, "The logical design of a one-microsecond adder using one-megacycle circuitry," *IRE Trans. Electron. Comput.*, June 1956, pp. 65–73.

Weiner, P., and T. F. Dwyer, "Discussions of some flaws in the classical theory of two level minimizations of multiple-output switching networks," *IEEE Trans. Comput.*, February 1968, pp. 184–186.

Weiner, P., and E. J. Smith, "On the number of state assignments for synchronous sequential machines," *IEEE Trans. Comput.*, April 1967, pp. 220–221.

Weisbecker, J., "A practical, low cost, home/school microprocessor system," *Computer*, August 1974, pp. 20–31.

Whitesitt, J. E., *Boolean Algebra and Its Applications*. Addison-Wesley, 1961, 182 pp.

References

Wilson, D. R., "Cell layout boosts speed of low-power 64K ROM," *Electronics*, Mar. 30, 1978, pp. 96–99.

Winder, R. O., "Threshold logic," Ph.D. Thesis, Mathematics Dept., Princeton Univ., 1962.

Winder, R. O., "The fundamentals of threshold logic, in *Applied Automata Theory*, edited by J. Tou, Academic Press, 1968a, pp. 235–318.

Winder, R. O., "Threshold logic will cut costs, especially with boost from LSI," *Electronics*, May 27, 1968b, pp. 94–103.

Wise, T. A., "IBM's $5,000,000,000 gamble," *Fortune*, September, pp. 118–123, 224–228; October, pp. 138–143, 199–212, 1966.

Wood, P. E., Jr., *Switching Theory*, McGraw-Hill, 1968, p. 65.

Wood, R. A., "High-speed dynamic programmable logic array chip," *IBM J. Res. Dev.*, July 1975, pp. 379–383.

Wooley, B. A., and C. R. Baugh, "An integrated m-out-of-n detection circuit using threshold logic," *IEEE J. Solid-State Circuits*, October 1974, pp. 297–300.

Wyland, D. C. (Monolithic Memories), "Shift registers can be designed using RAMs and counter chips," *EDN* Jan. 5, 1974, pp. 64–67.

Yau, S. S., and C. K. Tang, "Universal logic circuits and their modular realizations," *Spring Joint Computer Conference*, 1968, pp. 297–305.

Yoeli, M., and S. Rinon, "Application of ternary algebra to the study of static hazards," *J. ACM*, January 1964, pp. 84–97.

Index

Page numbers in italics indicate the more detailed explanation of entry.

Absolutely eliminable prime implicant, 130
Absolutely eliminable prime implicate, 152
Accumulator, 497
Alterm, *73*, 151
AND, 9
AND matrix (or array), 551
Antecedent, 93
Architectural simulation, 577
Associative logic array (ALA), 557
Asynchronous networks, 446

8-4-2-1 BCD code (binary coded decimal), 66, 537
Biform variable, 103
Bipolar transistor, 36, *47*
Boolean algebra, 111
Boolean difference, 82, *588*
Bouncing (of metal contact), 372
Bounding operation, 176
Branch-and-bound method, 175
Branching operation, 175
Breadboard, 577
Break-contact relay, 16

CAD (computer-aided design), 4, *106*
Canonical conjunctive form, 73
Canonical disjunctive form, 73
Canonical implicant, 73
Canonical implicate, 73
Canonical product, 73
Canonical sum, 73
Chattering (of metal contact), 372
Chip, 53
Circuit simulation, 577
Classification of functions, 327
Classification of networks, 329
Clock, 440

Clocked latch, 442
CMOS, 46, *77*, 84
CMOS/SOS, 46
Combinational network, 27
Comparable, 93
Comparator, 535
Compatible, 409
Completely specified flow table, 396
Completely specified function, 71
Complete product, 152
Complete set of gates (or gate types), 252
Complete set of primitive logic operations, 252
Complete sum, 101
Computer-aided design (CAD), 4, *106*
Conditionally eliminable prime implicant, 130
Conditionally eliminable prime implicate, 152
Conjunction, 9, 115
Conjunctive form, 73
Consensus, 97, 151
Consequent, 93
Counter, 511
Cover, 167
Covering problem, 168
Cricial race, 361
Custom design, 56
Custom-design IC package, 56
Cut set, 25
Cyclic table, 173

D-algorithm, *583,* 590
Data base management system, 117
Decimal specification, 13
Decision table, 117, 119
Decoder, 65

613

Degenerate function, 72
De Morgan's Theorem, 76
Denial, 115
Dependent variable, 72
Depletion type, 41
Design objectives, 124
Diagnosis, 578
Diffusion region, 37
Direct reset-input terminal, 454
Direct set-input terminal, 454
Discrete component, 53
Discrete gate, 53
Disjunction, 9, 116
Disjunctive form, 73
Distinguishable networks, 396
Distinguishable states, 397
Distinguished column, 169
Distinguished 1-cell, 140
Divide-by-N counter, *436*, 522
D latch, 459
Dominate column, 170
Dominate row, 169
Don't-care condition, *13,* 72
Double-rail input logic, *261,* 273
Double-rail (output) logic, 261
Drain (of MOSFET), 37
Driver (MOSFET), 41
D (Type) Master-slave flip-flop, *456,* 462
Dual, 69, *70,* 77
Duality theorem, 70
Dummy variable, 72
Duty cycle, 447
Dynamic hazard, 324
Dynamic MOS, 46, 445

EAROM (electrically alterable read-only memory), 548
Edge-triggered flip-flops, 457
Edge-triggered S-R flip-flop, 463
Effective gate delay, 128, *325*
Electrical switch, 20
Electronic (or electrical) performance, 59
Enchancement type, 37
EPROM (erasable PROM), 548
Essential prime implicant, 130
Essential prime-implicant loop, 140, 142
Essential prime implicate, 152
Essential row, 169

Excess-3 code, 494
Excitation (-output) table, 377, 378
Excitation variable, 376
EXCLUSIVE-NOR(gate), 256
EXCLUSIVE-OR, 116
EXCLUSIVE-OR(gate), 254
Exhaustive method, 250
Extrinsic hazard, 322

Fan-in, 60
Fan-out, 60
False (input) vector, 72
Fault table, 586
Field-programmable PLA (or FPLA), 554
Finite induction, 76
Firmware, 538, *561*
Flip-flop, 347, *459*
Flow chart, 564
Flow (output) table, 366, 367
Formation tree (or generalized-consensus formation tree), 198
Full adder, 7, 10
Functional test (of computers), 579
Functional testing (of components), 578
Fundamental mode, for asynchronous networks, 356
   for synchronous networks, 448
Fundamental product, 73
Fundamental sum, 73

Gate, logic gate, MOS gate, MOS cell, *37,* 42
   MOSFET gate, *37,* 42
Generalized consensus, *196,* 213
Generalized-consensus formation tree, 197
Generalized De Morgan's Theorem, 78
General model of sequential networks, 375
Generator of irredundant disjunctive forms, 204
GSI (grand scale integration), 55

Half adder, *7,* 67
Hard-core test, 579
Hard-wired control unit, 561
Hazardous race, 416
Hindrance, 31
Hybrid integrated circuit, 53

Index 615

IIL (integrated injunction logic), 340
Implicant, 95
Implicate, 149
Implication, 116
Implication relation, 92, 93
Imply (strictly), 92, 93
Inclusion function, 200
INCLUSIVE-OR, 116
Incomparable, 93
Incompletely specified flow table, 396
Incompletely specified function, 71
Independent variable, 72
Indistinguishability (equivalence) classes, 397
Indistinguishable networks, 396
Indistinguishable states, 397
Input state, 356
Input variable, 10
Integer programming logic design method, 250, 310
Integrated circuit, 53
Intermediate unstable state (in multiple transition), 366
Internal state, 356
Internal variables, 354
Intrinsic hazard, 320
Irredundant conjunctive form, *152*, 153
Irredundant disjunctive form, *129*, 140
Irredundant implication relation, 196
Irredundant network, 586
Iterated consensus method, 100

J-K latch, 459
J-K master-slave flip-flop, 452

Karnaugh map, 87

Latch, *347*, 459
Layout, 46
Library approach, *58*
Literal, *22*, 72
Loadable shift register, 535
Load (MOSFET), 41
Logic capability, 59
Logic design, 1, *4*, 11
Logic family, *46*, 49
Logic simulation, 577
Look-ahead adder, 518

Look-ahead comparator, 536
Look-ahead counter, 536
LSI (large scale integration), 55

Majority function, 259
Make-contact relay, 16
Malfunction, 28, 324, 359, 363
Map-factoring method, 281
Mask-programmable PLA, 554
Mask-programmable ROM, 548
Master-slave flip-flop, 452
Material implication, 116
Maximal, 409
Maxterm (expansion), 73
Mealy machine, 367
Metal oxide semiconductor field effect transistor (MOSFET), 36
Micro-instructions, 561
Minimal product (or minimal conjunctive form), *152,* 153
Minimal sum (or minimal disjunctive form), 131
Minimization of number of states, 393-413, 498-499
Minterm (expansion), 73
Modular structure, 511
Monoform variable, 103
Monolithic-integrated circuit, 53
Moore machine, 367
Moore reduction procedure, 498
MOS cell (or MOS gate), 42
m-out-of-n detector, 541
MSI (medium scale integration), 55
Multilevel network, 248, 276, 322
Multiple-contact relay, 19
Multiple-output prime implicant, *217*, 240
Multiplexer, 190-192
Multithreshold threshold gate, 260

NAND, 41
n-channel MOSFET, 37
Negation, 9
Negative edge-triggered flip-flop, 457
Negative function, 43, *260*
Negative logic, 41
N-equivalence, 329
Net gate delay, 326
Next state (internal) variables, 354

n-MOS, 37
Noncritical race, 361
Nondegenerate function, 72
NOR, 41
NOT, 9
NP-equivalence, 329
NPN-equivalence, 328
Number of levels (or gate delay) of network, 127
Number of phases of clock, 445

Off-the-shelf package, 2, *56*
OR, 8
OR matrix (or array), 551
Output table, 356
Output variable, 10

Package, 37
Parametric testing (of components), 578
Paramount prime implicant, 217
Parasitic capacitance, 54
Parity function, 12, 84, *254*
Partially symmetric function, 80
Path-sensitizing method, 579
pc board (printed circuit board), 44
p-channel MOSFET, 37
P-equivalence, 329
Perfect induction, 69
Permissible loop, 281
Petrick function, *173,* 223
Petrick method, 173
Petri net, 573
Pierce arrow, 253
p-MOS, 37
Polynomial expansion, 257
Positive edge-triggered flip-flop, 457
Positive function, *260*, 270
Positive logic, 41
Postanalysis, 423
Present state (internal) variables, 354
Prime implicant, 96
Prime-implicant loop, 140, 144, 147
Prime implicant table, *167*, 221
Prime implicate, *150*, 153
Prime-implicate loop, 153
Primitive D-cube, 590
Primitive form (of flow-output table), 386
Primitive gate (or gate type), 252

Primitive logic operation, 252
Product lines, 551
Programmable logic array (PLA), 549
PROM (programmable read-only memory), 548
Proposition, 114
Propositional logic, 114
Propositional variable, 114
Pulse duration, 447
Pulse mode, 468
Pulse repetition period, 447

Quine-McCluskey method, 163

Race, 361
Raceless flip-flop, 459
Random-logic network, 547
Relay, 16
Representative function, 328
Reset-dominant S-R latch, 352, *373*
Reset terminal, 348
Residue-threshold gate, 269
Resolution, 581
Right-shift register, 535
Ripple adder (or carry ripple adder) 8, *512*
Ripple comparator, 535
Ripple counter, *522*, 536
ROM (read-only memory), 538
Root (of formation tree), 197
R-S flip-flop, 352
RTL (resistor-transistor logic), 49

Secondary essential row, 171
Second source, 61
Self-dual function, 77
Sensitized path, 580
Sequential network, 27
Serial adder, 468
Series-parallel network, 22
Set-dominant S-R latch, 352, *373*
Set terminal, 348
Shannon's expansions, 83
Sheffer stroke, 253
Shift register, 511, *534*
Simulation, 576
Single-rail input logic, *261*, 276
Single-rail (output) logic, 261
Skew mode, 468

Sloppy minimal product, 152
Sloppy minimal sum, 131
Source (of MOSFET), 37
S-R flip-flop, 352
S-R latch, *348*, 459
SSI (small scale integration), 55
Stable state, 353
Standard networks, 2, 56, *510*
State-adjacency diagram, 416
State assignment, 380, *413-421*
State-comparison table, 399
State diagram, 365
State minimization, for completely specified networks, *399-405*, 498
   for incompletely specified networks, 405-413
State (output) table, 364, 365
State-splitting, 406
Static hazard, 324
Static MOS, 46, 446
Structural test, 579
Stuck-at-0 (or-1) fault, 579
Substrate, 44
Subsume (strictly), 95, 150
Switching algebra, 1, 68, 113
Switching expression, 10
Switching theory, 1
Switching time, 44, *50*
Switching variable, 10
Symmetric function, 80
Symmetric variables, 80
Synchronous networks, 446

Term, 72
Test, 577
Threshold function, 258

Threshold gate, 258
Tie set, 24
Tison function, 200
Tison method, derivation of all irredundant disjunctive forms, 198, 208
   derivation of all prime implicants, 103
Totally symmetric function, 80
Total state, 356
Transduction method, 251, 311
Transfer-contact relay, 20
Transient state, 492
Transistor-level logic design, 263
Transition diagram, 365
Transition (output) table, 354, 356
Transmission, 17
Tree network, 277
Triangular condition, *298*, 316, 345
True (input) vector, 72
Truth table, 12
T (Type) master-slave flip-flop, 455

Unate function, 260
Unstable state, 353
Up-down counter, 392, 433, *511*

Variable-entered Karnaugh map, 161
Veitch diagram, 88
VLSI (very large-scale integration), 55
VMOS, 273

Wired logic, 263
   wired-AND, *263*, 340
   wired-OR, 263
Word statement, 12, 374, *385*, 472

Yield, 55